Power and Electronic Engineering

Power and Electronic Engineering

Edited by **Marko Silver**

Willford Press

New York

Published by Willford Press,
118-35 Queens Blvd., Suite 400,
Forest Hills, NY 11375, USA
www.willfordpress.com

Power and Electronic Engineering
Edited by Marko Silver

International Standard Book Number: 978-1-68285-225-5 (Hardback)

Printed in China.

Contents

	Preface	**IX**
Chapter 1	**Measurement and Analysis Possibilities of Pulse Wave Signals** Stefan Borik and Ivo Cap	**1**
Chapter 2	**A Voltage-Mode First Order Allpass Filter Based on VDTA** Suwat Maneewan, Nawee Udorn, Danupat Duangmalai, Phamorn Silapan and Winai Jaikla	**9**
Chapter 3	**Linear Regression Based Real-Time Filtering** Misel Batmend and Daniela Perdukova	**16**
Chapter 4	**Reactive Power Optimization with SVC & TCSC using Genetic Algorithm** Biplab Bhattacharyya, Vikash Kumar Gupta and Sanjay Kumar	**23**
Chapter 5	**Generalized Radio Resource Management for Overlapping MBS Zones** Fang Luo, Yu-Feng Chou, Ray-Guang Cheng, Marek Kalika, Radek Holy and Jana Kalikova	**35**
Chapter 6	**Wearable Textile Electrodes for ECG Measurement** Lukas Vojtech, Radoslav Bortel, Marek Neruda and Milos Kozak	**42**
Chapter 7	**Modelling of H.264 MPEG2 TS Traffic Source** Stanislav Klucik and Martin Lackovic	**47**
Chapter 8	**Fault Tolerant Ethernet Based Network for Time Sensitive Applications in Electrical Power Distribution Systems** Leos Bohac and Jiri Vodrazka	**53**
Chapter 9	**Investigating and Simulating DGs to Improve Voltage Profile and Reducing Power Loss in Unbalanced Distribution Networks** Hassan Barati and Abdol Rasool Fatthi	**59**
Chapter 10	**Unsupervised Spelling Correction for the Slovak Text** Daniel Hladek, Jan Stas and Jozef Juhar	**68**
Chapter 11	**Three Input Single Output Voltage-Mode Multifunction Filter with Independent Control of Pole Frequency and Quality Factor** Witthaya Mekhum and Winai Jaikla	**74**
Chapter 12	**Power Quality Analysis in Off-Grid Power Platform** Stanislav Misak, Jindrich Stuchly, Jakub Vramba, Lukas Prokop and Marian Uher	**81**
Chapter 13	**Comparison of Standard and Fast Charging Methods for Electric Vehicles** Petr Chlebis, Martin Tvrdon, Ales Havel and Katerina Baresova	**89**

Chapter 14 **Test Frequency Selection Using Particle Swarm Optimization** **95**
Zdenek Kincl and Zdenek Kolka

Chapter 15 **Design of Two Feeder Three Phase Four Wire Distribution System Utilizing Multi Converter UPQC with Fuzzy Logic Controller** **102**
Chandra Babu Paduchuri, Subhransu Sekhar Dash and Subramani Chinnamuthu

Chapter 16 **Fuzzy Multi-Objective Optimal Power Flow Using Genetic Algorithms Applied to Algerian Electrical Network** **113**
Ahmed Salhi, Djemai Naimi and Tarek Bouktir

Chapter 17 **Direct Thrust Force and Flux Control of a PM-Linear Synchronous Motor Using Fuzzy Sliding-Mode Observer** **125**
Saeed Masoumi Kazraji and Mohammad Bagher Bannae Sharifian

Chapter 18 **Low-voltage Power Supply Subsystem for a Sub-Orbital Particle Physic Instrument** **134**
Hector Hugo Silva Lopez, Gustavo Medina-Tanco and Lauro Santiago Cruz

Chapter 19 **Optimal Placement of Non-Site Specific DG for Voltage Profile Improvement and Energy Savings in Radial Distribution Networks** **153**
Mudathir Funsho Akorede, Edris Pouresmaeil, Hashim Hizam, Ishak Aris and Mohd Zainal Ab-Kadir

Chapter 20 **A New Control Strategy Based Multi Converter UPQC Using Fuzzy Logic Controller to Improve the Power Quality Issues** **168**
Chandra Babu Paduchuri, Subhransu Sekhar Dash and Subramani Chinnamuthu

Chapter 21 **Trust Based Algorithm for Candidate Node Selection in Hybrid MANET-DTN** **180**
Jan Papaj and Lubomir Dobos

Chapter 22 **The Derivation of the Electrical Conductance/Temperature Dependency for Tin Dioxide Gas Sensor** **188**
Libor Gajdosik

Chapter 23 **On Ladder Diagrams Compilation and Synthesis to FPGA Implemented Reconfigurable Logic Controller** **196**
Adam Milik

Chapter 24 **An Experimental Study of Laboratory Hybrid Power System with the Hydrogen Technologies** **205**
Daniel Minarik, Bohumil Horak, Petr Moldrik and Zdenek Slanina

Chapter 25 **Comparative Method for Indirect Sensitivity Measurement of UHF RFID Reader with Respect to Interoperability and Conformance Requirements** **216**
Lukas Kypus and Lukas Vojtech

Chapter 26 **Increasing Robustness of Multi-homed Systems in Heterogeneous Environment** **222**
Matej Rohlik, Petr Chlumsky and Tomas Vanek

Chapter 27 **Extreme Production Conditions of Photovoltaic Power Plant Operated in Distribution Grid** **228**
Martin Smocek and Zdenek Hradilek

Chapter 28 **A Model-Based Fuzzy Control of an Induction Motor** **235**
Daniela Pedrukova and Pavol Fedor

Chapter 29 **Impact of H.264/AVC and H.265/HEVC Compression Standards on the Video Quality for 4K Resolution** **243**
Miroslav Uhrina, Jaroslav Frnda, Lukas Sevcik and Martin Vaculik

Chapter 30 **Speech Quality Measurement of GSM Infrastructure Built on USRP N210 and OpenBTS Project** **252**
Marcel Fajkus, Martin Mikulec, Miroslav Voznak, Martin Tomis and Peppino Fazio

Permissions

List of Contributors

Preface

This book has been a concerted effort by a group of academicians, researchers and scientists, who have contributed their research works for the realization of the book. This book has materialized in the wake of emerging advancements and innovations in this field. Therefore, the need of the hour was to compile all the required researches and disseminate the knowledge to a broad spectrum of people comprising of students, researchers and specialists of the field.

The various sub-fields of power and electronic engineering along with technological progress in this field are glanced at in this book. This book is a compilation of chapters that discuss the most vital concepts like solid state devices, power generation and transmission, analog and digital circuits, signal processing, etc. It strives to provide a fair idea about this discipline and to help develop a better understanding of the latest advances within this field. This book is a resource guide for experts as well as students.

At the end of the preface, I would like to thank the authors for their brilliant chapters and the publisher for guiding us all-through the making of the book till its final stage. Also, I would like to thank my family for providing the support and encouragement throughout my academic career and research projects.

Editor

1

Measurement and Analysis Possibilities of Pulse Wave Signals

Stefan BORIK, Ivo CAP

Department of Electromagnetic and Biomedical Engineering, Faculty of Electrical Engineering, University of Zilina, Univerzitna 1, 010 01 Zilina, Slovak Republic

stefan.borik@fel.uniza.sk, ivo.cap@fel.uniza.sk

Abstract. *This article deals with the implementation of wireless data transfer in biomedical applications concretely in photoplethysmography and pressure wave measurements. It deals with the application of a Bluetooth module for data transfer to the computer where further signal processing can be performed. The work describes a use of microcontroller serving for data flow control, communication and analog signal digitalization.*

Keywords

Bluetooth, microcontroller photoplethysmography, pulse wave, sensors.

1. Introduction

This article deals with design and realization of photoplethysmographic device serving for pulse wave investigation based on principle reflective photoplethysmography and with design of device prototype serving for pressure pulse wave measurements. The work describes sensor and hardware design, functionality principle, digitalization and data transfer using a Bluetooth module. A part of work involves analog signal processing such as signal amplification, filtering. The article is focused on data transfer to PC using selected microcontroller with integrated A/D converter and on the application of Bluetooth module serving for wireless data transfer.

2. Photoplethysmographic Device Design

2.1. Photoplethysmography

Suitable detection method for pulse wave investigation is photoplethysmography which serves as a diagnostic method for detection of blood volume changes in a selected part of the vessel or skin. It can serve with ECG measurements as useful diagnostic method for estimation of the cardiovascular system state. Basic principle is based on infrared radiation application to cutaneous area and on its reflection depending on blood volume changes. Reflected radiation corresponds to the propagation of pulse wave [1], [2], [3], [4], [5].

2.2. Sensor Design

Photoelectric detectors can be used for measurement. They detect either passing (transmission through tissue) or reflecting wave (reflection from tissue). Recorded pulse wave (see Fig. 3.) is similar to pressure wave curve [2].

Sensor was created from phototransistor and IR (infrared) diode (Fig. 1). Working wave length of this couple is 940 nm. Sensor is based on principle of radiation reflection.

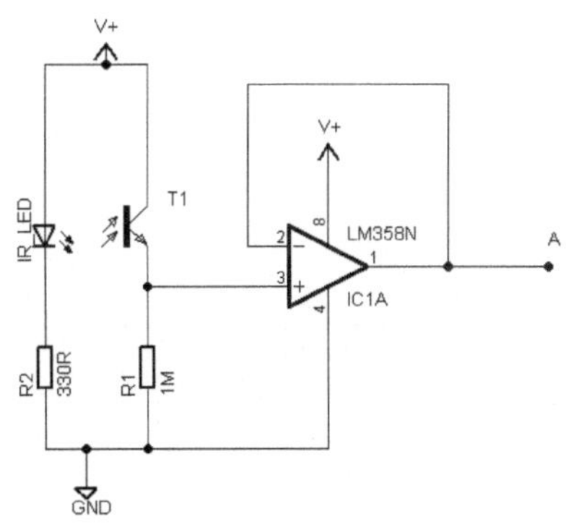

Fig. 1: Sensor circuit.

The used operational amplifier LM358 (shown on Fig. 1) serves as a voltage follower for an impedance matching. The signal from the voltage follower inputs to the first stage of two operational amplifiers serving for signal amplification because the signal from sensor output is small and it is needed to be amplified for further processing. A gain of the amplifier (Fig. 2) is set by resistors R_3 and R_4. It can be calculated by the next equation:

$$G = \frac{R_4}{R_3} = -100. \tag{1}$$

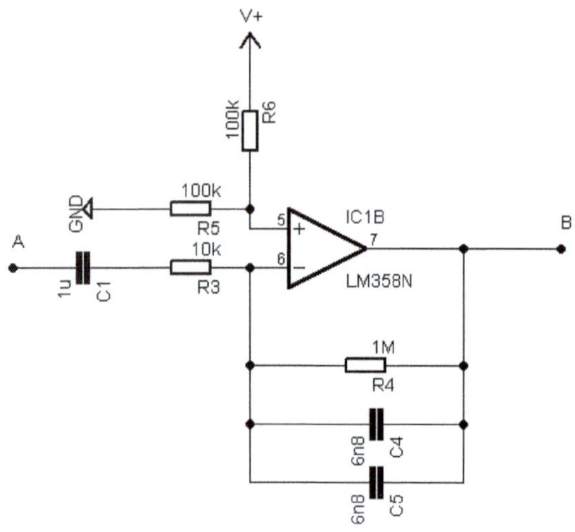

Fig. 2: First amplifier stage.

The capacitor C_1 and the resistor R_3 represent high pass filter with cutoff frequency f_{ch}:

$$f_{ch} = \frac{1}{2\pi R_3 C_1}. \tag{2}$$

The capacitors C_4, C_5 and the resistor R_4 represent low pass filter for filtering noise such as mains noise:

$$f_{cl} = \frac{1}{2\pi R_4 (C_4 + C_5)}. \tag{3}$$

Second amplifier stage has the same function as the first stage.

Both amplifiers have a single power supply. For that reason it is needed to create a virtual reference voltage on their non-inverting inputs. A half of power supply ensures that the signal amplitude swings around this value (+1,5 V). It is between supply voltage (+3 V) and ground [2], [6].

3. Microcontroller Based Signal Processing

The analog output signal is needed to be digitalized. For that reason integrated analog to digital converter (ADC) which is part of a selected microcontroller ATmega88A was used.

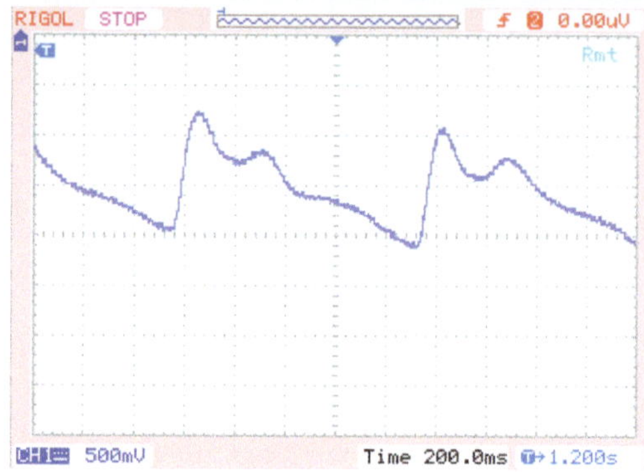

Fig. 3: Output signal measured by oscilloscope.

3.1. ADC Settings

The microcontroller ATmega88A offers the analog to digital converter with 10-bit resolution. A sampling frequency of the converter depends on a frequency of the connected crystal. In our case its value is 18,432 MHz. Regarding to the datasheet of manufacturer it is recommended to set the converter input frequency to range from 50 kHz to 200 kHz. In this way correct function of the converter is ensured. Dividing of the input frequency is done by setting of prescaler bits shown in the next table.

Tab. 1: ADC prescaler selections.

ADPS2	ADPS1	ADPS1	Division factor
0	0	0	2
0	0	1	2
0	1	0	4
0	1	1	8
1	0	0	16
1	0	1	32
1	1	0	64
1	1	1	128

By selecting of the division factor to 128 the input frequency is set to 144 kHz and it corresponds to the recommended value.

3.2. Conversion

After enabling of ADC (ADEN bit is set to the logical one) the function for conversion can be periodically called in a main loop of the program. In this function following parameters are set:

- Input channel selection (bits ADMUX),
- ADC left adjust result (bit ADLAR to 1).

By setting the bits MUX0 and MUX2 channel 5 is selected and 8-bit conversion result is selected by setting ADLAR bit to the logical one. The 8-bit resolution of the conversion was selected for higher transfer rate achievement using RS232 protocol. It follows that by using 3 V power supply the voltage resolution (VR) is about:

$$VR = \frac{3}{2^8 - 1} \cong 12\ mV. \qquad (4)$$

The conversion starts after setting these control bits. In a loop bit ADIF is compared to logical one. If its value changes from zero to one the conversion is complete and the conversion result can be read from the 8-bit register ADCH. Its value can be calculated by the following equation:

$$ADCH = \frac{V_{IN}}{V_{REF}} \times 2^8, \qquad (5)$$

where V_{IN} is an analog input voltage and V_{REF} is internal voltage reference 3 V.

3.3. Serial Communication

Digitalized data are sent using UART (Universal Asynchronous Receiver Transmitter) unit which is compatible with RS232 protocol. For correct function of this unit appropriate setting of baud rate generator is needed. Correct setting of UART baud rate generator register value allows baud rate generation without errors. For that reason the external crystal with the frequency 18,432 MHz was selected. The parameters of serial data transfer are shown in the next table:

Tab. 2: Parameters of the serial link.

Parameter	Value
Baud Rate	57600 B
Stop Bits	2
Parity	Even

3.4. Microcontroller Circuit Design

The used microcontroller ATmega88A requires for its function power supply voltage between 1,8 and 5,5 V. As the power supply a mobile phone battery was selected because the designed device should be serving for permanent patient monitoring and shouldn't limit his/her free movement with power supply cables. The battery output voltage is 3 V. The microcontroller circuit diagram is shown on the next figure:

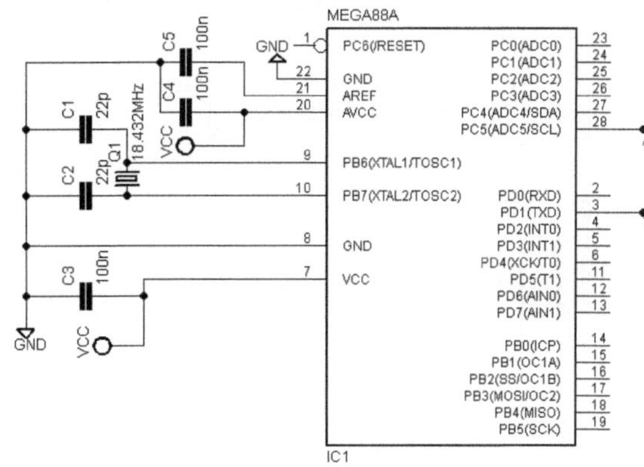

Fig. 4: Microcontroller circuit.

4. Investigation of Pressure Wave Propagation

There are two phenomena which are connected with a heart activity. By stroking function of heart the blood is ejected to circulation in a form of flow pulse wave propagation. The second phenomenon is distension of arteries propagating through whole systemic circulation. In this way a pressure pulse wave propagates to periphery and the arteries represent a medium for this propagation. Its velocity depends on many factors e.g. compliance, elasticity, diameter, thickness of arteries and intravascular pressure [1], [5], [6]. These parameters change with age. The velocities in the young subjects are shown in the next table [1]

Tab. 3: Pulse wave velocities.

Arterial type	Velocity $[\mathrm{m \cdot s^{-1}}]$
Aorta	3–5
Large arteries	7–10
Small arteries	15–35

The pressure wave can be measured by various methods. A measurement principle is based on use appropriate converter which converts mechanical energy to electrical energy. As converter piezoelectric transducer or pressure sensor can be used.

4.1. Device Prototype for Pressure Wave Measurement

For conversion of mechanical energy to electrical energy the pressure sensor SPD015G was selected. This sensor consists of four silicon resistors which are connected in bridge.

These resistors are in a plastic package which has two mechanical openings. The one serves for selected pressure measurement and the other is for reference.

A cuff is connected to the measurement input by rubber hose. This cuff is placed for example on the arm of subject and by using of pump the air is pumped to it. In this way the cuff can encircle arteria brachialis which runs through this place. Sensor, cuff and pump are in the next picture:

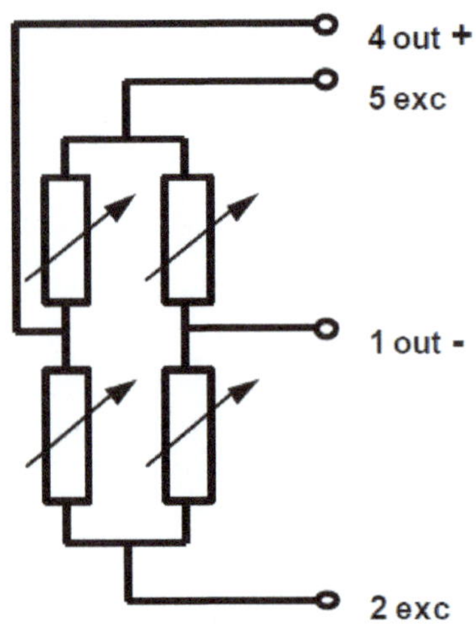

Fig. 5: Internal pressure sensor circuit (adapted from manufacturers datasheet).

Fig. 6: Sensor package (adapted from manufacturers datasheet).

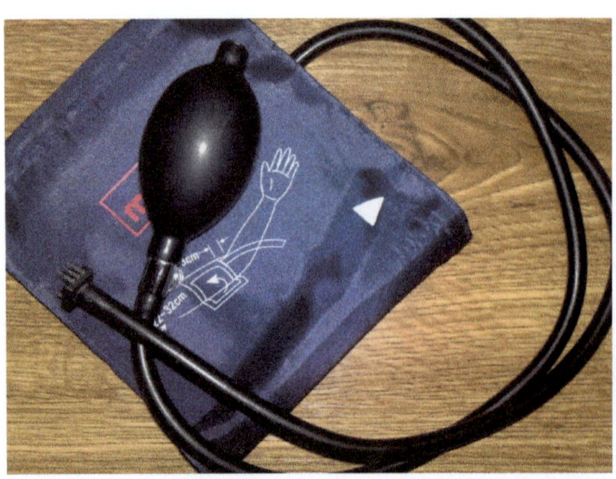

Fig. 7: Sensor, cuff and pump.

4.2. Electrical Circuit

Sensor output voltage corresponds to pressure wave propagation. Signal from the sensor has to be amplified. Primary amplification was realized by instrumentation amplifier INA 122 which serves for low-noise differential amplification of the signal. Manufacturer describes this amplifier that it provides excellent performance with very low quiescent current and is ideal for portable instrumentation and data acquisition systems. The INA122 can be operated with a single power supplies from 2,2 V to 36 V and quiescent current is a mere 60 mA. A single external resistor sets the gain from 5 to 10000.

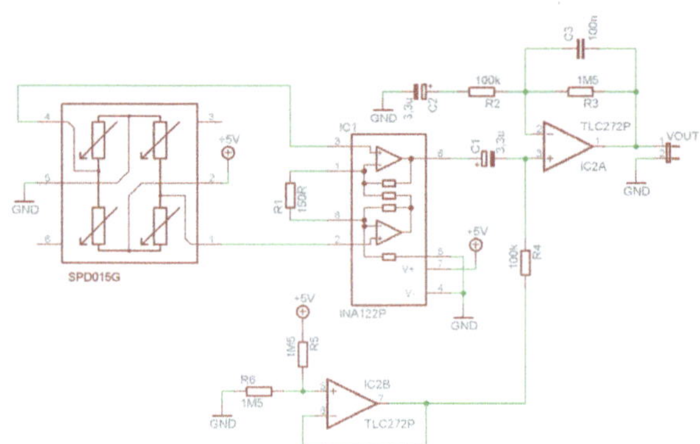

Fig. 8: Electrical circuit.

For filtering of the output signal operational amplifier TLC272 was used. The capacitor C_3 and resistor R_3 represent low-pass filter. This amplifier serves also for amplification of the signal. Selection of a single element values depends on the need of signal details.

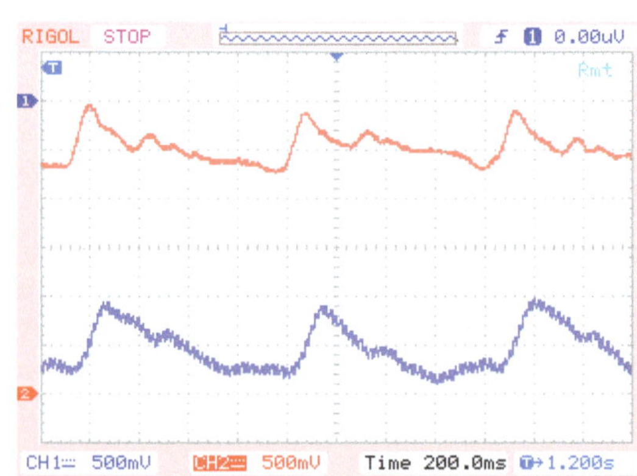

Fig. 9: Pressure wave measured by oscilloscope.

5. Wireless Data Transfer

For wireless data transfer Bluetooth® communication protocol was selected. From many devices module BTM-112 from Rayson was selected due to its performance ratio.

5.1. Bluetooth Module BTM-112

The selected module allows measured data transfer using serial communication protocol. For that reason another module is needed for the creation of a virtual serial port in the computer, which serves for further signal processing. It can be achieved by using conventional Bluetooth serial link adapter available on the market. The BTM-112 module has the following features:

- Bluetooth standard Ver. 2.0,
- Internal 1,8 V regulator,
- Low current consumption : average 46 mA,
- 3,0 V to 3,6 V operation,
- Support for up to seven slaves,
- Interface: USB, UART, PCM,
- SPP firmware with AT command sets,
- Small outline: 25 x 14.5 x 2.2 mm.

Fig. 10: BTM-112 module.

5.2. BTM-112 Settings

The settings of the selected module can be changed by using appropriate AT commands. All the AT commands except the "AT" test command use the below command structure:

¡header¿¡name¿¡ parameters¿ ¡cr¿

The module is shipped with the following default factory setting of UART:

- Baud rate: 19200 B,
- Data bits: 8,

Tab. 4: AT commands.

header	Each command will start with "AT" character sequence.
name	Command name.
parameters	The parameters are required for most of the commands. The parameter may be character, integer and character sequence (BD address, Pin code, Name, etc.) depending on the command operation.
cr	This character terminates the command packet and signals the device to proceed with command execution.

- Parity: none,
- Stop bits: 1,
- Flow control: H/W or none.

For that reason it is needed to change the parameters of the transfer. For our purposes we used the commands shown in the next table:

Tab. 5: Used AT commands.

Command	Name	Parameter
ATL	Baud Rate	4
ATK	Stop Bits	1
ATM	Parity	2

By using these commands desired serial link settings were achieved as it was described in Tab. 2 [7], [8].

5.3. Function Principle of BTM-112

If the conversion of the analog signal is done, microcontroller sends data to the RX pin of the module and the data are automatically sent to the PC by this module. The circuit diagram is shown in the Fig. 11.

6. Measured Data Analysis

Within this work we performed measurements on selected subjects and we tried to increase measured data worth by application of specific algorithms. There are many approaches to evaluate measured PPG curves. We can introduce comprehensive background research work of [9]. In this work we can read that it is possible to evaluate PPG signal curves in many ways. For our

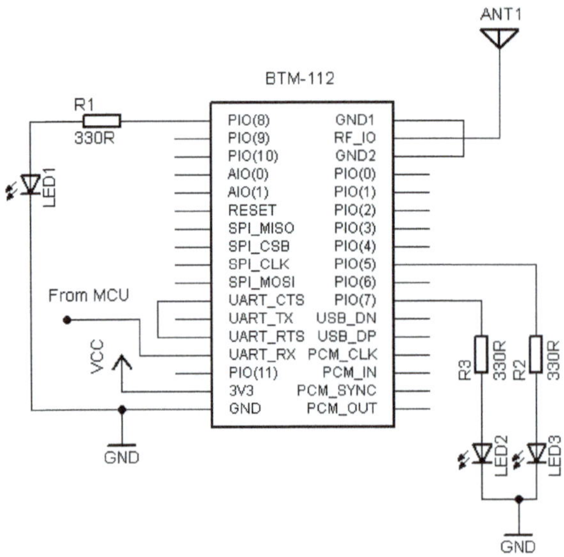

Fig. 11: BTM-112 – circuit diagram.

contribution we selected the second derivative of photoplethysmogram (SDPG) and we created the application for investigation SDPG curves parameters which are interconnected with state of the cardiovascular system parameter such as e.g. vascular compliance or elastic modulus [10], [11], [12], [13], [14].

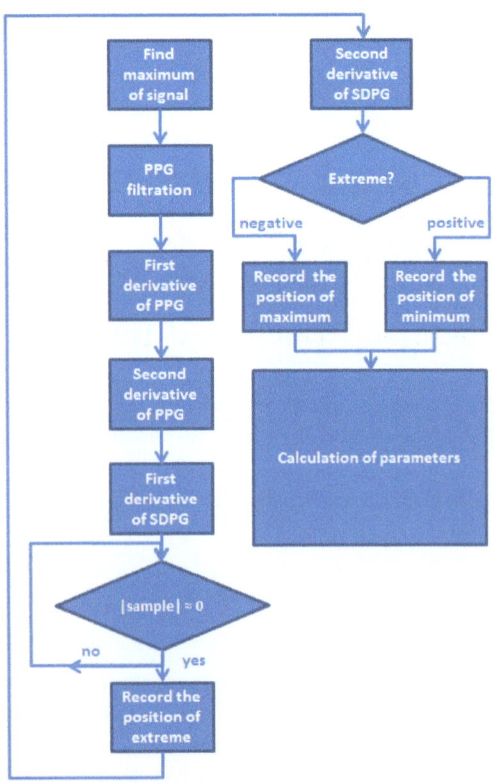

Fig. 12: Data analysis algorithm.

6.1. GUI for Evaluation of SDPG

An application for evaluation of SDPG was created in MATLAB programming environment. It is possible to evaluate measured PPG curves in this application and information about arterial elastic modulus can be obtained.

At first measured PPG curve has to be filtered because the next operation is a derivation of PPG signal. Without digital filtering of measured data noise is amplified by deriving of given signal. For data preprocessing was used moving average filter. Afterwards we applied first and second derivation to filtered signal and by that way we obtain SDPG curve. At this step it is needed to detect local maxima and minima of SDPG signal. If the absolute value of compared sample is around the zero then the position of local extreme is recorded. Afterwards the recorded local extremes are investigated. If the current recorded extreme has positive value then the position of local minimum is recorded and if the local extreme value is negative then the local maximum is detected. The data processing algorithm is shown in the Fig. 12.

Realized GUI for PPG analysis is shown in the Fig. 13. By using this application user can evaluate and determine the SDPG parameters such as peaks a, b, c, d and e. After determining of these peaks the single ratios are calculated which are interconnected with state of the cardiovascular system concretely with elastic modulus of arteries [10], [11], [12], [13], [14].

Measured b/a ratios are comparable to data presented by [10]. Our calculated slope of the trend line was 0,008 (see Fig. 14). This result is equal to result of the work [10].

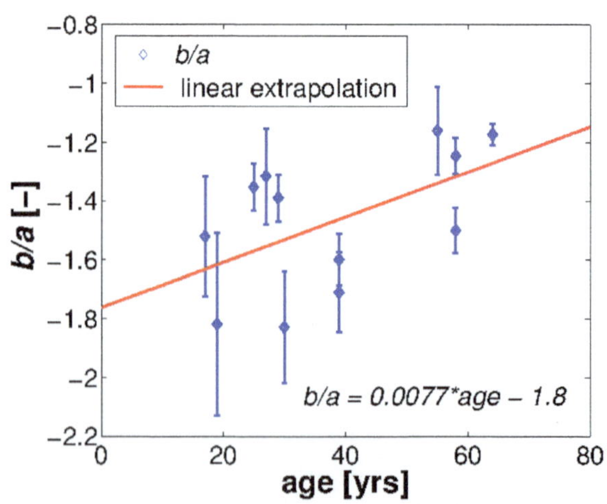

Fig. 14: Measurement results – b/a ratio dependence on age of measured subjects.

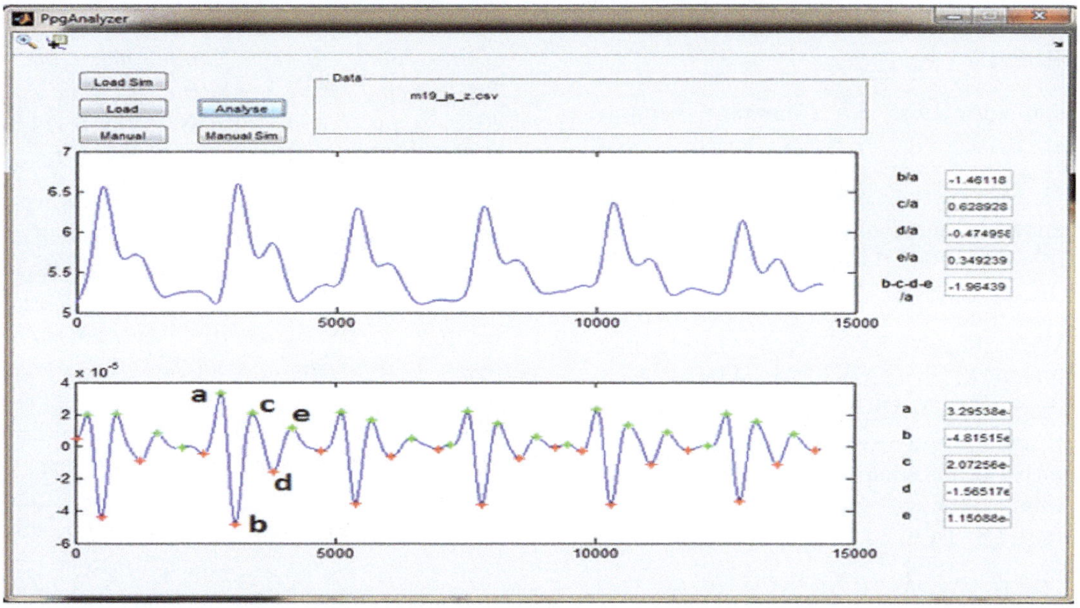

Fig. 13: Graphic user interface for PPG curve analysis.

7. Conclusion

Within the work device for scanning of the flow and pressure pulse wave was designed. Both types of pulse waves measured by these two principles can be compared and the comparison results could help to understand properties of the arterial system.The devices have two parts: analog and digital. The analog part serves for the analog signal amplification and filtering. The digital part is based on the use of the microcontroller which allows signal digitalization using integrated analog to digital converter. The microcontroller serves also for controlling of wireless data transfer. For this type of data transfer Bluetooth module was selected because standard mobile devices dispose of Bluetooth communication interface. The designed device should serve for patient monitoring and allows free patient movement without restrictions caused by communication and power supply cables. The patient can be monitored by a doctor for example by sending a message from his mobile phone. Within the work further analysis of measured and transferred data was presented. We created graphic user interface serving for PPG curve evaluation by using its second derivative SDPG. Calculated b/a ratios are interconnected with elastic modulus of arteries and they can give us information about vascular aging of measured subject. Measurement results could be combined for example with the existing model of the cardiovascular system based on electromechanical analogy. In this way the elastic modulus of arteries could be determined by comparing of measured and simulated data.

References

[1] JAVORKA, K. *Lekárska fyziológia: učebnica pre lekárske fakulty.* Martin: Osveta, 2001, ISBN 80-806-3023-2.

[2] BORIK, S, and I. CAP. Investigation of pulse wave velocity in arteries. In: *35th International Conference on Telecommunications and Signal Processing (TSP), 2012.* Prague: IEEE, 2012, pp. 562–565. ISBN 978-1-4673-1117-5. DOI: 10.1109/TSP.2012.6256358.

[3] BABUSIAK, B. and J. MOHYLOVA. The EEG Signal Prediction by Using Neural Network. *Advances in Electrical and Electronic Engineering.* 2008, vol 7, iss. 1–2, pp. 342–345. ISSN 1336-1376.

[4] BABUSIAK, B. and M. GALA. Detection of Abnormalities in ECG. In: *Information Technologies in Biomedicine, ITIB 2012.* Berlin: Springer Berlin Heidelberg, 2012, pp. 161–171. ISBN 978-3-642-31195-6. DOI: 10.1007/978-3-642-31196-3_17.

[5] CZIPPELOVA, B. and D. GOMBARSKA. The wave analysis of terminal segment effect on blood pressure and blood flow propagation. In: *8th international conference KRALIKY 2010.* Semily: Brno University of Technology, Faculty of Electrical Engineering and Communication, 2010, pp. 32–35. ISBN 978-80-214-4139-2.

[6] BORIK, S. Pulse wave velocity measurements. In: *XIV International PhD Workshop OWD 2012.* Gliwice: Conference Archives PTETIS, 2012, pp. 324–327. ISBN 978-83-935427-0-3.

[7] Rayson. *Class2 BC04-ext Module BTM-112.* [Datasheet] 2012.

[8] LM Technologies Ltd. *AT Command Manual.* [Datasheet] 2010.

[9] ELGENDI, M. On the analysis of fingertip photoplethysmogram signals. *Current cardiology reviews.* 2012, vol. 8, iss. 1, pp. 14–25. ISSN 1573-403X. DOI: 10.2174/157340312801215782.

[10] TAKAZAWA, K., N. TANAKA, M. FUJITA, O. MATSUOKA, T. SAIKI, M. AIKAWA, S. TAMURA and Ch. IBUKIYAMA. Assessment of vasoactive agents and vascular aging by the second derivative of photoplethysmogram waveform. *Hypertension.* 1998, vol. 32, iss. 2, pp. 365–370. ISSN 0194-911X. DOI: 10.1161/01.HYP.32.2.365.

[11] USMAN, S. B., M. A. B. M. ALI, M. M. B. I. REAZ and K. CHELLAPAN. Second derivative of photoplethysmogram in estimating vascular aging among diabetic patients, In: *International Conference for Technical Postgraduates 2009, TECHPOS 2009.* Kuala Lumpur: IEEE, 2009, pp. 1–3. ISBN 978-142445223-1. DOI: 10.1109/TECHPOS.2009.5412099.

[12] MOHAMAD R., S. USMAN, M. A. M. ALI and M. M. B. I. Reaz. Second derivatives of photoplethysmography (PPG) for estimating vascular aging of atherosclerotic patients. In: *Biomedical Engineering and Sciences (IECBES), 2012 IEEE EMBS Conference on.* Langkawi: IEEE, 2012, pp. 256–259. ISBN 978-1-4673-1664-4. DOI: 10.1109/IECBES.2012.6498064.

[13] JANG, Dae-Geun, Jang-Ho PARK, Seung-Hun PARK and M. HAHN. A morphological approach to calculation of the second derivative of photoplethysmography. In: *IEEE 10th International Conference on Signal Processing (ICSP), 2010.* Beijing: IEEE, 2010, pp. 1–4. ISBN 978-1-4244-5897-4. DOI: 10.1109/ICOSP.2010.5656472.

[14] BAEK, H. J., J. S. KIM, Y. S. KIM, H. B. LEE and K. S. PARK. Second Derivative of Photoplethysmography for Estimating Vascular Aging. In: *6th International Special Topic Conference on Information Technology Applications in Biomedicine, 2007. ITAB 2007.* Tokyo: IEEE, 2007, pp. 70–72. ISBN 978-1-4244-1868-8. DOI: 10.1109/ITAB.2007.4407346.

About Authors

Stefan BORIK was born in 1986. He received his M.Sc. from Faculty of Electrical Engineering, Department of Electromagnetic and Biomedical Engineering, University of Zilina in 2011. In the present time he is a Ph.D. student at the University of Zilina. He especially deals with analog and digital hardware design, biological signal processing and modeling.

Ivo CAP was born in 1948. He graduated from the Czech Technical University in Prague, Faculty of Electrical Engineering, CSc. (eq received at the Slovak Academy of Sciences, Institute of Electrical Engineering, professor of Physics of condensed matter and Acoustics at the University in Zilina. The present appointment is professor of Biomedical Engineering at the Faculty of Electrical Engineering of the University of Zilina. He is specialized in modelling of physiological processes, influence of EM field on living organisms and EM non-destructive testing of conducting materials.

2

A Voltage-Mode First Order Allpass Filter Based on VDTA

Suwat MANEEWAN[1], *Nawee UDORN*[2], *Danupat DUANGMALAI*[2], *Phamorn SILAPAN*[3], *Winai JAIKLA*[4]

[1]Department of Technology Electronics, Faculty of Industrial Technology, Buriram Rajabhat University, Buriram, 310 00, Thailand
[2]Department of Electronic Technology, Faculty of Industrial Technology, Nakhon Phanom University, Nakhon Phanom, 480 00, Thailand
[3]Department of Electrical Computer and Industrial, Faculty of Industrial Technology, Uttaradit Rajabhat University, Uttaradit, 530 00, Thailand
[4]Department of Engineering Education, Faculty of Industrial Education, King Mongkut's Institute of Technology Ladkrabang, Bangkok, 105 20, Thailand

suwat_m@hotmail.com, ornavy@yahoo.com, danupat@npu.ac.th, phamorn@mail.uru.ac.th, kawinai@kmitl.ac.th

Abstract. *This article presents a new voltage-mode first order allpass filter (APF) employing voltage differencing transconductance amplifier (VDTA). The advantages of the circuit are that: the circuit description is very compact, consists of merely a VDTA and a capacitor: the phase shift can be electronically adjusted by current bias: it provides the lower THD of the output signal. Without any component matching conditions, the proposed circuit is very appropriate to further develop into an integrated circuit. Moreover, the proposed APF can provide the output current with high output impedance without modification of the circuit topology. The PSpice simulation results are depicted. The given results agree well with the theoretical anticipation. The maximum power consumption is 400μW at ±1.25V power supplies.*

Keywords

Allpass filter, VDTA, voltage-mode.

1. Introduction

A phase shifter or first order allpass filter (APF) is widely employed in analog signal processing system such as, oscillators, high-Q band-pass filters and phase shifters [1], [2], [3], [4], [5], [6]. The literature surveys display that the first-order all-pass filter circuit using different high-performance active building blocks such as, current conveyors (CCIIs) [7], [8], [9], [10], [11], [12], OTAs [13], current controlled current conveyors (CCCIIs) [14], [15], [16], [17], differential voltage current conveyor (DVCC) [18], differential difference current conveyors (DDCCs) [19], [20], current differencing buffered amplifier (CDBA) [4], [21] and operational transresistance amplifiers (OTRAs) [22], [23], [24], voltage differencing-differential input buffered amplifier (VD-DIBA) [25], voltage differencing inverting buffered amplifier (VDIBA) [26], have been reported. However, these reported circuits suffer from one or more of the following weaknesses:

- use more one active element [9], [13],
- requirement of an external resistor [4], [7], [13], [18], [19], [20], [21], [22], [23], [24],
- absence of electronic adjustability [4], [7], [8], [9], [10], [11], [12], [18], [19], [20],
- absence of the output current with high output impedance in the same circuit topology [4], [7], [8], [9], [10], [11], [12], [13], [14], [15], [16], [17], [18], [19], [20], [21], [22], [23], [24], [25], [26].

In 2011, the voltage differencing transconductance amplifier is presented [27]. It consists of 2 ports input voltage, 2 port output current and 2 external current bias. The voltage differencing of V_n with V_n is transferred to current at the terminal z by first transconductance gain and the voltage at z port is transferred to current at the seems to be a versatile component in the realisation of a class of analog signal processing circuits. The fact is that the device can operate in both current and voltage-modes, provides flexibility and enables a variety of circuit designs. In addition, output current of VDTA can be electronically adjusted. The aim of this paper is to propose a voltage-mode first order allpass filter, emphasizing on the use of the VDTA. The features of the proposed circuit are that: the phase shift can be tuned by current bias: the circuit description is very simple, it uses 1 VDTA and a capacitor as passive elements, which is suitable for fabricating in monolithic chip or off-the-shelf implementation: phase shift can be independently adjusted. The performances of proposed circuit are illustrated by PSpice simulations, they show good agreement as mentioned.

2. Theory and Principle

2.1. Basic Concept of VDTA

Since the proposed APF is based on VDTA, it is realized by CMOS technology, a brief review of VDTA is given in this Section. The characteristics of the ideal VDTA are presented by the following hybrid matrix:

$$\begin{bmatrix} I_z \\ I_{x+} \\ I_{x-} \end{bmatrix} = \begin{bmatrix} g_{m1} & -g_{m1} & 0 \\ 0 & 0 & g_{m2} \\ 0 & 0 & -g_{m2} \end{bmatrix} \begin{bmatrix} V_p \\ V_n \\ V_z \end{bmatrix}, \tag{1}$$

g_{m1} and g_{m2} are the transconductances of the VDTA, I_{b1} and I_{b2} are currents bias employed to adjusted g_{m1} and g_{m2}, respectively. They can be found to be:

$$g_{m1} = \sqrt{I_{b1}\mu_i C_{ox}(W/L)}, \tag{2}$$

and

$$g_{m2} = \sqrt{I_{b2}\mu_i C_{ox}(W/L)}, \tag{3}$$

where μ_i is the mobility of the carrier for NMOS and PMOS transistors, C_{ox} is the gate-oxide capacitance/unit area, W and L are the effective channel width and length, respectively. The symbol of VDTA is shown in Fig. 1, where V_p and V_n are input terminals and z, $x+$ and x- are output terminals.

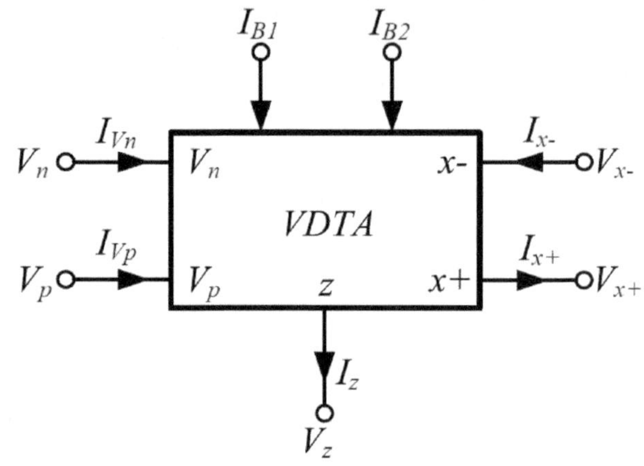

Fig. 1: Symbol of VDTA.

2.2. Proposed Voltage-Mode First-Order All-Pass Filter

The proposed voltage-mode first-order APF is illustrated in Fig. 2. The proposed circuit consists of only one VDTA and a capacitor. Not only the output voltage is achieved, the output current with high output impedance is also achieved which is well-known as transconductance-mode. Considering the circuit in Fig. 2, the current of z terminal can be found to be:

$$0 = I_c + I_z + I_{x-}. \tag{4}$$

From the properties of VDTA, Eq. (4) can be rewritten as:

$$0 = sC(V_{in} - V_0) - g_{m1}V_{in} - V_0 g_{m2}. \tag{5}$$

From Eq. (5), the voltage and transconductanc transfer functions can be express to be:

$$\frac{V_0}{V_{in}} = \frac{sC - g_{m1}}{sC + g_{m2}}, \tag{6}$$

and

$$\frac{I_0}{V_{in}} = g_{m2}\left(\frac{sC - g_{m1}}{sC + g_{m2}}\right). \tag{7}$$

For easy consideration, if g_{m2}=g_{m1}=g_m, the transfer functions can be rewritten to:

$$\frac{V_0}{V_{in}} = \frac{sC - g_m}{sC + g_m}, \tag{8}$$

and

$$\frac{I_0}{V_{in}} = g_m\left(\frac{sC - g_m}{sC + g_m}\right). \tag{9}$$

From Eq. (8), the pole frequency, voltage gain and phase response of the proposed circuit are:

$$\omega_p = \frac{g_m}{C}, \tag{10}$$

$$\left|\frac{V_0}{V_{in}}\right| = 1, \tag{11}$$

and

$$\phi(\omega_p) = \pi - 2\tan^{-1}\left(\frac{\omega_p C}{g_m}\right). \tag{12}$$

For transconductance-mode, the pole frequency and phase response are same to the voltage-mode APF. But the transconductance-gain is written as:

$$\left|\frac{I_0}{V_{in}}\right| = g_m. \tag{13}$$

From properties of VDTA as shown in Eq. (2) and Eq. (3), the pole frequency and phase response for both mode can be modified to be:

$$\omega_p = \frac{C}{\sqrt{I_B \mu_i C_{ox}(W/L)}}, \tag{14}$$

and

$$\phi(\omega_p) = \pi - 2\tan^{-1}\left[\frac{\omega_p C}{\sqrt{I_B \mu_i C_{ox}(W/L)}}\right]. \tag{15}$$

It can be seen that the circuit gives a phase shift from 180° –0°. Moreover, the angle pole frequency can be electronically controlled by I_B. For transconductance-mode, the transconductance gain is rewritten as:

$$\left|\frac{I_O}{V_{in}}\right| = \sqrt{I_B \mu_i C_{ox}(W/L)}. \tag{16}$$

The ω_p sensitivities of the filter can be written to:

$$S_C^{\omega_p} = -1, S_{I_B}^{\omega_p} = 1.5. \tag{17}$$

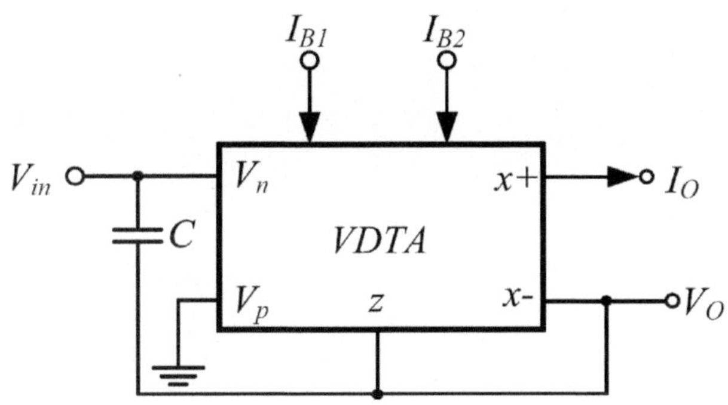

Fig. 2: Proposed voltage-mode first-order APF.

2.3. Non Ideal Case

In practice, the influences of voltage and current tracking errors and also the parasitic terminal impedances of VDTA will affect the filter performance. In this Section, these parameters will be taken into account. For non-ideal the VDTA can be respectively characterized with the following equations:

$$\begin{bmatrix} I_z \\ I_{x+} \\ I_{x-} \end{bmatrix} = \begin{bmatrix} \alpha_P g_{m1} & \alpha_N g_{m1} & 0 \\ 0 & 0 & \beta g_{m2} \\ 0 & 0 & -\beta g_{m2} \end{bmatrix} \begin{bmatrix} V_P \\ V_N \\ V_Z \end{bmatrix}, \tag{18}$$

where α_P and α_N are the transconductance error gains from P and N ports to z port. β is the transconductance error gain from z port to x port. The influences of parasitic impedances are resistive and capacitive parts affecting the P, N, Z and X ports of VDTA. Let us denote them R_P, C_P, R_N, C_N, R_z, C_Z, and R_X, C_X, respectively. Considering into these effects, the voltage and transconductane transfer functions will be modified to the more general forms:

$$\frac{V_O}{V_{in}} = \frac{sC - \alpha_N g_{m1}}{s(C+C^*) + G^* + \beta g_{m2}}, \tag{19}$$

and

$$\frac{I_O}{V_{in}} = \beta g_{m2}\left(\frac{sC - \alpha_N g_{m1}}{s(C+C^*) + G^* + \beta g_{m2}}\right), \tag{20}$$

where $C^* = C_N + C_Z + C_X$ and $G^* = G_N + G_Z + G_X$. In this case, the pole frequency, voltage gain and phase response are modified to:

$$\omega_p = \frac{G^* + \alpha_N g_m}{C + C^*}, \tag{21}$$

$$\left|\frac{V_O}{V_{in}}\right| = \frac{\sqrt{(\omega_p C)^2 - (\alpha_N g_{m1})^2}}{\sqrt{[\omega_p (C+C^*)]^2 + (G^* + \beta g_{m2})^2}}, \tag{22}$$

and

$$\begin{aligned} \phi(\omega_p) &= \pi - \tan^{-1}\left(\frac{\omega_p C}{\alpha_N g_{m1}}\right) \\ &- \tan^{-1}\left[\frac{\omega_p (C+C^*)}{G^* + \beta g_{m2}}\right]. \end{aligned} \tag{23}$$

The transconductance-gain from Eq. (20) is written as:

$$\left|\frac{I_O}{V_{in}}\right| = \beta g_{m2}\frac{\sqrt{(\omega_p C)^2 - (\alpha_N g_{m1})^2}}{\sqrt{[\omega_p (C+C^*)]^2 + (G^* + \beta g_{m2})^2}}. \tag{24}$$

It should be mentioned that the non-ideal parameters of the VDTA affect the pole frequency, voltage gain and phase response.

3. Simulation Results

The performances of the proposed voltage-mode first order allpass filter have been tested by PSpice simulation. This work employed a VDTA realized by a CMOS technology. The NMOS and PMOS transistors employed in the proposed circuit as shown in Fig. 2, were simulated by respectively using the parameters of the 0.25 µm TSMC CMOS technology (level 7) with ±1.25 V supply voltages. Fig. 3 depicts the schematic description of VDTA used in the simulations. The aspect ratios of PMOS and NMOS transistor are W/L = 8 µm/0.25 µm and W/L = 5 µm/0.25 µm, respectively. C = 10 pF, I_B = 80 µA. Simulated gain and phase responses of the APF are given in Fig. 4. It can be found that the simulated gain is slightly deviated from ideal responses due to the error terms as expressed in Eq. (22). Phase response for different I_B is shown in Fig. 5. This result confirms that the angle natural frequency can be electronically controlled by setting IB as shown in Eq. (15). The time-domain response of the proposed APF is shown in Fig. 6, where a sine wave of 150 mV/5 MHz was applied as the input to the filter. Fig. 7 shows the total harmonic distortion (THD) variation with respect to amplitude of the applied sinusoidal input voltage at the pole frequency of the all-pass filter. The tuning of pole frequency by I_B is confirmed by the result in Fig. 8.

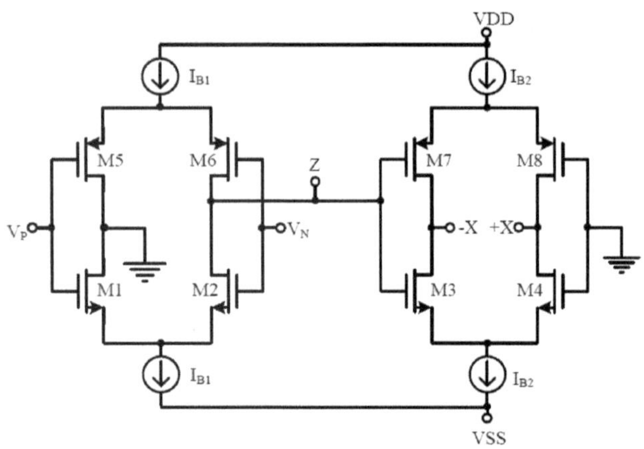

Fig. 3: Internal construction of VDTA.

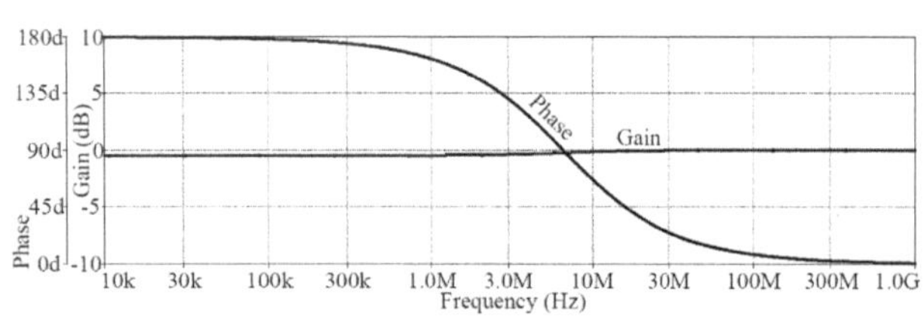

Fig. 4: Output phase and gain response.

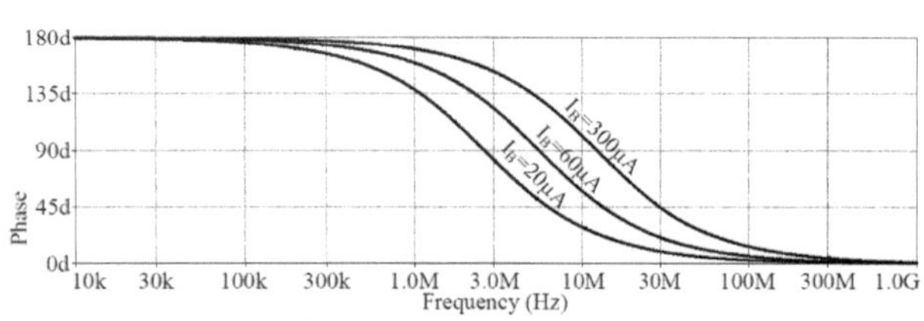

Fig. 5: Output phase and gain responses.

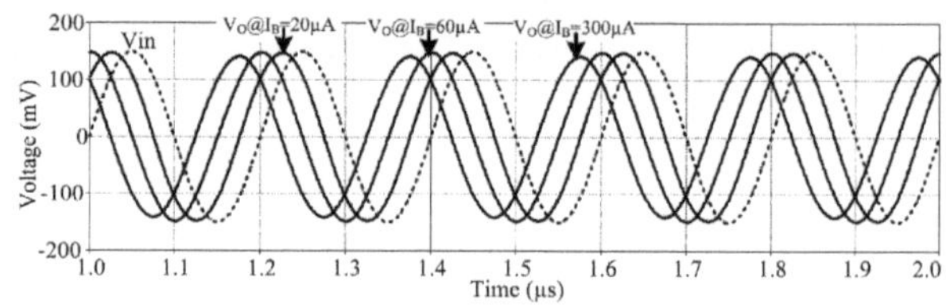

Fig. 6: Time domain response of the circuit in Fig. 2.

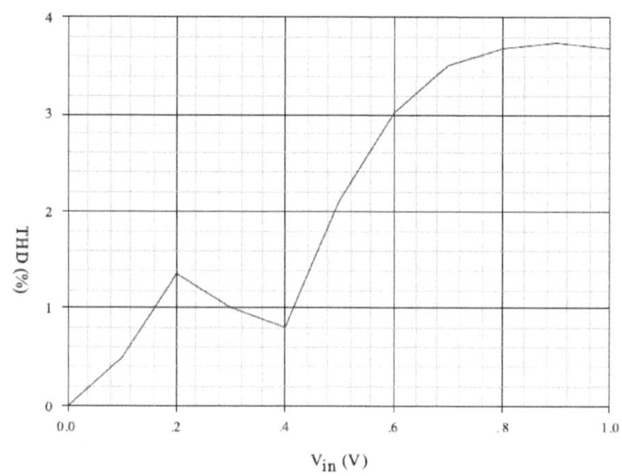

Fig. 7: THD variations versus amplitudes of the applied sinusoidal input voltages at f_0= 6.75 MHz.

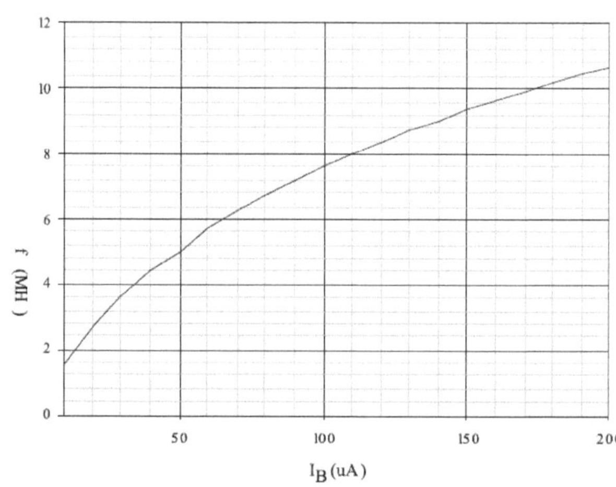

Fig. 8: Pole frequency vs I_B.

4. Conclusion

An electronically tunable voltage-mode first-order all-pass filter has been introduced via this paper. The proposed configuration is very simple and can be electronically controlled. It consists of single VDTA and single capacitor. So it is easy to fabricate in IC form to use in battery-powered or portable electronic equipments such as wireless communication devices. In addition, the output current with high output impdenace is achieved. The PSpice simulation results were depicted, and agree well with the theoretical anticipation.

References

[1] SCHAUMANN, R. and E. VALKENBURG. *Design of Analog Filters.* New York: Oxford University Press, 2001. ISBN 0-19-973046-6.

[2] METIN, B., N. HERENCSAR and K. PAL. Supplementary first-order all-pass filters with two grounded passive elements using FDCCII. *Radioengineering.* 2011, vol. 20, iss. 2, pp. 433–437. ISSN 1210-2512.

[3] COMER, D. and J. MCDERMID. Inductorless Bandpass Characteristics Using All-Pass Networks. *IEEE Transactions on Circuit Theory.* 1968, vol. 15, iss. 4, pp. 501–503. ISSN 0018-9324. DOI: 10.1109/TCT.1968.1082852.

[4] TOKER, A., S. OZOGUZ, O. CICEKOGLU and C. ACAR. Current-mode all-pass filters using current differencing buffered amplifier and a new high-Q bandpass filter configuration. *IEEE Transactions on Circuits and Systems II: Analog and Digital Signal Processing.* 2000, vol. 47, iss. 9, pp. 949–954. ISSN 1057-7130. DOI: 10.1109/82.868465.

[5] HORNG, J.-W. DVCCs based high input impedance voltage-mode first-order allpass, highpass and lowpass filters employing grounded capacitor and resistor. *Radioengineering.* 2010, vol. 19, iss. 4, pp. 653–656. ISSN 1210-2512.

[6] CICEKOGLU, O., H. H. KUNTMAN a S. BERK. All-Pass Filters Using A Single Current Conveyor. In: *Proceedings of the 6th International Conference on Optimization of Electrical and Electronic Equipments.* Brasov: IEEE, 1998, pp. 649–654. ISBN 973-98511-2-6. DOI: 10.1109/OPTIM.1998.708015.

[7] PAL, K. and S. RANA. Some New First-Order All-Pass Realizations Using CCII. *Active and Passive Electronic Components.* 2004, vol. 27, iss. 2, pp. 91–94. ISSN 0882-7516. DOI: 10.1080/0882751031000116188.

[8] SOLIMAN, A. M. Generation of current conveyor-based all-pass filters from op amp-based circuits. *IEEE Transactions on Circuits and Systems II: Analog and Digital Signal Processing.* 1997, vol. 44, iss. 4, pp. 324–330. ISSN 1057-7130. DOI: 10.1109/82.566650.

[9] HORNG, J.-W. Current conveyors based all-pass filters and quadrature oscillators employing grounded capacitors and resistors. *Computers & Electrical Engineering.* 2005, vol. 31, iss. 1, pp. 81–92. ISSN 0045-7906. DOI: 10.1016/j.compeleceng.2004.11.006.

[10] IBRAHIM, M. A., H. KUNTMAN, S. OZCAN, O. SUVAK and O. CICEKOGLU. New first-order inverting-type second-generation current conveyor-based all-pass sections including canonical forms. *Electrical Engineering.* 2004, vol. 86, iss. 5, pp. 299–301. ISSN 0948-7921. DOI: 10.1007/s00202-003-0205-3.

[11] HIGASHIMURA, M. and Y. FUKUI. Realization of all-pass and notch filters using a single current conveyor. *International Journal of Electronics.* 1988, vol. 65, iss. 4, pp. 823–828. ISSN 0020-7217. DOI: 10.1080/00207218808945282.

[12] HORNG, J.-W., C.-L. HOU, C.-M. CHANG, W.-Y. CHUNG, H.-L. LIU and C.-T. LIN. High output impedance current-mode first-order allpass networks with four grounded components and two CCIIs. *International Journal of Electronics.* 2006, vol. 93, iss. 9, pp. 613–621. ISSN 0020-7217. DOI: 10.1080/00207210600711580.

[13] PSYCHALINOS, C. and K. PAL. A novel all-pass current-mode filter realized using a minimum number of single output OTAs. *Frequenz.* 2010, vol. 64, iss. 1–2, pp. 30–32. ISSN 0016-1136. DOI: 10.1515/FREQ.2010.64.1-2.30.

[14] KUMAR, P. A., U. KESKIN and K. Pal. Wide-band resistor less all-pass sections with single element tuning. *International Journal of Electronics.* 2007, vol. 94, iss. 6, pp. 597–604. ISSN 0020-7217.

[15] TOKER, A., E. O. GUNE and S. OZOGUZ. New high-Q band-pass filter configuration using current controlled current conveyor based all-pass filters. In: *Proceedings of ICECS 2001. 8th IEEE International Conference on Electronics, Circuits and Systems.* Malta: IEEE, 2001, pp. 165–168. ISBN 0-7803-7057-0. DOI: 10.1109/ICECS.2001.957706.

[16] MINAEI, S. and O. CICEKOGLU. A Resistorless realization of the first-order all-pass filter. *International Journal of Electronics.* 2006, vol. 93, iss. 3, pp. 177–183. ISSN 0020-7217. DOI: 10.1080/00207210600562173.

[17] PRASAD, D. Electronically Controllable Fully-Uncoupled Explicit Current-Mode Quadrature Oscillator Using VDTAs and Grounded Capacitors. *Circuits and Systems.* 2013, vol. 04, iss. 02, pp. 169–172. ISSN 2153-1285. DOI: 10.4236/cs.2013.42023.

[18] IBRAHIM, M. A., S. MINAEI and H. KUNTMAN. DVCC based differential-mode all-pass and notch filters with high CMRR. *International Journal of Electronics.* 2006, vol. 93, iss. 4, pp. 231–240. ISSN 0020-7217. DOI: 10.1080/00207210600562181.

[19] IBRAHIM, M. A., H. KUNTMAN and O. CICEKOGLU. First-Order All-Pass Filter Canonical in the Number of Resistors and Capacitors Employing a Single DDCC. In: *Circuits System and Signal Processing*. 2003, vol. 22, iss. 5. pp. 525–536. ISSN 0278-081X. DOI: 10.1007/s00034-003-1111-7.

[20] HORNG, J.-W., C.-L. HOU, C.-M. CHANG, Y.-T. LIN, I.-C. SHIU and W.-Y. CHIU. First-order allpass filter and sinusoidal oscillators using DDCCs. *International Journal of Electronics*. 2006, vol. 93, iss. 7, pp. 457–466. ISSN 0020-7217. DOI: 10.1080/00207210600711481.

[21] MAHESHWARI, S. Voltage-Mode All-Pass Filters Including Minimum Component Count Circuits. *Active and Passive Electronic Components*. 2007, vol. 2007, pp. 1–5. ISSN 1555-029X. DOI: 10.1155/2007/79159.

[22] KILINC, S. and U. CAM. Realization of allpass filters using operational transresistance amplifier (OTRA). In: *Proceedings of the IEEE 12th Signal Processing and Communications Applications Conference, 2004*. Kusadasi: IEEE, 2004, pp. 133–136. ISBN 0-7803-8318-4. DOI: 10.1109/SIU.2004.1338276.

[23] KILINC, S. and U. CAM. Operational transresistance amplifier based first-order allpass filter with an application example. In: *The 2004 47th Midwest Symposium on Circuits and Systems, 2004*. Piscataway: IEEE, 2004, pp. 65–68. ISBN 0-7803-8346-X. DOI: 10.1109/MWSCAS.2004.1353898.

[24] CAKIR, C., U. CAM and O. CICEKOGLU. Novel allpass filter configuration employing single OTRA. *IEEE Transactions on Circuits and Systems II: Express Briefs*. 2005, vol. 52, iss. 3, pp. 122–125. ISSN 1549-7747. DOI: 10.1109/TCSII.2004.842055.

[25] BIOLEK, D. and V. BIOLKOVA. First-order voltage-mode all-pass filter employing one active element and one grounded capacitor. *Analog Integrated Circuits and Signal Processing*. 2010, vol. 65, iss. 1, pp. 123–129. ISSN 0925-1030. DOI: 10.1007/s10470-009-9435-2.

[26] HERENCSAR, N., S. MINAEI, J. KOTON, E. YUCE and K. VRBA. New resistorless and electronically tunable realization of dual-output VM all-pass filter using VDIBA. *Analog Integrated Circuits and Signal Processing*. 2013, vol. 74, iss. 1, pp. 141–154. ISSN 0925-1030. DOI: 10.1007/s10470-012-9936-2.

[27] YESIL, A., F. KACAR and H. KUNTMAN. New simple CMOS realization of voltage differencing transconductance amplifier and its RF filter application. *Radioengineering*. 2011, vol. 20, iss. 3, pp. 632–637. ISSN 1210-2512.

About Authors

Suwat MANEEWAN was born in Buriram, Thailand in 1974. He has been with Department of Technology Electronics, Faculty of Industrial Technology, Buriram Rajabhat University since 1998. His research interests include electronic communications, analog signal processing and analog integrated circuit.

Nawee UDORN was born in Nakhom Phanom, Thailand in 1972. He received his Higher Diploma in Technology (Industrial Education) in Electrical-Telecommunication from Patumwan Technical College, Thailand in 1994, M. Ed. in Education Technology from Mahasarakham University, Thailand in 2002. He has been with department of Electronics Technology, Faculty of Industrial Technology, Nakhon Phanom University, Nakhon Phanom, Thailand since 1994. His research interests include electronic communications, analog signal processing and analog integrated circuit.

Danupat DUANGMALAI was born in Nakhon Phanoom, Thailand in 1974. He received his B. Eng. degree in electronic and communications engineering from Pathumwan institute of Technology, Thailand in 2004, M. Tech. Ed. in electrical technology from King Mongkut's University of Technology North Bangkok (KMUTNB) in 2006. He has been with department of Electronicsl technology, Faculty of Industrial Technology, Nakhon Phanoom University, Nakhon Phanoom, Thailand. His research interests include electronic communications, analog signal processing and analog integrated circuit.

Phamorn SILAPAN was born in Phaitsanulok, Thailand in 1977. He received his B. Eng. degree in electrical engineering from Mahanakorn University of Technology, Thailand in 2002, M. Tech. Ed. in electrical technology and Ph.D. in electrical education from King Mongkut's University of Technology North Bangkok (KMUTNB) in 2005 and 2011, respectively. He has been with department of Electrical Computer and Industrial technology, Faculty of Industrial Technology, Uttaradit Rajabhat University, Uttaradit, Uttaradit, Thailand since 2006. His research interests include electronic communications, analog signal processing and analog integrated circuit.

Winai JAIKLA was born in Buriram, Thailand. He received the B. S. I. Ed. degree in telecommunication engineering from King Mongkut's Institute of Technology Ladkrabang (KMITL), Thailand in 2002,

M. Tech. Ed. in electrical technology and Ph.D. in electrical education from King Mongkut's University of Technology North Bangkok (KMUTNB) in 2004 and 2010, respectively. From 2004 to 2011 he was with Electric and Electronic Program, Faculty of Industrial Technology, Suan Sunandha Rajabhat University, Bangkok, Thailand. He has been with Department of Engineering Education, Faculty of Industrial Education, King Mongkut's Institute of Technology Ladkrabang, Bangkok, Thailand since 2012. His research interests include electronic communications, analog signal processing and analog integrated circuits. He is a member of ECTI, Thailand.

3

Linear Regression Based Real-Time Filtering

Misel BATMEND, Daniela PERDUKOVA

Department of Electrical Engineering and Mechatronics, Faculty of Electrical Engineering and Informatics, Technical University of Kosice, Letna 9, 042 00 Kosice, Slovak Republic

misel.batmend@tuke.sk, daniela.perdukova@tuke.sk

Abstract. *This paper introduces real time filtering method based on linear least squares fitted line. Method can be used in case that a filtered signal is linear. This constraint narrows a band of potential applications. Advantage over Kalman filter is that it is computationally less expensive. The paper further deals with application of introduced method on filtering data used to evaluate a position of engraved material with respect to engraving machine. The filter was implemented to the CNC engraving machine control system. Experiments showing its performance are included.*

Keywords

Filtering, filtering based on model, linear least squares method, linear regression.

1. Introduction

The problem of real-time signal filtering is addressed extensively in technical society. Variety of filters has been created. As an illustration, we can mention simple averaging or median filter; highpass, lowpass, bandpass filters [1] or widely used Kalman filter [2], [3]. The superiority of the latter comes out of fact that it relies not only on measured data, but also on prior knowledge of a measured signal in the form of a mathematical model. Moreover, the model is iteratively updated based on previous measurements.

In this paper, we introduce filtering method, which is also using prior knowledge of the measured signal. We will show that under certain conditions linear regression can be used to filter a signal with known mathematical description. The idea of filtering lies in fitting a curve to already measured data, given the mathematical description of a curve. This basically means that if we know what kind of signal we expect we can interpolate a measured data by a curve of expected shape in real-time, thus filtering the noised data.

In the theoretical part of this paper, we describe mathematical fundamentals of linear regression (LR) filter for a line fit.

The practical part of the paper is focused on implementation of LR filter in a CNC engraving machine. For a correct functionality, a constant distance between engraving tool and the material has to be held. To accomplish a constant distance, proximity measurements are taken and the tool's position in Z axis is controlled accordingly. The proximity sensor data are corrupted by noise and need to be filtered. To filter the data and to estimate the position of material surface with reference to the machine's frame, we use LR filter.

2. Least Squares Filter

Least squares (LS) error fitting is widely used method in statistics and other fields of math. Usually it is used to fit a curve to already measured set of data.

Given a set of n points (data pairs) (x_i,y_i), $i = 1,\dots,n$, where x_i is an independent variable and y_i is dependent variable, we can fit a line $y = ax + b$, using following equations:

$$a = \frac{n\sum_{i=1}^{n} x_i y_i - \sum_{i=1}^{n} x_i \sum_{i=0}^{n} y_i}{n\sum_{i=1}^{n} x_i^2 - \left(\sum_{i=1}^{n} x_i\right)^2}, \tag{1}$$

$$b = \frac{n\sum_{i=1}^{n} x_i^2 - a\sum_{i=1}^{n} x_i}{n}. \tag{2}$$

This is a common least squares line fitting problem solution and complete derivation of above equations can be found in [4]. Theoretically, it can be used to filter any mathematically described signal which is a linear combination of functions in form:

$$f(x_i, \beta) = \sum_{j=1}^{m} \beta_j \phi_j(x_i), \tag{3}$$

where β_j are coefficients and ϕ_j are functions of independent variable xi. The signal can also be nonlinear, but there is no closed-form solution for such LR problem. Instead, numerical algorithms must be used [5]. Computing LR for more complex functions requires more sophisticated algorithms such as LU (lower-upper) decomposition, or singular value decomposition, which are computationally more expensive and real time filtering would require more powerful environment.

If we know that measured signal is linear, we can use this method for real time signal filtering. With each measured point, we can compute a new fitted line. The more points there are the more accurate fit we get. Pseudo code of LS real time filter would look as follows:

```
sumx = 0; sumy = 0; sumxx = 0; sumxy = 0; n=0
//initialize variables
while (true)
{
x = determine_x();
y = measure_y();
n++;  //iterate number of points
sumx += x;  //update sum of x 's
sumy += y; //update sum of y 's
sumxy += x * y; //update sum of (x * y)
sumxx += x * x; //update sum of (x * x)
a = (n * sumxy - sumx * sumy) / (n * sumxx -
sumx * sumx);    //calculate parameter a
b = (sumy - a * sumx) / n; //calculate
paramenter b
}
```

Advantage of this code is, that it is very fast and consumes a small amount of memory. There is no need to remember all measured points, only sums of x and y need to be stored. Figure 1 shows filtering of noisy signal with LS filter. In Fig. 2 averaging filter was used on the same dataset. At the beginning, performance of LS filter and averaging filter is comparable. After having more than 20 data points ($x = 0$) measured, LS filter clearly outperforms averaging filter.

We can compare the complexity of LR filter to Kalman filter, since it also relies on a model of measured signal. The Kalman filter model assumes the true state at time k is evolved from the state at $(k-1)$ according to:

$$x_k = F_k x_{k-1} + B_k u_K + w_k, \tag{4}$$

where F_k is the state transition model which is applied to the previous state x_{k-1}; B_k is the control input model which is applied to the control vector u_k; w_k is the process noise which is assumed to be drawn from a zero mean multivarate normal distribution with covariance Q_k [2].

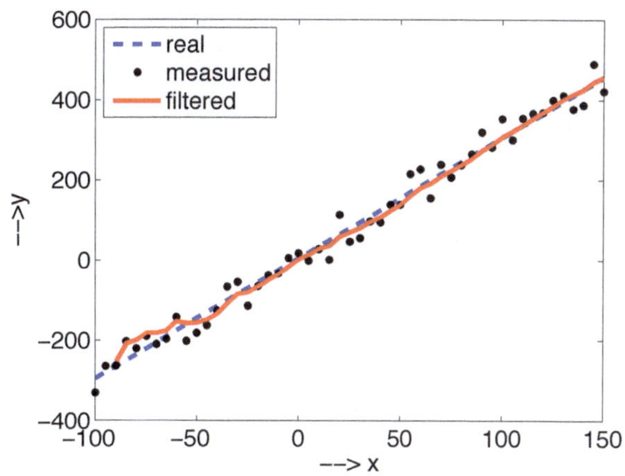

Fig. 1: Least squares filter.

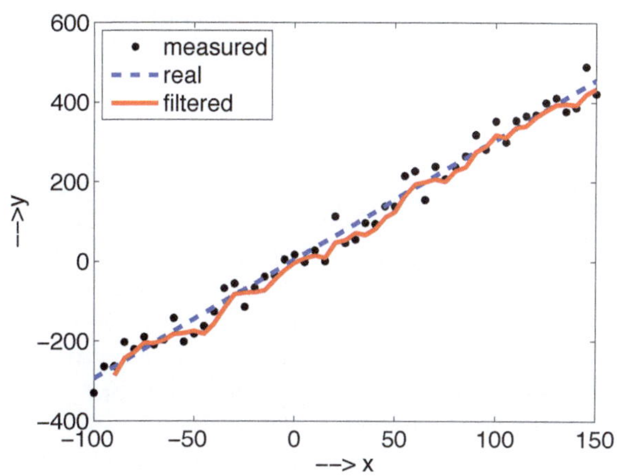

Fig. 2: Average filter using 3 samples.

At time k an observation (or measurement) z_k of the true state x_k is made according to:

$$z_k = H_k x_k + v_k, \tag{5}$$

where H_k is the observation model which maps the true state space into the observed space and v_k is the observation noise which is assumed to be zero mean Gaussian white noise with covariance R_k [2].

The filtered value $\widehat{x}_{k|k}$ of measured signal in time k is computed in two stages: predict and update.

Predict:

$$\widehat{x}_{k|k-1} = F_k \widehat{x}_{k-1|k-1} + B_{k-1} u_{k-1}, \tag{6}$$

$$P_{k|k-1} = F_k P_{k-1|k-1} F_k^T + Q_k. \tag{7}$$

Update:

$$K_k = P_{k|k-1} H_k^T (H_k P_{k|k-1} H_k^T + R_k)^{-1}, \tag{8}$$

$$\widehat{x}_{k|k} = \widehat{x}_{k|k-1} + K_k(z_k - H_k \widehat{x}_{k|k-1}), \tag{9}$$

$$P_{k|k} = (I - K_k H_k) P_{k|k-1}. \tag{10}$$

Practical implementation of Kalman filter requires a model of the system in the form of matrices F_k, B_k, H_k, and process noise and sensor noise covariances Q_k and R_k. The more precise model and estimate of covariances are, the better the performance of filter is achieved. From a computing point of view, in the general case, Kalman filter requires multiplications, additions and inversion of matrices. Computationally most expensive is the inversion of matrix in Eq. (8).

3. Application of LS Filter

In our application, LS filter is used to filter a signal from a proximity sensor which is measuring a distance of material that is engraved by CNC machine. The machine was built for engraving planar solid surfaces, such as stone blocks. The engraved picture is not 3D, but only planar (similar to drawing). Engraving is performed by hammering on the material with pointed diamond. This would theoretically require $X-Y$ plotter which has a working plane in parallel with material. This parallelism must be rather accurate, so that a constant distance between the tool and material is held. In our case, required distance is $1 \pm 0,5$ mm. Bedding the material with such accuracy is often a difficult task. For this reason, a degree of freedom in Z-axis was added to the machine. The only purpose of Z-axis movement is to compensate for non-parallelism of material surface with respect to working plane of the machine. A deflection from parallelism is evaluated by measuring the distance between the diamond tool tip and the material surface in certain points. Measurements are performed by rather unconventional proximity sensor which uses touching of a tool tip with material to evaluate the distance. A detailed description of the sensor can be found in [6]. From the nature of touch sensor principle, it is obvious that the bigger the distance, the longer it takes to measure it. In our machine small distances up to 3 mm can be measured in tens of milliseconds, but a 15 mm distance can take up to 3 seconds (!) to measure. This constraint comes out of fact, that small distances are measured only by moving the tool tip which has low mass. In contrast, for longer distances, whole engraving tool with considerably bigger mass has to be moved. Taking in account above constraints, sensor allows to measure distance with each and every stroke of hammering tool while engraving (the engraving distance is 1 mm). In addition, the measurement is taken during the time the tool is performing its stroke, so it does not introduce any additional time delay.

On the other hand, initial evaluating of material surface plane might take longer, because every point in which we measure distance might take 3 seconds to measure. That is why amount of initial measurement points is limited.

4. Estimation of Material Surface Plane Position

To evaluate the parallelism deflection of material surface and machine working plane, we measure distance in a few representative points and fit a plane on these points. Based on this equation, a Z-axis would be driven, so that a constant distance between the tool and material is kept. To define a plane, at least 3 points are necessary. To make sure, that engraved picture lies completely on a material surface, we picked from 4 to 6 representative measurement points depending on outer shape of the picture. Since the sensor has some error, measured points do not lie exactly on a plane. So we came to over-determined task of fitting a plane to $4 \div 6$ points. To do this, we used a multiple linear least squares regression [7], [8], [9], [10]. The plane equation is:

$$ax + by + cz + d = 0. \tag{11}$$

z can be calculated as:

$$z = -\frac{d}{c} - \frac{a}{c}x - \frac{b}{c}y. \tag{12}$$

Since there is an additional sensor error in z direction we can write:

$$z = k_0 + k_1 x + k_2 y + e, \tag{13}$$

where

$$k_0 = -\frac{d}{c},\ k_1 = -\frac{a}{c},\ k_2 = -\frac{b}{c}. \tag{14}$$

As we have more measured points, we can write the Eq. (13) in matrix form:

$$\mathbf{z} = \mathbf{J}\vec{k} + \mathbf{E}, \tag{15}$$

where

$$\mathbf{J} = \begin{pmatrix} 1 & x_1 & y_1 \\ \vdots & \vdots & \vdots \\ 1 & x_n & y_n \end{pmatrix}. \tag{16}$$

n is the number of measured points and $\vec{k}$ is a vector of unknown parameters.

The square residual error $\mathbf{E}^2$ is calculated by rearranging and then squaring $\mathbf{E}$ as is shown below:

$$\begin{aligned}\mathbf{E}^2 &= (\mathbf{z} - \mathbf{J}\vec{k})^2 = (\mathbf{z} - \mathbf{J}\vec{k})^T(\mathbf{z} - \mathbf{J}\vec{k}) \\ &= \mathbf{z}^T\mathbf{z} + \vec{k}^T(\mathbf{J}^T\mathbf{J})\vec{k} - 2\mathbf{z}^T\mathbf{J}\vec{k}.\end{aligned} \tag{17}$$

At the minimum square residual error, the derivatives of $\mathbf{E}^2$ with respect to the unknown $\vec{k}$ are equal to zero:

$$\frac{\partial(\mathbf{E}^2)}{\partial\vec{k}} = 2\mathbf{J}^T\mathbf{J}\vec{k} - 2\mathbf{J}^T\mathbf{z} = 0. \tag{18}$$

This gives the equation:

$$\mathbf{J}^T\mathbf{J}\vec{k} = \mathbf{J}^T\mathbf{z}. \tag{19}$$

By solving the equation we get the vector of parameters $\vec{k}$:

$$\vec{k} = \mathbf{J}^T\mathbf{J}\backslash\mathbf{J}^T\mathbf{z}. \tag{20}$$

The equation of fitted plane ρ is then:

$$\rho:\ z = k_0 + k_1 x + k_2 y. \tag{21}$$

Estimation of material surface position is depicted in Fig. 3. The origin of coordinate system is placed into the tip of the tool. After measuring five representative points (pink) a fitted plane ρ is calculated.

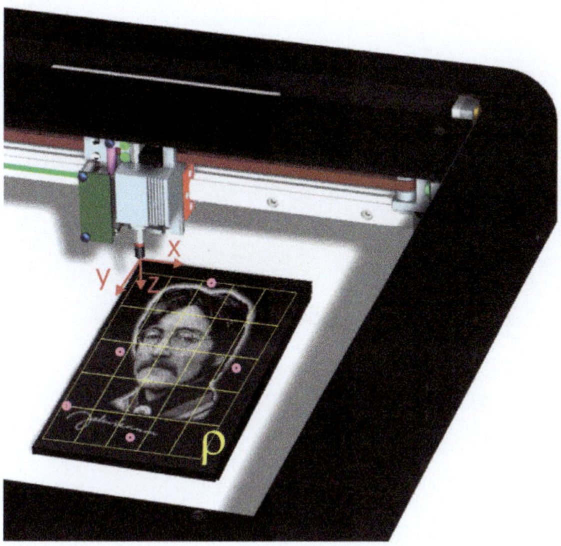

Fig. 3: CNC machine mb2300; Red - origin of coordinate system is placed on the tip of diamond tool. Yellow- fitted plane ρ is calculated out of five measured points (pink).

5. Z-Axis Position Control

The computed fitted plane equation is fetched to the machine control system in the form of coordinates of image corners.

$$\mathbf{C} = \begin{pmatrix} x_1 & y_1 & z_1 \\ \vdots & \vdots & \vdots \\ x_4 & y_4 & z_4 \end{pmatrix}.$$

$$\mathbf{C} = \begin{pmatrix} 0 & 0 & z_1 \\ img\_width & 0 & z_2 \\ 0 & img\_width & z_3 \\ img\_width & img\_width & z_4 \end{pmatrix}. \tag{22}$$

Matrix $\mathbf{C}$ basically represents a rectangle that belongs to plane ρ. Each row of $\mathbf{C}$ is a corner of engraved image. We denote them $C1$, $C2$, $C3$ and $C4$ where $C1 = (x_1, y_1, z_1)$ and so on.

Subsequently, two lines l and r, identical with left and right side of the rectangle are computed using parametric equations:

$$l:\ L_L = C1 + (C1 - C3)t, \tag{23}$$

$$r:\ L_R = C2 + (C2 - C4)t, \tag{24}$$

where parameter $t \in< 0, 1 >$. Finally, equation of the middle line m, intersecting left and right line, perpendicular to both lines is computed:

$$m:\ L_M = L_L + (L_L - L_R)s. \tag{25}$$

where parameter $s \in< 0, 1 >$. The engraved image is the result of thousands of strokes of the tool. When projected to $X - Y$ plane (Fig. 4), the tool trajectory starts at the top of the image and continues row by row, until it reaches the bottom of the image. We calculate the Z-axis tool position in any 2D point p with coordinates (x, y) as follows:

1. Calculate points L_L and L_R letting $t = y/img\_height$.
2. Calculate point L_M letting $s = x/img\_width$.

Z coordinate of point L_M is a desired Z-axis tool position in point p. Since coordinates (x, y) are known, it is sufficient to calculate only z-coordinate of vectors L_L, L_R and L_M.

6. On the Fly Parameters Tuning

A few experiments have shown that driving a Z-axis tool position based only on the fitted plane ρ is not accurate enough. Every measured point involves some error. Fitted plane minimizes squared sum of errors. In adverse cases it might happen that summed error of fitted plane is much smaller than measurement error, which concludes to rather big error in coincidence of a real and fitted plane. In worst cases, the Z axis error was bigger than 1 mm. However, the fitted plane gives a good estimate of where the real material surface is.

Computing z-coordinate of the tool is based on lines l, r and m. By adjusting the parameters of these lines, more accurate Z-axis control could be achieved. Refinement of parameters of these lines is accomplished by carrying out further measurements of surface distance during engraving.

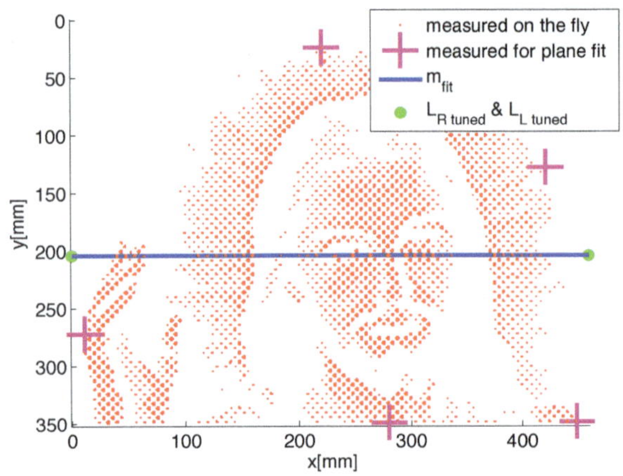

Fig. 4: $X-Y$ projection of measured stroke lengths of engraved image.

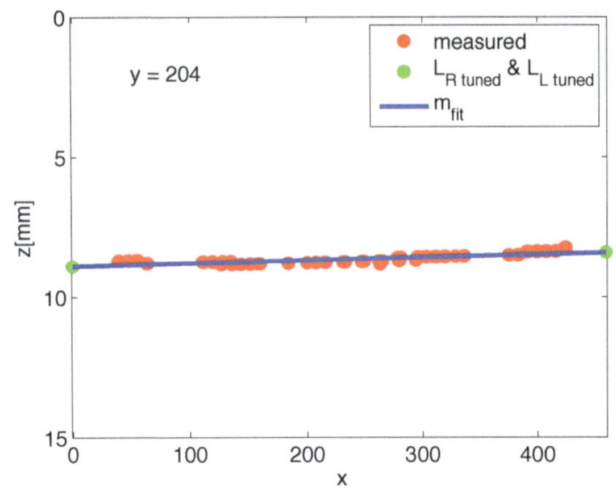

Fig. 5: $X-Z$ projection of one row of engraved image.

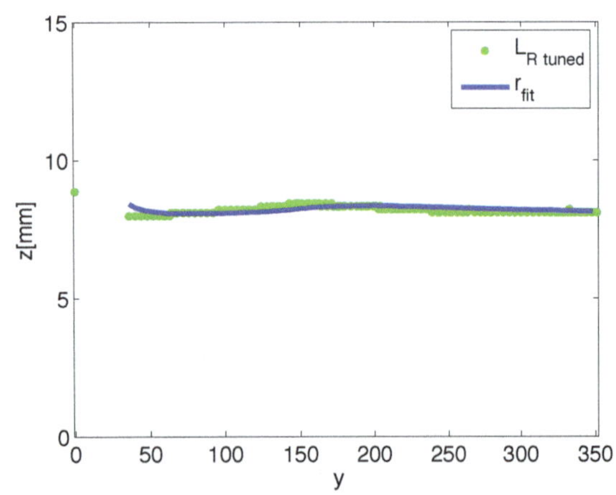

Fig. 6: $Y-Z$ projection of line r_{fit} (blue), fitted on points $L_{R\ tuned}$ (green) which are computed for each row of engraved image.

Proximity sensor allows measuring distance with each stroke of hammering tool. Regular image consists of tens to hundreds of strokes in each row (Fig. 4).

By fitting a line m_{fit} to points measured in a row, we get more accurate estimate of surface in given row (Fig. 5). In order to handle cases with a few measurement points (< 20), we also include marginal points (intersections of lines m, l and m, r) to set of points fitted by m_{fit}. The equation of m_{fit} is calculated at the end of each row. Based on this equation, we calculate a z-coordinate of new point $L_{L\ tuned}(0, y_{curr}, z)$ and point $L_{(R\ tuned}(img\_width, y_{curr}, z)$, where y_{curr} is y-coordinate of the current row. Points $L_{L\ tuned}$ and $L_{R\ tuned}$ serve for tuning parameters of lines l and m, respectively. This is done by calculating new lines l_{fit} and r_{fit}. Line l_{fit} is fitted on set $\{C1, C3, L_{L\ tuned\ i}\}$, line r_{fit} is a fitted on set $\{C2, C4, L_{R\ tuned\ i}\}$, where $i = 0, \ldots, y_{curr}$ (Fig. 6). Finally, we update a z-coordinate of corners $C1$, $C2$, $C3$ and $C4$ by calculating it using starting and ending points of l_{fit} and r_{fit}. In the next image row, new values of corners are used to calculate lines l, r and m.

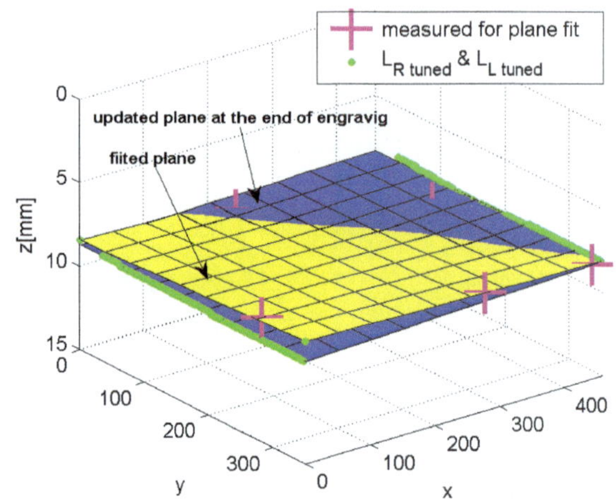

Fig. 7: Estimated plane ρ (yellow) and its last adjustment (blue).

Figure 4 to Fig. 7 were created by plotting a measured data from real engraved picture. In Fig. 7, fitted plane ρ (yellow) computed out of five measured points (pink crosses) is depicted. Blue plane is the final update of ρ at the end of engraving. As can be seen, two planes are slightly displaced. That confirms the fact,

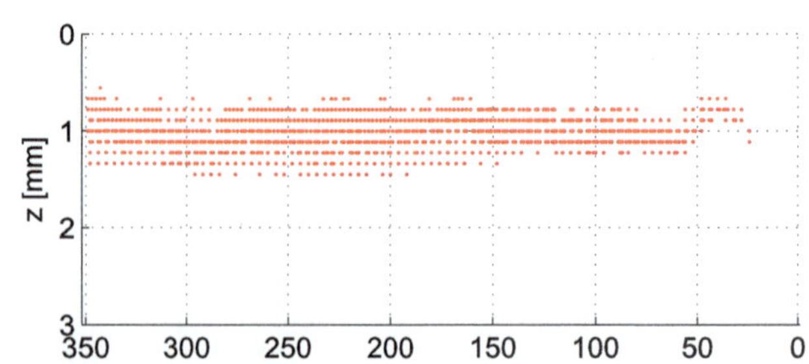

Fig. 8: Accuracy of Z-axis position control.

that estimating a surface position only by five measurements is not very accurate.

In Fig. 8, there is a $X - Z$ projection of engraved image from Fig. 4. The measurement shows that the distance is kept on 1 mm with an error $< 0,5$ mm which comprises with initial requirements of engraving machine.

7. Conclusion

In this paper the real time filtering method based on linear regression has been introduced. The filter is based on the idea of fitting a curve on a set of already measured data, supposing that a mathematical model of the signal is available. The experimental measurements were focused on practical application of filter in CNC machine, to estimate a position of machined material, where a filtered signal has the shape of a line. Therefore, computing of fit is simple and fast. LR filter in comparison to Kalman filter is less expensive from a computing point of view because in the general case Kalman filter in practical implementation requires multiplications, additions and inversion of matrices.

In this application, a main goal was to keep a constant distance of engraving tool from engraved material. Numerous portraits have been engraved using proposed system, while measuring and saving the tool position data, which confirm a correct functionality of the system, i.e. that the system is capable of keeping the tool in required distance. The control system of the machine is based on an ATmega162, 8-bit, fixed-point microcontroller.

The advantage of our method is in computational simplicity since it requires only a few summations, multiplications and divisions of floating point variables so it is well suited for a real time use in computationally non-powerful environments e.g. with 8-bit microcontrollers.

Acknowledgment

This work was supported by project KEGA 011TUKE-4/2013.

References

[1] IFEACHOR, E. C. and B. W. JERVIS. *Digital Signal Processing: a Practical Approach.* Essex: Pearson Education Limited, 2002. ISBN 987-0-201-59619-9.

[2] CHUI, C. K. and CHEN, G. *Kalman Filtering with Real-Time Applications.* New York: Springer-Verlag, 1999. ISBN 978-3-540-87848-3.

[3] LEE, J. H., K. PARK, K. H. YOU and J. W. JEON. Kalman filter applications to laser interferometer measurements in position-servo systems. *Electronics Letters.* 2006, vol. 42, iss 17, pp. 969–970. ISSN 0013-5194. DOI: 10.1049/el:20061572.

[4] FANY, CH. L. and M. Y. PANGZ. A Method for Filtering Noise Data by Blending Local Least Squares Fitting Curves. In: *Second International Workshop on Computer Science and Engineering, 2009. WCSE '09.* Qingdao: IEEE, 2009, pp. 538–542. ISBN 978-0-7695-3881-5. DOI: 10.1109/WCSE.2009.727.

[5] BJORCK, A. *Numerical Methods for Least Squares Problems.* Amsterdam: SIAM, 1996. ISBN 978-0-89871-360-2.

[6] LICHEN, G., K. XUEQIN and J. JIA. Distance measurement for tower crane obstacle based on multi-ultrasonic sensors. In: *IEEE International Conference on Automation Science and Engineering (CASE).* Seoul: IEEE, 2012, pp. 1028–1032. ISBN 978-1-4673-0429-0. DOI: 10.1109/CoASE.2012.6386344.

[7] MYERS, R. H. *Classical and Modern Regression with Applications.* Duxbury: Thomson Learning, 1990. ISBN 134-0-78134-5.

[8] PEDHAZUR, E. J. *Multiple regression in behavioral research: Explanation and prediction.* New York: Harcourt Brace, 1997. ISBN 978-0030417603.

[9] GUIDRY, T. Linear and multiple linear least squares regression. In: *Trent F Guidry* [online]. 2009. Available at: `http://www.trentfguidry.net/post/2009/07/19/Multiple-Linear-Least-Squares-Regression.aspx`.

[10] STOCK, J. H. and M. W. WATSON. *Introduction to Econometrics.* Boston: Pearson Addison-Wesley, 2012. ISBN 978-0138009007.

About Authors

Misel BATMEND was born in Ulanbatar, Mongolia in 1987. He received his M.Sc. degree in electrical engineering in 2010 from Faculty of Electrical Engineering and Informatics at Technical University in Kosice. Now, he is a Ph.D. student. His research interest included automation of engraving process for specific applications in whole range of its portfolio from hardware level up to image processing.

Daniela PERDUKOVA was born in Spisska Sobota, Slovakia in 1960. She received the M.Sc. degree in technical cybernetics and Ph.D. in electrical drives and electrical traction at the Technical University of Kosice, Slovak Republic. She is currently a Professor at the Department of Electrical Engineering and Mechatronics at the Technical University of Kosice. Her research and educational activities are focused mostly on AI techniques and their applications in the field of non-linear systems; a special attention is also paid to modelling of technological processes, their monitoring and technological process visualization.

4

Reactive Power Optimization with SVC & TCSC using Genetic Algorithm

Biplab BHATTACHARYYA , Vikash Kumar GUPTA , Sanjay KUMAR

Department of Electrical Engineering, Indian School of Mines, Dhanbad, 826004 Jharkhand, India

biplabrec@yahoo.com, vikash1146@gmail.com, sanjayism2012@gmail.com

Abstract. *In this paper Genetic Algorithm (GA) is used as an evolutionary tecthniques for the optimal placement of flexible AC transmission systems (FACTS) devices in an interconnected power system. Here two types of FACTS devices has been discussed nemely, Thyristor Controlled Series Capacitor (TCSC) and Static Var Compensator (SVC) for the economic operation and to reduce the transmission loss. Reactively loading of the system is taken from base to 200 % of base loading and the system performance is observed without and with FACTS devices. Optimal placement of FACTS devices in the system is determined by calculating active and reactive power flow in lines. FACTS devices along with reactive generation of generators and transformer tap setting are used for the power transfer capacity using GA. The proposed approach is applied on IEEE 14 and IEEE 30-bus test systems. Finally the effectiveness of the proposed GA based method of placement of FACTS devices is established by comparing the results with another standard method of optimization like Particle Swarm Optimization (PSO) technique.*

Keywords

FACTS devices, genetic algorithm, line power Flow, optimal location of FACTS devices, particle swarm optimization.

1. Introduction

Due to increase in power demand, restriction on the construction of new lines, environment, unscheduled power flows in lines creates congestion in the transmission network and increases transmission loss. Maintenance of bus voltages and line loads within predefined limits is one of the challenging tasks in an interconnected power system. Effective control of reactive compensation on weak nodes improves voltage profile, reduces power loss and improves both steady state & dynamic performance of the system.

Power flow through an ac transmission line is a function of line impedance, the magnitude and the phase angle between the sending end and the receiving end voltages. With the development of FACTS devices both the active and reactive power flow in the lines can be controlled. The concept of flexible AC transmission system (FACTS) was first introduced by Hingorani [1]. FACTS devices are solid-state converters having the capability of control of various electrical parameters in transmission circuits.

Sensitivity analysis and linear programming technique for the optimal location and size of Static Var Compensator (SVC) in a power system is discussed in [2]. Optimization techniques are widely used in the field of technology. Optimal placement of Thyristor Controlled Series Capacitor (TCSC) for increasing loadability and minimizing transmission loss by Genetic Algorithm (GA) is discussed in [3]. Optimal reactive power dispatch along with the setting of switchable series & shunt FACTS devices is presented in [4]. Optimal placement of Var sources by loss sensitivity based method is presented in [5]. A hybrid Genetic Algorithmic approach with FACTS devices for optimal power flow is dealt in [6].

Solution of optimal power flow using GA is presented by Osman et al. in [7]. Authors have discussed Genetic Algorithm based approach for the placement of different types of FACTS devices in [8]. Computational Intelligence based algorithm is presented in [9] to determine the optimal placement and parameter setting of TCSC for enhancing the security of power system under single line contingency.

Concept of comutational intelligence technique using FACTS controller is applied in [10] for the loadability enhancement in a restructured power system. In [11] GA based technique is discussed for the placement of FACTS devices in some test systems. Das et al. in [12] applied GA to minimize active power loss in a radial distribution network using SVC. About the modelling and selection of possible locations for the installation of FACTS devices have been discussed in [13]. An optimization method is used in [14] that combines the reliability and the efficiency of radial power distribution systems to reduce the active power loss, through a process of network reconfiguration. Effect of implementation of Genetic Algorithm for the determination of locations and size of the FACTS controller is discussed in [15].

Nomenclature:

- X_{Line}: reactance of line,
- S: operating range of FACTS devices,
- C_{Total}: total cost of system operation,
- $C_1(E)$: cost due to energy loss,
- $C_2(F)$: total investment cost of the FACTS Devices,
- P_{ni}^{min}, P_{ni}^{max}: lower and upper limit of nodal active power in the i-th bus respectively,
- P_{ni}, Q_{ni}: nodal active and reactive power output of the i-th bus respectively,
- Q_{ni}^{min}, Q_{ni}^{max}: lower and upper limit of nodal reactive power in the i-th bus respectively,
- Q_{gi}^{min}, Q_{gi}^{max}: lower and upper limit of existing nodal reactive capacity in the i-th bus respectively,
- Q_{gi}: output of existing nodal reactive capacity in the i-th bus,
- P_{Gi}, Q_{Gi}: active and reactive power generation in the i-th bus respectively,
- P_{Di}, Q_{Di}: active and reactive power consumed by load in the i-th bus respectively,
- P_i, Q_i (inj): real and reactive power flow change takes place at the node i due to TCSC connected to a particular line between the nodes i & j,
- Q_{iL} (inj): reactive power injection due to SVC,
- V_i, V_j: voltage of i-th and j-th bus respectively,
- N: number of lines,
- G_{ij}, B_{ij}: real and imaginary part of admittance between buses i & j respectively,
- θ_{ij}: phase angle between V_i & V_j,
- V_i^{gen-1}: current velocity of agent i at previous generation,
- w: weight function for velocity of agent i,
- rand: is the random number between 0 and 1,
- S_i^{gen-1}: current position of agent i at previous generation,
- C_i: weight coefficient for each term,
- p_{besti}: pbest of agent i,
- g_{besti}: gbest of agent i.

2. FACTS Devices

2.1. FACTS Devices & Cost Functions

Two types of FACTS devices namely thyristor controlled series capacitors (TCSC) and static VAr compensators (SVC) are used in the transmission network:

- TCSC: By modifying the line reactance TCSC acts as either inductive or capacitive compensator. The maximum value of the capacitance is fixed at $-0.8X_{Line}$ and $0.2X_{Line}$ is the maximum value of the inductance.
- The SVC can be operated as either inductive or capacitive compensation. It can be modeled as a fixed capacitor and a thyristor controlled reactor. So the function of the SVC is either to inject reactive power to the bus or to absorb reactive power from the bus where it is connected.

According to [18], cost functions for SVC and TCSC are given below:

- TCSC:

$$C_{TCSC}\,(\text{US\$/kVar}) = \\ = 0.0015 \cdot S^2 - 0.7130 \cdot S + 153.75, \quad (1)$$

- SVC:

$$C_{SVC}\,(\text{US\$/kVar}) = \\ = 0.0003 \cdot S^2 - 0.3051 \cdot S + 127.38. \quad (2)$$

Here, S is the operating range of the FACTS devices.

3. Optimal Placement of FACTS Devices

Having made the decision to install a FACTS device in the system, there are three main issues that are to be considered: types of device, its capacity and location. The decision where they are to be placed is largely dependent on the desired effect and the characteristics of the specific system. SVC's are mostly suitable when reactive power flow or voltage support is necessary. Also the costs of the devices play an important role for the choice of a FACTS devices. There are two distinct means of placing a FACTS device in the system for the purpose of increasing the system's ability to transmit power, thereby allowing for the use of more economic generating units. That is why FACTS devices are placed in the more heavily loaded lines to limit the power flow in that line. This causes more power to be sent through the remaining portions of the system while protecting the line with the device from being overloaded. This method which sites the devices in the heavily loaded line is the most effective. If reactive power flow is a significant portion of the total flow of the limiting transmission line, either a TCSC device in the line or a SVC device located at the end of the line that receives the reactive power, may be used to reduce the reactive power flow, thereby increasing the active power flow capacity.

4. The Proposed Approach

Here the main objective is to minimize the transmission loss by incorporating FACTS devices at suitable locations of the transmission network. Inclusion of FACTS controllers also increase the system cost. So, optimal placement of FACTS devices is required such that the gain obtained by reducing the transmission loss is significant even after the placement of costly FACTS devices. Installation costs of various FACTS devices and the cost of system operation, namely, energy loss cost are combined to form the objective function to be minimized. The optimal allocation of FACTS devices can be formulated as:

$$\mathrm{C_{total}} = \mathrm{C_1}\,(\mathrm{E}) + \mathrm{C_2}\,(\mathrm{F}), \tag{3}$$

where $\mathrm{C_1}\,(\mathrm{E})$ is the cost due to energy loss and $\mathrm{C_2}\,(\mathrm{F})$ is the total investment cost of the FACTS devices. Subject to the nodal active and reactive power balance:

$$\mathrm{P_{ni}^{min}} \le \mathrm{P_{ni}} \le \mathrm{P_{ni}^{max}}, \tag{4}$$

$$\mathrm{Q_{ni}^{min}} \le \mathrm{Q_{ni}} \le \mathrm{Q_{ni}^{max}} \tag{5}$$

and voltage magnitude constraints:

$$\mathrm{V_i^{min}} \le \mathrm{V_i} \le \mathrm{V_i^{max}} \tag{6}$$

and the existing nodal reactive capacity constraints:

$$\mathrm{Q_{gi}^{min}} \le \mathrm{Q_{gi}} \le \mathrm{Q_{gi}^{max}}. \tag{7}$$

Superscripts min, max, are the minimum and maximum limits of the variables. The power flow equations between the nodes i–j after incorporating FACTS devices would appear as:

- TCSC:

$$\mathrm{P_{Gi}} - \mathrm{P_{Di}} + \mathrm{P_i} - \\ - \textstyle\sum_{j=1}^{N-1} \mathrm{V_i V_j}\,(\mathrm{G_{ij}} cos\theta_{ij} + \mathrm{B_{ij}} sin\theta_{ij}) = 0, \tag{8}$$

$$\mathrm{Q_{Gi}} - \mathrm{Q_{Di}} + \mathrm{Q_{i(inj)}} - \\ - \textstyle\sum_{j=1}^{N-1} \mathrm{V_i V_j}\,(\mathrm{G_{ij}} sin\theta_{ij} + \mathrm{B_{ij}} cos\theta_{ij}) = 0, \tag{9}$$

$$\mathrm{Q_{Gj}} - \mathrm{Q_{Dj}} + \mathrm{Q_j} - \\ - \textstyle\sum_{j=1}^{N-1} \mathrm{V_i V_j}\,(\mathrm{G_{jj}} sin\theta_{jj} + \mathrm{B_{jj}} cos\theta_{jj}) = 0, \tag{10}$$

$$\mathrm{Q_{Gj}} - \mathrm{Q_{Dj}} + \mathrm{Q_{j(inj)}} - \\ - \textstyle\sum_{j=1}^{N-1} \mathrm{V_i V_j}\,(\mathrm{G_{jj}} sin\theta_{jj} + \mathrm{B_{jj}} cos\theta_{jj}) = 0, \tag{11}$$

- SVC:

$$\mathrm{Q_{Gi}} - \mathrm{Q_{Di}} + \mathrm{Q_{iL(inj)}} - \\ - \textstyle\sum_{j=1}^{N-1} \mathrm{V_i V_j}\,(\mathrm{G_{ij}} sin\theta_{ij} + \mathrm{B_{ij}} cos\theta_{ij}) = 0. \tag{12}$$

These changes in the power flow equations are taken into consideration by appropriately modifying the bus admittance matrix for execution of load flow in evaluating the objective function for each individual population of generation of both Genetic Algorithm (GA) and Particle Swarm Optimization (PSO) based optimization technique.

In this approach, first the locations of FACTS devices are defined by calculating the power flow in the transmission lines. Here we choose only four locations in IEEE 14 bus and eight locations in IEEE 30 bus system for the placement of FACTS devices. SVC's positions are selected by choosing the lines carrying largest reactive power. In IEEE 14 bus system, 10th, 13th & 14th buses and in IEEE 30 bus system, 21st, 7th, 17th & 15th buses are found as buses where suitable reactive injection by SVC's could improve the system performance. Line number 7th in IEEE 14 bus and lines 25th, 41^{thst}, 28th & 5th in IEEE 30 bus system are found as the lines for TCSC's placement and simultaneously series reactance of these lines are controlled.

In the proposed approach combined effect of SVC & TCSC is tested. Simultaneous use of shunt (SVC) and series (TCSC) FACTS controller has a better effect than if either of the FACTS controller used singly. We only can connect SVC at buses where reactive injections are required and also there is a limit of a number of SVC's that can be connected to a particular system. Similarly, modifying line reactance helps greatly in reducing line loss & improving overall system performance. Application of series & shunt combination of FACTS controller on a standard system is analyzed in the present work. We have taken only one number of TCSC and three number of SVC's in IEEE 14 bus

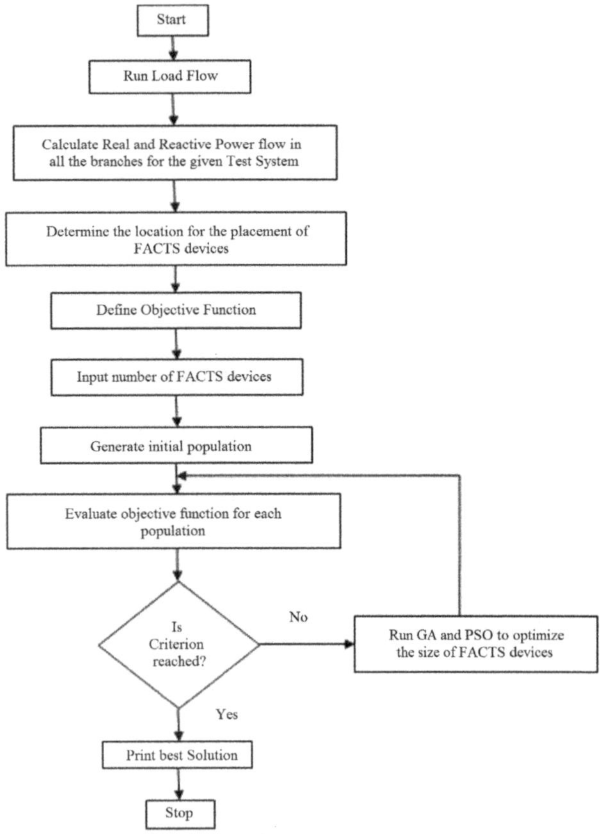

Fig. 1: Flowchart for the proposed approach.

system & four numbers of TCSC's and four numbers of SVC's in IEEE 30 bus systemfor the purpose of analysis in the present problem. The number of FACTS controller depends also on the size of the system.

Maximum value of SVC is taken as 50 MVAr & maximum value of TCSC is taken as 10 MVAr in the present problem. Fig. 1 shows a flowchart for the proposed approach.

4.1. Genetic Algorithm in the Proposed Method

The function of the GA is to find the optimum value of the different FACTS devices. Here two different types of FACTS devices are used. The locations for the placement of different FACTS devices are determined on the basis of power flow analysis. TCSC's modifies reactance of the lines and SVC's are to control reactive injection at buses.

In addition transformer tap positions along with reactive generations of the generators are controlled. In IEEE 14 bus system there are three transformer tap positions and four generator buses while in IEEE 30 bus system there are four transformer tap positions and five generator buses. So, as a whole all the controlling parameters are to be optimized by Genetic Algorithm [19]. These controlling parameters is represented within a string. This is shown in Tab. 1 for IEEE 14 bus system and in Tab. 2 for IEEE 30 bus system. Initially a population of N strings is randomly created in their limits. Then the objective function is computed for every individual of the population.

A biased roulette wheel is created such a way so that the parameter values are selected according to their fitness obtained after computing the objective function for all the individuals of the current population. Thereafter the usual Genetic operation such as Reproduction, Crossover & Mutation takes place. Two individual are randomly selected from the current population for reproduction. Then crossover takes place with a probability close to one (here 0.8).

Finally mutation with a specific probability (very low) completes one Genetic cycle and individuals of the same population with improved characters are created in the next generation. The objective function is then again calculated for all the individuals of the new generation and all the genetic operations are again performed and the second generation of the same population size is produced. This procedure is repeated till the final goal is achieved. The population size is taken as 80 & GA is run for 100 generation.

4.2. PSO Approach in Brief

Particle swarm optimization (PSO) is a population based stochastic optimization technique developed by Dr. Eberhart and Dr. Kennedy in 1995, inspired by social behavior of bird flocking or fish schooling.

A population is initialized of random feasible solutions and searches for optima by updating generations. In PSO, the potential solutions, called particles have their own positions and velocities move in the search space of an optimization problem by following the current optimum particles. Each particle tracks its own best position found so far in the exploration and each particle searches for better positions in the search space by updating its velocity. The movement of each particle naturally evolves to an optimal or near-optimal solution.

The position of each agent is represented by XY-axis position and the velocity (displacement vector) is expressed by Vx (the velocity along X-axis) and Vy (the velocity along Y-axis). Modification of the agent position is realized by using the position and the velocity information. The behavior of particles in PSO is shown in Fig. 2.

Each agent or particle knows its best value so far (pbest) and its x, y position. Each agent knows the best value so far in the group (gbest) among pbests.

Tab. 1: String representing the control variables in IEEE 14 bus system.

TCSC	SVC			Transformer Tap			Reactive Generations of Generators			
1 Nos.	3 Nos.			3 Nos.			4 Nos.			
1	1	2	3	1	2	3	1	2	3	4

Tab. 2: String representing the control variables in IEEE 30 bus system.

TCSC				SVC				Transformer Tap				Reactive Generations of Generators				
4 Nos.				4 Nos.				4 Nos.				5 Nos.				
1	2	3	4	1	2	3	4	1	2	3	4	1	2	3	4	5

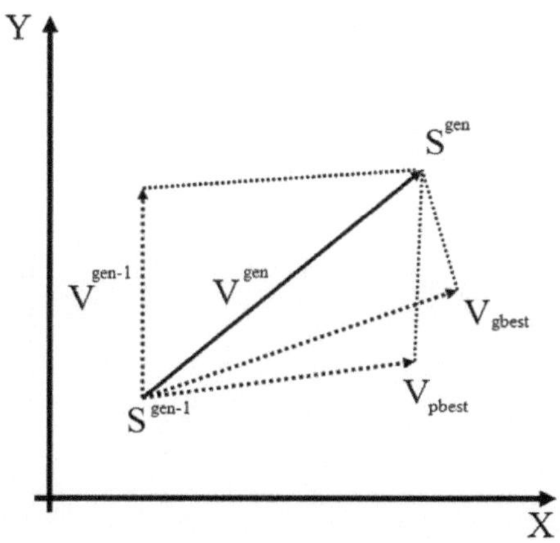

Fig. 2: Behavior of particles in PSO.

Each particle tries to modify their position using the following information:

- the current positions (x, y),
- the current velocities (Vx, Vy),
- the distance between the current position and pbest,
- the distance between the current position and gbest.

The basic equation for the optimization of nonlinear functions using particle swarm optimization technique is:

$$V_i^{gen} = w \cdot V_i^{gen-1} + \mathrm{C_1 rand}\left(\mathrm{p_{besti}} - S_i^{gen-1}\right) + +\mathrm{C_2 rand}\left(\mathrm{g_{besti}} - S_i^{gen-1}\right), \quad (13)$$

$$S_i^{gen} = S_i^{gen-1} + V_i^{gen}, \quad (14)$$

where w is updated at each iteration:

$$w = w_{max} - \frac{w_{max} - w_{min}}{gen_{max}} gen. \quad (15)$$

Tab. 3: Locations of different FACTS Devices in the Transmission Network.

IEEE 14 bus		IEEE 30 bus	
TCSC in line	SVC in buses	TCSC in line	SVC in line
7	10; 13; 14	25; 41; 28; 5	21; 7; 17; 15

Here $w_{max} = 0.9$; $w_{min} = 0.4$; $gen_{max} = 500$ and gen = current iteration; C_1 and C_2 are set to 2.0.

In PSO, the gbest particle always improves its position and finds the optimum solution and the rest of the population follows it. String representing control variables using PSO are shown in Tab. 1 and Tab. 2.

5. Results and Discussions

After detecting the locations of FACTS devices by power flow analysis it becomes necessary to determine their magnitudes. GA & PSO based optimization technique is run to serve this purpose. IEEE 14 & IEEE 30 bus system is taken as standard test system. Both test systems are loaded (reactive loading is considered) from it's base value to 200 % of it's base value. The locations where different FACTS devices are placed is shown in Tab. 3.

Active power loss without and with FACTS devices using GA & PSO technique is shown in Tab. 4 & Tab. 5 respectively for both systems. The magnitude and phase angle of the voltages of weak nodes without & with FACTS devices for highest reactive loading i.e. for 200 % is shown in Tab. 6 & Tab. 7. Phase angles are given in radian. Here, we see that after connecting FACTS devices, voltage profile of all buses of the IEEE 14 & IEEE 30 bus system improves, though question may arise why we connect FACTS devices if the voltages are in the acceptable range as seen from Tab. 6 & Tab. 7. Our main objective is to reduce overall system loss with the aid of the FACTS controller. In doing so, the voltage profile improves and transmission loss reduces significantly. A comparative study of the operating cost of the system without and with FACTS devices using GA & PSO is given in Tab. 8 & Tab. 9 for both the systems. From Tab. 8 & Tab. 9, we see that large economic gain is achieved using GA & PSO based placement of FACTS devices in all cases of loading. Here, it is clearly observed from the result, there is a great saving in the system cost in various system loading condition. Table 10 & Tab. 11 shows the amount of FACTS devices in p.u and controlled reactive sources present in the network under different cases of loading using both the techniques and for both the systems. Reactive power flows in different lines before and after the placement of FACTS devices for 200 % loading are shown in Tab. 12 & Tab. 13 for IEEE 14 and IEEE 30 bus system respectively.

It is observed from Tab. 3, that SVC's are connected at the buses 10th, 13th & 14th those are at the finishing ends of lines 13th, 19th & 20th respectively in IEEE 14 bus system, while buses 21st, 7th, 17th & 15th are at the finishing ends of the lines 27th, 26th, 9th & 18th respectively in IEEE 30 bus system, since these are the lines very high reactive power without FACTS devices. After connecting SVC's at theses buses, voltage profile at these buses are improved, also reactive power flow is reduced in a large amount in the lines 13th, 19th & 20th in IEEE 14 bus and in the lines 27th, 26th, 9th & 18th in IEEE 30 bus for all cases of loading. The placement of TCSC's in lines using both GA & PSO approaches reduces reactive power flow in lines significantly for both the test systems. As transmission line congestion is directly related with reactive power flows in different lines, we may conclude that line congestion reduces heavily with the reduction of reactive power flow in different lines.

It is also to be noticed that no FACTS device is connected in line 1 because of the fact that it is in between bus 1 and bus 2 though it carries very large active power. Bus 1 is the slack bus and already a FACTS device regulates the voltage of the bus 2. Again in any line or in a bus connected to the line, only one FACTS device can be placed. Tab. 4 & Tab. 5 shows that transmission loss is reduced in a considerable amount with the FACTS devices connected in different loacations of the network, as a result operating cost reduces significantly. This effect is observed with different sets of loading values. Hence benefit in terms of saving using both GA & PSO is observed in each cases of loading for both systems as shown in Tab. 8 & Tab. 9. Here, energy cost is taken as 0.06 dollar/kWh.

Figure 3 and Fig. 4 shows the variations of operating cost with generation for 200 % of reactive loading of the system with GA based approach in IEEE 14 & IEEE 30 bus system respectively, while Fig. 5 and

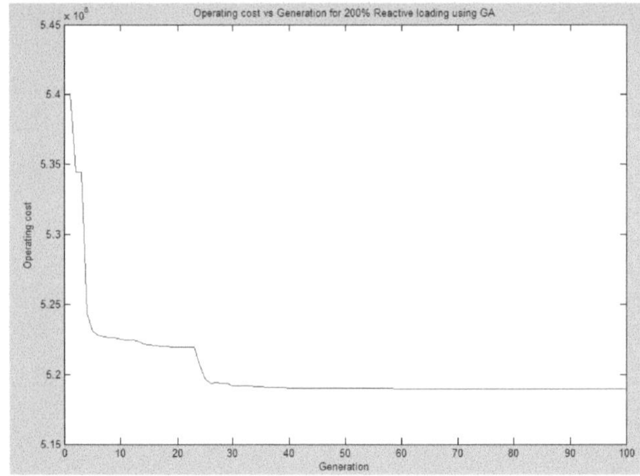

Fig. 3: Variations of operating cost with generation for 200 % of base reactive loading with GA in IEEE 14 bus system.

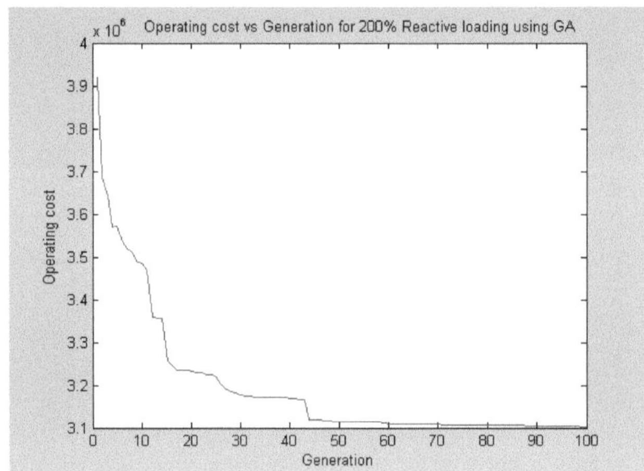

Fig. 4: Variations of operating cost with generation for 200 % of base reactive loading with GA in IEEE 30 bus system.

Fig. 6 shows the variations of operating cost with generation for 200 % of reactive loading of the system with PSO based approach in IEEE 14 & IEEE 30 bus system respectively. Fig. 7 and Fig. 8 shows a single line diagram for IEEE 14 & IEEE 30 bus test system respectively. The results obtained by the GA based method is compared with other standard optimization approach like PSO. It is observed that though PSO yields comparable solution with GA in minimizing transmission loss and transmission cost in the case of IEEE 14 bus system, but in larger test system as in IEEE 30 bus system PSO fails to yield a satisfactory result as GA in all cases of loading.

6. Conclusion

Here Genetic Algorithm (GA) based approach for the placement of FACTS devices and simultaneous control of existing reactive sources is presented. Two different types of FACTS devices are considered. It is

Tab. 4: Active power loss without & with FACTS devices in IEEE 14 bus system.

Reactive Reactive	Active Power Loss without FACTS (p.u.)	Active Power Loss with FACTS using GA (p.u.)	Active Power Loss with FACTS using PSO (p.u.)
100 %	0.1339	0.0666	0.0668
150 %	0.1356	0.0734	0.0734
175 %	0.1368	0.0763	0.0763
200 %	0.1384	0.0983	0.0984

Tab. 5: Active power loss without & with FACTS devices in IEEE 30 bus system.

Reactive Reactive	Active Power Loss without FACTS (p.u.)	Active Power Loss with FACTS using GA (p.u.)	Active Power Loss with FACTS using PSO (p.u.)
100 %	0.0711	0.0406	0.0445
150 %	0.0742	0.0433	0.0478
175 %	0.0765	0.0448	0.0497
200 %	0.0795	0.0573	0.0637

Tab. 6: Bus voltages & phase angles without and with FACTS devices for 200 % reactive loading in IEEE 14 bus system.

Bus. No.	Bus Voltage without FACTS	Bus Angle without FACTS	Evolutionary Methods with FACTS devices	Bus Voltage with FACTS	Bus Angle with FACTS
10	1.0232	−0.2612	GA	1.0454	−0.2633
			PSO	1.0444	−0.2631
13	1.0386	−0.2612	GA	1.0510	−0.2691
			PSO	1.0513	−0.2689
14	1.0066	−0.2764	GA	1.0361	−0.2830
			PSO	1.0334	−0.2818

Tab. 7: Bus voltages & phase angles without and with FACTS devices for 200 % reactive loading in IEEE 30 bus system.

Bus. No.	Bus Voltage without FACTS	Bus Angle without FACTS	Evolutionary Methods with FACTS devices	Bus Voltage with FACTS	Bus Angle with FACTS
7	1.0014	−0.1391	GA	1.0082	−0.1413
			PSO	0.9952	−0.1383
15	1.0036	−0.1797	GA	1.0747	−0.1746
			PSO	1.0574	−0.1711
17	1.0050	−0.1775	GA	1.0797	−0.1733
			PSO	1.0662	−0.1696
21	0.9959	−0.1816	GA	1.0771	−0.1800
			PSO	1.0684	−0.1773

Tab. 8: Operating cost analysis without and with FACTS devices using GA & PSO pproach in IEEE 14 bus system.

Reactive Loading	**Operating Cost due to Energy Loss (A)** (dollar)	**Evolutionary Methods with FACTS devices**	**Operating Cost (B)** $\times 10^6$ (dollar)	**Net Saving (A–B)** (dollar)
100 %	7037784	GA	3.5046	3533184
		PSO	3.5147	3523084
150 %	7127136	GA	3.9029	3224236
		PSO	3.9135	3213636
175 %	7190208	GA	4.0431	3147108
		PSO	4.0527	3137508
200 %	7274304	GA	5.1896	2084704
		PSO	5.1927	2081604

Tab. 9: Operating cost analysis without and with FACTS devices using GA & PSO pproach in IEEE 30 bus system.

Reactive Loading	**Operating Cost due to Energy Loss (A)** (dollar)	**Evolutionary Methods with FACTS devices**	**Operating Cost (B)** $\times 10^6$ (dollar)	**Net Saving (A–B)** (dollar)
100 %	3737016	GA	2.1786	1558416
		PSO	2.4052	1331816
150 %	3899952	GA	2.3429	1557052
		PSO	2.6080	1291952
175 %	4020840	GA	2.4745	1546340
		PSO	2.7693	1251640
200 %	4178520	GA	3.1024	1076120
		PSO	3.4460	732520

Tab. 10: Amount of FACTS devices and other reactive sources in the transmission network by GA & PSO in IEEE 30 bus system.

Reactive Loading	**SVC amount** (p.u.)		**TCSC amount in lines** (p.u.)		**Reactive Generation Qg** (p.u.)		**Transformer Tap Position** (p.u.)	
	GA	**PSO**	**GA**	**PSO**	**GA**	**PSO**	**GA**	**PSO**
100 %	0.0003 0.0060 0.0327	0.0000 0.0000 0.0389	0.0001	0.0000	0.2356 0.1605 0.2370 0.1812	0.0000 0.2055 0.0023 0.1655	0.9641 0.9516 0.9998	0.9527 1.0113 0.9087
150 %	0.0000 0.3245 0.0355	−0.0572 0.2976 0.0515	0.0249	0.0254	0.1829 0.2305 0.2049 0.2234	0.2157 0.4271 0.1682 0.0939	0.9615 0.9750 0.9977	0.9695 1.0525 0.8802
175 %	0.0934 0.0604 0.1105	0.1564 0.0434 0.1319	0.0025	0.0009	0.1128 0.1669 0.2189 0.2133	0.4559 0.3337 0.1719 0.0384	0.9799 0.9433 0.9652	0.9785 0.9196 1.0363
200 %	0.0351 0.0663 0.0643	0.0273 0.0770 0.0492	0.0011	0.0000	0.2230 0.2282 0.2212 0.2141	0.5208 0.3454 0.2380 0.1696	0.9896 0.9567 0.9378	0.9774 1.0132 0.8748

Tab. 11: Amount of FACTS devices and other reactive sources in the transmission network by GA & PSO in IEEE 14 bus system.

Reactive Loading	SVC amount (p.u.)		TCSC amount in lines (p.u.)		Reactive Generation Qg (p.u.)		Transformer Tap Position (p.u.)	
	GA	PSO	GA	PSO	GA	PSO	GA	PSO
100 %	0.0892 0.0511 0.0398 0.0621	0.0000 0.0000 0.0000 0.0000	0.0001 0.0419 0.0002 0.0515	0.1463 0.0419 0.1049 0.1388	0.3409 0.1815 0.1911 0.1975 0.1023	0.6000 0.0000 0.0000 0.4000 0.0000	0.9099 0.9859 0.9133 0.9344	0.9000 0.9000 0.9248 0.9000
150 %	0.1586 0.1172 0.0714 0.1036	0.0869 0.0000 0.0000 0.1510	0.0010 0.0117 0.0002 0.0545	0.1463 0.0419 0.1049 0.1388	0.3899 0.1818 0.3185 0.2371 0.0971	0.6000 0.0000 0.0000 0.2474 0.0000	0.9431 1.0109 0.9331 0.9081	0.9000 0.9000 0.9358 0.9000
175 %	0.3351 0.2194 0.1877 0.1350	0.3202 0.0063 0.2336 0.2018	0.0008 0.0419 0.0008 0.0501	0.1463 0.0419 0.1049 0.1388	0.3630 0.2073 0.2158 0.1606 0.2500	0.0672 0.0183 0.4656 0.2370 0.1942	0.9601 0.9004 0.9993 0.9482	0.9195 0.9308 0.9673 0.9006
200 %	0.2399 0.1673 0.1149 0.1579	0.1457 0.0000 0.0000 0.1089	0.0011 0.0051 0.0004 0.0500	0.1463 0.0419 0.1049 0.1388	0.3318 0.2240 0.2751 0.2145 0.1357	0.6000 0.0000 0.0000 0.3168 0.0000	0.9366 0.9880 0.9189 0.9001	0.9000 0.9000 0.9483 0.9000

Tab. 12: Comparative study of rective power flow in lines using GA & PSO based proposed approach for 200 % of base loading in IEEE 14 bus.

Lines	For reactive loading of 200 % (before) (p.u.)	For base reactive loading of 200 % using GA (p.u.)	For base reactive loading of 200 % using PSO (p.u.)
7	0.1417	0.0139	0.0503
13	0.0414	0.0538	0.0604
19	0.0222	−0.0013	−0.0020
20	0.0638	0.0194	0.0277

Tab. 13: Comparative study of rective power flow in lines using GA & PSO based proposed approach for 200 % of base loading in IEEE 30 bus.

Lines	For reactive loading of 200 % (before) (p.u.)	For base reactive loading of 200 % using GA (p.u.)	For base reactive loading of 200 % using PSO (p.u.)
5	0.0384	0.0388	0.0380
25	0.0664	0.0649	0.0879
28	0.0883	0.0115	0.0495
41	0.0751	0.0662	0.0388
9	0.1032	0.0100	0.0714
18	0.1365	−0.0227	−0.0034
26	0.0860	0.0173	0.1544
27	0.1925	0.0070	0.0923

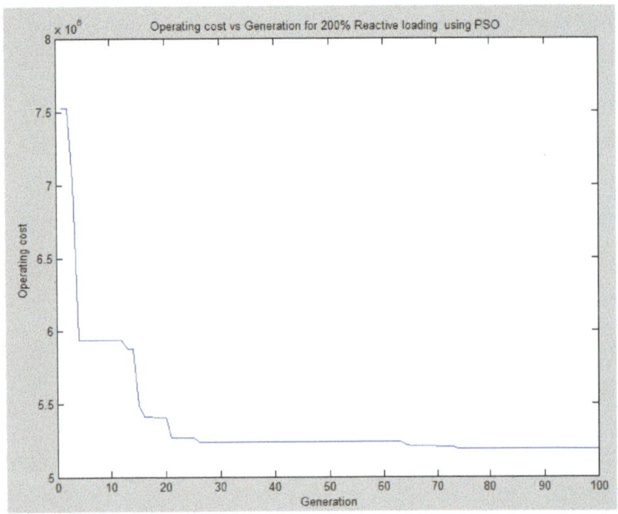

Fig. 5: Variations of operating cost with generation for 200 % of base reactive loading with PSO in IEEE 14 bus system.

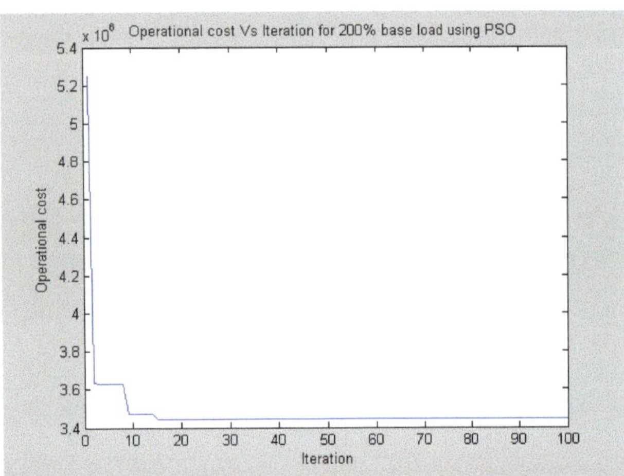

Fig. 6: Variations of operating cost with generation for 200 % of base reactive loading with PSO in IEEE 30 bus system.

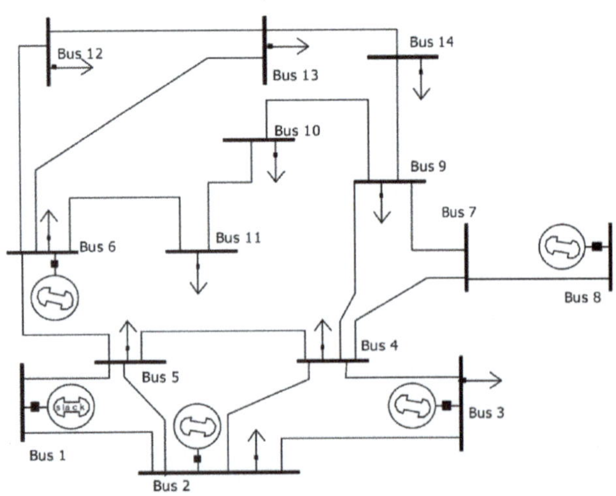

Fig. 7: Single line diagram of IEEE 14 bus system.

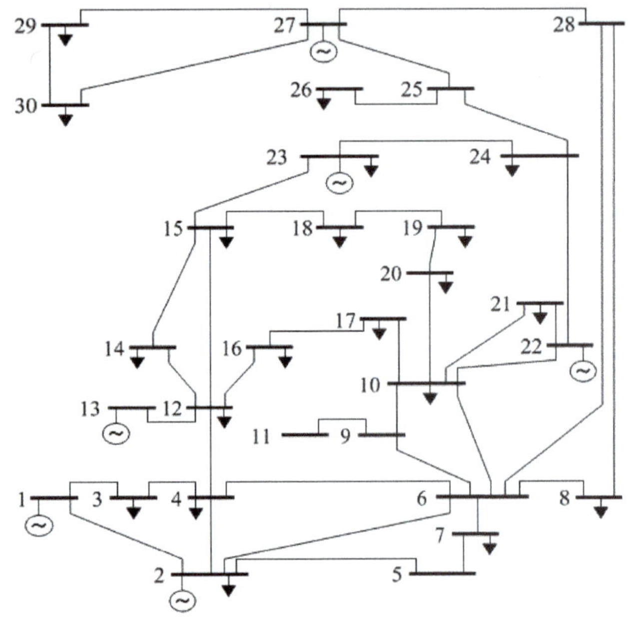

Fig. 8: Single line diagram of IEEE 30 bus system.

clearly evident from the results that effective placement of FACTS devices in proper locations along with the proper planning of existing reactive sources by using suitable optimization technique can significantly improve system performance. Also it is significantly noticeable from the results that the FACTS devices can be an alternative to reduce the flows in heavily loaded lines, resulting in an increased loadabilty, low system loss, improved stability of the network. Particle Swarm Optimization (PSO) based algorithm is developed for the purpose of comparison with the GA based approach. But GA based approach is more effective in minimizing the total operating cost, transmission loss than PSO based approach. Hence this GA based approach could be a new technique for planning of the existing reactive sources and efficient utilization of FACTS devices for improved power transfer.

References

[1] HINGORANI, N. G. High Power Electronics and flexible AC Transmission System. *Power Engineering Review.* 1988, vol. 8, iss. 7, pp. 3–4. ISSN 0272-1724. DOI: 10.1109/MPER.1988.590799.

[2] IBRAHIM, E. Optimal Allocations of SVCs for Improvement of Power System Performance. *Electric Power Components and Systems.* 2003, vol. 31, iss. 1, pp. 27–46. ISSN 1532-5008. DOI: 10.1080/15325000390112044.

[3] ABDELAZIZ, A. Y., M. A. EL-SHARKAWY and M. A. ATTIA. Optimal Location of Thyristor-

controlled Series Compensators in Power Systems for Increasing Loadability by Genetic Algorithm. *Electric Power Components and Systems*. 2011, vol. 39, iss. 13, pp. 1373–1387. ISSN 1532-5008. DOI: 10.1080/15325008.2011.584108.

[4] PREEDAVICHIT, P. and S. C. SRIVASTAVA. Optimal reactive power dispatch considering FACTS devices. *Electric Power Systems Research*. 1998, vol. 46, iss. 3, pp. 251–257. ISSN 0378-7796. DOI: 10.1016/S0378-7796(98)00075-3.

[5] BHATTACHARYYA, B., S. K. GOSWAMI and R. C. BANSAL. Loss Sensitivity Approach in Evolutionary Algorithms for Reactive Power Planning. *Electric Power Components and Systems*. 2009, vol. 37, iss. 3, pp. 287–299. ISSN 0378-7796. DOI: 10.1080/15325000802454468.

[6] CHUNG, T. S. and Y. Z. LI. A hybrid GA approach for OPF with consideration of FACTS devices. *Power Engineering Review*. 2000, vol. 20, iss. 8, pp. 54–57. ISSN 0272-1724. DOI: 10.1109/39.857456.

[7] OSMAN, M. S., M. A. ABO-SINNA and A. A. MOUSA. A solution to the optimal power flow using genetic algorithm. *Applied Mathematics and Computation*. 2004, vol. 155, iss. 2, pp. 391–405. ISSN 0096-3003. DOI: 10.1016/S0096-3003(03)00785-9.

[8] GERBEX, S., R. CHERKAOUI a A. J. GERMOND. Optimal location of multi-type FACTS devices in a power system by means of genetic algorithms. *Transactions on Power Systems*. 2001, vol. 16, iss. 3, pp. 537–544. ISSN 0885-8950. DOI: 10.1109/59.932292.

[9] RASHED, G. I. and Y. SUN. Optimal Placement of Thyristor Controlled Series Compensation for Enhancing Power System Security Based on Computational Intelligence Techniques. *Procedia Engineering*. 2011, vol. 15, pp. 908–914. ISSN 1877-7058. DOI: 10.1016/j.proeng.2011.08.168.

[10] NAGALAKSHMI, S. and N. KAMARAJ. Comparison of computational intelligence algorithms for loadability enhancement of restructured power system with FACTS devices. *Swarm and Evolutionary Computation*. 2012, vol. 5, pp. 17–27. ISSN 2210-6502. DOI: 10.1016/j.swevo.2012.02.002.

[11] IWARI, P. K. and Y. R. SOOD. Optimal location of FACTS devices in power system using Genetic Algorithm. In *World Congress on Nature & Biologically Inspired Computing*. Coimbatore: IEEE, 2009, pp. 1034–1040. ISBN 978-1-4244-5053-4. DOI: 10.1109/NABIC.2009.5393860.

[12] DAS, D. Reactive power compensation for radial distribution networks using genetic algorithm. *International Journal of Electrical Power*. 2002, vol. 24, iss. 7, pp. 573–581. ISSN 0142-0615. DOI: 10.1016/S0142-0615(01)00068-0.

[13] LIE, T. T. and W. DENG. Optimal flexible AC transmission systems (FACTS) devices allocation. *International Journal of Electrical Power*. 1997, vol. 19, iss. 2, pp. 125–134. ISSN 0142-0615. DOI: 10.1016/S0142-0615(96)00036-1.

[14] VITORINO, R. M., H. M. JORGE and L. P. NEVES. Loss and reliability optimization for power distribution system operation. *Electric Power Systems Research*. 2013, vol. 96, pp. 177–184. ISSN 0378-7796. DOI: 10.1016/j.epsr.2012.11.002.

[15] GHAHREMANI, Esmaeil a Innocent KAMWA. Optimal placement of multiple-type FACTS devices to maximize power system loadability using a generic graphical user interface. *Transactions on Power Systems*. 2013, vol. 28, iss. 2, pp. 764–778. ISSN 0885-8950. DOI: 10.1109/TPWRS.2012.2210253.

[16] RASHED, G. I., Y. SUN a H. I. SHAHEEN. Optimal Location and Parameter Setting of TCSC for Loss Minimization Based on Differential Evolution and Genetic Algorithm. *Physics Procedia*. 2012, vol. 33, pp. 1864–1878. ISSN 1875-3892. DOI: 10.1016/j.phpro.2012.05.296.

[17] ABUL'WAFA, A. R. Optimal capacitor allocation in radial distribution systems for loss reduction: A two stage method. *Electric Power Systems Research*. 2013, vol. 95, pp. 168–174. ISSN 0378-7796. DOI: 10.1016/j.epsr.2012.09.004.

[18] CAI, L. J., I. ERLICH and G. STAMTSIS. Optimal choice and allocation of FACTS devices in deregulated electricity market using genetic algorithms. In: *Power Systems Conference and Exposition*. New York: IEEE, 2004, pp. 815–821. ISBN 0-7803-8718-X. DOI: 10.1109/PSCE.2004.1397562.

[19] GOLDBERG, D. E. *Genetic algorithms in search, optimization, and machine learning*. Boston: Addison-Wesley, 1989. ISBN 02-011-5767-5.

About Authors

Biplab BHATTACHARYYA born in 9th January, 1970. He is currently working as Associate Professor of Electrical Engineering in the Indian School of Mines University, Dhanbad, India. He has joined in the Electrical Engineering department as Assistant Professor in the year of 2007. Then he promoted to the post of Associate Professor on 2010. He had served department of Electrical Engineering of National Institute of Technology, Durgapur, India for six years as senior Lecturer. He was in the position of Lecturer in the Electrical Engineering department of BITS, Pilani, Rajasthan, India for nearly one year. He worked as Assistant Engineer (Electrical Test), in a reputed cable industry for nearly three years. He obtained B.Sc (Hons) in Physics from Calcutta University, India in 1990. He obtained his B-Tech and M-Tech degree in the field of Electrical machines and Power system from Calcutta University in 1993 and 1995 respectively. He obtained his Phd degree in Engineering from the department of Electrical Engineering, Jadavpur University, India in 2006. He has published several technical papers in international/national journals and conference proceedings. His research area mainly includes Evolutionary approaches, Power system optimization, Planning, Dispatch, FACTS devices.

Vikash Kumar GUPTA born in 9th January 1984. He received B. E. degree Anna University, Chennai in 2009, and the M.Tech degree from B.I.T. Sindri, Dhnabad, Jharkhand in 2011. Currently, he is the Research Scholar in the Electrical Engineering Department, Indian School of Mines, Dhanbad, Jharkhand. His area of interest is Power System.

Sanjay KUMAR born in 5th June 1980. He received B.Tech degree from N.I.T Jamshedpur in 2006, M.Tech from N.I.T Patna in 2010. He has having two years of teaching experience and currently, he is the Research Scholar in the Electrical Engineering Department, Indian School of Mines, Dhanbad, Jharkhand.

5

Generalized Radio Resource Management for Overlapping MBS Zones

Fang LUO[1], *Yu-Feng CHOU*[1], *Ray-Guang CHENG*[1], *Marek KALIKA*[2], *Radek HOLY*[2], *Jana KALIKOVA*[2]

[1]Department of Electronic and Computer Engineering, College of Electrical and Computer Engineering, National Taiwan University of Science and Technology, No. 43, Sec. 4, Keelung Road, Taipei 106, Taiwan
[2]Department of Applied Informatics in Transport, Faculty of Transportation Sciences, Czech Technical University in Prague, Konviktska 20, 110 00 Prague, Czech Republic

fang.lou1@facebook.com, ecomno@cht.com.tw, crg@ieee.org, kalika@vc.cvut.cz, holy@vc.cvut.cz, kalikova@fd.cvut.cz

Abstract. *Multicast and broadcast service (MBS) is a point-to-multipoint service where data packets are transmitted simultaneously from a single source to multiple destinations. In MBS, some base stations (BSs) may form an MBS zone and transmit identical MBS contents simultaneously using the same modulation and coding scheme. Hence, the network has to coordinate the transmission of BSs such that BSs belonging to multiple MBS zones may utilize non-conflict resources to transmit different MBS contents. This paper extends the work in [1] for accommodating MBS zones with different service presence probabilities and various bandwidth requirements. A continuous allocation algorithm and a non-continuous allocation algorithm are presented to allocate resource units for overlapping MBS zones. Simulations were conducted to verify the effectiveness of the proposed algorithms.*

Keywords

Multicast and broadcast services, multiple MBS zones, resource allocation.

1. Introduction

Multimedia broadcast/multicast service (MBMS), which is also known as multicast and broadcast service (MBS) in IEEE 802.16m, is one of the important services defined by 3GPP Long Term Evolution (LTE) for the 4th generation cellular systems. MBMS is a point-to-multipoint service where data packets are transmitted simultaneously from a single source to multiple destinations [2]. In MBMS, the same data is transmitted via a common broadcast or multicast channel to multiple MBMS subscribers. Two service architectures are supported by 3GPP for delivering MBMS services: single cell-point to multi-points (SC-PTM) and multicast broadcast single frequency network (MBSFN). In SC-PTM, each base station (BS) transmits MBS content using its own modulation coding scheme (MCS).

In MBSFN, the MBS content is transmitted over a geographical area identified as a zone. A cluster of base stations (BSs) that transmit the same content in a zone is referred as an MBS zone. In general, each BS may belong to one or more MBS zones. Two MBS zones are overlapped if there is at least one BS belonging to both zones [3]. The network should allocate separate radio resource units for overlapping MBS zones such that BSs belonging to different MBS zones may transmit individual MBS content without confliction [4]. Normally, the service interval of an MBS service is much longer than that of a voice call. Hence, the coverage area of an MBS zone may be changed during the MBS service interval due to the movement of mobile stations (MSs). A BS may decide to join a nearby MBS zone if an MS requests for a new MBS service that is not existed in this cell but is available at this MBS zone. A BS may also leave an MBS zone if there is no MS demanding the MBS service. In response to the topology change of the MBS zones, the network should reallocate the radio resource units during the service interval. Hence, the network should minimize the signalling exchanged in the air interface and the core network.

The radio resource allocation for overlapping MBS zones is similar to the channel assignment problem discussed in traditional cellular networks. However, they are different in nature. In the channel assignment problem, the network normally allocates a number of static channels to a cluster of BSs at the cell planning phase to prevent from co-channel interference among neighboring BSs [5], [6]. The network may further reserve some dynamic channels to support extra traffic demanding at a specific BS during a particular time interval [5], [6]. The radio resource allocation for overlapping MBS zones is similar to the static channel assignment problem. Both of them can be solved based on graph theory [7]. However, each BS has the same number of neighboring cells in channel assignment but an MBS zone may be overlapped with different number of MBS zones. The radio resource re-allocation for overlapping MBS zones is different to the dynamic channel assignment problem. It is because that the traffic demanding of two BSs are uncorrelated but the topologies between two overlapping MBS zones are highly correlated. Moreover, the number of channels is a constant in dynamic channel assignment problem but the number of radio resource units used by MBS zones is a variable. Hence, existing channel assignment techniques cannot be directly applied here.

The coverage area of an MBS zone can be in regular shape or irregular shape [8]. The MBS zone with regular shape consists of a center cell and one or more surrounding rings of cells. The total number of cell in an MBS zone can be determined based the number of surrounding rings. The MBS zone with irregular shape is a group of neighboring cells with arbitrary shape. For example, the number of cells in an MBS zone was modeled by a binomial distribution with a given service presence probability [9]. An MBS zone with a higher service presence probability contains more cells.

The radio resource management for overlapping MBS zones has been studied in [1]. The paper assumed that each non-isomorphic graph of overlapped MBS zones occur with equal probabilities and each MBS service consumes only one radio resource unit [1]. Based on the two assumptions, a radio resource allocation algorithm is proposed to assign radio resources to each MBS zone based on graph theory; a radio resource re-allocation algorithm is then proposed to minimize the number of re-allocated radio resource units by utilizing the correlation between successive MBS zone topologies; and a simple resource estimation model is presented for estimating the number of radio resource units required by a given number of MBS zones.

This paper considers a generalized radio resource management problem for overlapping MBS zones. We extend the work in [1] by relaxing the assumption of identical bandwidth requirement in each MBS zone. The rest of the paper is organized as follows. In Section 2, a system model is presented. A continuous allocation algorithm and a non-continuous allocation algorithm are elaborated. Numerical results are shown in Section 3. Section 4 summarizes the conclusions.

2. System Model

This paper considers a hexagonal cellular network with M cells supporting N MBS services ($M > N$) as shown in Fig. 1. Let $\vec{B} = [B_1 \ldots B_N]$ be the bandwidth requirement of the MBS services, where B_i is the number of radio resource units required by the ith MBS zone and $B_1 \geq B_2 \geq \ldots \geq B_N$. The ith MBS zone contains a center cell and R_i surrounding-ring of cells. For example, $R_1 = R_2 = 1$ and $R_3 = 2$ in Fig. 1.

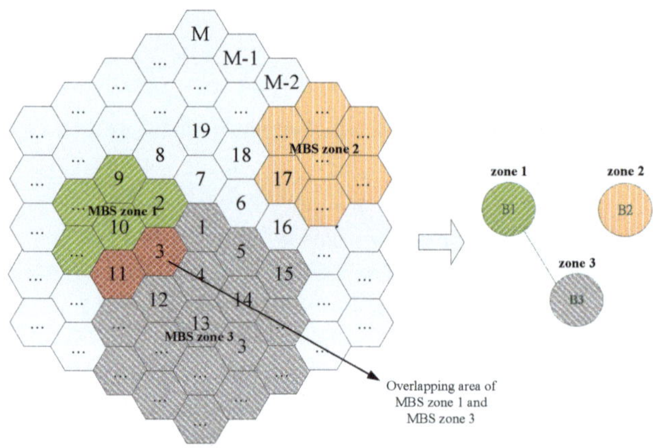

Fig. 1: System model.

We use an undirected labeled graph to represent the topology of the MBS zones. In this graph, each vertex represents an MBS zone; the number shown on the vertex represents the number of radio resource units required by the MBS service; and the edge connecting two vertexes indicates the overlapping of two MBS zones. Similar to [1], non-conflict radio resource units should be allocated to two vertexes connected by an edge. An M-by-N matrix $\mathbf{Q}$ ($\mathbf{Q}_{M\times N}$) is used herein to indicate the relationship among cells and their MBS zones. The ith row and the jth column element of $\mathbf{Q}$, $q_{i,j}$, is set as:

$$q_{i,j} = \begin{cases} 1 & \text{if cell } j \text{ belongs to MBS zone } i, \\ 0 & \text{otherwise.} \end{cases} \tag{1}$$

An N-by-N adjacent matrix $\mathbf{X}$ ($\mathbf{X}_{N\times N}$) presented in [1] is used to represent the interconnectivity of the network containing N MBS zones. Let $x_{i,j}$ be the ith row and the jth column element of $\mathbf{X}$, where

$$x_{i,j} = \begin{cases} 1 & \text{if MBS Zones } i \text{ and } j \text{ are overlaped,} \\ 0 & \text{otherwise.} \end{cases} \tag{2}$$

$\mathbf{X}$ can be obtained from $\mathbf{Q}$. That is,

$$x_{i,j} = \begin{cases} \text{sign}\left(\sum_{m=1}^{M} (q_{i,m} \times q_{j,m})\right), & i \neq j \\ 0, & i = j, \end{cases} \tag{3}$$

where sign(.) is the sign function.

For MBS zones with various bandwidth requirements, the network may allocate radio resource units for the MBS services in either continuous or non-continuous manner. In continuous resource allocation, a block of adjacent radio resource units are assigned to each MBS service. In non-continuous resource allocation, non-adjacent radio resource units are assigned to the MBS services [3].

2.1. Continuous Allocation (CA)

In continuous allocation, a block of radio resource units is first assign to MBS zone which has the maximum bandwidth requirement.

CA Algorithm

- Step 1: For a given $N \times N$ adjacent matrix $\mathbf{X}$, set $x_{i,j} = 0$.
- Step 2: Find a degree vector $\vec{d_x} = [d_1\, d_2\, d_3 \ldots d_N]$ of matrix $\mathbf{X}$, which is defined as:

$$d_i = \sum_{j=1}^{N} x_{i,j}. \tag{4}$$

- Step 3: Assign a block of resource unit to each MBS zone.
  - Step 3.1: Set all vertexes as uncolored. Let $C_0 = -\mathbf{X}$.
  - Step 3.2: Assign resource unit 1 to MBS zone n, (i.e., set $c_{n,n} = 1$) if MBS zone n has the maximum resource unit requirement (i.e., $B_n = \arg\max_i B_i, 1 \leq i \leq N, B_i \in \vec{B}$). Note, assign resource unit 1 to MBS zone n if $B_n = B_m$ and $d_n > d_m$. Construct C_1 by replacing the following elements in C_0,

$$\begin{cases} c_{i,n} = c_{n,n}, & \text{if } c_{i,n} = -1, i \in \{1, N\}, \\ c_{n,j} = c_{n,n}, & \text{if } c_{n,j} = -1, j \in \{1, N\}. \end{cases} \tag{5}$$

  - Step 3.3: Assign a used resource unit to MBS zone m (set $c_{m,m} = 1$ if resource unit 1 is assigned) if the unassigned MBS zone m has the maximum bandwidth requirement and its neighbor does not use this used color. Construct C_2 by replacing the following elements in C_1

$$\begin{cases} c_{i,m} = c_{m,m}, & \text{if } c_{i,m} = -1, i \in \{1, N\}, \\ c_{m,j} = c_{m,m}, & \text{if } c_{m,j} = -1, j \in \{1, N\}. \end{cases} \tag{6}$$

  - Step 3.4: Repeat Step 3.3 using an extra resource unit until each vertex is assigned a resource unit. The allocating resource block is conflict if $c_{q,i} \leq c_{q,q} \leq c_{q,i} + B_i - 1$ or $c_{q,i} \leq B_q \leq c_{q,i} + B_i - 1$. In this case, let $c_{q,q} = c_{q,q} + 1$. Repeat this process until no conflict is found. Construct C_k by replacing the following elements in C_{k-1},

$$\begin{cases} c_{i,q} = c_{q,q}, & \text{if } c_{i,q} = -1, i \in \{1, N\}, \\ c_{q,j} = c_{q,q}, & \text{if } c_{q,j} = -1, j \in \{1, N\}. \end{cases} \tag{7}$$

- Step 4: The resource unit vector set is $\vec{c}_X = diag(C_N)$.

2.2. Non-Continuous Allocation (NCA)

The concept proposed in [10] is used to transform a multi-coloring problem to a single-coloring problem by splitting each node i into B_i nodes.

Let $\overline{I_n}$ be the complement of the identity matrix of size n (i.e., an n-by-n square matrix with zeros on the diagonal and ones elsewhere). That is,

$$\overline{\mathbf{I_n}} = \begin{pmatrix} 0 & 1 & \cdots & 1 \\ 1 & 0 & \cdots & 1 \\ \vdots & \vdots & \ddots & \vdots \\ 1 & 1 & \cdots & 0 \end{pmatrix}. \tag{8}$$

Let $\mathbf{A}_{m \times n}$ be an m-by-n matrix with all elements equal to one. That is,

$$\mathbf{A_{m \times n}} = \begin{bmatrix} 1 & 1 & \cdots & 1 \\ 1 & 1 & \cdots & 1 \\ \vdots & \vdots & \ddots & \vdots \\ 1 & 1 & \cdots & 1 \end{bmatrix}. \tag{9}$$

The adjacent matrix $\mathbf{X}$ can be transformed to the extended adjacent matrix $\mathbf{E}$ (i.e., a square matrix of size W, $W = \sum_{i=1}^{N}(B_i)$ by Eq. (10), where $x_{i,j}$ is the ith row and the jth column element of $\mathbf{X}$, B_i is the required radio resource units of the ith MBS zone. For example, $\mathbf{X} = \begin{bmatrix} 0 & 1 & 0 \\ 1 & 0 & 0 \\ 0 & 0 & 0 \end{bmatrix}$ and $\vec{B} = [\ 3 \quad 2 \quad 1\ ]$. We can have Eq. (11).

After splitting the adjacent matrix $\mathbf{X}$ to the extended adjacent matrix $\mathbf{E}$, the algorithm proposed in [1] can directly apply as below.

NCA Algorithm

- Step 1: For a given $N \times N$ adjacent matrix $\mathbf{X}$, find the extended adjacent matrix $\mathbf{E}$.

$$\mathbf{E}=\begin{bmatrix} \overline{I_{B_1}} & x_{1,2}A_{B_1\times B_2} & \cdots & x_{1,N}A_{B_1\times B_N} \\ x_{2,1}A_{B_2\times B_1} & \overline{I_{B_2}} & \cdots & \vdots \\ \vdots & \vdots & \ddots & \vdots \\ x_{N,1}A_{B_N\times B_1} & x_{N,2}A_{B_N\times B_2} & \cdots & \overline{I_{B_N}} \end{bmatrix}. \tag{10}$$

$$\mathbf{E}=\begin{bmatrix} \overline{I_3} & 1\times\begin{bmatrix}1&1\\1&1\\1&1\end{bmatrix} & 0\times\begin{bmatrix}1\\1\\1\end{bmatrix} \\ 1\times\begin{bmatrix}1&1&1\\1&1&1\end{bmatrix} & \overline{I_2} & 0\times\begin{bmatrix}1\\1\end{bmatrix} \\ 0\times\begin{bmatrix}1&1&1\end{bmatrix} & 0\times\begin{bmatrix}1&1\end{bmatrix} & \overline{I_1} \end{bmatrix}=\begin{bmatrix}0&1&1&1&1&0\\1&0&1&1&1&0\\1&1&0&1&1&0\\1&1&1&0&1&0\\1&1&1&1&0&0\\0&0&0&0&0&0\end{bmatrix}. \tag{11}$$

- Step 2: Find a degree vector $\vec{d_x}=\begin{bmatrix} d_1 & d_2 & \dots & d_W \end{bmatrix}$ of $\mathbf{E}$, which is defined as:

$$d_i=\sum_{j=1}^{W} e_{i,j}. \tag{12}$$

- Step 3: Assign one resource unit to each vertex.
  - Step 3.1: Set all vertexes as uncolored. Let $C_0=-\mathbf{E}$.
  - Step 3.2: Assign the first resource unit to vertex n, (i.e., set $c_{n,n}=1$) if vertex n has the maximum degree (i.e. $d_n=\arg\max_i d_i$, $1\le i\le W$, $d_i\in\vec{d_E}$). Construct C_1 by replacing the following elements in C_0,

$$\begin{cases} c_{i,n}=c_{n,n}, & \text{if } c_{i,n}=-1, i\in\{1,W\},\\ c_{n,j}=c_{n,n}, & \text{if } c_{n,j}=-1, j\in\{1,W\}.\end{cases} \tag{13}$$

  - Step 3.3: Assign a used resource unit to an unassigned vertex m (set $c_{m,m}=1$) if the vertex m has the maximum degree and does not have a neighbor with the used resource unit. Construct C_2 by replacing the following elements in C_1,

$$\begin{cases} c_{i,m}=c_{m,m}, & \text{if } c_{i,m}=-1, i\in\{1,W\},\\ c_{m,j}=c_{m,m}, & \text{if } c_{m,j}=-1, j\in\{1,W\}.\end{cases} \tag{14}$$

  - Step 3.4: Repeat Step 3.3 using extra resource unit until each vertex is assigned by a non-conflict resource unit. Note that color p can be assigned to vertex q if and only if:

$$\begin{cases} c_{q,q}\neq c_{i,q}, & \text{if } c_{i,q}\neq 0 \text{ or } -1, i\in\{1,W\},\\ c_{q,q}\neq c_{q,j}, & \text{if } c_{q,j}\neq 0 \text{ or } -1, j\in\{1,W\}.\end{cases} \tag{15}$$

  Contruct C_k by replacing the following elements in C_{k-1},

$$\begin{cases} c_{i,q}=c_{q,q}, & \text{if } c_{i,q}=-1, i\in\{1,W\},\\ c_{q,j}=c_{q,q}, & \text{if } c_{q,j}=-1, j\in\{1,W\}.\end{cases} \tag{16}$$

- Step 4: The resource unit vector set is $\vec{c_E}=diag(C_W)$.

Note that the NCA algorithm is expected to achieve a better performance than CA algorithm does but at the cost of a higher computational complexity. The computational complexity of the resource allocation algorithm depends on the dimension of the adjacent matrix or extended adjacent matrix. Hence, the complexity of the continuous resource allocation algorithm $O(N^2)$ and the complexity of the non-continuous resource allocation algorithm is $O(W^2)$, where $W=\sum_{i=1}^{N} B_i$. For all MBS zones have identical bandwidth requirements (i.e., B_i is a constant), the complexity of the non-continuous resource allocation algorithm is $O((B_iN)^2)$.

3. Simulation Results

Simulations were conducted on top of a C-based simulation platform to verify the effectiveness of the proposed algorithms. In the simulation, each sample was obtained by averaging 10000 outcomes. The results of a random allocation algorithm, CA algorithm, and NCA algorithm were evaluated. The random allocation algorithm randomly assigns a block of continuous radio resource units to each MBS zone. In the simulation, the number of radio resource units required by the ith MBS zone, Bi, was generated from a Gaussian distribution with parameter (μ,σ^2).

Figure 2 and Fig. 3 showed the mean number of allocated resource units required in three algorithms for different number of MBS zones with identical R_i and different R_i, respectively. In Figure 2 and Fig. 3, the number of cell M was 169 and N varied from 3 to 10. As defined in Sec. 2, R_i is the number of surrounding-ring of cells in the ith MBS zone. A larger R_i implies a larger MBS zone size. It can be found in Fig. 2 that the mean number of allocated resource units increases as the number of MBS zones increases; both CA and NCA algorithms require less resource units than the random allocation algorithm does; and the improvement of NCA over CA becomes noticeable when N reaches 7. Compare Fig. 2 and Fig. 3, it was found that the effect of R_i on the mean number of allocated resource units required in three algorithms is not significant. It is because that the size of the MBS zone and its bandwidth requirement (i.e, R_i and B_i) were independently generated.

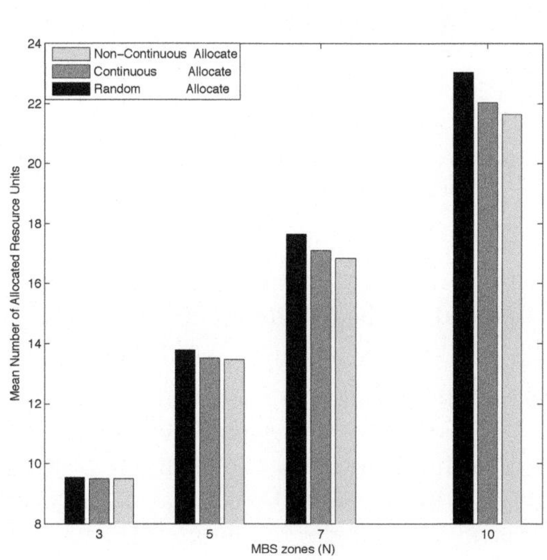

Fig. 2: Performance of radio resource allocation algorithm with identical R_i.

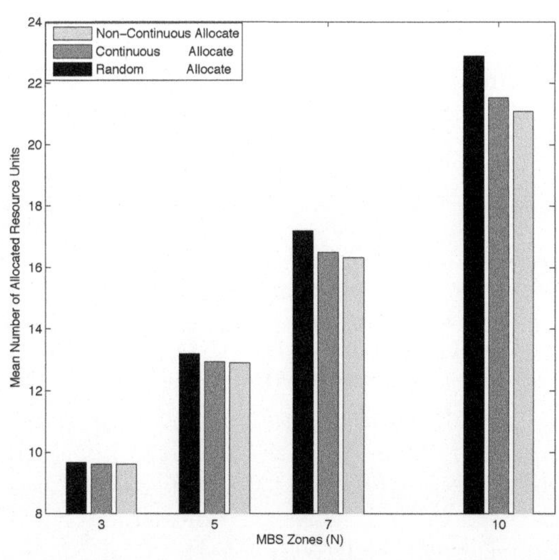

Fig. 3: Performance of radio resource allocation algorithms with different R_i.

Figure 4 showed the running time of the CA and NCA algorithms for MBS zones with different R_i and $M = 169$. A constant of $B_i = 5$ was chosen in order to verify the time complexity analysis. It can be found from Fig. 4 that NCA requires less resource units than CA does but at the cost of extra computational complexity. From the simulation results, it was found that the average running time of NCA algorithm is about B_i^2 times than that of CA algorithm. Note that the running time of the cases shown in Fig. 2 and Fig. 3 had similar results as that in Fig. 4. Therefore, the results are not demonstrated herein.

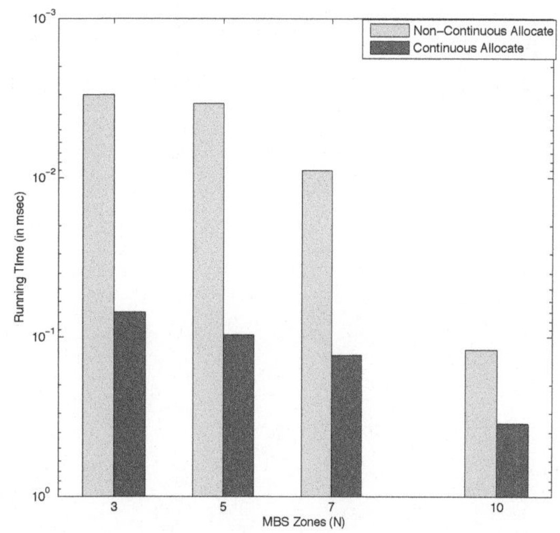

Fig. 4: Running time of CA and NCA algorithms.

Figure 5 and Fig. 6 showed the required resource units of the three resource allocation algorithms for $M = 169$ and 91, respectively. In Fig. 5 and Fig. 6, $B_i = 5$ and $R_i = 2$ were considered and the effect of the bandwidth requirement variation on the required number of resource units was investigated. The case of standard deviation equal to 0 is equivalent to the single resource allocation algorithm considered in [1]. In this case, CA and NCA are identical. Figure 5 and Fig. 6 showed that the mean number of required resource units increases as the bandwidth requirement variation increases. It can be found in Fig. 5 that the benefit of NCA was ignorable when the bandwidth requirement variation increases. Figure 6 demonstrated the effect of reducing M on the mean number of required resource units. Reducing M is equivalent to increasing R_i since both cases increase the overlapping probabilities of MBS zones. Figure 6 showed that the performance of CA and NCA was similar if the MBS zones were highly overlapped. Figure 7 showed the required resource units of the three resource allocation algorithms for $M = 37$ and $R_i = 1$. It can be found from Figure 6 and Fig. 7 that the performance of the three algorithms was similar in the two cases since the effect of reducing M was compensated by the effect of decreasing R_i.

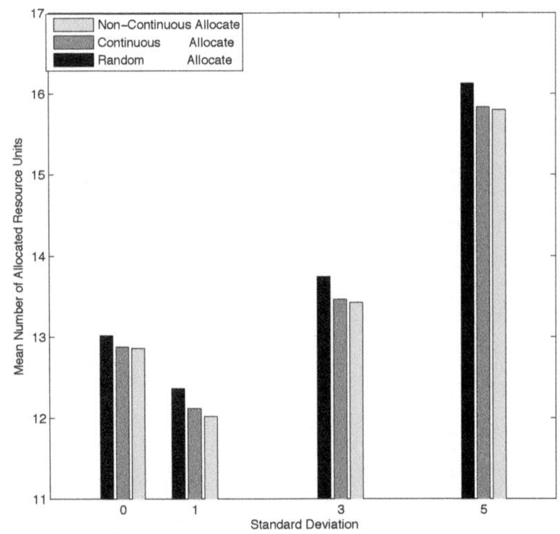

Fig. 5: Performance of radio resource allocation algorithm with different standard deviation $M = 169$.

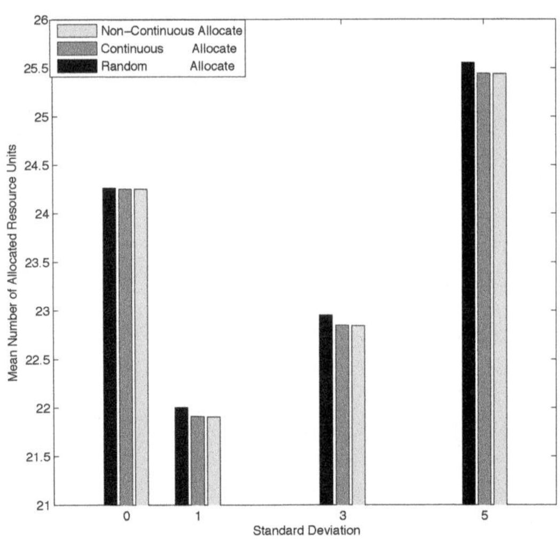

Fig. 6: Performance of radio resource allocation algorithm with different standard deviation $M = 91$.

4. Conclusion

This paper presents two algorithms to allocate radio resource units for overlapping MBS zones requiring different radio resource units. A continuous allocation algorithm and a non-continuous allocation algorithm were presented to allocate radio resource units to M overlapping MBS zones. Simulations were conducted to evaluate the effectiveness of the proposed

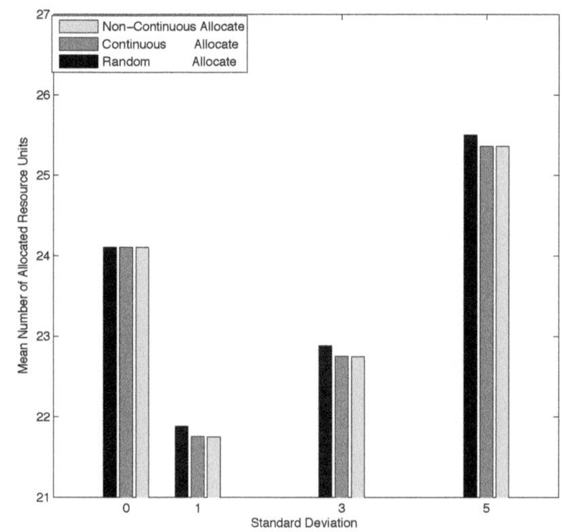

Fig. 7: Performance of radio resource allocation algorithm with different standard deviation $M = 37$.

algorithms. It was found that the proposed allocation algorithms may reduce the required number of resource units.

Currently, a radio resource re-allocation is under development. The radio resource re-allocation aims to re-allocate resource in response to the changing of MBS zone topology. It may utilize the correlation between the old and the new topologies to minimize the number of re-allocated zones. In addition, we are developing a resource estimation model which can accurately estimate the number of radio resources required by the MBS services.

Acknowledgment

This work was supported in part by National Science Council, Taiwan, under Contract NSC 101-2219-E-011-005, NSC 102-2218-E-011-002, NSC 102-3113-P-006-015, and NSC 102-2911-I-011-506.

References

[1] CHENG, R. G. and K. J. HUANG. Resource allocation for overlapping MBS zones. *EURASIP Journal on Wireless Communications and Networking.* 2011, vol. 2011, no. 205612, pp. 1–10. ISSN 1687-1499. DOI: 10.1155/2011/205612.

[2] JIANG, T., W. XIANG, H.-H. CHEN and Q. NI. Multicast broadcast services support in OFDMA-Based WiMAX systems. *IEEE Communications Magazine.* 2007,

vol. 45, iss. 8, pp. 78–86. ISSN 0163-6804. DOI: 10.1109/MCOM.2007.4290318.

[3] ALVARION AND STARENT NETWORKS. *MCBCS synchronous transmission support over WiMAX network: WiMAX Forum Contribution 01328_r000*. 2008.

[4] MCBCS_DSX_STAGE2_WORKING_BASELINE. *WiMAX Forum Network Working Group Contribution 01358_r003*. 2008.

[5] COX, D. C. and D. O. REUDINK. A comparison of some channel assignment strategies in Large-scale mobile communications system. *IEEE Transactions on Communications*. 1972, vol. 20, iss. 2, pp. 190–195. ISSN 0090-6778. DOI: 10.1109/TCOM.1972.1091134.

[6] COX, D. C. and D. O. REUDINK. Increasing channel occupancy in large-scale mobile radio Systems: dynamic channel reassignment. *IEEE Transactions on Communications*. 1973, vol. 21, iss. 11, pp. 1302–1306. ISSN 0090-6778. DOI: 10.1109/TCOM.1973.1091564.

[7] MALAGUTI, E. and P. TOTH. A survey on vertex coloring problems. *International Transactions in Operational Research*. 2009, vol. 17, iss. 1, pp. 1–34. ISSN 1475-3995. DOI: 10.1111/j.1475-3995.2009.00696.x.

[8] KAMBLE, V., S. KALYANASUNDARAM, V. RAMACHANDRAN and R. AGRAWAL. Efficient resource allocation strategies for multicast/broadcast services in 3GPP Long Term Evolution single frequency networks. In: *IEEE Wireless Communications and Networking Conference. WCNC 2009*. Budapest: IEEE, 2009, pp. 1–6. ISBN 978-1-4244-2947-9. DOI: 10.1109/WCNC.2009.4917701.

[9] JIANG, A., C. Y. FENG and T. K. ZHANG. Research on resource allocation in multi-cell MBMS single frequency networks. In: *IEEE Conference on Wireless and Optical Communication Network (WOCN), 2010*. Colombo: IEEE, 2010, pp. 1–5. ISBN 978-1-4244-7203-1. DOI: 10.1109/WOCN.2010.5587312.

[10] LIM, A., X. ZHANG and Y. ZHU. A hybrid method for the graph coloring and its related problems. In: *Andrew* [online]. 2003. Available at: `http://144.214.237.85/andrew/images/course/PAPERS/mic2003lzz.pdf`.

About Authors

Fang LUO was born in 1986 in Taipei, Taiwan. He received his M.Sc. from National Taiwan University of Science and Technology in 2011. His research interests include wireless resource allocation and social network.

Yu-Feng CHOU was born in 1984 in Taipei, Taiwan. He received his M.Sc. from National Chiao-Tung University in 2008. He is currently working toward the Ph.D. degree at National Taiwan University of Science and Technology. His research interests include game theory and social network.

Ray-Guang CHENG was born in 1969 in Keelung, Taiwan. He received his M.Sc. and Ph.D. from National Chiao-Tung University, Taiwan, in 1993 and 1996, respectively. He is currently a Professor with the Department of Electronic and Computer Engineering, National Taiwan University of Science and Technology (NTUST), Taiwan. His research interests include multi-hop wireless networks and machine-to-machine communications.

Marek KALIKA was born in 1970 in Prague, Czech Republic. He received his M.Sc. and Ph.D. from Czech Technical University in Prague (CTU) in 1993 and 2000, respectively. He is currently an Assistant Professor with the Department of Applied Informatics in Transport, Faculty of Transportation Sciences, CTU. His research interests include process systems engeneering, secure, ontologies and smart cards.

Raedk HOLY was born in 1976 in Prague, Czech Republic. He received his M.Sc. from Czech Technical University in Prague (CTU) in 1994. He is currently an Assistant Professor with the Department of Applied Informatics in Transport, Faculty of Transportation Sciences, CTU. His research interests include secure data transfer, hardware and smart cards.

Jana KALIKOVA was born in 1970 in Jablonec n. N., Czech Republic. She received her M.Sc. and Ph.D. from Czech Technical University in Prague (CTU) in 1993 and 2000, respectively. She is currently an Assistant Professor with the Department of Applied Informatics in Transport, Faculty of Transportation Sciences, CTU. Her research interests include secure data transfer, biometrics, and smart cards.

6

Wearable Textile Electrodes for ECG Measurement

Lukas VOJTECH[1], Radoslav BORTEL[2], Marek NERUDA[1], Milos KOZAK[1]

[1]Department of Telecommunication Engineering, Faculty of Electrical Engineering, Czech Technical University in Prague, Technicka 2, 166 27 Prague, Czech Republic
[2]Department of Circuit Theory, Faculty of Electrical Engineering, Czech Technical University in Prague, Technicka 2, 166 27 Prague, Czech Republic

vojtecl@fel.cvut.cz, bortelr@fel.cvut.cz, nerudmar@fel.cvut.cz, kozakmil@fel.cvut.cz

Abstract. *The electrocardiogram (ECG) is one of the most important parameters for monitoring of the physiological state of a person. Currently available systems for ECG monitoring are both stationary and wearable, but the comfort of the monitored person is not at a satisfactory level because these systems are not part of standard clothing. This article is therefore devoted to the development and measurement of wearable textile electrodes for ECG measurement device with high comfort for the user. The electrode material is made of electrically conductive textile. This creates a textile composite that guarantees high comfort for the user while ensuring good quality of ECG measurements. The composite is implemented by a carrier (a T-shirt with flame retardant) and sensing electrodes embroidered with yarn based on a mixture of polyester coated with silver nanoparticles and cotton. The electrodes not only provide great comfort but are also antibacterial and antiallergic due to silver nanoparticles.*

Keywords

Conductive textile, ECG, wearable electronics.

1. Introduction

The electrocardiogram (ECG) is one of the most important parameters for monitoring of the physiological state of a person. Currently available systems for ECG monitoring are both stationary and wearable, but the comfort of the monitored person is not at a satisfactory level because these systems are not part of standard clothing. Conventional stationary systems often use suction cup electrodes or adhesive electrodes. The wearable monitoring systems use metallic, textile or conductive polymeric electrodes or contactless sensing. Both solutions can be inconvenient and poorly applicable for long term ECG monitoring of patients and athletes. The use of existing methods for electrical potential sensing are not at a satisfactory level of comfort in applications of preventive ECG measurement, especially in monitoring of stress factors of rescue workers, fireman or soldiers. The low comfort of the existing solutions therefore limits the applicability of these methods.

One of the promising applications of ECG monitoring is also monitoring of excitation level in persons working under great physical and mental stress. This includes soldiers, firefighters, police officers and rescue workers in mines. Stress in these professions has a major impact on the quality of their work, but also of their life [1]. High levels of stress factors have an impact on the rescue itself – they can prolong the response time and increase the chance of mistakes. This directly increases the danger to health, life or property. It is, therefore, advantageous for the rescue operation management to react quickly to increased stress of workers and to stop or reduce their exposure to stress.

The means for measuring ECG can be divided into two basic groups. The first one is the technology for ambulatory ECG measurements that need to perform measurements with high accuracy and QoS (Quality of Services) of these systems. The second category can be described as a device for mobile measurement, especially designed for athletes [2] or for long-term ECG monitoring during a specified period of examination of patients [3]. The second category also includes the systems for optimization of the dosage of drugs that affect the activity of the patients.

The currently available wearable ECG monitoring systems use several types of electrodes. Adhesive electrodes are commonly used in commercial Holter monitors. Also, many systems use textile electrodes embedded in clothing [4], [5], [6], [7]. Other approaches use hand held electrodes [8] or contactless sensing [9].

This article is therefore devoted to the development and measurement of wearable textile electrodes for

ECG measurement device with high comfort for the user. The electrode material is made of electrically conductive textile. This creates a textile composite that guarantees high comfort for the user while ensuring good quality of ECG measurements. The composite is implemented by a carrier (a T-shirt with flame retardant) and sensing electrodes embroidered with yarn based on a mixture of polyester coated with silver nanoparticles and cotton. The electrodes not only provide great comfort but are also antibacterial and antiallergic (due to silver nanoparticles).

The rest of the paper is organized as follows: Section 2 describes wearable electronics, Section 3 denotes construction of T-shirt with textile electrodes, Section 4 presents experimental results, Section 5 describes evaluation and the conclusion is presented in Section 6.

2. Wearable Electronics

With the development of electrically conductive textile materials included in the various types of clothing, a new area of applications called wearable electronics is formed. Heated textile structures [10] or textile antenna structures directly implemented in the clothes [11] can be therefore realized. Similar technology can be used for realization of wearable textile electrodes for ECG sensor, which can be part of standard clothing.

Following characteristics are preferred for the realization of these electrodes:

- It is necessary to use a textile material with sufficiently high conductivity, which can be pre-defined and propose suitable composition of textile material [12] (providing sensing of electrical signal with a sufficient quality).
- Textile material must be cotton based fabric (high comfort on contact fabric and skin, thanks to the use of cotton blended yarn).
- Textile electrode should be a fixed part of the clothes (easy application of the system and position stability of the electrode).
- Complete textile product should allow standard maintenance (washing and ironing).
- Fabric with galvanic plating surface cannot be used (implementation on non-cotton substrates causes skin irritation because of synthetic fibers, upper material is copper, which oxidizes, or nickel, which is allergenic).
- Electrodes require sufficient adhesion to the pectorals (to ensure sufficient difference of electrical potentials between electrodes).

3. Construction of T-shirt with Textile Electrode

For ensuring the above requirements the T-shirt was designed and patterned from a knitted fabric modified for reducing flammability (used for rescue workers and firefighters with sustained high wearing comfort). Rectangles of electrically conductive textile fabric in knitted design were sutured on the position of the pectoral muscles. The material has the following textile and electrical parameters:

- Fineness of yarn 29,5 tex.
- Silverstat 60 %, 40 % PES.
- Surface conductivity of 1,2 Ω/square, measured using the method [13].

The knitted fabric is realized from a blended yarn of the composite containing silver nanoparticles, which provide electrical conductivity of the yarn and the resultant knitted fabric. The content of silver nanoparticles provides corrosion resistance of textile electrodes, antibacterial and antiallergic properties, mechanical and electrical stability when exposed to sweat. The design of blended fabric made from conductive yarn enables traditional maintenance of T-shirts (washing, ironing) and long-term stability of surface conductivity of the electrodes with a high number of wash cycles.

The textile electrodes were rectangular in shape, with dimensions 70 × 100 mm. They were positioned so that they contacted the upper chest immediately below the clavicles, Fig. 1. In order to provide a sufficient potential difference, the electrodes were positioned 100 mm apart.

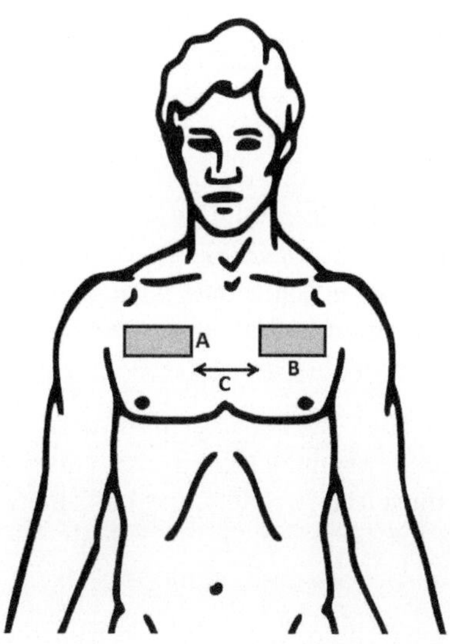

Fig. 1: Placement of textile electrodes (with using [14]).

4. Experimental Verification

The suitability of the textile electrodes for ECG recording was verified by experimental measurements. Specifically, we compared the ECG recorded using the textile electrodes with the ECG measured by a high quality reference measurement system.

During the measurement the subject was lying in the supine position, and was instructed to relax his muscles. This position minimizes the presence of muscular artifacts, and allows evaluating the quality of a pure ECG.

The signal was first recorded using two metallic electrodes positioned on the subject's fingers (the right and left index finger). The measurement system also used the third electrode that was positioned on the right middle finger, and was driven by a noise suppression circuit [16], [17]. The placement of the electrodes is shown in Fig. 2. From our previous measurements we know that this methodology provides a high quality ECG recording; therefore, we used it as a reference for the evaluation of the ECG recorded with the textile electrodes.

Once the reference record was obtained, we proceeded with the ECG recording using the textile electrodes. This measurement used only 2 leads without any active noise suppression. The electrode placement is shown in Fig. 4.

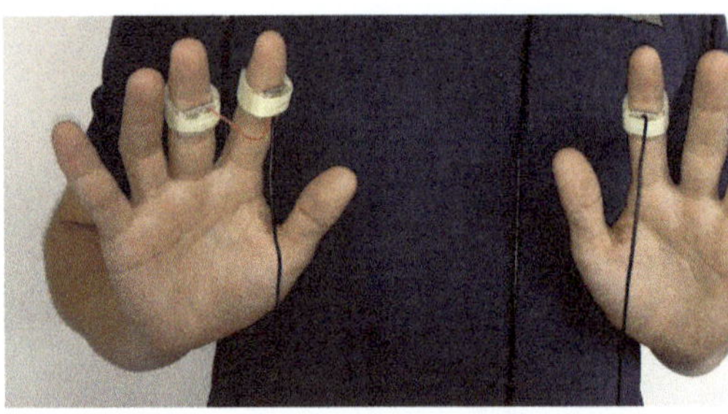

Fig. 2: Electrode placement for three-lead ECG measurement.

The measurement electronics implemented high pass and low pass filters with cut-off frequencies 0,1 Hz and 200 Hz, respectively. Once the signals were digitized (the sampling frequency was 800 Hz) they were further filtered with a set of notch filters with the center frequencies at multiples of 50 Hz. This filtering efficiently suppressed the power line interference.

It is expected that during the movement the ECG signal will contain various kinds of movement artifacts (especially the muscular artifacts and baseline wonder). However, these artifacts can be well dealt with using the digital signal processing techniques [15].

Measurement results of the reference ECG signal are shown in Fig. 3. The ECG record has clearly recognizable P waves, QRS complexes and T waves.

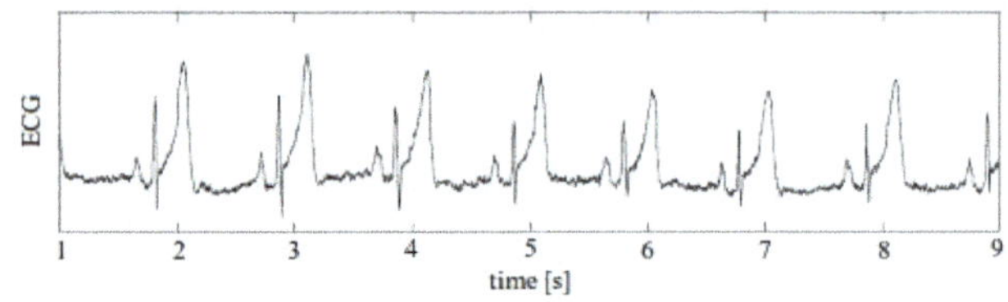

Fig. 3: The waveform of the three-lead measurement method. This record serves as a reference for the evaluation of the measurement performed with the textile electrodes.

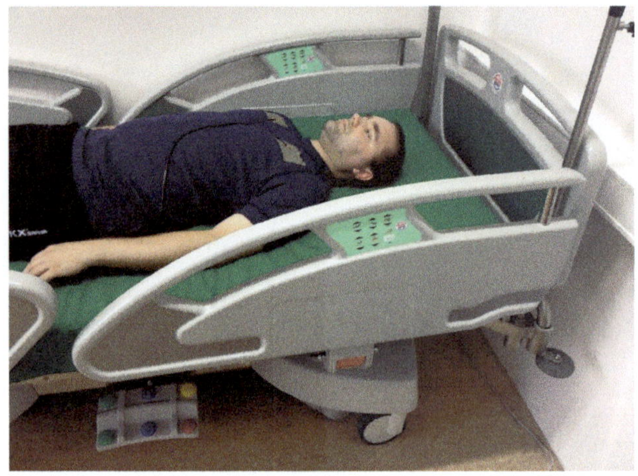

Fig. 4: Placement of textile electrodes.

The ECG record obtained using the textile electrodes is shown in Fig. 5. Similarly to the reference record the signal appears to be clear with well recognizable P waves, QRS complexes and T waves.

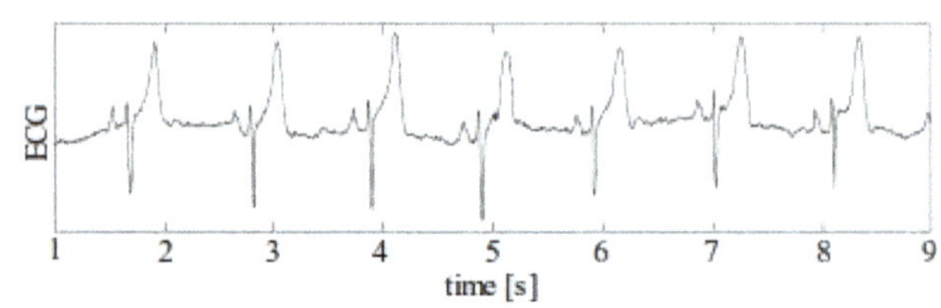

Fig. 5: Signal measured with two lead arrangement using textile electrodes.

The comparison of measured signals is better shown in Fig.6. The amplitude of the individual waves of ECG signal obtained using the textile electrodes is smaller than ECG reference signal, which is caused by different electrode placement. Both measurements clearly show P waves, QRS complexes and T waves, which results in the ECG measurement with textile electrodes is applicable method and it can be advantageously used in applications for ECG measurement.

5. Conclusion

Test arrangement of wearable textile electrodes for ECG measurement showed the suitability of this plan

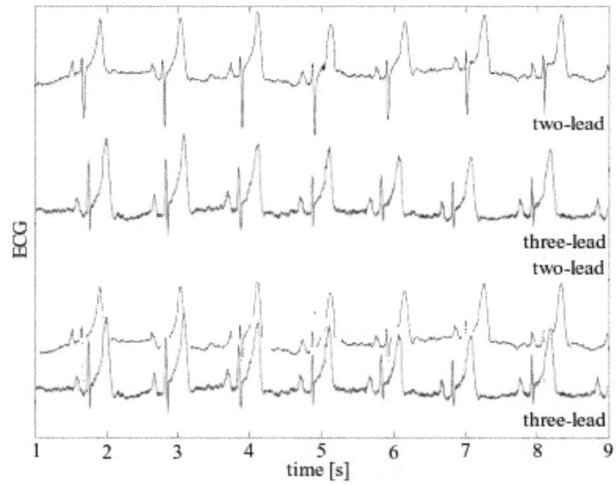

Fig. 6: Comparison of the measured signal with two-lead (using textile electrodes) and three-lead (ECG reference signal) arrangement.

in the Proof of Concept phase. It is possible to clearly distinguish the P wave, QRS complex and T wave from received signals and it is possible to measure the heart rate. Furthermore, it was observed only insignificant baseline wonder which indicates stable contact between the conductive textile electrode and the skin.

The result is the prototype of textile product designed as a T-shirt enabling capture ECG signals using two leads directly on the body of the monitored person who is not troubled by the presence of the electrodes. Implementation of this type of wearable textile electrodes allows standard methods of maintenance (washing, ironing), keeps high user comfort and simultaneously ensures sensing of electrical potentials for ECG measurement in sufficient quality. As shown, the quality of the measured signal using wearable textile electrodes for ECG sensor with two leads is comparable to the signals measured by the reference method, which incorporates active noise suppression using a third electrode.

Acknowledgment

This work was supported by the RFID Locator VG3VS/068 - Research and development of a new type of locator, especially for active RFID tags and by the Kompozitex FR–TI4/202 - Composite textile materials for humans and technology protection from the effects of electromagnetic and electrostatic fields.

References

[1] PRATI, G., L. PIETRANTONI and E. CICOGNANI. Self-efficacy moderates the relationship between stress appraisal and quality of life among rescue workers. *Anxiety, Stress & Coping.* 2010, vol. 23, iss. 4, pp. 463–470. ISSN 1061-5806. DOI: 10.1080/10615800903431699.

[2] VARNAUSKAS, E., P. BJORNTORP, M. FAHLEN, I. PREROVSKY and J. STENBERG. Effects of physical training on exercise blood flow and enzymatic activity in skeletal muscle. *Cardiovascular Research.* 1970, vol. 4, iss. 4, pp. 418–422. ISSN 0008-6363. DOI: 10.1093/cvr/4.4.418.

[3] ACHILLI, A, M. SASSARA, S. FICILI, D. PONTILLO, P. ACHILLI, C. ALESSI, S. DE SPIRITO, R. GUERRA, N. PATRUNO and F. SERRA. Long-term effectiveness of cardiac resynchronization therapy in patients with refractory heart failure and "narrow" QRS. *Journal of the American College of Cardiology.* 2003, vol. 42, iss. 12, pp. 2117–2124. ISSN 0735-1097. DOI: 10.1016/j.jacc.2003.08.024.

[4] LEE, Y.D. and W.Y. CHUNG. Wireless sensor network based wearable smart shirt for ubiquitous health and activity monitoring. *Sensors Actuators B Chemical.* 2009, vol. 140, iss. 2, pp. 390–395. ISSN 0925-4005. DOI: 10.1016/j.snb.2009.04.040.

[5] LOPEZ, G., V. CUSTODIO, J.I. MORENO. LOBIN: E-textile and wireless-sensor-network-based platform for healthcare monitoring in future hospital environments. *IEEE Transactions on Information Technology in Biomedicine.* 2010, vol. 14, iss. 6, pp. 1446–1458. ISSN 1089-7771. DOI: 10.1109/TITB.2010.2058812.

[6] YOO, J., Y. LONG, L. SEULKI, K. HYE-JUNG and Y. HOI-JUN. Awearable ECG acquisition system with compact planar-fash-ionable circuit board-based shirt. *IEEE Transactions on Information Technology in Biomedicine.* 2009, vol. 13, iss. 6, pp. 897–902. ISSN 1089-7771. DOI: 10.1109/TITB.2009.2033053.

[7] MERRITT, C. R., H. T. NAGLE and E. GRANT. Fabric-based active electrode design and fabrication for health monitoring clothing. *IEEE Transactions on Information Technology in Biomedicine.* 2009, vol. 13, iss. 2, pp. 274–280. ISSN 1089-7771. DOI: 10.1109/TITB.2009.2012408.

[8] POLLONINI, L., N. O. RAJAN, S. XU, S. MADALA and C. C. DACSO. A novel handheld device for use in remote patient monitoring of heart failure patients-design and preliminary validation on healthy subjects. *Journal of Medical Systems.* 2012, vol. 36, iss. 2, pp. 653–659. ISSN 0148-5598. DOI: 10.1007/s10916-010-9531-y.

[9] OTTENBACHER, J., L. JATOBA, U. GROSSMANN, W. STORK and K. MULLER-GLASER. ECG electrodes for a context-aware cardiac permanent monitoring system. In: *World Congress on Medical Physics and Biomedical Engineering 2006.* Seoul: IFMBE Proceedings, 2007, vol. 14. pp. 672–675. ISSN 1433-9277. ISBN 978-3-540-36839-7.

[10] VOJTECH, L. and M. NERUDA. Self-drying planar textile materials with electrically conductive structures. In: *15th International Symposium MECHATRONIKA 2012.* Prague: IEEE, 2012. pp. 1–5, ISBN 978-1-4673-0979-0.

[11] DAHAL, R., D. MERCAN, L. VOJTECH AND M. NERUDA. Textile Antenna for 50 ohm Applications. *Advances in Electrical and Electronic Engineering.* 2012, vol. 10, no. 4, pp. 229–234. ISSN 1336-1376.

[12] NERUDA, M. and L. VOJTECH. Verification of Surface Conductance Model of Textile Materials. *Journal of Applied Research and Technology.* 2012, vol. 10, no. 4, pp. 579–585. ISSN 1665-6423.

[13] VOJTECH, L. and M. NERUDA. Modelling of Surface and Bulk Resistance for Wearable Textile Antenna Design. *Przeglad Elektrotechniczny.* 2013, vol. 89, iss. 2b, pp. 217–222. ISSN 0033-2097.

[14] HAGGSTROM, M. Pioneer plaque man upper body as diagram template [online]. 2009 [cit. 2013-07-07]. Available at: `http://commons.wikimedia.org/wiki/`.

[15] BORTEL, R. and P. SOVKA. *Real-Time Robust Heart Rate Estimation Based on Bayesian Framework and Grid Filters.* Medical Applications of Intelligent Data Analysis: Research Advancements. IGI Global, 2012, pp. 67–90. ISBN 978-1-4666-1803-9. DOI: 10.4018/978-1-4666-1803-9.

[16] HOSPODKA, J. and R. BORTEL. Amplifier for Biological Signal Sensing with Efficient Common Mode Rejection. In: *2011 International Conference on Applied Electronics.* Plzen: University of West Bohemia, 2011, pp. 161–164. ISSN 1803-7232. ISBN 978-80-7043-987-6.

[17] HOSPODKA, J. and R. BORTEL. *System for Biological Signal Measurement with Noise Suppression* [patent]. Czech Republic. Urad prumysloveho vlastnictvi, 302454. 2011.

About Authors

Lukas VOJTECH was born in 1977 in Nachod. He received the M.Sc. degree in electrical engineering from the Czech Technical University in Prague, Faculty of Electrical Engineering, in 2003. In 2005, he received the bachelor degree engineering pedagogy from the Masaryk Institute of Advanced Studies in Prague. In 2010, he received the Ph.D. degree from Czech Technical University in Prague, Faculty of Electrical Engineering. Currently, he works as an assistant professor at the Department of telecommunication engineering, CTU in Prague. His research interests include wireless technologies, technology RFID and mainly EMC in area of shielding materials.

Radoslav BORTEL was born in 1979 in Martin, Slovakia. He received the M.Sc. degree in electrical engineering from the Czech Technical University in Prague, Faculty of Electrical Engineering, Czech Republic in 2005. In 2010, he received the Ph.D. degree from Czech Technical University in Prague, Faculty of Electrical Engineering. Currently, he works as an assistant professor at the Department of circuit theory, CTU in Prague. His research interests focuses on processing and measurement of biological signals.

Marek NERUDA was born in Hradec Kralove in Czech Republic in 1983. He received the M.Sc. degree in electrical engineering from the Czech Technical University in Prague, Faculty of Electrical Engineering, Czech Republic in 2007. In 2009, he received the bachelor degree engineering pedagogy from the Masaryk Institute of Advanced Studies in Prague, Czech Republic. In 2010, he received the M.Sc. degree in entrepreneurship and commercial engineering in industry from the Masaryk Institute of Advanced Studies in Prague, Czech Republic. Currently, he is a student of Ph.D. degree.

Milos KOZAK was born in Rakovnik in Czech Republic in 1984. He received the M.Sc. degree in electrical engineering from the Czech Technical University, Prague, in 2009. Since 2009 until 2012, he had been teaching optical communication systems and data networks with the Czech Technical University in Prague. His research interest is on the application of high-speed optical transmission systems in a data network. Particularly regenerators placement in all optical networks.

Modelling of H.264 MPEG2 TS Traffic Source

Stanislav KLUCIK, Martin LACKOVIC

Institute of Telecommunications, Faculty of Electrical Engineering and Information Technology, Slovak University of Technology in Bratislava, Ilkovicova 3, 812 19 Bratislava, Slovakia

klucik@ut.fei.stuba.sk, lackovic@ut.fei.stuba.sk

Abstract. *This paper deals with IPTV traffic source modelling. Traffic sources are used for simulation, emulation and real network testing. This model is made as a derivation of known recorded traffic sources that are analysed and statistically processed. As the results show the proposed model causes in comparison to the known traffic source very similar network traffic parameters when used in a simulated network.*

Keywords

H.264, IPTV, MPEG2 TS, packetization, traffic generation.

1. Introduction

The aim of this work is to define a precise VBR (Variable Bit Rate) HD (High Definition) H.264 MPEG2 TS (Moving Pictures Experts Group Transport Stream) IPTV (Internet Protocol Television) traffic generator that can be used as a replacement to real traffic sources. Generators are often used for network stress testing, by network delivered QoS testing [1] or as a help for prediction of QoS parameters like one-way delay, packet losses and jitter [2].

Traffic generators have some major advantages over real traffic in creating network load. The load can be created with a small amount of effort, no content streaming rights are needed, and preparing a large test with hundreds of streams can be performed in several seconds instead of hours. More information about IPTV networks and signal delivery can be seen in [3].

Our proposed model uses the following packetization stack: MPEG2 TS streamed over RTP (Real-time Transport Protocol), UDP (Used Datagram Protocol), IPv4 (Internet Protocol version 4), and Ethernet (Ethernet II or 802.3an), [4]. This stack is often used for several reasons. MPEG2 TS is used as a container format for streaming multiplexed video, audio and data. TS has several control mechanisms that are used for stream monitoring at end stations. RTP [5] ensures a better QoS supervision and quality management in IP network. The Ethernet II or 802.3ab are considered because these types of Ethernet technologies were used in our measurements. Measurements were made at the Ethernet layer, so L2 (layer 2 of the Reference Model Open System Interconnection) headers and minimal Ethernet frame sizes are also considered.

The rest of this paper is organized as follows. The second chapter describes the current state of art, IPTV stream analysis and evaluations of measurements. The third chapter is devoted to the model of the IPTV generator. The fourth and last chapter is the conclusion.

2. State of Art

IPTV packet generation can be a quite complex problematic. IPTV traffic itself is a time-variant self-similar process. These traffic properties can be partially described by several modeling approaches. Some of these IPTV traffic models are working only at the packet level. In [6] an analytical two level Markovian model is described. These types of generators are often not compute intensive but their can lack of accuracy.

Another approach is in the levelization of the process of IPTV packets generation. This means, that generation does not occur only at the packet level but also higher levels of creating MPEG video are involved. We created a four level traffic model were the MPEG GoP (Group of Pictures) length and structure, I (Intracoded), B (Bi-directional) and P (Predicted) frame sizes and MPEG2 TS packetization are taken into account.

Also other mathematical approaches like semi-Markovian [7], wavelets [8] or multifractal analyses [9] are applicable. These models are often aiming at a special field in the overall process of generators.

2.1. IPTV Stream Parameters and Measurements

IPTV streams can be distinguished by a number of parameters such as:

- VBR or CBR (Constant Bit Rate),
- GoP length,
- static or variable GoP structure,
- MPEG profile (e.g. Baseline, Main),
- number of streams carried within a multiplex,
- packetization.

Our compressed video stream uses a common frame rate of 25 frames/s with 188 bytes long MPEG2 TS Media Frames. 7 Media Frames are encapsulated into RTP/UDP/IP/Ethernet packets. Only 7 frames are encapsulated because of Ethernet data field length limitation.

2.2. Measurements

We want to thank to Slovak Telekom Inc. which provided to us IPTV traffic records. Without these records our research could not be successful. We made measurements on this real IPTV traffic with variable and fixed GoP structures. The first measurement is packet based where times between generation of packets and packet lengths were observed. The next measurement is aimed to the H.264 video structure where GoP lengths and frame sizes are inspected. From these two investigations we are able to create IPTV statistics. For packet based measurements we used the well known Wireshark program that can also run on Ubuntu Linux. We modified the settings of Wireshark so it is able to display the interarrival time with a nanosecond time resolution. This nanosecond resolution is not the same as the precision of measurement. Because normal non real time operating systems are not very usable for time sensitive signal processing or measurements, measured results are affected by a variable operating system load dependent error. Also when a real-time kernel would be used the error wouldn't be much smaller because of minimal guaranteed system response time which is commonly bigger than 20 ms. But this error can be reduced because the processed signal is a MPEG2 TS with constant interarrival time of packets between two consecutive PCR (Program Clock Reference) transmissions.

1) IPTV Packets Interarrival Times

Figure 1 shows the measured interarrival time of IPTV packets with time errors. These errors can be effectively filtered with a 1 dimensional median filter which can be used multiple times in tandem (Fig. 2). Times T1 and T2 equals. The picture shows them as two nonequal intervals but the X axes displays the packet number and the Y axes shows the interarrival time. When interarrival times are smaller, more packets have to be sent out because the 100 ms PCR interval can't be broken. On the other hand with a higher interarrival time fewer packets need to be transmitted.

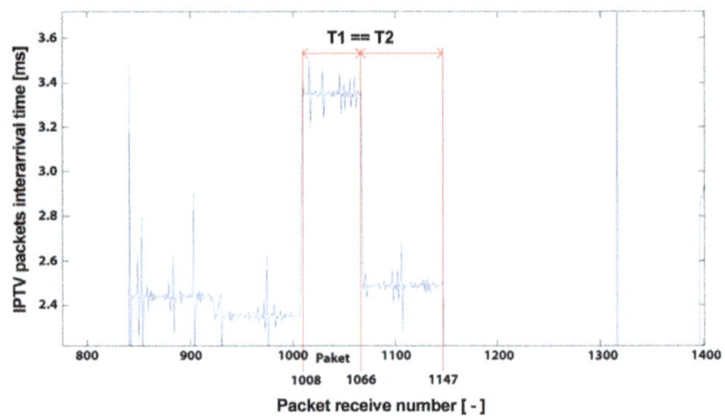

Fig. 1: Measured IPTV packets interarrival time.

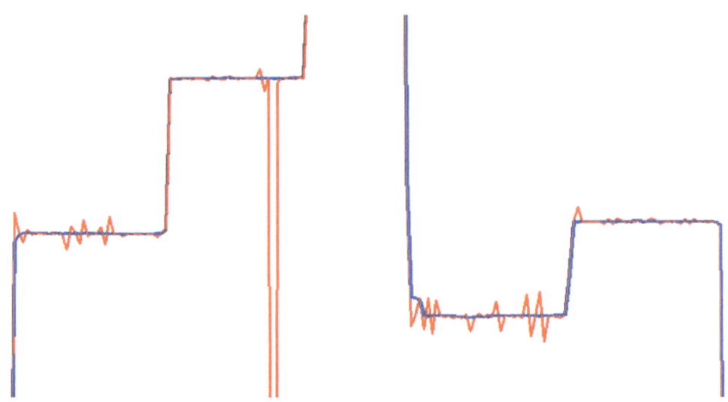

Fig. 2: Filtration of IPTV packets interarrival time using a 1D median filter. Red line - original measured interarrival time, blue line – filtered curve.

Figure 3 shows the PDF (Probability Distribution Function) of measured interarrival time of a chosen IPTV HD stream. As the figure shows the PDF can be approximated by the Inverse Gaussian PDF. Other measurements fit better to other distribution functions e.g. Johnson SB, Log-Logistic, Gamma, Burr or Lognormal. The best fitted distribution function depends on many factors when creating an IPTV stream. Therefore it cannot be said that the best fitted distribution function is one of the mentioned above. We have also made excessive measurements to evaluate an overall PDF for a mean IPTV HD stream which is made from 20 IPTV HD streams. We choose as the best fitting PDF the Burr distribution function. The best fitting PDF depends on the content and video coding.

Therefore another IPTV provider can use other video coding settings. The results would vary but the selected Burr distribution function can lead to similar behavior when IPTV services provider's streams uses VBR and GoP structure that matches to the proposed model.

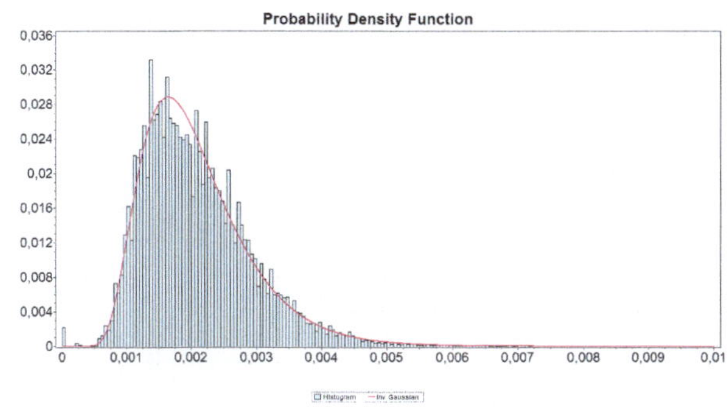

Fig. 3: PDF of packets interarrival time of a one chosen IPTV HD stream. Red curve shows an approximation by the Inverse Gaussian distribution.

2) IPTV GoP Structure and Frame Sizes

The next section describes results obtained from disassembled IPTV HD H.264 streams. For purposes of analysis, we used 4,8 hours long IPTV records with a predefined I B B B P B B B P .... GoP structure with a variable length. The overall number of used frames is 429 993 that are arranged into 16 388 GoP intervals with the mean GoP length of 26,24 frames.

Figure 4 shows the distribution of GoP structures in the used movies. In most cases it uses the GoP of 30 frames. Smaller GoPs are results of scene cuts in video content. The average length of GoP structure is 26,238 frames. This variable cannot be described by a known discrete PDF. Therefore we model this PDF manually and automatically.

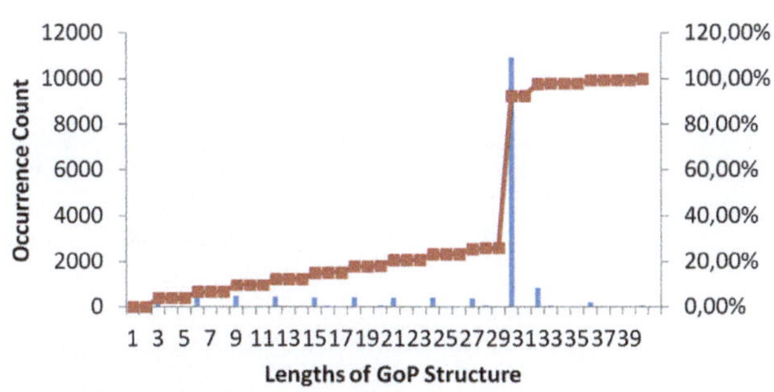

Fig. 4: Histogram of GoP lengths.

Figure 5 displays the PDF of I frame sizes. This distribution can be precisely approximated by the Burr PDF. Also B and P frames can be approximated by this PDF. The Burr CDF (Cumulative Distribution Function) is defined in Eq. (1), where α and k are con-

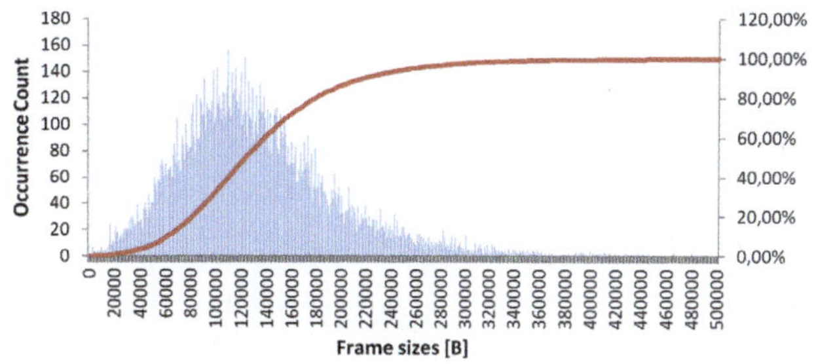

Fig. 5: Histogram of I frame sizes.

tinuous shape parameters and β is a continuous scale parameter:

$$\mathrm{CDF}\left(x\right) = 1 - \left(1 + \left(\frac{x}{\beta}\right)^{\alpha}\right)^{-k}. \tag{1}$$

PDF of I, B and P frame sizes are approximated with Burr PDF with the following parameters:

- I: $k = 2{,}3478$, $\alpha = 2{,}8266$ and $\beta = 180\ 166$,
- B: $k = 1{,}0607$, $\alpha = 3{,}2577$ and $\beta = 18\ 576$,
- P: $k = 1.7858$, $\alpha = 3{,}5889$ and $\beta = 78\ 457$.

3. H.264 HD Generator

This section describes the generator design and results. The results are compared to real IPTV H.264 HD traffic.

3.1. Generator Design

The generator defines four levels of creating content. At the first the generator defines the GoP lengths. This can be considered as the fourth level. The lengths are then replaced by a defined static GoP structure of the H.264. Main profile which is the third level. These frames are after the creation of GoP structure replaced by frame sizes in bytes which is the second level. The first level is the packetization of generated frames into RTP, UDP, IP and Ethernet frames. The packetization process takes the MPEG2 TS PCR into account.

Manually generated GoP lengths are using an uniform random number generator in $R \in \langle 0, 1\rangle$ which drives an ON/OFF model. A generated real number is used as an input to the ON/OFF model. For example if the generated value is equal to 0,24015 and defined ranges for the twenty first state of the ON/OFF model are between 0,240125 and 0,24025 the generated GoP length equals to 21.

The automated generation of GoP lengths is defined by the measured GoP PDF. The generator code uses a uniform random numbers generator in $R \in \langle 0,1\rangle$ which uses the recorded discrete GoP PDF value. The output value is used for searching the recorded GoP PDF in reverse to discover the best position of the current bar of the discrete GoP PDF. After this search a GoP length output value is generated.

The generated number of the manual or automated generators is replaced by a defined GoP structure of I, B and P frames. This structure is defined as static, so it does not change during generation of IPTV packets. We used the I B B B P B B B P ... structure. Every generated stream begins with I frames. When generating more streams in parallel they will generate at the beginning of this tests a high amount of data because of this fact. Therefore delayed stream generation should be used.

Frame sizes are generated using the Burr CDF. The inverse function of the Burr CDF formula Eq. (1) is showed in Eq. (2) which is used as a generating function for I, B and P frame sizes in bytes. The CDF (x) argument is replaced by a uniform random number generator in $R \in \langle 0,1\rangle$. The generated GoP structure is then replaced by frame sizes in bytes. These sizes take also the MPEG2 TS overhead into account. We also define minimum and maximum frame sizes in bytes for generated frames: I $\langle 782, 750000\rangle$, B $\langle 188, 115000\rangle$ and P $\langle 188, 252000\rangle$ and the occurrence of B and P frames that could be bigger than I frames to 0,5 %:

$$x = \beta\left(\left(1 - \mathrm{CDF}\left(x\right)\right)^{\frac{1}{-k}} - 1\right)^{\frac{1}{\alpha}}. \tag{2}$$

Packets are generated from the sequence of frame sizes that consist of payload and a MPEG2 header. These data have to be split into 188 bytes long chunks. Seven of these chunks are then encapsulated into packets with the 12 B RTP, 8 B UDP, 20 B IP and 38 B Ethernet overhead (Ethernet overhead includes Inter Frame Gap, Preamble, Destination Address, Source Address, Type field and Forward Error Correction). Because PCR is sent every 100 ms the interarrival time of IPTV packets have to be constant in this time. When the frame rate of an IPTV stream is 25 frames per second the 100 ms time interval equals to 2,5 frames. So the number of generated packets is a function of the number of bytes that have to be sent out and the 100 ms time interval. Then the number of packets is converted to the time between generation of packets in this manner: (100/number of packets) in milliseconds. The packet size (Ethernet frame size) of 1 394 bytes remain always the same.

The comparison of measured and generated IPTV HD streams is discussed in several ways. Figure 6 shows the comparison of CDFs of measured and generated times between generations of packets. The solid line shows the CDF of measured IPTV traffic as a function of itself. The dashed line shows the CDF of generated IPTV traffic as a function of the CDF of measured traffic. As can be seen from these results the slope of lines is approximately the same and the maximal absolute difference between them is 0,1.

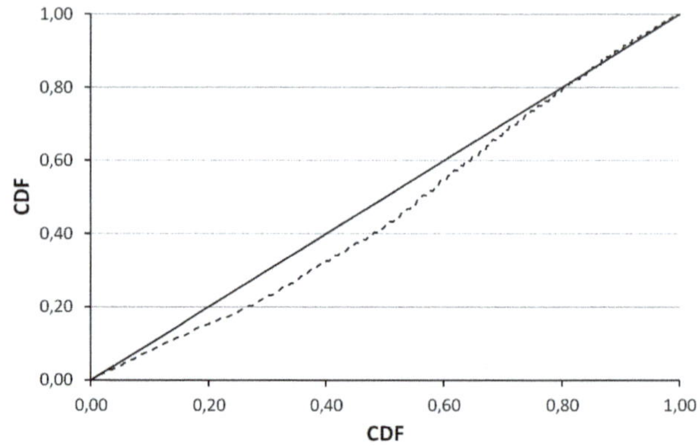

Fig. 6: Comparison of Cumulative Distribution Functions where: solid line shows the CDF of measured IPTV HD stream and the dashed line shows the CDF of generated IPTV HD stream.

Figure 7 and Fig. 8 show Lag-plots of measured and generated interarrival times of IPTV packets. The generated traffic is more concentrated around the base line as the generated traffic. This is the result of a non perfect median filtering of measured times. These times vary a bit around the mean value within a PCR interval. On the other side the generated traffic remains a constant time between generation of IPTV packets within a PCR interval.

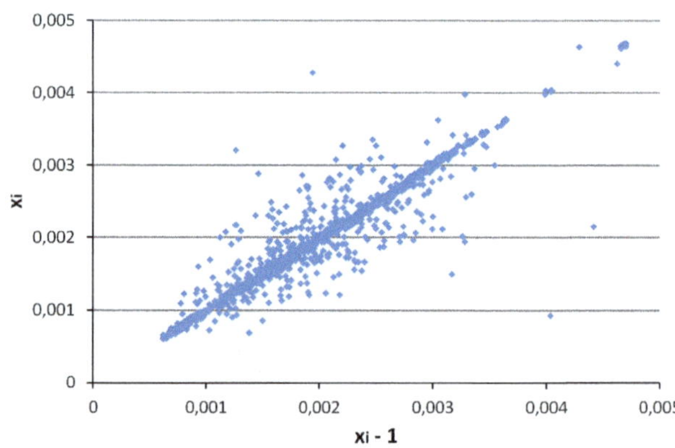

Fig. 7: Lag plot of interarrival times of chosen measured IPTV HD traffic stream with applied median filter.

For the purpose of testing the model we used an in MatLab developed network simulator with QoS (Quality of Service) mechanisms applied. The testing scenario uses IPTV streams multiplexed to 366 Mb·s^{-1} using a FIFO switch at Head End. These streams are then transported to a switch where VoIP and Data traffic classes are merged. This switch uses the WRR (Weighted Round-Robin) method for preferring VoIP and IPTV classes. Every test uses the same VoIP and

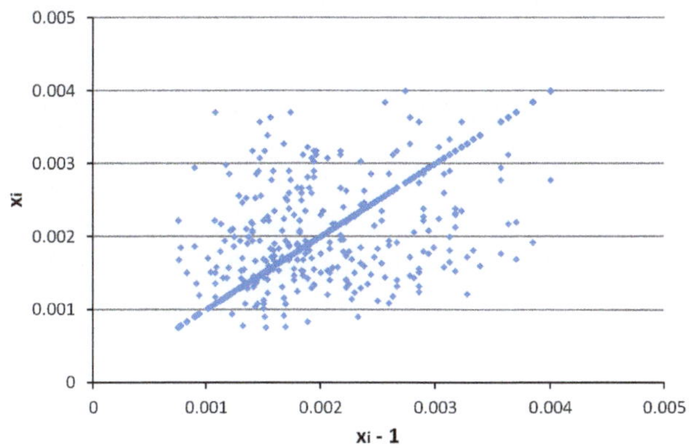

Fig. 8: Lag plot of times between generation of IPTV HD traffic packets.

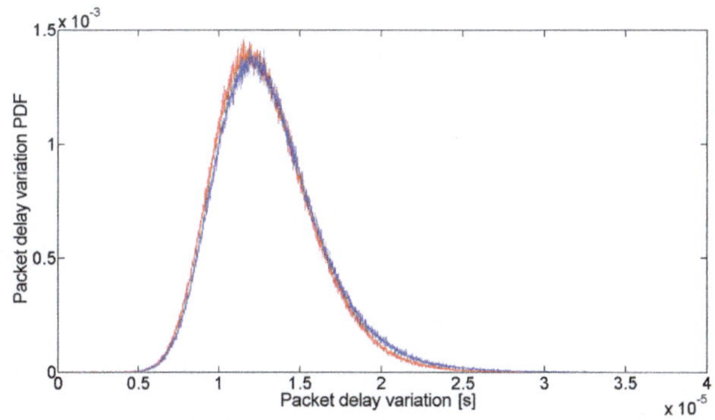

Fig. 9: RFC 3550 packet delay variation PDF of multiplexed 336 Mb·s^{-1} generated (red) and real (blue) IPTV H.264 HD traffic.

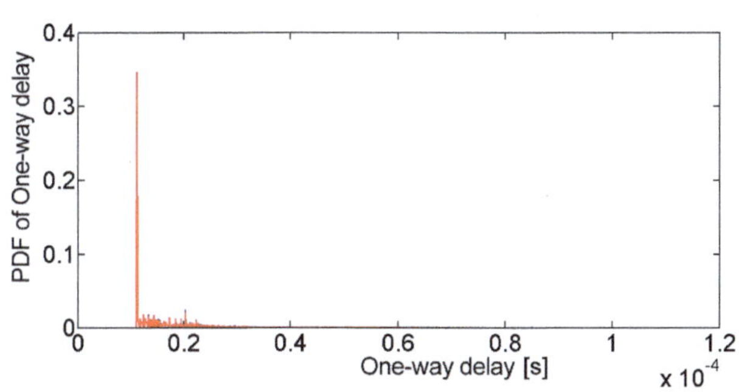

Fig. 10: One-way delay PDF of generated (red) and real (blue) IPTV traffic that occurred in the test network.

Data classes' sequences with different IPTV records (generated and measured). Figure 9 shows the RFC 3550 packet delay variation PDF of generated and real traffic measured between the input of the first switch in the simulated network and the output of the last switch in network. The correlation coefficient of these two curves equals to 0,99987. Also one way delay has equivalent properties with a correlation coefficient of 0,99987 (Fig. 10). In the test scenarios no IPTV packets were lost so the lost rate is not discussed. In tests where the IPTV traffic suffers transfer losses in simulation network these losses are about the same.

4. Conclusion

In this paper we propose a four-level HD H.264 VBR IPTV stream generator. Real world video streams are analyzed and results are used for the model definition. Realized simulations and measurements have shown that the generator is precise and can be used in simulations and in real word testing.

Acknowledgment

This work is a part of research activities conducted at Slovak University of Technology Bratislava, Faculty of Electrical Engineering and Information Technology, Institute of Telecommunications, within the scope of the projects "Grant programme to support young researchers of STU - Modeling of Traffic and Traffic Parameters in IPTV Networks", "Support of Center of Excellence for SMART Technologies, Systems and Services II, ITMS 26240120029, co-founded by the ERDF" and project VEGA no. 1/0106/11 "Analysis and Proposal for Advanced Optical Access Networks in the NGN Converged Infrastructure Utilizing Fixed Transmission Media for Supporting Multimedia Services".

References

[1] CHROMY, E., M. JADRON and T. BEHUL. Admission Control Methods in IP Networks. *Advances in Multimedia.* 2013, vol. 2013, no. 2, pp. 1–7. ISSN 1687-5699. DOI: 10.1155/2013/918930.

[2] KOVAC, A. and M. HALAS. E-Model MOS Estimate Precision Improvement and Modelling of Jitter Effects. *Advances in Electrical and Electronic Engineering.* 2012, vol. 10, no. 4, pp. 276–281. ISSN 1336-1376.

[3] KOCKOVIC, L. and I. BARONAK. Alternatives of Providing IPTV Using IP Multimedia Subsystem. *International Journal of Computers and Technology.* 2012, vol. 3, no. 2, pp. 188–192. ISSN 2277-3061.

[4] O'DRISCOLL, G. *Next generation IPTV services and technologies.* Hoboken: Wiley-Interscience, 2008. ISBN 978-0-470-16372-6.

[5] RFC 3550. *RTP: A Transport Protocol for Real-Time Applications.* Columbia University: ITU-T Network Working Group, 2003. Available at: `http://datatracker.ietf.org/doc/rfc3550/`.

[6] WAN, F., L. CAI and T. A. Gulliver. A Simple, Two-Level Markovian Traffic Model for IPTV Video Sources. In: *2008 IEEE Global*

Telecommunications Conference. New Orleans: IEEE, 2008, pp. 1–5. ISBN 978-1-4244-2324-8. DOI: 10.1109/GLOCOM.2008.ECP.288.

[7] KEMPKEN, S. and W. LUTHER. Modeling of H.264 high definition video traffic using discrete-time semi-markov processes. In: *Proceedings of the ITC 2007, Lecture Notes in Computer Science (LNCS)*. Ottawa: Springer-Verlag, 2007, pp. 42–53. ISBN 978-3-540-72990-7. DOI: 10.1007/978-3-540-72990-7_8.

[8] DAI, M., Y. ZHANG and D. LOGUINOV. A unified traffic model for MPEG-4 and H.264 video traces. *IEEE Transactions on Multimedia*. 2009, vol. 11, iss. 5, pp. 1010–1023. ISSN 1520-9210. DOI: 10.1109/TMM.2009.2021802.

[9] RELJIN, I., A. SAMCOVIC and B. RELJIN. H.264/AVC video compressed traces: multifractal and fractal analysis. *EURASIP Journal on Applied Signal Processing*. 2006, vol. 2006, no. 1, pp. 1–13. ISSN 1687-6180. DOI: 10.1155/ASP/2006/75217.

About Authors

Stanislav KLUCIK was born in Bratislava, Slovakia, in 1984. He received the Master degree in telecommunications in 2009 from Faculty of Electrical Engineering and Information Technology of Slovak University of Technology in Bratislava (FEI, STU). In 2012 he submitted a Ph.D. work from the field of Source and QoS traffic parameters modeling in NGN networks and his scientific research is focused on packet generators and modeling of packet networks. Nowadays he works as research at the Institute of Telecommunications of FEI STU Bratislava.

Martin LACKOVIC was born in Trnava, Slovakia, in 1989. He received the Master degree in telecommunications in 2013 from Faculty of Electrical Engineering and Information Technology of Slovak University of Technology (FEI, STU) Bratislava. His scientific research is focused on modeling of packet generators, traffic parameters and networks. Nowadays he works as technician in an IT company.

8

Fault Tolerant Ethernet Based Network for Time Sensitive Applications in Electrical Power Distribution Systems

Leos BOHAC, Jiri VODRAZKA

Department of Telecommunications Engineering, Faculty of Electrical Engineering, Czech Technical University in Prague, Technicka 2, 160 00 Prague, Czech Republic

bohac@fel.cvut.cz, vodrazka@fel.cvut.cz

Abstract. *The paper analyses and experimentally verifies deployment of Ethernet based network technology to enable fault tolerant and timely exchange of data among a number of high voltage protective relays that use proprietary serial communication line to exchange data in real time on a state of its high voltage circuitry facilitating a fast protection switching in case of critical failures. The digital serial signal is first fetched into PCM multiplexer where it is mapped to the corresponding E1 (2 Mbit/s) time division multiplexed signal. Subsequently, the resulting E1 frames are then packetized and sent through Ethernet control LAN to the opposite PCM demultiplexer where the same but reverse processing is done finally sending a signal into the opposite protective relay. The challenge of this setup is to assure very timely delivery of the control information between protective relays even in the cases of potential failures of Ethernet network itself. The tolerance of Ethernet network to faults is assured using widespread per VLAN Rapid Spanning Tree Protocol potentially extended by 1+1 PCM protection as a valuable option.*

Keywords

E1, PCM, protective switch, RSTP, smart grid, time sensitive applications.

1. Introduction

The foreseen concept of Smart Grid technology, which promises highly efficient and dynamic energy systems in Europe, has to be backed by suitable information and communication system. From this perspective it is more than obvious to take already acquired knowledge from well-developed ordinary information and communication systems and apply it to the realm of the power energy system.

However, the energy systems like the power electrical grids have more stringent requirements on the performance factors of the communication system, at least in terms of reliability, than is generally required and expected from ordinary information systems.

There has been a growing interest in recent couple of years to apply already existing ICT technology in an electrical power industry. The main stimulus behind these activities can be primarily seen in the lower and continually decreasing capital costs, strong technology standardization and its maturity. What is not also negligible is that one can select the particular devices from a broader range of manufacturers avoiding potential vendor lock as often seen in the power distribution industry today. Ethernet technology, originally developed and targeted specifically only for the local area network segment, is nowadays very mature, low cost, scalable and reliable communication network platform that is easily upgradable to speeds of tens of gigabits. Thus, it is not surprising there is a great effort to lay this network as a foundation for the future integrated communication system of the power electricity engineering industry [1], [11].

However, Ethernet technology is not alone as the candidate considered replacing the legacy systems in an electrical power industry. Its combination with IP and MPLS (Multiprotocol Label Switching) is also taken into account [2] to specifically cover communication needs in an extended geographical domain rather than a small area of only one substation as is most common in Ethernet's deployment today [5], [10]. This extended communication coverage is very typical for deployment in large SCADA systems (Supervisory Control and Data Acquisition) [8].

The basic Ethernet technology is well suited for supporting the data transfer for classical information technology applications running in the campus or office environment or accessing Internet resources, but not so much for the process automation or device monitoring in real time. Also, though the reliability of the basic Ethernet technology is sufficient for aforementioned office applications it is not as well for the control data exchange in the electrical power industry. Thus, the efforts have been made to alter or supplement the function of basic Ethernet technology to provide a better reliability figures particularly in the area of the substation deployment [4], [6].

As was already stated, one of the most important applications in the electrical power distribution systems is that of providing the protection relay intercommunication that is very demanding as regards the guarantee of reliability and minimum delay. Several papers [3], [7], [9], [12], [13] have already addressed these issues in different ways. It is possible either to define new protocol to deal with a potential node or link failure or use one from already existing set of standardized protocols. Implementing new algorithms is quite complicated and also in the very pragmatic world of an electrical power industry challenging to enforce. In our work we target first choice and analyze a situation where there in Ethernet based network is deployed the per VLAN (Virtual Local Area Network) Spanning Tree Protocol (RSTP) in combination with 1+1 PCM (Pulse-code Modulation) protection scheme in addition to the dedicated primary communication line among protection relays themselves. This protocol selected was well in line with one of our practical projects we have dealt with where there was a requirement to use already existing protocol (if possible) in the network (in this case the per VLAN RSTP) and validate its practicability for the protection relay traffic in the network with multiple VLANs.

2. Network Test Arrangement

The laboratory test interconnection of the network devices is depicted on Fig. 1. In order to stay more aligned with real practical situations, our intention was to use such a network arrangement that would be more relevant to a similar network topology deployed in one electrical power distribution company in Czech republic. All devices in our testbed are manufactured by Cisco and the network is strictly based on Ethernet L2 switching of data frames. The laboratory network architecture also follows the standard layered approach recommended by Cisco for enterprise/campus-like networks where switches SW11 and SW21 represent the core or distribution layer and all others are part of the access layer.

In practice, each electrical substation site is connected to the network core through two identical local Cisco switches (e.g. SW12 and SW13 in our lab) to ensure both communication device and medium redundancy needed in case of the fiber or access device failure. The issues related to other potential single points of failures, like common installation of both uplink fiber cables in the same duct or feeding both redundant switches from a single power supply source were not considered.

In the paper we have strictly concentrated only on studying the effects of the dynamic network reconfiguration on the data transport interruption when one of the uplink-fiber fails (e.g. fiber cut). At the very bottom of the Fig. 1 there are two boxes named Siprotec. They represent the real high voltage remote differential protection relays from Siemens communicating in real time between each other at the electrical power distribution network to protect the electrical grid from detrimental failures that may cause the potential damage to the power lines or other power grid devices. The data communication between these relays is very important in the power distribution engineering field and as such it makes very sensitive issue to deal with. The communication requirement of the protection relays belongs to the low-bandwidth high-response real time category that is generally difficult to support unless the network is able to guarantee some sort of the priority dispatching schema of frames corresponding to this type of service.

In our case the electrical power distribution company has been using historically for ensuring the protection relays communication two independent lines, one primary optical fiber based line (blue arrowed line in Fig. 1) and second redundant PCM E1 based line (yellow line connecting PCM muldexes). If primary optical fiber line is not interrupted the relays communicate only through them. When the primary line fails the traffic is automatically rerouted to the secondary line hosted by PCM transmission system. The serial data of relays are first mapped in PCM muldex to the corresponding E1 frame channel of E1 and then are sent to the opposite PCM muldex through another pair of fibers. The direct primary and secondary optical fibers (via PCM) pass along the different pathways to tackle forming a single point of failure at common duct/cable level.

Our goal was to check whether the PCM can be potentially replaced by Ethernet based network without incurring unacceptable transit delay and decreased reliability. The laboratory testbed corresponding to the Ethernet network is also depicted in Fig. 1. Four Ethernet switches form a representative part of entire Ethernet metropolitan network. This network is also used for other applications, particularly SCADA, to move data from Remote Terminal Units (RTU's) installed in appropriate substations (the pair of switches SW12/SW13 is in practice installed in the place of one substation and SW22/SW23 in another) to the central SCADA servers connected to the core switches SW11 and SW21. In other words, the interconnection of switches is not purpose built only for the relays, but the network as a whole must support also hub-and-spoke type of communication. A key design requirement was to preserve currently running resilience protocol in the network, which is per VLAN RSTP, and check its reconfiguration/interruption time in the case of link failure. As can be seen from network topology, there is potentially a number of possible failures, both on the side of the devices and connecting fibers.

However, most probable failures come from the uplink fibers (e.g. from SW11 to SW12) as they are relatively long (several kilometers) and pass through various pathways that may be cut or another way interrupted. Other fiber segments in the network (e.g. SW11-SW21, SW12-13 and SW22-23) are only short local interconnection of switches not posing a great reliability issue if properly treated. The device failures were not extensively studied here thought in some cases they can be simulated by the node's concurrent multiple fiber failure model.

3. RSTP Convergence Time

Ethernet technology historically supported only active tree topology. This requirement resulted in only one active pass permissible for any pair of communicating in the entire bounded switched network. If this rule was not adhered to, the potential loop in the network would leave network totally inoperable causing endless circulation of the data frames and inducing excessive load on the switches that forced very quickly reindex station's position in the network. Sooner or later this way looped Ethernet network is sentenced to its own load collapse. However, the impossibility to intentionally introduce the loops in the network topology significantly affects its resilience as there is only one path available between any two communicating devices and when any component of this path fails the path fails fully disabling any subsequent data transfer. In this sense the loops are very vital providing standby paths in case the active ones fail.

The ability of the Ethernet to resist the line or node failures is very significant for applications that operate in the real time and do not tolerate long traffic disruption. For the classical office and campus network applications Ethernet was supplemented by the distributed algorithm called Spanning Tree Protocol (STP) that enabled design of Ethernet network topology with multiple physical loops. However, in this case STP dynamically builds up logical tree topology above the physical one by blocking specific segments between switches and thus disconnecting loops holding operational topology again as the tree. If the failure of the switch or any segment occurs in the active logical tree, the STP progressively builds up a new and different active logical tree out of still operating parts (i.e. switches and segments) of the network.

The STP has been standardized and implemented for years in accordance with the document known as IEEE 802.1D. However, the main drawback of this historically first standard has been a low convergence speed of the algorithm leading in the event of the failure to a very long traffic disconnection time in the order of tens of seconds, which is not acceptable for mission critical applications like the ones used in the process control of utility systems.

RSTP protocol was specifically designed to tackle slow convergence of its predecessor. In our project we have used this protocol because it can support very short outage of data packet streams in case of network failures.

4. RSTP Convergence Measurement

In our situation of the laboratory tests we have used aforementioned network topology in Fig. 1. The core switches were Cisco Catalyst C3560 and all access switches Cisco Catalyst C2960. The software images we have used ordinarily supported the rapid spanning tree protocol. Though better choice was to implement multiple spanning tree protocol (MSTP), we have rather concentrated on the tests and verification of per VLAN rapid spanning tree protocol (per VLAN RSTP) as this one was required to be checked in our project. The slowest convergence of RSTP protocol is when the root switch fails and hence the remaining switches are forced to elect new root switch to continue to support a logical spanning tree in the network. However, this event was not considered for it has a lower probability to occur than the uplink fiber failures.

We have configured the network in such a way that SW11 played a role of active root switch in our topology and SW21 as a standby root bridge for the case of the primary is lost. The star-type topology was dictated by the design of the production network, traffic patterns in the network and the primary requirement to have minimum devices between substations and the substations and the network center, here represented by SW11 and SW21 switches. Thought this double-star like topology is very demanding as the number of the optical fibers and lengths required it does not make a big trouble in our case as the corresponding distribution company already had its own very rich optical network with a plenty of the optical fibers already installed. Alternative topologies can be considered,

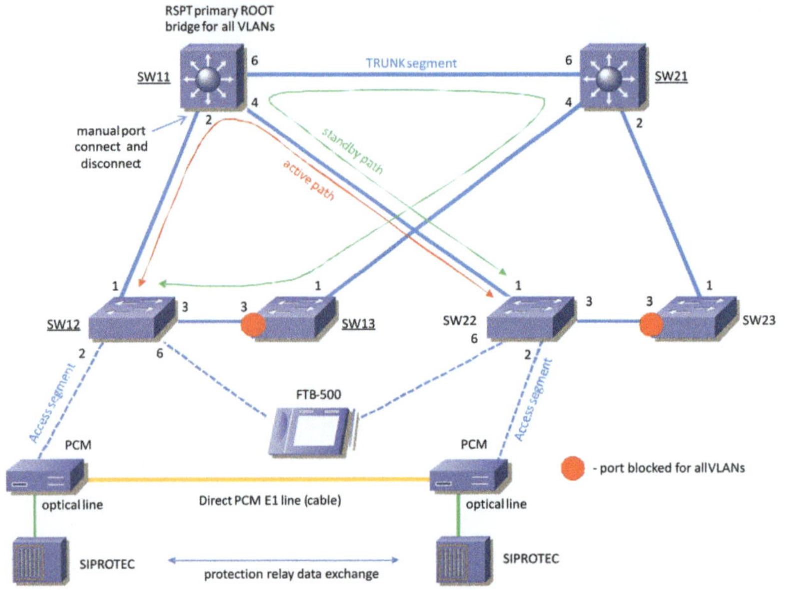

Fig. 1: The laboratory test topology; the core and access switches from Cisco were used (C3560 and C2960 respectively); for the measurement of the traffic disruptions we used professional network analyzer form EXFO, FTB 500; the network was configured to support the increasing number of VLANs; port #2 on SW11 was repeatedly disconnected and connected to simulate break of fiber between switches SW11 a SW12; the red line represents the active path and green standby path respectively that is activated by RSTP in case of disconnection of port#2 on SW11.

of course (like the ring), but redesigning this critical network was not welcomed at all by the distribution company as it is very pragmatic to any considerable changes in any system if these are not absolutely necessary bringing clear benefit for the potential failure of the network is in stake anytime when any configuration and hardware change are made. These type of networks and systems fall under strict rules of reconfiguration, testing and management procedures and thus it is not easy to do some experiments in live there.

The test procedure of our measurement was quite simple. We used a professional EXFO analyzer FTB 500 to inject repeated train of Ethernet frames of 200 bytes long to the port #6 of SW12 every 1 ms in time and at the SW22 this sequence was received and lost frames counted. The disruption time is then given by the number of lost frames multiplied by frames sending period at the transmitter site using the formula

$$t_{dis} = N_{lost} T_p, \tag{1}$$

where t_{dis} is time of traffic disruption in the case of the port#6 connect or disconnect event, N_{lost} is the number of lost Ethernet frames and T_p is the period of time forframes are sent to the port#6 on SW12 (in our case 1 ms).

We used the physical access speed of 100 Mb·s^{-1} on all ports in the network in our test. The processing delay at the transmitter and receiver site of the analyzer and the length of frames were neglected as they are in the order of microseconds and we are dealing here with disruption times in the order of tens of milliseconds and even more, see further.

The precision of our measurement is primarily dictated by a finite time of sending frames in the network, in this case it is ±2 ms. Again it can be neglected in comparison with the average disruption times measured.

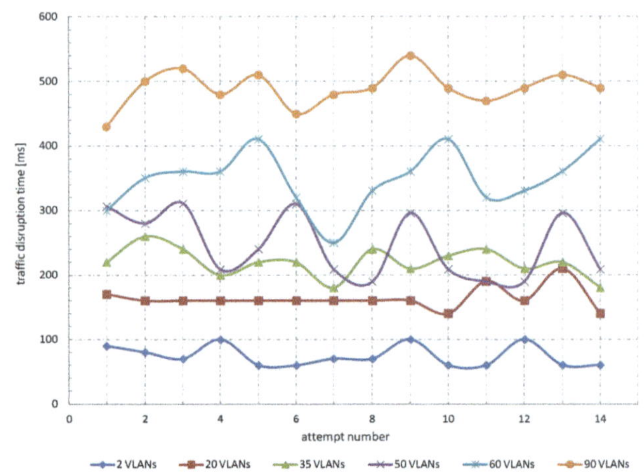

Fig. 2: The plot of traffic disruption time in case of port#2 disconnection versus the number of the attempt tried and number of VLANs.

The number of VLANs was changed in each step on the switch SW11 and using VTP protocol the list of VLANs was distributed to remaining switches in the network followed by other cycle of the experiment. One experiment consisted of 14 cycles of port#2 disconnect/reconnect events at SW11 and results were plotted in graphs shown in Fig. 2 and Fig. 3. Fig. 2 corresponds to disconnect events and Fig. 3 reconnect events. As can be seen when a number of VLANs is kept low the disruption time is also low and has even small variance.

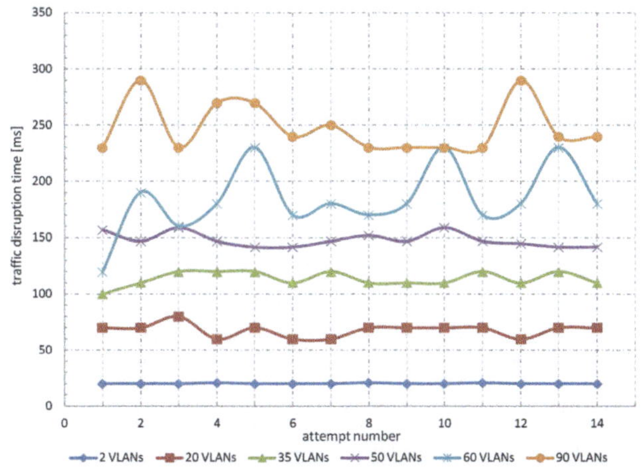

Fig. 3: The plot of traffic disruption time in case of port#2 reconnection versus the number of the attempt tried and number of VLANs.

However, as the number of VLANs increases the disruption time as well and also its variance. The reason for that behavior can be attributed to how multiple control messages of RSTP are processed inside the switches. The similar curves corresponding to the reconnection events are shown in Fig. 3. As can be seen, the reconnection disruption times are shorter than disconnection ones, and again a larger variance is evident when the number of VLANs is increasing. In order to

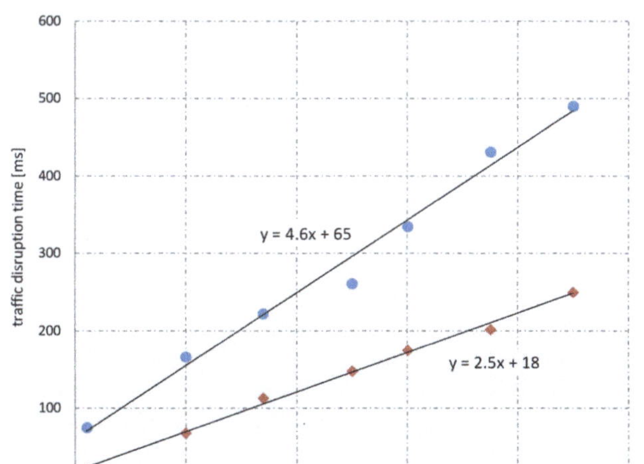

Fig. 4: The plot of traffic disruption time in case of the port #2 disconnect (blue) and connect (red) events versus the number of VLANs in the network.

better understand dependence of the disruption time on the number of VLANs we have plotted the average values of disruption times for all cycles in other plot shown in Fig. 4, where a nearly linear dependence of the disruption time on the number of VLANs can be seen.

5. Conclusion

In the paper we have summarized measurements done in a laboratory environment to check the feasibility of RSTP and Ethernet network for deployment in demanding electrical power distribution network. We have found that performance of per VLAN RSTP protocol as implemented by Cisco switched is linearly dependent on the number of VLANs configured in the network and that time of traffic disruption grows faster in the case of uplink disconnect than reconnect event. Thus, to use per Vlan RSTP the number of VLANs has to be considered even in a very simple hub-and-spoke topology. Also, the more VLANs in the network the larger variance in disruption traffic was found.

When the protective relay's communication is considered in this scenario, it is evident that in the moment of the uplink fiber cut (disconnect) the communication line is not available at least for 80 ms at the average time and with only one VLAN in the network configured that is not typical situation in practice at all. In many case such a long disruption time is not tolerated by most differential protection relays where the required ranges are in the fraction up to the order of a few milliseconds (0,5–5 ms) depending on the type of protection. Thus, in the failure scenario the relays are not able to communicate among each other. In order to make communication more robust, it is possible to send relay's data along two separate paths and hence use 1+1 protection or to strictly keep communication fiber's or equipment's failures independent of those in the electrical power system to ensure a break in power distribution system will not cause at the same time a failure at communication network.

Acknowledgment

This work was supported by the Grant of the Ministry of the Interior of the Czech Republic, No. VG20132015104, "Research and development of secure and reliable communications network equipments to support the distribution of electric energy and other critical infrastructures", and was researched in cooperation with TTC Telecomunikace.

References

[1] JANCA, T. R. Utility Ethernet network architecture: Networked Electrical eXchange Topology-

NEXT. In: *Sarnoff Symposium (SARNOFF)*. Newark: IEEE, 2012, pp. 1–6. ISBN 978-1-4673-1465-7. DOI: 10.1109/SARNOF.2012.6222721.

[2] FUJIKAWA, F., K. KUWABARA, Y. KODA and M. KIUCHI. Examination of electric power utility network applying IP router/MPLS router/wide-area Ethernet. In: *Power Engineering Society General Meeting*. Denver: IEEE, 2004, pp. 901–904. ISBN 0-7803-8465-2. DOI: 10.1109/PES.2004.1372956.

[3] BRUNELLO, G., R. SMITH and C. B. CAMPBELL. An application of a protective relaying scheme over an ethernet LAN/WAN. In: *Transmission and Distribution Conference and Exposition*. Atlanta: IEEE, 2001, pp. 522–526. ISBN 0-7803-7285-9. DOI: 10.1109/TDC.2001.971288.

[4] MIDANCE, R. and D. IADONIS. Ethernet networks redundancy with focus on IEC 61850 applications. In: *20th International Conference and Exhibition on Electricity Distribution*. Prague: Curran, 2009, pp. 1–4. ISBN 978-184919126-5. DOI: 10.1049/cp.2009.0637.

[5] QUERASHI, M., A. RAZA and D. KUMAR. A survey of communication network paradigms for substation automation. In: *Power Line Communications and Its Applications*. Jeju City: IEEE, 2008, pp. 310–315. ISBN 978-1-4244-1975-3. DOI: 10.1109/ISPLC.2008.4510445.

[6] HEINE, H. and O. KLEINEBERG. The High-Availability Seamless redundancy protocol (HSR): Robust fault-tolerant networking and loop prevention through duplicate discard. In: *Factory Communication Systems (WFCS)*. Lemgo: IEEE, 2012, pp. 213–222. ISBN 978-1-4673-0693-5. DOI: 10.1109/WFCS.2012.6242569.

[7] WARD, S. and W. HIGIBHOTAN. Network errors and their influence on current differential relaying. In: *Protective Relay Engineers*. College Station: IEEE, 2011, pp. 79–90. ISBN 978-1-4577-0494-9. DOI: 10.1109/CPRE.2011.6035607.

[8] KWOK-HONG, M. and B. L. HOLLAND. Migrating electrical power network SCADA systems to TCP/IP and Ethernet networking. *Power Engineering Journal*. 2002, vol. 16, iss. 6, pp. 305–311. ISSN 0950-3366. DOI: 10.1049/pe:20020604.

[9] DEMETER, E., S. FARIED and T. S. SIDHU. Power System Protective Functions Performance Over an Ethernet-based Process Bus. In: *Electrical and Computer Engineering*. Vancouver: IEEE, 2007, pp. 264–267. ISBN 1-4244-1020-7. DOI: 10.1109/CCECE.2007.72.

[10] VASEL, J. One plant, one system: Benefits of integrating process and power automation. In: *Protective Relay Engineers*. College Station: IEEE, 2012, pp. 215–250. ISBN 978-146731841-9. DOI: 10.1109/CPRE.2012.6201235.

[11] POZZUOLI, M. P. and R. MOORE. Ethernet in the substation. In: *Power Engineering Society General Meeting*. Montreal: IEEE, 2006, pp. 1–7. ISBN 1-4244-0493-2. DOI: 10.1109/PES.2006.1709165.

[12] MOXLEY, R., K. FODERO and H. J. ALTUVE. Updated transmission line protection communications. In: *Power Systems Conference*. Austin: IEEE, 2009, pp. 394–401. ISBN 978-1-4244-4182-2. DOI: 10.1109/CPRE.2009.4982528.

[13] CHLUMSKY, P., Z. KOCUR and J. VODRAZKA. Comparison of Different Scenarios for Path Diversity Packet Wireless Networks. *Advances in Electrical and Electronic Engineering*. 2012, vol. 10, iss. 4, pp. 199–203. ISSN 1336-1376.

About Authors

Leos BOHAC received the M.Sc. and Ph.D. degrees in electrical engineering from the Czech Technical University, Prague, in 1992 and 2001, respectively. Since 1992, he has been teaching optical communication systems and data networks with the Czech Technical University, Prague. His research interest is on the application of high-speed optical transmission systems in a data network. He has also participated in the optical research project Czech Education and Scientific NETwork (CESNET) - the academic data network provider to help implement a long-haul high-speed optical research network. Currently, he has been actually involved in and led some of the projects on optimal protocol design, routing, high speed optical modulations and industrial network design.

Jiri VODRAZKA was born in Prague, Czech Republic in 1966. He joined the Department of Telecommunication Engineering, Faculty of Electrical Engineering, Czech Technical University, Prague in 1996 as a research assistant and received his Ph.D. degree in electrical engineering in 2001. He has been the head of the Transmission Media and Systems scientific group since 2005 and became Associate Professor in 2008. He participates in numerous projects in cooperation with external bodies. Currently he acts also as vice-head of the Department.

Investigating and Simulating DGs to Improve Voltage Profile and Reducing Power Loss in Unbalanced Distribution Networks

Hassan BARATI, Abdol Rasool FATTHI

Department of Electrical Engineering, Dezful Branch, Islamic Azad University, Dezful, Iran

barati216@gmail.com, rasool_425@yahoo.com

Abstract. *Development of Distributed generations (DGs) has changed the nature of power networks from passive to active. In order to specify a proper model for DG units, function and connection style of DGs to the network (direct or indirect) must be clarified. In this paper, operation of DGs in an unbalanced condition (due to unbalanced loads) will be studied. A three-phase power flow model is used to confirm the certainty of the proposed model. Generally DG modelling, depend on its type and controlling circuit, is like either PV or PQ bus. In this paper a three-phase unbalanced power flow model using Newton-Raphson method over a 4 buses power system is implemented. Based on the type of bus (PV or PQ), Simulation results show that the DGs can improve voltage profile (NEME and IEEE unbalance factor) and reduce network losses.*

Keywords

DG, network loss, unbalance 3-phase power flow, unbalance factor, voltage profile.

1. Introduction

Increasing in energy consumption and expansion of nonlinear and sensitive loads (like drive systems, heavy single phase loads), continuous variation in network power consumption and unpredicted change in load lead to unbalance in power systems. In most of developed countries evolution of power generation and transmission industry eliminates all technical, academic and commercial requirements. Since operators of power system realized the need of different energy sources, the distributed generation was formed. Distributed generation concept includes small power plants with a capacity between 15 kilowatts to 25 megawatts to supply nearby consumers e.g. small power plants like wind, solar, fuel cells and etc.

Nowadays using distributed generation due to its many advantages like balancing power flow, balancing steady state condition, backup P and Q, reduction of investing in the transmission system, easy installation and startup, from consumers and power companies' point of view is unavoidable. Development of Distributed generations (DGs) has changed the nature of power networks from an active to a reactive one. A distribution company market, which is operated by Distribution System Operator (DSO) is proposed, and Effects of unbalance in power systems has been studied in [1], [2]. In [3], [4], [5], [6], [7], [8], [9], models of generators, transformers, lines and capacitors are presented to study three-phase power flow. Many studies normally focus on finding procedures to introduce DGs as PV and PQ buses in power flow calculations [10], [11], [12], [13], [14], [15], [16], [17]. To indicate a proper model for each DG, function and its connecting style to networks (direct or indirect) different kinds of DG like synchronous and inductive has been studied in [18]. In [19] loads models are presented. To show DG effects in power systems, it is necessary to update most of analyzing tools that used by power system engineers. Probabilistic computer calculations and usual power flow solution to analyze steady state of the network are commonly used. Trustful power flow solution in power networks due to its real and applied nature is challenging. Too many calculating methods have been suggested for this subject, from which the Newton-Raphson method with its converging specification is known as the best method. Also it is a popular method in the industry [19]. In [20], [21], [22], about operation management of DGs power sources connected to distribution networks worked.

As the goal of this paper is the study of DG's impact on voltage profile and power system losses in an unbalanced condition, so despite the DG's type all of DGs' are modeled as PV or PQ and their location is selected randomly based on trial and error method. DG's impact on three-phase power flow in an unbalanced condition based on the NEMA and IEEE indexes are presented.

2. Voltage Unbalance

2.1. The Definitions of Unbalance

Increase of the nonlinear consumers that are switched by power electronic elements is one of the main reasons for unbalancing. On the other hand, these consumers have taken part in power market sensitive to the unbalance. These factors double the importance of unbalancing phenomenon and emphasize on its control. The definitions of voltage unbalance are stated below:

1) First Definition IEEE Std. 936-1987

The difference between the highest and the lowest rms voltage referred to the average of the three-phase voltages.

2) Second Definition IEEE Std. 112-1991

The maximum deviation from the average phase voltage, referred to the average of the phase voltage.

3) Third Definition NEMA [1], [2]

The definition of voltage unbalance, also known as the line voltage unbalance rate (LVUR), is given by:

$$\%LVUR = \\ = \frac{max\ voltage\ deviation\ from\ the\ avg.\ line\ voltage}{avg.\ line\ voltage}. \quad (1)$$

The true definition of voltage unbalance is defined as the ratio of negative sequence voltage component to the positive sequence voltage component. The percentage voltage unbalance factor (%VUF), or the true definition, is given by:

$$\%VUF = \\ = \frac{negative\ sequence\ voltage\ component}{positive\ sequence\ voltage\ component} \cdot 100. \quad (2)$$

2.2. Methods of Unbalance Reduction

- Equally distributing single phase loads between all phases.
- Reducing unbalance caused by impedance of the system.
- Using single phase regulators.
- Using passive and active compensators.

3. DG Model for Power Flow Study

3.1. DG Units Connecting Styles to Power Network are Categorized as Below

1) Wind Turbines

Wind turbines are divided to constant and variable/floating speed. In the first group rotor of an inductive squirrel cage generator is rotated through a gearbox and the generator is directly connected to the power network, but in the second group a synchronic generator or a reference model is used. Output of these units is converted to the proper AC power by means of rectifiers and power electronic inverters for the network.

2) Fuel Cells

Fuel cells directly transform potential energy stored in the fuel to heat and electric energy without any electrical interface machine. Then Produced DC power is converted to AC power adapted to the network using an inverter.

3) Photovoltaic Systems

In photovoltaic systems solar energy is converted to DC current, then like fuel cells their DC output is converted to AC power that is adaptive with the network using an inverter.

4) Internal Combustion Engines

In an internal combustion engine the expansion of the high-temperature and high-pressure gases produced by combustion apply direct force to some component of the engine. The force is applied typically to pistons, turbine blades, or a nozzle. This force moves the component over a distance, transforming chemical energy into useful mechanical energy. Internal combustion engine driven generators are commonly used to produce electricity.

5) Gas Turbines

These units convert chemical energy stored in fossil fuels into heat and to mechanical energy finally synchronous or inductive generator that is directly connected to the network is rotated.

6) Microturbine

These units operate the same way as gaseous turbines. The only difference is that a permanent magnet instead of rotating synchronous generator is rotated. So the generator is connected to the network via power electronic interfaces.

Concerning the points above, primary produced energy by DGs (generated by either of synchronous or asynchronous electric machines that are directly or indirectly connected to the network) would be injected to the network, using a combination of electric machines electronic interfaces or only the electronic interfaces may be involved.

3.2. DG Model and Its Control Circuit Interface Characteristic

DGs model and converter control circuit characteristic are categorized as follow:

1) Induction Generator Model

Generally, in an induction generator both active and reactive powers are functions of slip.

$$\begin{cases} P = P(V, S) \\ Q = Q(V, S) \end{cases}, \tag{3}$$

where P and Q active and reactive power respectively, S is the slip of induction generator speed, and V is the bus voltage. Assuming P to be constant and neglecting the very low dependency of reactive power to the slip, the Eq. (3) can be expressed as follows:

$$\begin{cases} P = Ps = \mathrm{Cons}\tan t \\ \quad Q = F(V) \end{cases}. \tag{4}$$

The expressed model by Eq. (4), which is called SVCM, is an appropriate model of squirrel cage induction generator for power flow studies.

Since the bus voltages are near 1.0 pu in steady-state cases, squirrel cage induction generator can be modeled as a PQ bus for simplicity.

2) Synchronous Generator Model

Depending on the excitation system, synchronous generators are divided into two categories:

- (a) with adjustable excitation voltage,
- (b) with constant excitation voltage.

The first case can be divided into two separate groups:

- (a1) voltage control mode or constant terminal voltage,
- (a2) power factor control mode or constant power factor.

DGs in subgroup (a1) and (a2) are modeled with PV and PQ buses respectively.

3) Power Electronic Interface:

Power generated by photovoltaic, fuel cells, microturbines, and some wind farms are injected to network through power electronic converters. In this case power flow depends on controlling scheme that is used in converter control circuit. As a general rule, if the converter control circuit is designed to control P and V independently, the DG model shall be as a PV bus and when it is designed to control P and Q independently, the DG model shall be considered as a PQ bus.

3.3. Incorporating DG Units in Power Flow Algorithm

The DG units, which are modeled as PQ nodes can be treated as negative PQ loads in power flow studies without any problem. However, handling PV nodes in power flow studies requires some additional processes. It should be noted that the generator terminal voltage is typically controlled by the specification of positive sequence component. So for a PV node, the three-phase active power output and positive sequence voltage of the generator are specified.

4. Three-Phase Power Flow

Start point for developing node power equations that is suitable for solving three-phase power flow using Newton-Raphson method, describes the relation between injected current to a bus and its voltage. See transmission circuit shown in Fig. 1.

4.1. Power Flow Equations

Equations for injected active and reactive power to three-phase buses k and m is obtained from following complex power equations:

$$\begin{bmatrix} S_k^{\mathrm{abc}} \\ S_m^{\mathrm{abc}} \end{bmatrix} = \begin{bmatrix} P_k^{\mathrm{abc}} + \mathrm{j}Q_k^{\mathrm{abc}} \\ P_m^{\mathrm{abc}} + \mathrm{j}Q_m^{\mathrm{abc}} \end{bmatrix} = \begin{bmatrix} E_k^{\mathrm{abc}} I_k^{\mathrm{abc}^*} \\ E_m^{\mathrm{abc}} I_m^{\mathrm{abc}^*} \end{bmatrix}, \tag{5}$$

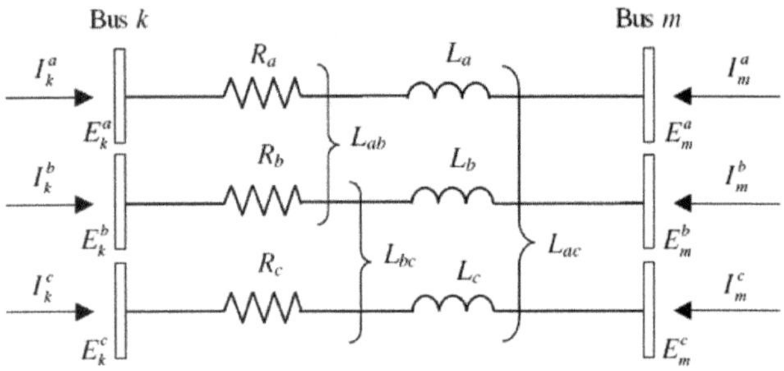

Fig. 1: Transmission lines model for three-phase power flow studies.

Following some complicated algebraic operations, Equations to express injected active and reactive power to phases a, b, c of bus k is acquired by:

$$P_k^\rho = V_k^\rho \left\{ \sum_{i=k,m} \sum_{j=a,b,c} V_i^j \left[G_{\mathrm{ki}}^{\rho j} \cos\left(\theta_k^\rho - \theta_i^j\right) + B_{\mathrm{ki}}^{\rho j} \sin\left(\theta_k^\rho - \theta_i^j\right) \right] \right\}, \quad (6)$$

$$Q_k^\rho = V_k^\rho \left\{ \sum_{i=k,m} \sum_{j=a,b,c} V_i^j \left[G_{\mathrm{ki}}^{\rho j} \sin\left(\theta_k^\rho - \theta_i^j\right) - B_{\mathrm{ki}}^{\rho j} \cos\left(\theta_k^\rho - \theta_i^j\right) \right] \right\}, \quad (7)$$

where ρ represents phases a, b, c.

As predicted phrases that are applied to calculate injected active and reactive power to bus m are like Eq. (6) and Eq. (7) with the difference that m replaces k and vice versa:

$$P_m^\rho = V_m^\rho \left\{ \sum_{i=m,k} \sum_{j=a,b,c} V_i^j \left[G_{\mathrm{mi}}^{\rho j} \cos\left(\theta_m^\rho - \theta_i^j\right) + B_{\mathrm{mi}}^{\rho j} \sin\left(\theta_{\mathrm{mk}}^\rho - \theta_i^j\right) \right] \right\}, \quad (8)$$

$$Q_m^\rho = V_m^\rho \left\{ \sum_{i=m,k} \sum_{j=a,b,c} V_i^j \left[G_{\mathrm{mi}}^{\rho j} \sin\left(\theta_m^\rho - \theta_i^j\right) - B_{\mathrm{mi}}^{\rho j} \cos\left(\theta_m^\rho - \theta_i^j\right) \right] \right\}. \quad (9)$$

4.2. Newton-Raphson Power Flow Algorithm

Solving three-phase node power equations using Newton-Raphson method presents good certainty in converging.

Power Eq. (6) and Eq. (9) are linearized around a reference working point. In three phases application power difference and state variables are 3×1 vectors and each of them are independent terms of 3×3 Jacobean matrixes. Acquired linear equations that are suitable for solving iterative solution are as below:

$$\begin{bmatrix} \Delta P_l^\rho \\ \Delta Q_l^\rho \end{bmatrix}^{(i)} = \begin{bmatrix} \frac{\partial P_l^\rho}{\partial \theta_j^\rho} & \frac{\partial P_l^\rho}{\partial V_j^\rho} V_j^\rho \\ \frac{\partial Q_l^\rho}{\partial \theta_j^\rho} & \frac{\partial Q_l^\rho}{\partial V_j^\rho} V_j^\rho \end{bmatrix}^{(i)} \begin{bmatrix} \Delta \theta_j^\rho \\ \frac{\Delta V_j^\rho}{V_j^\rho} \end{bmatrix}^{(i)}, \quad (10)$$

where $L = k$, m and $j = k$, m and I is the number of iteration.

Vector statement is supposes like the below one:

$$\Delta P_l^\rho = \begin{bmatrix} \Delta P_k^a & \Delta P_k^b & \Delta P_k^c & \Delta P_m^a & \Delta P_m^b & \Delta P_m^c \end{bmatrix}^t, \quad (11)$$

$$\Delta Q_l^\rho = \begin{bmatrix} \Delta Q_k^a & \Delta Q_k^b & \Delta Q_k^c & \Delta Q_m^a & \Delta Q_m^b & \Delta Q_m^c \end{bmatrix}^t, \quad (12)$$

$$\Delta \theta_i^\rho = \begin{bmatrix} \Delta \theta_k^a & \Delta \theta_k^b & \Delta \theta_k^c & \Delta \theta_m^a & \Delta \theta_m^b & \Delta \theta_m^c \end{bmatrix}^t, \quad (13)$$

$$\frac{\Delta V_j^\rho}{V_j^\rho} = \begin{bmatrix} \frac{\Delta V_k^a}{V_k^a} & \frac{\Delta V_k^b}{V_k^b} & \frac{\Delta V_k^c}{V_k^c} & \frac{\Delta V_m^a}{V_m^a} & \frac{\Delta V_m^b}{V_m^b} & \frac{\Delta V_m^c}{V_m^c} \end{bmatrix}^t. \quad (14)$$

Jacobean array comprises:

$$\frac{\partial P_l^\rho}{\partial \theta_j^\rho} = \begin{bmatrix} \frac{\partial P_l^a}{\partial \theta_j^a} & \frac{\partial P_l^a}{\partial \theta_j^b} & \frac{\partial P_l^a}{\partial \theta_j^c} \\ \frac{\partial P_l^b}{\partial \theta_j^a} & \frac{\partial P_l^b}{\partial \theta_j^b} & \frac{\partial P_l^b}{\partial \theta_j^c} \\ \frac{\partial P_l^c}{\partial \theta_j^a} & \frac{\partial P_l^c}{\partial \theta_j^b} & \frac{\partial P_l^c}{\partial \theta_j^c} \end{bmatrix}, \quad (15)$$

$$\frac{\partial P_l^\rho}{\partial V_j^\rho} V_j^\rho = \begin{bmatrix} \frac{\partial P_l^a}{\partial V_j^a} V_j^a & \frac{\partial P_l^a}{\partial V_j^b} V_j^b & \frac{\partial P_l^a}{\partial V_j^c} V_j^c \\ \frac{\partial P_l^b}{\partial V_j^a} V_j^a & \frac{\partial P_l^b}{\partial V_j^b} V_j^b & \frac{\partial P_l^b}{\partial V_j^c} V_j^c \\ \frac{\partial P_l^c}{\partial V_j^a} V_j^a & \frac{\partial P_l^c}{\partial V_j^b} V_j^b & \frac{\partial P_l^c}{\partial V_j^c} V_j^c \end{bmatrix}, \quad (16)$$

$$\frac{\partial Q_l^\rho}{\partial \theta_j^\rho} = \begin{bmatrix} \frac{\partial Q_l^a}{\partial \theta_j^a} & \frac{\partial Q_l^a}{\partial \theta_j^b} & \frac{\partial Q_l^a}{\partial \theta_j^c} \\ \frac{\partial Q_l^b}{\partial \theta_j^a} & \frac{\partial Q_l^b}{\partial \theta_j^b} & \frac{\partial Q_l^b}{\partial \theta_j^c} \\ \frac{\partial Q_l^c}{\partial \theta_j^a} & \frac{\partial Q_l^c}{\partial \theta_j^b} & \frac{\partial Q_l^c}{\partial \theta_j^c} \end{bmatrix}, \quad (17)$$

$$\frac{\partial Q_l^\rho}{\partial V_j^\rho} V_j^\rho = \begin{bmatrix} \frac{\partial Q_l^a}{\partial V_j^a} V_j^a & \frac{\partial Q_l^a}{\partial V_j^b} V_j^b & \frac{\partial Q_l^a}{\partial V_j^c} V_j^c \\ \frac{\partial Q_l^b}{\partial V_j^a} V_j^a & \frac{\partial Q_l^b}{\partial V_j^b} V_j^b & \frac{\partial Q_l^b}{\partial V_j^c} V_j^c \\ \frac{\partial Q_l^c}{\partial V_j^a} V_j^a & \frac{\partial Q_l^c}{\partial V_j^b} V_j^b & \frac{\partial Q_l^c}{\partial V_j^c} V_j^c \end{bmatrix}. \quad (18)$$

It should be noted that linear Eq. (10) is only applied to one three-phase transmission line between m and k buses. However, the result can simply be generalized to a more practical one than includes nl transmission lines between nb buses l and j where $l = 1, \ldots, k, m, \ldots n_b - 1$ and $j = 1, \ldots k, m, \ldots n_b - 1$, Also when slack bus is not stated in linear equations, there will be only $n_b - 1$ buses. It is supposed that element number l in Eq. (10) is connected between k and m buses. Self and mutual Jacobean statement are simply calculated as below, where $\rho 1 = \rho 2$ are used for phases a, b, c respectively.

For $k = m$ and $\rho 1 = \rho 2$:

$$\frac{\partial P_{k,l}^{\rho 1}}{\partial \theta_{k,l}^{\rho 1}} = -Q_k^{\rho 1 \mathrm{cal}} - \left(V_k^{\rho 1}\right)^2 B_{\mathrm{kk}}^{\rho 1 \rho 1}, \quad (19)$$

$$\frac{\partial P_{k,l}^{\rho 1}}{\partial V_{k,l}^{\rho 1}} V_{k,l}^{\rho 1} = P_k^{\rho 1 \mathrm{cal}} + \left(V_k^{\rho 1}\right)^2 G_{\mathrm{kk}}^{\rho 1 \rho 1}, \quad (20)$$

$$\frac{\partial Q_{k,l}^{\rho 1}}{\partial \theta_{k,l}^{\rho 1}} = P_k^{\rho 1 \mathrm{cal}} - \left(V_k^{\rho 1}\right)^2 G_{\mathrm{kk}}^{\rho 1 \rho 1}, \quad (21)$$

$$\frac{\partial Q_{k,l}^{\rho 1}}{\partial V_{k,l}^{\rho 1}} V_{k,l}^{\rho 1} = Q_k^{\rho 1 \text{cal}} - \left(V_k^{\rho 1}\right)^2 B_{\text{kk}}^{\rho 1 \rho 1}. \quad (22)$$

For $k = m$ and $\rho 1 \neq \rho 2$:

$$\frac{\partial P_{k,l}^{\rho 1}}{\partial \theta_{k,l}^{\rho 2}} = V_k^{\rho 1} V_k^{\rho 2} \left[G_{\text{kk}}^{\rho 1 \rho 2} \sin\left(\theta_k^{\rho 1} - \theta_k^{\rho 2}\right) - B_{\text{kk}}^{\rho 1 \rho 1} \cos\left(\theta_k^{\rho 1} - \theta_k^{\rho 2}\right)\right], \quad (23)$$

$$\frac{\partial P_{k,l}^{\rho 1}}{\partial V_{k,l}^{\rho 2}} V_{k,l}^{\rho 2} = V_k^{\rho 1} V_k^{\rho 2} \left[G_{\text{kk}}^{\rho 1 \rho 2} \cos\left(\theta_k^{\rho 1} - \theta_k^{\rho 2}\right) + B_{\text{kk}}^{\rho 1 \rho 1} \sin\left(\theta_k^{\rho 1} - \theta_k^{\rho 2}\right)\right], \quad (24)$$

$$\frac{\partial Q_{k,l}^{\rho 1}}{\partial \theta_{k,l}^{\rho 2}} = -V_k^{\rho 1} V_k^{\rho 2} \left[G_{\text{kk}}^{\rho 1 \rho 2} \cos\left(\theta_k^{\rho 1} - \theta_k^{\rho 2}\right) + B_{\text{kk}}^{\rho 1 \rho 1} \sin\left(\theta_k^{\rho 1} - \theta_k^{\rho 2}\right)\right], \quad (25)$$

$$\frac{\partial Q_{k,l}^{\rho 1}}{\partial V_{k,l}^{\rho 2}} V_{k,l}^{\rho 2} = V_k^{\rho 1} V_k^{\rho 2} \left[G_{\text{kk}}^{\rho 1 \rho 2} \sin\left(\theta_k^{\rho 1} - \theta_k^{\rho 2}\right) - B_{\text{kk}}^{\rho 1 \rho 1} \cos\left(\theta_k^{\rho 1} - \theta_k^{\rho 2}\right)\right]. \quad (26)$$

For $k \neq m$:

$$\frac{\partial P_{k,l}^{\rho 1}}{\partial \theta_{m,l}^{\rho 2}} = V_k^{\rho 1} V_m^{\rho 2} \left[G_{\text{km}}^{\rho 1 \rho 2} \sin\left(\theta_k^{\rho 1} - \theta_m^{\rho 2}\right) - B_{\text{km}}^{\rho 1 \rho 1} \cos\left(\theta_k^{\rho 1} - \theta_m^{\rho 2}\right)\right], \quad (27)$$

$$\frac{\partial P_{k,l}^{\rho 1}}{\partial V_{m,l}^{\rho 2}} V_{m,l}^{\rho 2} = V_k^{\rho 1} V_m^{\rho 2} \left[G_{\text{km}}^{\rho 1 \rho 2} \cos\left(\theta_k^{\rho 1} - \theta_m^{\rho 2}\right) + B_{\text{km}}^{\rho 1 \rho 1} \sin\left(\theta_k^{\rho 1} - \theta_m^{\rho 2}\right)\right], \quad (28)$$

$$\frac{\partial Q_{k,l}^{\rho 1}}{\partial \theta_{m,l}^{\rho 2}} = -V_k^{\rho 1} V_m^{\rho 2} \left[G_{\text{km}}^{\rho 1 \rho 2} \cos\left(\theta_k^{\rho 1} - \theta_m^{\rho 2}\right) + B_{\text{km}}^{\rho 1 \rho 1} \sin\left(\theta_k^{\rho 1} - \theta_{\text{mk}}^{\rho 2}\right)\right], \quad (29)$$

$$\frac{\partial Q_{k,l}^{\rho 1}}{\partial V_{m,l}^{\rho 2}} V_{k,l}^{\rho 2} = V_k^{\rho 1} V_m^{\rho 2} \left[G_{\text{km}}^{\rho 1 \rho 2} \sin\left(\theta_k^{\rho 1} - \theta_m^{\rho 2}\right) - B_{\text{km}}^{\rho 1 \rho 1} \cos\left(\theta_k^{\rho 1} - \theta_m^{\rho 2}\right)\right]. \quad (30)$$

Iterative power flow solution using Newton-Raphson method needs observation of points applied for positive sequence solutions i.e. specify primary value to the state variable and checking required reactive power for the generator. In three-phase application voltages a, b, c primary phase angles are valued 0, $-2\pi/3$ and $2\pi/3$ respectively.

5. Three-phase Power Flow Computer Simulation Results of a 4 Buses IEEE Distribution Network

IEEE 4 buses distribution network includes an infinite bus, load, a transformer and two lines that its single line diagram is illustrated in Fig. 2 and its data is collected in [20]. In the next section, outputs resulted from Newton-Raphson method for three-phase power flow in an unbalanced condition with and without DG are presented. As the goal of this paper is the study of DG's impact on voltage profile and power system losses in an unbalanced condition, so despite the DG's type all of DGs' are modeled as PV or PQ and their location is selected randomly based on trial and error method.

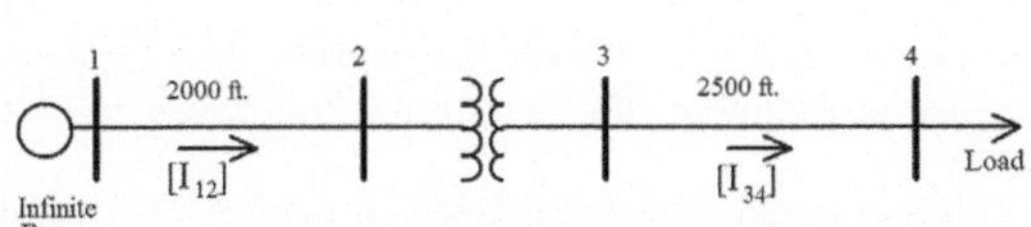

Fig. 2: IEEE four buses single line diagram.

5.1. Three-phase Power Flow Without DG

To evaluate the certainty of the computer program and algorithm, results of some of three-phase power flow with different transformer connections are added and are compared with results of reference [9] as shown in Tab. 1.

Tab. 1: Comparison between three-phase power flow and reference network.

Transformer Connection		Gr Y − Gr Y		Gr Y − D	
		Suggested Algorithm Result	IEEE [9]	Suggested Algorithm Result	IEEE [9]
Bus Voltage 2					
Va	(V)	7108.2	7107	7115.8	7113
	deg	−0.34	−0.3	−0.32	−0.3
Vb	(V)	7143.3	7140	7129.6	7132
	deg.	−120.33	−120.3	−120.31	−120.3
Vc	(V)	7121.4	7121	7130.6	7123
	deg.	119.65	119.6	119.61	119.6
Bus Voltage 4					
Va	(V)	1919.4	1918	3534.4	3437
	deg.	−9.19	−9.1	−5.8	−7.8
Vb	(V)	2073	2061	3269.8	3497
	deg.	−128.16	−128.3	−128.97	−129.3
Vc	(V)	1977.9	1981	3566	3388
	deg.	111	110.9	108.55	110.6

Table 1 represent voltage magnitude and phase angle in buses of distribution network. For example in bus four; voltage amplitude and phase angle of phase a are 1919.4 volts and −9.19°, these values have been recorded 1918 volts with −9.1° in reference [9]; comparison between these two validate suggested algorithm. Voltage profile and network losses will be studied. In Fig. 3 voltages magnitudes are shown, and it is observed that bus four has the worst condition from phase's voltage profile point of view.

Figure 4 demonstrates active power losses in phases A, B, C of lines 1-2, transformer and lines 3-4 as well as total network active and reactive power losses which are 0.32157 and 0.88962 respectively.

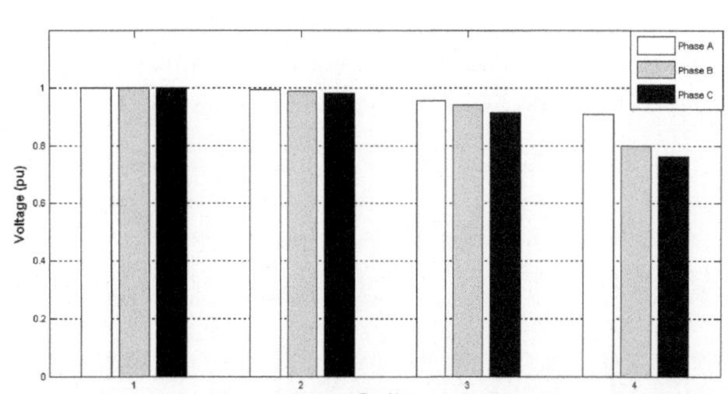

Fig. 3: Voltage magnitude in absence of DG.

In Tab. 2, four buses simulation results without DG are shown. By observing Tab. 2, it can be seen that NEMA and IEEE unbalance factor in bus 4 has the

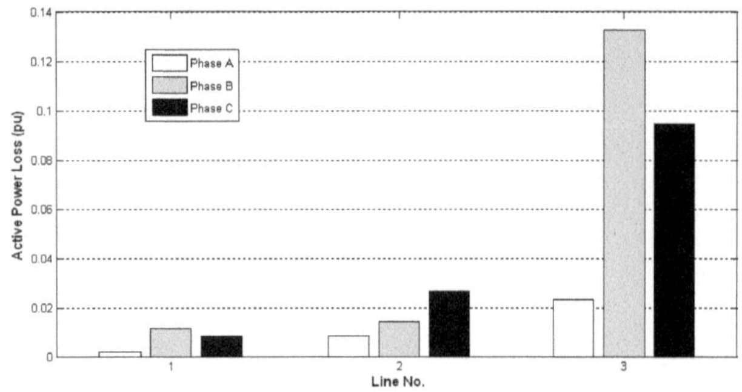

Fig. 4: Power losses in absence of DG.

most value; that means voltage unbalance in the worst case is 0.10124 and 0.064413 respectively.

Tab. 2: 4 buses network results – without DG.

Bus Number	1	2	3	4
Phase 'A' Voltage Mag. (pu)	1	0.99562	0.95604	0.90742
Phase 'B' Voltage Mag. (pu)	1	0.98735	0.94217	0.80043
Phase 'C' Voltage Mag. (pu)	1	0.98377	0.91608	0.76415
Phase 'A' Voltage Angle (Deg)	0.0	−0.12	−2.40	−4.04
Phase 'B' Voltage Angle (Deg)	−120	−120.13	−123.52	101.83
Phase 'C' Voltage Angle (Deg)	120	119.21	114.81	111
Zero Sequence	1.67×10^{-16}	0.0066923	0.016967	0.093352
Neg. Sequence	1.11×10^{-16}	0.0030186	0.018371	0.052771
NEMA Unbala. Factor	0	0.0067791	0.023472	0.10124
IEEE Unbala. Factor	1.11×10^{-16}	0.0030525	0.019588	0.064413

5.2. Three-phase Power Flow with PQ Model of DG

If DG is modeled like PQ and installed in bus 3 with the value of $0.3 + j0.4$ pu for each phase, implementing three-phase power flow will lead to results shown in Fig. 5 for bus voltage amplitude in phases A, B, C. In compare to previous scheme phases voltage unbalance specially in bus four has been reduced and voltage profile has been improved in the entire network.

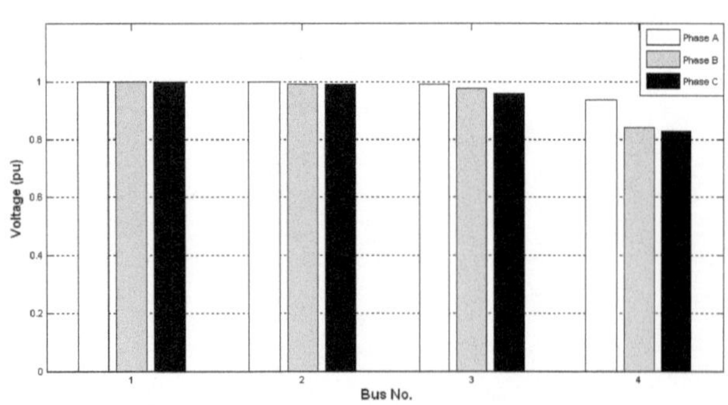

Fig. 5: Voltage amplitude with PQ model of DG.

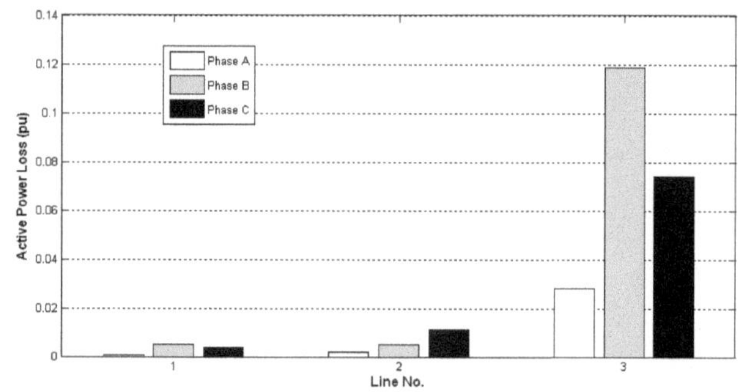

Fig. 6: Active power losses with PQ model of DG.

Tab. 3: 4 buses network results with PQ model of DG.

Bus Number	1	2	3	4
Phase 'A' Voltage Mag. (pu)	1	1.0006	0.99033	0.93808
Phase 'B' Voltage Mag. (pu)	1	0.9922	0.97733	0.83997
Phase 'C' Voltage Mag. (pu)	1	0.99153	0.96025	0.82803
Phase 'A' Voltage Angle (Deg)	0.0	−0.07	−1.48	−2.72
Phase 'B' Voltage Angle (Deg)	−120	−120.13	−122.51	−125.27
Phase 'C' Voltage Angle (Deg)	120	119.29	116.01	104.68
Zero Sequence	1.67×10^{-16}	0.0061001	0.015253	0.08411
Neg. Sequence	1.11×10^{-16}	0.0022112	0.015113	0.045399
NEMA Unbala. Factor	0	0.0058741	0.016107	0.079875
IEEE Unbala. Factor	1.11×10^{-16}	0.0022228	0.015488	0.052495

Figure 6 shows phases A, B, C active power losses in lines 1-2, transformer and lines 3-4 as well as total active and reactive network losses which are 0.24673 and 0.60588 respectively.

In Tab. 3, four buses simulation results with the PQ model of DG are shown. By observing Tab. 4, it can be seen that NEMA and IEEE unbalance factors in all buses are improved; that means voltage unbalance in the worst case is 0.079875 and 0.052495 respectively. These are better factors in compared to without DG condition.

5.3. Three-phase Power Flow in Presence of PV Model of DG

If DG is modeled like PV, installed in bus 3 and its active power for each phase and voltage amplitude set to 0.3 pu and 1 pu respectively a better improvement will be observed in voltage profile in compare to that for the PQ model. Buses voltage amplitude in phases A, B, C is plotted in Fig. 7.

Figure 8 shows active power losses in phases a, b, c in lines 1-2, transformer and lines 3-4. Network total active and reactive power losses are 0.22577 and 0.55086 respectively.

In Tab. 4, four buses simulation results with PV model of DG are shown. By observing Tab. 5, it can be seen that NEMA and IEEE unbalance factors in all buses are improved; that means voltage unbalance in the worst case is 0.050001 and 0.0444418 respectively. These factors in comparison to the case without DG are better.

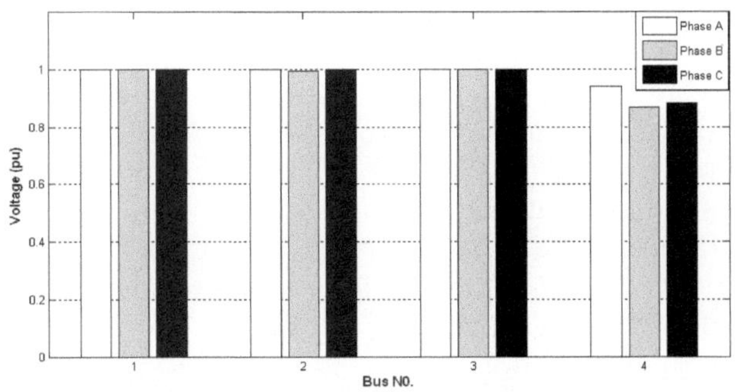

Fig. 7: Voltage amplitude with PV model of DG.

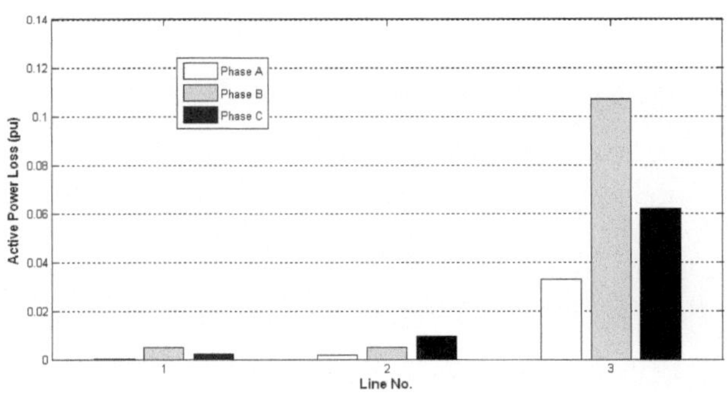

Fig. 8: Active power loses with PV model of DG.

Tab. 4: 4 buses network results with PQ model of DG.

Bus Number	1	2	3	4
Phase 'A' Voltage Mag. (pu)	1	0.99967	1	0.94178
Phase 'B' Voltage Mag. (pu)	1	0.9947	1	0.8678
Phase 'C' Voltage Mag. (pu)	1	0.99945	1	0.88119
Phase 'A' Voltage Angle (Deg)	0.0	−0.04	−1.55	−2.70
Phase 'B' Voltage Angle (Deg)	−120	−120.38	−122.85	−125.83
Phase 'C' Voltage Angle (Deg)	120	119.25	115.87	105.81
Zero Sequence	1.67×10^{-16}	0.0051727	0.012913	0.072333
Neg. Sequence	1.11×10^{-16}	0.00019602	0.013082	0.039696
NEMA Unbala. Factor	0	0.0032301	0	0.050001
IEEE Unbala. Factor	1.11×10^{-16}	0.0019643	0.013084	0.044418

5.4. Comparison Between Three-phase Power Flow in Different Cases

Table 5 shows minimum and maximum voltage amplitude, total active and reactive power losses and NEMA and IEEE unbalance factor in different conditions with and without DG. These results showed that in the presence of DGs, losses and voltage unbalance are reduced.

Tab. 5: Comparing results of four buses network.

Different Cases of three-phase power flow	Whitout DG	Whit DG PQ model	Whit DG PV model
Total active power losses (pu)	0.32157	0.24673	0.22577
Total reactive power losses (pu)	0.88963	0.60588	0.55086
Maximum voltage amplitude (pu)	1	1.0006	1
Minimum voltage amplitude (pu)	0.76415	0.82803	0.8678
NEMA Unbalance Factor	0.10124	0.079875	0.050001
IEEE Unbalance Factor	0.064413	0.052495	0.044418

6. Conclusion

A useful list of DG models with their method of connection to the network (direct or indirect) for power flow studies is considered in this paper. DG's model for power flow studies selected based on initial energy of DG unit that is injected to the network through an electrical machine(which is directly connected to the network). Also, DGs are connected to the network through a power electronic interface, or through combination of the electric machine beside power electronic interface.

With the suggested newton-Raphson three-phase power flow algorithm it is possible to study DGs in PV and PQ models and good divergence of this algorithm makes it to be the most successful method accepted in the industry. DGs reduce network losses and improve voltage profile.

References

[1] JOUANNE, A. and B. BANERJEE. Assessment of Voltage Unbalance. *IEEE Trans. on power delivery*. 2001, vol. 16, no. 4, pp. 782–790. ISSN 0885-8977. DOI: 10.1109/61.956770.

[2] PILLAY, P. and M. MANYAGE. Definitions of voltage unbalance. *IEEE Power Engineering Review*. 2001, vol. 21, no. 5, pp. 49–51. ISSN 0272-1724. DOI: 10.1109/MPER.2001.4311362.

[3] SANTOS VALOIS, P. V., C. M. VIEIRA TAHAN, N. KAGAN and H. ARANGO. Voltage unbalance

in low voltage distribution networks. In: *International Conference and Exhibition on Electricity Distribution.* Amsterdam: IET, 2001, pp. 1–5. ISBN 0-85296-735-7. DOI: 10.1049/cp:20010770.

[4] CHEN, T. H., M. S. CHEN, T. INOUE, P. KOTAS, and E. A. CHEBLI. Three-phase cogenerator and transformer models for distribution system analysis. *IEEE Transactions On Power Delivery.* 1991, vol. 6, no. 4, pp. 1671–1681. ISSN 0885-8977. DOI: 10.1109/61.97706.

[5] CHOQUE, J. L., D. RODAS and A. PADILHA-FELTRIN. Distribution Transformer Modeling for Application in Three-phase Power Flow Algorithm. *IEEE Latin America Transaction.* 2009, vol. 7, no. 24, pp. 196–202. ISSN 1548-0992. DOI: 10.1109/TLA.2009.5256829.

[6] CORCOLES, F., L. SAINZ, J. PEDRA, J. SANCHEZ-NAVARRO and M. SALICHS. Three-phase Transformer Modelling for Unbalanced Conditions. *IET Electric Power Applications.* 2008, vol. 2, no. 2, pp. 99–112. ISSN 1751-8660. DOI: 10.1049/iet-epa:20070290.

[7] PALMA-BEHNKE, R., J. L. A. CERDA, L. S. VARGAS and A. JOFRE. A distribution company energy acquisition market model with integration of distributed generation and load curtailment options. *IEEE Trans. on Power Systems.* 2005, vol. 20, no. 4, pp. 1718–1727. ISSN 0885-8950. DOI: 10.1109/TPWRS.2005.857284.

[8] LI, H., Y. LI and Z. LI. A multi-period energy acquisition model for a distribution company with distributed generation and interruptible load. *IEEE Trans. on Power Systems.* 2007, vol. 22, no. 2, pp. 588–596. ISSN 0885-8950. DOI: 10.1109/TPWRS.2007.894862.

[9] IEEE PES. *Distribution Test Feeders* [online]. 2013. Available in `http://ewh.ieee.org/soc/pes/dsacom/testfeeders/index.html`.

[10] CHEN, M.-S. and T.-H. CHEN. Application of Three-phase Load Flow to Power System Distribution Automation. In: *International Conference on Advances in Power System Control, Operation and Management.* Hong Kong: IEEE, 1991, pp. 472–478. ISBN 0-86341-246-7.

[11] LUO, G. X. and A. SEMLYEN. Efficient load flow for large weakly meshed networks. *IEEE Trans. on Power Systems.* 1990, vol. 5, no. 4, pp. 1309–1316. ISSN 0885-8950. DOI: 10.1109/59.99382.

[12] CHENG, C. S. and D. SHIRMOHAMMADI. A three-phase power flow method for real-time distribution system analysis. *IEEE Trans. on Power Systems.* 1995, vol. 10, no. 2, pp. 671–679. ISSN 0885-8950. DOI: 10.1109/59.387902.

[13] KHUSHALANI, S. and N. SCHULZ. Unbalanced distribution power flow with distributed generation. In: *Transmission and Distribution Conference and Exhibition.* Dallas: IEEE, 2006, pp. 301–306. ISBN 0-7803-9194-2. DOI: 10.1109/TDC.2006.1668508.

[14] KHUSHALANI, S., J. M. SOLANKI and N. N. SCHULZ. Development of three-phase unbalanced power flow using PV and PQ models for distributed generation and study of the impact of DG models. *IEEE Trans. on Power Systems.* 2007, vol. 22, no. 3, pp. 1019–1025. ISSN 0885-8950. DOI: 10.1109/TPWRS.2007.901476.

[15] AGALGAONKAR, A. P., S. V. KULKARNI and S. A. KHAPARDE. Impact of wind generation on loss and voltage profile in a distribution system. In: *Conference on Convergent Technologies for Asia-Pacific Region.* Bangalore: IEEE, 2003, pp. 775-–779. ISBN 0-7803-7651-X.

[16] TENG, J.-H. Modelling for Distributed Generation in Three-Phase Distribution Load Flow. *IET Generation, Transmission & Distribution.* 2008, vol. 2, no. 3, pp. 330–340. ISSN 1751-8687. DOI: 10.1049/iet-gtd:20070165.

[17] SHIQIONG, T. and K. N. MIU. A Network-Based Distributed Slack Bus Model for DGs in Unbalanced Power Flow Studies. *IEEE Trans. on Power Systems.* 2005, vol. 20, no. 2, pp. 835–842. ISSN 0885-8950. DOI: 10.1109/TPWRS.2005.846056.

[18] MOGHADDAS-TAFRESHI, S. M. and E. MASHHOUR. Distributed Generation Modeling for Power Flow Studies and a Three-Phase Unbalanced Power Flow Solution for Radial Distribution Systems Considering Distributed Generation. *Electric Power Systems Research.* 2009, vol. 79, no. 4, pp. 680–686. ISSN 0378-7796. DOI: 10.1016/j.epsr.2008.10.003.

[19] CHEN, H., J. CHEN, D. SHI and X. DUAN. Power flow study and voltage stability analysis for distribution systems with distributed generation. In: *Power Engineering Society General Meeting.* Montreal: IEEE, 2006, pp. 1–8. ISBN 1-4244-0493-2. DOI: 10.1109/PES.2006.1709198.

[20] NIKNAM, T., H. ZEINODDINI-MEYMAND, H. DOAGOU-MOJARRAD and J. AGHAEI. Multi-objective Daily Operation Management of distribution network considering Fuel Cell Power Plants. *IET Renewable Power Generation.* 2011, vol. 5, no. 5, pp. 356–367. ISSN 1752-1416. DOI: 10.1049/iet-rpg.2010.0190.

[21] NIKNAM, T., A. KAVOUSIFARD and J. AGHAEI. Optimal operation management of fuel cell/wind/photovoltaic power sources connected to distribution networks. *Journal of Power Sources*. 2001, vol. 196, no. 20, pp. 8881–8896. ISSN 0378-7753. DOI: 10.1016/j.jpowsour.2011.05.081.

[22] NIKNAM, T., A. KAVOUSIFARD and J. AGHAEI. Scenario-Based Multi-objective Distribution Feeder Reconfiguration considering Wind Power Using Adaptive Modified PSO. *IET Renewable Power Generation*. 2012, vol. 6, no. 4, pp. 236-247. ISSN 1752-1416. DOI: 10.1049/iet-rpg.2011.0256.

About Authors

Hassan BARATI was born in Dezful, Iran, in 1969. He received his B.Sc degree in electronic engineering from the Isfahan University of Technology, Isfahan, Iran in 1992. He received his M.Sc degree in electrical engineering from the Tabriz University in 1996. He obtained the Ph.D. degree in Electrical Engineering from the Islamic Azad University, Science and Research Branch, Tehran, Iran, in 2007. His research interests are power system analysis, restructured power system, FACTS devices, DGs and renewable energies.

Abdol Rasool FATHI was born in Larestan, Iran. He received his B.Sc degree in electronic engineering from the Islamic Azad University, Kazeroun Branch, Kazeroun, Iran in 1999. He received his M.Sc degree in electrical engineering from the Islamic Azad University, Dezful Branch in 2011. His research interests are power system analysis and FACTS devices.

10

UNSUPERVISED SPELLING CORRECTION FOR THE SLOVAK TEXT

Daniel HLADEK, Jan STAS, Jozef JUHAR

Department of Electronics and Multimedia Communications,
Faculty of Electrical Engineering and Informatics, Technical University of Kosice,
Park Komenskeho 13, 042 00 Kosice, Slovakia

daniel.hladek@tuke.sk, jan.stas@tuke.sk, jozef.juhar@tuke.sk

Abstract. *This paper introduces a method to automatically propose and choose a correction for an incorrectly written word in a large text corpus written in Slovak. This task can be described as a process of finding the best matching sequence of correct words to a list of incorrectly spelled words, found in the input. Knowledge base of the classification system - statistics about sequences of correctly typed words and possible corrections for incorrectly typed words can be mathematically described as a hidden Markov model. The best matching sequence of correct words is found using Viterbi algorithm. The system will be evaluated on a manually corrected testing set.*

Keywords

Automatic spelling correction, hidden Markov model, natural language processing.

1. Introduction

Important part of the natural language processing, statistical language modeling and information retrieval is a preparation of text data. The problem is that the text data often contain typographical and grammatical errors that decrease its information value. This paper will focus on the problem of correcting possibly incorrectly typed sentence. Spell-checking refers to the task of identifying and marking incorrectly spelled words in a document written in a natural language [1].

As it is known from the world of word processors, the spell-checker containing a dictionary of correct words is probably able to find incorrect word form and propose a list of corrections. Choosing the best matching correct word depends on the surrounding context. The user then can manually check the selected word, choose one of the proposed word forms, correct the word manually or proclaim that the correction is not necessary, because the word is truly correct.

The algorithm for correction of the text using an incorporated spell-checker then can be described as:

- check if the highlighted word is really incorrect,
- if it is, check if the correction is in the list of the proposed corrections,
- if a correction is in the list, choose the best possible one,
- if a correction is not available, then manually rewrite the word to its correct form.

In the case of a very large textual data, such as training corpora or web search indices it is simply not possible to use this method that requires human intervention. In order to improve the data it is necessary to reduce human work and do it as much as it is possible in an unsupervised way. One part of the work can be done using a rule based system that can solve the simplest correction by replacing by its usual correction. In the case when there are more corrections possible, a statistical approach is necessary in order to choose the best correction from the list according to the surrounding context.

2. The State of the Art

The technique of finding misspelled word and providing a list of possible corrections is known for a long time in common word processing software, such as Microsoft Word or OpenOffice Writer. It is common that these applications also provide other tools, checking not only spelling, but also the grammar using a rule-based system. However, proposing a correction autonomously is still not common in the office area.

According to [1], [2], errors related to the misspelled words can be categorized into two basic classes:

- **non-word errors** - where the misspelled word is not a valid word in a language,
- **real-word errors** - where the word in question is valid yet inappropriate in the context, and hence not giving the intended meaning.

Paper [1] states that human typing leads to non-word errors that can arise due to three major factors:

- **typographic errors** - a result of motor coordination slips and are related to keyboard mis-punches (e.g. "the" - "teh", "spell" - "speel"),
- **cognitive errors** - caused by the writer's misconceptions (e.g. "receive" - "recieve", "conspiracy" - "conspiricy"),
- **phonetic errors** - a result of substituting a phonetically equivalent sequence of letters (e.g. "seperate" - "separate").

Grobbelaar and Kinyua [3] proposed a system capable of providing corrections for the South African language. Sirts [4] deals with spelling errors, caused by learners of Estonian. This approach utilizes hidden Markov model (HMM) and part-of-speech (POS) tagging in order to choose the best correction of a word. In [5], Li, Duan and Zhai focused on the correction of web search queries and used HMMs with discriminative training. Lund and Ringer [6] used a decision lists in order to improve OCR recognition accuracy. Rodphon et al. [7] used bigram and trigram probability to select the best word in the OCR recognition result in the Thai language. In [8], Zhou et al. used a tribayes method to propose the best correction.

3. The Proposed Approach

The most simple correction mechanism possible is to use search and replace, where a manually created a list of incorrect expressions and their corrections are used to search and replace common errors. However, this approach have its limitations and have to be supplemented by some more sophisticated methods.

If the word is out-of-vocabulary (OOV), it can be one of:

- **word tokens** - regular word, proper name, foreign word,
- **non-word tokens** - numbers, web links, incorrectly typed words.

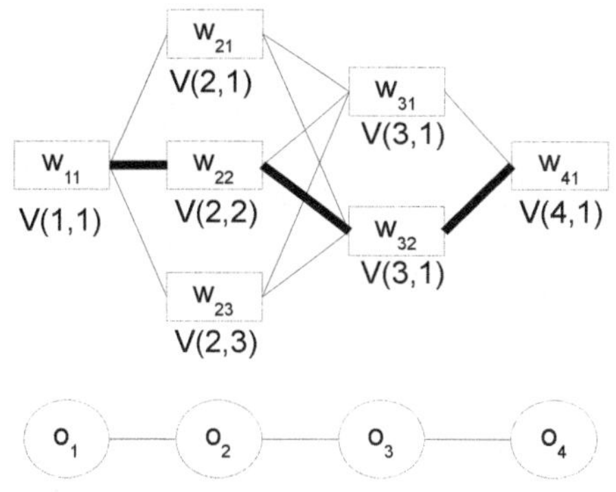

Fig. 1: Viterbi trellis for a sentence.

The first step of the automatic correction should decide, if the OOV word is feasible for automatic word correction. This step ensures that possibly useful OOV word are not "destroyed" and incorrectly replaced by one of the vocabulary words. The following heuristic rules are proposed to identify an incorrect word. The word probably contains typing error if:

- it is not in the vocabulary of the correct words,
- it is lowercase,
- it does not contain features characteristic for foreign word, such as strange letters, numerals, uncommon letter combinations.

After the words qualified for automatic correction are identified, the spell-checking can be described as a process of finding the best possible sequence of correct words, given to the list of possibly incorrect words.

The most common method for classification of the time dependent (sequence) data is hidden Markov model and the Viterbi algorithm. In this case, words distorted by errors in the sentence can be described as a sequence of observation o (bottom part of Fig. 1). Possible corrections w for these words o are hidden states. The Viterbi algorithm then can assign the most probable sequence of the corrected words to the given list of possibly incorrect words (marked by the bold lines).

To make correct representation of this problem using the HMM framework in a way similar to [9], it is necessary to correctly express basics components of the HMM: observation matrix $P(o|w)$ and state transition matrix $P(w|h)$.

Observation matrix is a probability of the occurrence of the incorrect form of the word according to its correct form. The first problem is that the number of possible incorrect word forms is infinite. Then the method of maximum likelihood estimation is impossible to use and have to be estimated in a different way. A heuristic procedure for probability $P(o|w_j)$ of observation o, according to the state w_j is then estimated as

$$P(o|w_j) = \frac{C - j}{\sum_{i=0}^{C}(C - i)}, \quad (1)$$

where j is order of the word proposed by the spell-checking module. C is a number of proposed corrections by the spell-checking module and $\sum_{i=0}^{C}(C - i)$ is a normalizing constant, so that the first proposed correction of the spell-checker has the highest probability. It assumes that the first proposed word by the spell-checking module has the highest probability of a match with current observation. Also note that this expression is not mathematically correct, because it does not ensure that sum of probabilities for all possible bad forms for the inspected state is one.

State set of a HMM is a list of all correct forms of words, given by the manually checked dictionary. The state-transition matrix is a language model of the target language and expresses probability of occurrence of a word according to the given context. The Slovak language is characterized by many possible morphological word forms. As a consequence, every operation that considers statistical information about sequences of words, such as word spelling correction, needs to have a proper method of training to improve the performance of the system [10].

In the case of n-gram language model, the maximum likelihood estimation $P(w|h)$ of the word w according to the context h is given by

$$P(w|h) = \frac{C(h, w)}{C(h)}, \quad (2)$$

where $C(h, w)$ is count of the sequence (h, w) and $C(h)$ is count of the context h in the training corpus.

However, this formula cannot be used in practice, because in the case of insufficient training data, counts of n-grams are often zero. This type of language model then incorrectly gives zero probability also for those situations that are perfectly valid in the given language. From this reason a method of adjusting resulting probability (smoothing) have to be chosen to estimate the probability of events that had not been observed in the training corpus as it is in [11].

If incorrect words, state and observation probabilities can be estimated, the most probable sequence of the corrections can be calculated using Viterbi algorithm. For every possible correct word form a trellis is constructed and every node has assigned a certain value $V(t, i)$ that is used to find the best possible sequence as a path in this trellis.

The Viterbi value of a node, representing a transition a correct word form to another (a state-transition) can

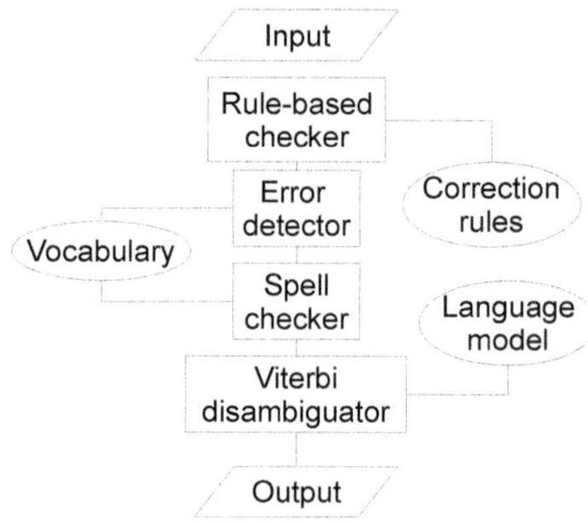

Fig. 2: The proposed system structure.

be calculated using a recursive formula:

$$V(t, i) = P(o_t|w_{t_i}) \max_{j \in S_{t-1}} V(t - 1, j)P(w_{t_i}|w_{t-1,j}), \quad (3)$$

where $V(t, i)$ is Viterbi value for word w_i in time t, $P(o_t|w_{t_i})$ is observation probability similar to the Eq. (1), $V(t - 1, j)$ is Viterbi value of word w_j in time $(t - 1)$ and $P(w_{t_i}|w_{t-1,j})$ is transition probability given by the language model, as it is described by the Eq. (2). The whole process is depicted in the Fig. 1, where each word o_t in a given sentence has some possible corrections $w_i(t)$. Then a value $V_i(t)$ is assigned for every possible correction and using backtracking from the last value it is possible to derive the best sequence of corrections for the given sentence. After all nodes in the Viterbi trellis are evaluated, the best path (the best sequence of correct words) can be found using backtracking, taking paths with the best Viterbi evaluation.

To summarize, the automatic correction algorithm, processes a given sentence:

1. applies rule based corrections in order to deal with the most simple errors,
2. identifies possibly incorrect words,
3. for every incorrect word proposes a list of corrections,
4. constructs a Viterbi trellis, evaluating all possible transitions between correct states,
5. using backtracking proposes the best sequence of correct states.

4. The Experimental Part

The most important part of any natural language processing task is a proper training corpus preparation.

Tab. 1: Evaluation of bad words detection on the testing set.

Result/Class	Word correct	Contains error
true	28 699	128
false	99	393
	Prec: 0,24	*Rec*: 0,56

Tab. 2: Evaluation of detected errors and their corrections.

Correct corrections	93	72,6 %
Bad correction proposal	15	11,7 %
Bad classifier decision	20	15,6 %
Classifier corrects	73	78,4 %

It is required the language model for the unattended spelling corrector is composed of the grammatically and typographically correct text. As a training set, a corpus created from Slovak fiction books have been used, as it is considered to be manually checked and unaffected by the typographic errors. The training corpus contains 49 592 554 tokens in 2 910 180 sentences.

The training corpus preparation step includes word tokenization, sentence boundary identification, sentence and token filtering, transcription and training sentences correction as it is noted in [12]. The list of words have been constrained to the biggest manually checked the dictionary of the Slovak words (6, 5 mil. of unique words) [13].

It is important to say that both training text for the classifier's language model and input text containing errors have very similar structure - both should be a result of the same text preparation process. For this purpose, the same method for the testing set preparation has been used on a selection of the Slovak blog web pages [12], because it is considered to contain typing errors and its style is close to the fiction, used in the training corpus. Testing part of the blog corpus has 29 319 tokens and 1 507 sentences.

In the first experiment, success rate of the bad word detection has been evaluated. Results summarized in the Tab. 2 show that the weakest part of the proposed system is identification of words where spell-check should be applied. Heuristic procedure presented above seems to be insufficient. In order to improve efficiency of the system it is required to extend dictionary to reduce the number of words that should not have proposed corrections. Otherwise, too large number of words that are good but have incorrectly assigned correction is present in the output of the system.

The second evaluation (see Tab. 1) focused on the words that were truly incorrect and the system found a correction for them. Results are comparable to the ones presented in [4], [8] that claim accuracy of spelling correction around 85 %.

It is possible to say that in some cases correction failed, because the spell-check module was not able to propose correct form and classiffier had to choose only from incorrect forms. If a correct form had been available, sometimes Viterbi algorithm chosen wrong. Anyway, it is possible to conclude that the described classifier is sufficient and the biggest room for improvement is in the dictionary, that contains a list of words, that are not qualified for correction.

The results of the experiments show that error in the assigning a correction can be made if: a) the given rules and dictionaries contain errors; b) the heuristic rules for recognition of possibly incorrect word fail; c) the estimated observation probability is not sufficiently correct; d) the given language model is not good enough; e) the spell-checking module propose wrong possible corrections. If all of these weak points are implemented in a plausible way, then the result of the human unattended correction process might be acceptable. If the output of the automatic correction is gathered, processed and manually checked by human supervisor, the whole system can be improved by analyzing the result, updating the list of good words and correction rules.

5. Conclusion

This approach can utilize human effort as much as possible and large amounts of text can be corrected in an unsupervised way. Alternatively, this approach can be used in a semi-automatic way, where the system autonomously proposes a correction and human operator can approve it. Decision then can be remembered and used in latter cases. As a result, probability of occurrence of OOV word is reduced and the well corrected training corpus should produce a better results.

This contribution is the first for the Slovak language and thus it can not be directly compared to other approaches. Still it can be used as a starting point for future research and it seems to be valuable for other languages. The presented results can be improved by better detection of the incorrect word and by improving the vocabulary. It also might be useful to try other classification techniques.

Acknowledgment

The research presented in this paper was supported by the Ministry of Education, Science, Research and Sport of the Slovak Republic under the research project MS SR 3928/2010-11 (50 %) and Research and Development Operational Program funded by the ERDF under the project ITMS-26220220141 (50 %).

References

[1] JAYALATHARACHCHI, E., A. WASALA and R. WEERASINGHE. Data-driven spell checking: The synergy of two algorithms for spelling error detection and correction. In: *International Conference on Advances in ICT for Emerging Regions (ICTer), 2012.* Colombo: IEEE, 2012, pp. 7–13. ISBN 978-1-4673-5529-2. DOI: 10.1109/ICTer.2012.6422063.

[2] KUKICH, K. Techniques for automatically correcting words in text. *ACM Computing Surveys.* 1992, vol. 24, iss. 4, pp. 377–439. ISSN 0360-0300. DOI: 10.1145/146370.146380.

[3] GROBBELAAR, L. and D. J. M. KINYUA. A spell checker and corrector for the native South African language, South Sotho. In: *Proceedings of 2009 Annual Conference of the Southern African Computer Lecturers' Association, SACLA 2009.* Mpekweni Beach Resort: ACM, 2009, pp. 50–59. ISBN 978-1-60558-683-0. DOI: 10.1145/1562741.1562747.

[4] SIRTS, K. Noisy-channel spelling correction models for Estonian learner language corpus lemmatisation. In: *The 5th International Conference Human Language Technologies–The Baltic Perspective, HLT 2012.* Tartu: IOS Press, 2012, pp. 213–220. ISBN 978-1-61499-132-8. DOI: 10.3233/978-1-61499-133-5-213.

[5] LI, Y., H. DUAN and C.-X. ZHAI. A generalized hidden Markov model with discriminative training for query spelling correction. In: *Proceedings of the 35th International ACM SIGIR Conference on Research and Development in Information Retrieval, SIGIR 12.* Portland: ACM, 2012, pp. 611–620. ISBN 978-1-4503-1472-5. DOI: 10.1145/2348283.2348365.

[6] LUND, W. B. and E. K. RINGGER. Error correction with in-domain training across multiple OCR system outputs. In: *11th International Conference on Document Analysis and Recognition, ICDAR 2011.* Beijing: IEEE, 2011, pp. 658–662. ISBN 978-1-4577-1350-7. DOI: 10.1109/ICDAR.2011.138.

[7] RODPHON, M., K. SIRIBOON and B. KRUATRACHUE. Thai OCR error correction using token passing algorithm. In: *IEEE Pacific Rim Conference on Communications, Computers and Signal Processing, PACRIM 2001.* Victoria: IEEE, 2001, pp. 599–602. ISBN 0-7803-7080-5. DOI: 10.1109/PACRIM.2001.953704.

[8] ZHOU, Y., S. JING, G. HUANG, S. LIU, and Y. ZHANG. A correcting model based on tribayes for real-word errors in English essays. In: *Fifth International Symposium on Computational Intelligence and Design (ISCID), 2012.* Hangzhou: IEEE, 2012, pp. 407–410. ISBN 978-1-4673-2646-9. DOI: 10.1109/ISCID.2012.108.

[9] HLADEK, D., J. STAS and J. JUHAR. Dagger: The Slovak morphological classiffer. In: *Proceedings of 54th International Symposium ELMAR 2012.* Zadar: IEEE, 2012, pp. 195–198. ISBN 978-1-4673-1243-1.

[10] STAS, J., D. HLADEK, M. PLEVA and J. JUHAR. Slovak language model from internet text data. In: *Proceedings of the Third COST 2102 international training school conference on Toward autonomous, adaptive, and context-aware multimodal interfaces: theoretical and practical issues.* Berlin: Springer, 2011, pp. 340–346. ISBN 978-3-642-18183-2. DOI: 10.1007/978-3-642-18184-9_29.

[11] STAS, J., D. HLADEK and J. JUHAR. Language model adaptation for Slovak LVCSR. In: *Proceedings of the International Conference on Applied Electrical Engineering and Informatics, AEI 2010.* Venice: ACM, 2010, pp. 101–106. ISBN 978-80-553-0519-6.

[12] HLADEK, D. and J. STAS. Text gathering and processing agent for language modeling corpus. In: *12th International Conference on Research in Telecommunication Technologies, RTT 2010.* Velke Losiny: VSB–Technical University of Ostrava, 2010, pp. 137–140. ISBN 978-80-248-2261-7.

[13] KRAJCI, S., M. MATI and R. NOVOTNY. Morphonary: A Slovak language dictionary, tools for acquisition, organisation and presenting of information and knowledge. In: *Proceedings in Informatics and Information Technologies.* Bratislava: STU, 2006, pp. 162–165. ISBN 80-227-2468-8.

About Authors

Daniel HLADEK was born in Kosice, Slovakia in 1982. He received his M.Sc. (Ing.) from Artificial Intelligence in 2007 and Ph.D. in 2009. He is currently working as a post-doctoral researcher at the Department of Electronics and Multimedia Communications, Faculty of Electrical Engineering and Informatics, Technical University of Kosice with a focus on natural language processing, speech and audio processing and intelligent decision methods. He is an author and co-author of more than 30 conference and journal papers from this area.

Jan STAS was born in Bardejov, Slovakia in

1984. He received his M.Sc. from Telecommunications in 2007 and Ph.D. in 2011. He is currently working as a post-doctoral researcher at the Department of Electronics and Multimedia Communications, Faculty of Electrical Engineering and Informatics, Technical University of Kosice with a focus on natural language processing and understanding and language modeling. He is an author and co-author of more than 30 conference and journal papers in this area.

Jozef JUHAR was born in Poproc, Slovakia in 1956. He received his M.Sc. (Ing.) in Radioelectronics from in 1980 and Ph.D. in 1991. He is currently working as a Full Professor at the Department of Electronics and Multimedia Communications, Faculty of Electrical Engineering and Informatics, Technical University of Kosice. He is an author and co-author of more than 220 scientific papers. His research interests include digital speech and audio processing, speech synthesis and development in spoken dialogue and speech recognition systems.

Three Input Single Output Voltage-Mode Multifunction Filter with Independent Control of Pole Frequency and Quality Factor

Witthaya MEKHUM[1], *Winai JAIKLA*[2]

[1]Department of Industrial Management, Faculty of Industrial Technology, Suan Sunandha Rajabhat University, Ratchasima 295, 103 00 Bangkok, Thailand
[2]Department of Engineering Education, Faculty of Industrial Education, King Mongkut's Institute of Technology Ladkrabang, Chalongkrung Road, 105 20 Bangkok, Thailand

mekhum@yahoo.com, kawina@kmitl.ac.th

Abstract. *This paper presents a voltage-mode biquadratic filter performing completely standard functions: low-pass, high-pass, band-pass, band-reject and all-pass functions, based on single voltage differencing transconductance amplifier (VDTA). The proposed filter has three input voltage and a single output voltage. The features of the circuit are that; the quality factor and natural frequency can be tuned independently; the circuit description is very simple, consisting of merely one VDTA, one resistor and two capacitors; the pole frequency can be electronically adjusted. Additionally, each function response can be selected by suitably selecting input signals with digital method; the double input voltage is not required. Using only single active element, the proposed circuit is very suitable to further develop into an integrated circuit. The PSPICE simulation results are depicted. The given results agree well with the theoretical anticipation.*

Keywords

Filter, integrated circuit, VDTA, voltage-mode.

1. Introduction

Filters are important blocks in many integrated circuit applications such as radio frequency applications, data conversion in A/D and D/A systems and many others [1], [2], [3]. One of most popular analog filters is a multiple-input single-output biquadratic filter (MISO) which different output filter functions can be realized simply by different combinations of switching on or off the input currents where the selection can be done digitally using a microcontroller.

The synthesis and design of analog filters using modern electronically controllable active building blocks (ABBs) give flexibility and convenience for designer. These filters can be easily controlled by microcomputer or microcontroller. Also some filter circuits which use active building block can avoid the use of the external resistors. This will reduce the cost and chip area. Yesil et al [4] proposed several circuit ideas of building blocks for voltage-, current- and mixed mode applications. One of them is the voltage differencing transconductance amplifier (VDTA). This device allows applications with interesting features, especially those providing the electronic controllability.

Numerous techniques using active building block for designing the MISO voltage-mode filters have been developed in the past two decades [5], [6], [7], [8], [9], [10], [11], [12], [13], [14], [15], [16], [17], [18], [19], [20], [21], [22], [23], [24], [25], [26], [27], [28], [29], [30], [31]. These MISO filters are based on current conveyor (CCII) [5], [6], [7], [8], [9], [10], [11], current feedback amplifier (CFA) [12], [13], [14], differential difference current conveyor (DDCC) [15], [16], operational transconductance amplifier (OTA) [17], [18], fully differential current conveyor (FDCCII) [19], [20], [21], operational transresistance amplifier (OTRA) [22], current conveyor transconductance amplifier (CCTA) [23], current follower tansconductance amplifier (CFTA) [24], dual X second-generation current conveyor (DXCCII) [25], current controlled CCII (CCCII) [26], differential voltage current conveyor (DVCC) [27], [28], current differencing buffered amplifier (CDBA) [29], and voltage differencing differential input buffered amplifier (VD-DIBA) [30] and VDTA [31]. The advantages of the proposed MISO voltage-mode filter are compared with several previous MISO voltage-mode filters which the results are shown in Tab. 1.

Tab. 1: Comparsion between various MISO voltage-mode filters.

Ref	ABB	No. of ABB	No. of R+C	Electronic control	Independent tune of ω_0 and Q	No need of double input	No need of matching condition
[5]	CCII	2	3+2	No	Yes	Yes	Yes
[6]	CCII	3	5+2	No	Yes	Yes	Yes
[7]	CCII	2	2+2	No	No	Yes	Yes
[8]	CCII	3	3+3	No	No	Yes	No
[9]	CCII	2	2+2	No	No	Yes	Yes
[10]	CCII	2	3+2	No	Yes	Yes	No
[11]	CCII	3	2+2	No	No	Yes	No
[12]	CFA	1	3+2	No	No	Yes	Yes
[13]	CFA	1	3+2	No	Yes	Yes	No
[14]	CFA	1	2+2	No	No	Yes	Yes
[15]	DDCC	2	2+2	No	No	Yes	Yes
[16]	DDCC	3	2+2	No	No	Yes	Yes
[17]	OTA	2	0+2	Yes	No	Yes	Yes
[18]	OTA	6	0+2	Yes	No	Yes	Yes
[19]	FDCCII	1	3+2	No	Yes	Yes	Yes
[20]	FDCCII	2	2+2	No	No	Yes	Yes
[21]	FDCCII	2	2+2	No	No	Yes	Yes
[22]	OTRA	1	4+4	No	No	Yes	No
[23]	CCTA	1	2+2	Yes	Yes	Yes	Yes
[24]	CFTA	1	2+2	Yes	No	Yes	No
[25]	DXCCII	1	4+2	No	No	Yes	No
[26]	CCCII	2	0+2	Yes	No	Yes	Yes
[27]	DVCC	1	2+2	Yes	No	Yes	No
[28]	DVCC	3	4+2	No	Yes	Yes	No
[29]	CDBA	2	4+2	No	Yes	Yes	Yes
[30]	VD-DIBA	1	1+2	Yes	No	Yes	Yes
[31]	VDTA	1	0+2	Yes	No	No	No
Proposed filter	VDTA	1	1+2	Yes	Yes	Yes	Yes

The aim of this paper is to propose a MISO voltage-mode filter, emphasizing on the use of single voltage differencing transconductance amplifier (VDTA) as the active element. The features of the proposed circuit are that:

- The natural frequency and the quality factor are orthogonally tunable.
- Obtains a minimum number of components required to achieve a second-order transfer function; low pass (LP), high pass (HP), band pass (BP), band stop (BS) and all pass (AP) functions are obtained without any change in the filter topology.
- Passive and active elements enjoy low sensitivities.
- No need of matching condition of passive elements.
- No need of double input voltage to achieve all pass function.

2. Principle of Operation

There are two topics in this section as follows:

2.1. VDTA Overview

The circuit symbol of VDTA is shown in Fig. 1, where V_P and V_N are the input terminals, Z and X are the output ones. Hence, Z is the current output terminal; current through Z terminal follows the difference of the voltages at V_P and V_N terminals by transconductances g_{m1}. The voltage V_Z on Z terminal is transferred into current using transconductance g_{m2}, which flows into output terminal X. The g_{m1} and g_{m2} are tuned by I_{B1} and I_{B2}, respectively. In general, VDTA can contain an arbitrary number of x terminals, providing currents IX of both directions. All terminals of VDTA exhibit high impedance values [4]. The characteristics of VDTA are represented by the following hybrid matrix:

$$\begin{bmatrix} I_Z \\ I_{X+} \\ I_{X-} \end{bmatrix} = \begin{bmatrix} g_{m1} & -g_{m1} & 0 \\ 0 & 0 & g_{m2} \\ 0 & 0 & -g_{m2} \end{bmatrix} \begin{bmatrix} V_P \\ V_N \\ V_Z \end{bmatrix}. \quad (1)$$

If the VDTA is realized using CMOS technology, gm1 and gm2 can be respectively written as

$$g_{m1} = \sqrt{kI_{B1}}, \quad (2)$$

and

$$g_{m2} = \sqrt{kI_{B2}}. \quad (3)$$

Here k is the physical transconductance parameter of the CMOS transistor. I_{B1} and I_{B2} are the bias current used to control the g_{m1} and g_{m2}, respectively. The internal construction of CMOS VDTA is shown in Fig. 2.

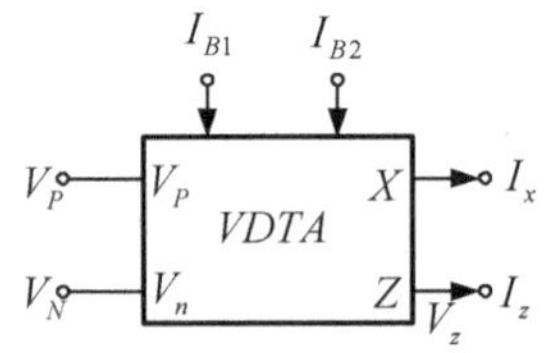

Fig. 1: Circuit symbol of VDTA.

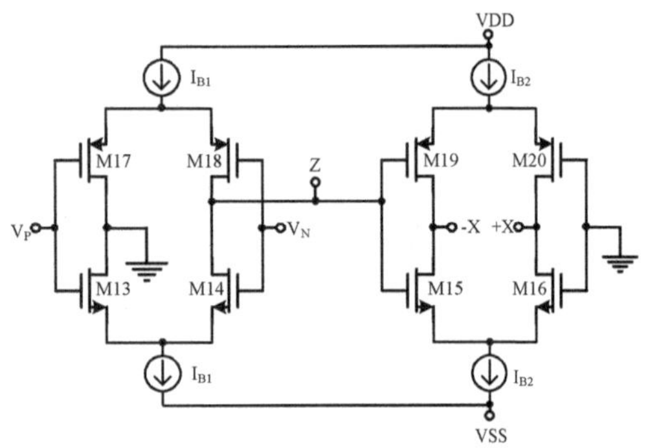

Fig. 2: Internal construction of VDTA.

2.2. Proposed MISO Voltage-Mode Filter

The proposed biquad filter is illustrated in Fig. 3. It consists of one VDTA, one resistor and two capacitors. Considering the circuit in Fig. 3 and using VDTA properties in section 2.1, the output voltage of the proposed filter can be written as

$$V_o = \frac{s^2C_1C_2V_1 + sC_1GV_2 + g_{m1}g_{m2}V_3}{s^2C_1C_2 + sC_1G + g_{m1}g_{m2}}, \tag{4}$$

where G=1/R. The following relations are valid for the pole frequency and the quality factor:

$$\omega_0 = \sqrt{\frac{g_{m1}g_{m2}}{C_1C_2}}, \tag{5}$$

and

$$Q = R\sqrt{\frac{g_{m1}g_{m2}C_2}{C_1}}. \tag{6}$$

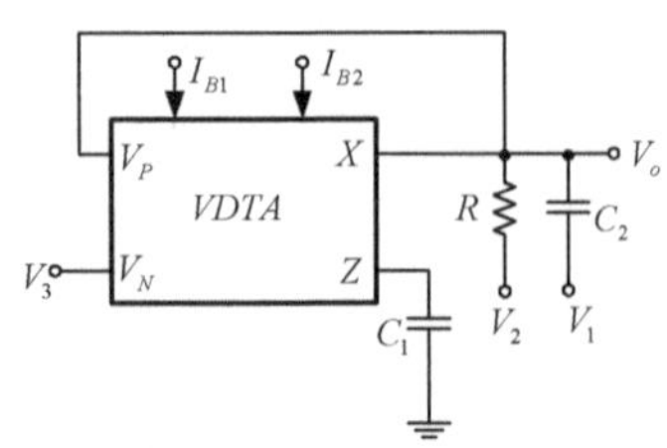

Fig. 3: Proposed filter.

From Eq. 5 and Eq. 6, if the g_{m1} and g_{m2} are equal to Eq. 2 and Eq. 3, the pole frequency and quality factor are re-written as

$$\omega_0 = \sqrt{\frac{k(I_{B1}I_{B2})^{\frac{1}{2}}}{C_1C_2}}, \tag{7}$$

and

$$Q = R\sqrt{\frac{C_2k(I_{B1}I_{B2})^{\frac{1}{2}}}{C_1}}. \tag{8}$$

It is apparent from Eq. 8 that the quality factor could be controlled by R without affecting the pole frequency. Also the pole frequency can be electronically tuned by I_{B1} and I_{B2}. It should remark that the parameter k is proportional to mobility and mobility falls with increasing temperature. Then the temperature variation will affect the ω_0 and Q_0.

Depending on the voltage status of V_1, V_2, and V_3 in the numerator of Eq. 4, one of the following five filter functions is obtained.

- If V_1=V_2=0 (grounded), and V_3=V_{in} (connected to the input voltage source), a second-order low pass filter can be obtained.
- If V_2=V_3=0 (grounded) and V_1=V_{in} (connected to the input voltage source) a second-order high pass filter can be obtained.
- If V_1=V_3=0 (grounded) and V_2=V_{in} (connected to the input voltage source), a second-order inverting band pass filter can be obtained.
- If V_3=0 and V_1=V_2=V_{in}, a second-order notch filter can be obtained.
- If V_1=V_{-2}=V_3=V_{in} a second-order all pass filter can be obtained.

From above conditions, the filter respond selections can be shown in Tab. 2, where number 1 is the magnitude of input voltage and it also represent the digital logic that means the proposed filter can be digitally controlled. It should be remarked that for all pass response, the inverting unit gain amplifier is required. The ω_0-sensitivity analysis with respect to the parameters of the active and passive element used can be given by:

$$S^{\omega_0}_{C_1,C_2} = -\frac{1}{2}, S^{\omega_0}_{g_{m1},g_{m2}} = \frac{1}{2}. \tag{9}$$

3. Analysis of Non-Ideal Case

In practice, the influences of voltage and current tracking errors and also the parasitic terminal impedances

$$\begin{bmatrix} I_Z \\ I_{X+} \\ I_{X-} \end{bmatrix} = \begin{bmatrix} \alpha_P g_{m1} & \alpha_N g_{m1} & 0 \\ 0 & 0 & \beta g_{m2} \\ 0 & 0 & -\beta g_{m2} \end{bmatrix} \begin{bmatrix} V_P \\ V_N \\ V_Z \end{bmatrix}. \tag{10}$$

$$V_o = \frac{\alpha_N \beta g_{m2} g_{m1} V_3 + (sC_1^* + G_Z)GV_2 + (sC_1^* + G_Z)sC_1V_2}{[s^2(C_e^*C_1^* + C_1^*C_2) + s(C_1^*G_e^* + G_ZC_e^* + GC_1^* + C_2G_Z) + G_ZG_e^* + GG_Z + \alpha_P\beta g_{m1}g_{m2}]}. \tag{11}$$

of VDTA will affect the filter performance. In this Section, these parameters will be taken into account. For non-ideal voltage buffer, the VDTA can be respectively characterized with the Eq. 10, where α_P and α_N are the transconductance error gains from P and N ports to z port. β is the transconductance error gain from z port to x port. The influences of parasitic impedances are resistive and capacitive parts affecting the P, N, Z and X ports of VDTA. Let us denote them RP, CP, RN, CN, Rz, CZ, and RX, CX, respectively.

Considering into these effects, the output voltage will be modified to the more general forms in the Eq. 11, where $C_1^* = C_1 + C_Z$, $C_e^* = C_X + C_P$, and $G_e^* = G_X + G_P$. In this case, the pole frequency and quality factor is modified to

$$\omega_0^* = \sqrt{\frac{G_ZG_e^* + GG_Z + \alpha_P\beta g_{m2}g_{m1}}{C_e^*C_1^* + C_1^*C_2}}, \tag{12}$$

and

$$Q_0^* = \frac{C_e^*C_1^* + C_1^*C_2}{C_1^*G_e^* + G_ZC_e^* + GC_1^* + C_2G_Z} \cdot \sqrt{\frac{G_ZG_e^* + GG_Z + \alpha_P\beta g_{m2}g_{m1}}{C_e^*C_1^* + C_1^*C_2}}. \tag{13}$$

It should be mentioned that the stray/parasitic z-terminal capacitances are absorbed by C_1 as it appears in shunt with them. However, the parasitic resistance R_z and capacitance C_z not only affect the pole frequency and Q but they also add parasitic zeros to all transfer functions (except LP). The parameters α_P, α_N and β of the VDTA affect the gain of low pass filter.

4. Simulation Results

To prove the performances of the proposed filter, the PSPICE simulation program was used for the examination. Internal construction of VDTA used in the simulation is from [4]. The PMOS and NMOS transistors have been simulated by respectively using the parameters of a 0,25 μm TSMC CMOS technology [32]. The aspect ratios of PMOS and NMOS transistor are W/L = 8 μm/0,25 μm and W/L = 5 μm/0,25 μm, respectively. The circuit was biased with ±1, 25 V supply voltages, $C_1 = C_2 = 20$ pF, $I_{B1} = I_{B2} = 40$ μA, R = 4,7 kΩ. It yields the natural frequency of 2,376 MHz. The results shown in Fig. 4 are the gain and phase responses of the proposed biquad filter from Fig. 3.

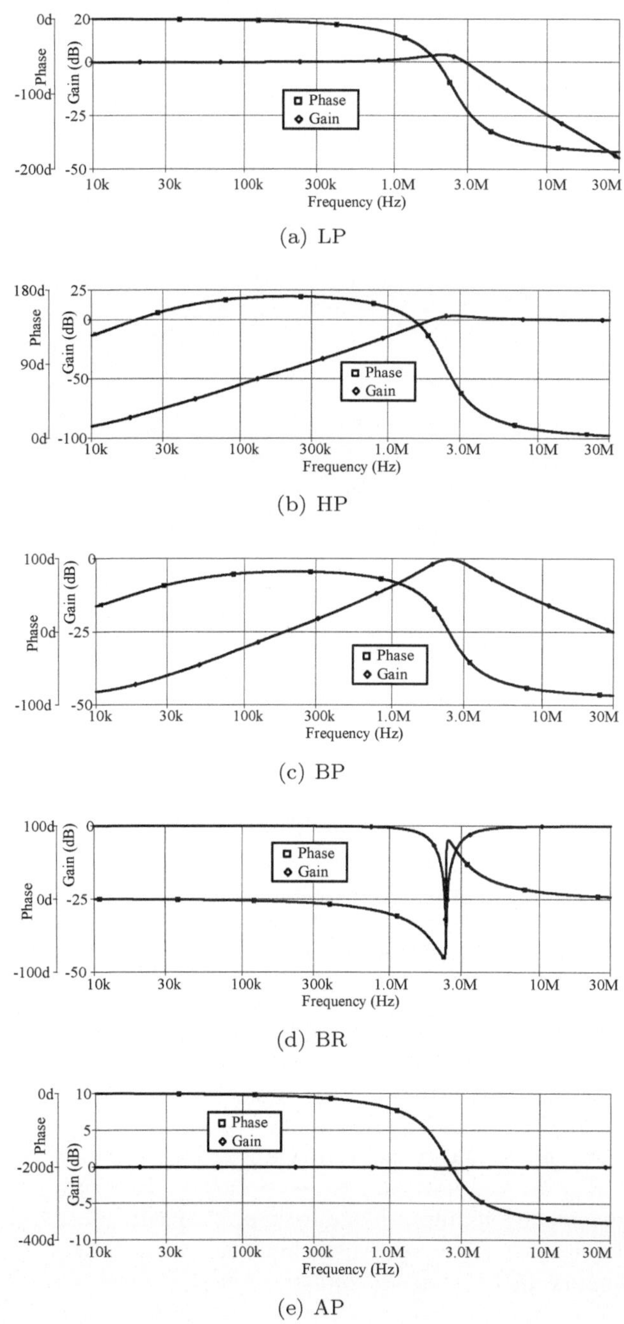

Fig. 4: Gain and phase responses of proposal VM biquad filter.

There are seen that the proposed filter can provide low-pass, high-pass, band-pass, band-reject and all-pass functions dependent on selection as shown in Tab. 2, without modifying circuit topology. Figure 5 shows gain responses of band-pass function where R is set for several values. It is found that quality factor can be adjusted without affecting pole frequency as shown in Eq. 6. The transient response of the proposed filter from band-pass function for the center frequency of 2,376 MHz can be seen in Fig. 6. The total harmonic distortion (THD) is about 0,0786 %.

Tab. 2: The V_1, V_2 and V_3 values selection for each filter function response.

Filter Responses	Input selections		
	V_1	V_2	V_3
LP	0	0	1
HP	1	0	0
BP	0	1	0
BR	1	0	1
AP	1	-1	1

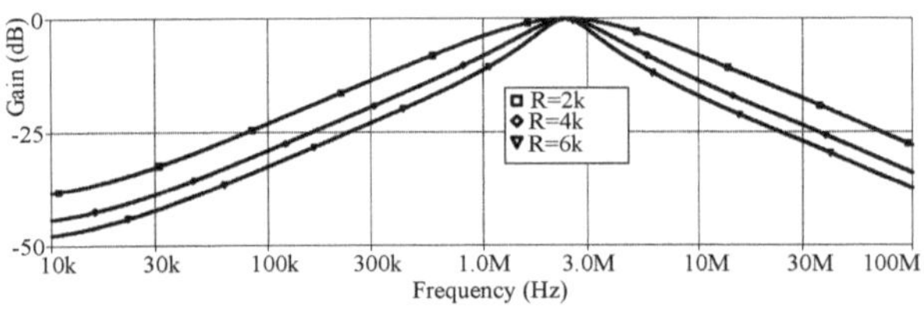

Fig. 5: Voltage-mode band-pass response for different values of R.

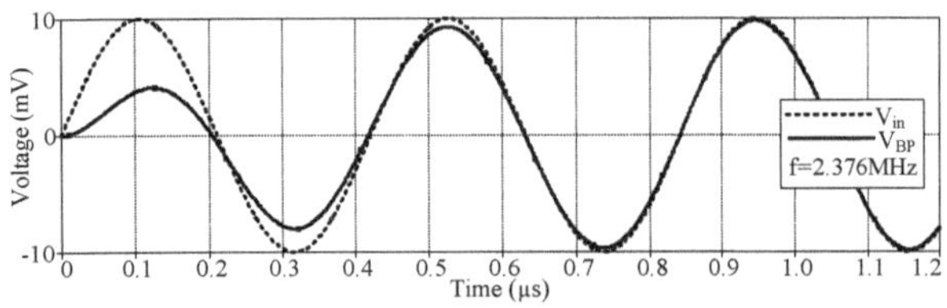

Fig. 6: Time domain response for BP filter.

5. Conclusion

The multiple-inputs single-output biquadratic filter based on single VDTA is presented. The advantages of the proposed circuit are that: it performs low-pass, high-pass, band-pass, band-reject and all-pass functions dependent on an appropriate selection of three input signals: the natural frequency can be electronically controlled via input bias currents, it is easily modified to use in control systems using a microcontroller: the quality factor can be adjusted without affecting natural frequency. The circuit description comprises only one VDTA, one resistor two capacitors. With mentioned features, it is very suitable to realize the proposed circuit in monolithic chip to use in battery-powered, portable electronic equipment such as wireless communication system devices.

References

[1] SMITH, K. C. and A. S. SEDRA. *Microelectronic Circuits: 3rd edition.* Florida: Holt, Rinehart and Wiston, 1991. ISBN 0195116631.

[2] IBRAHIM, M. A., S. MINAEI and H. A. KUNTMAN. A 22.5 MHz current-mode KHN-biquad using differential voltage current conveyor and grounded passive elements. *International Journal of Electronics and Communication (AEU).* 2005, vol. 59, iss. 5, pp. 311–318. ISSN 1434-8411. DOI: 10.1016/j.aeue.2004.11.027.

[3] PANDEY, R., N. PANDEY, S. K. PAUL and A. SINGH. Voltage mode OTRA MOS-C single input multi output biquadratic universal filter. *Advances in Electrical and Electronic Engineering.* 2012, vol. 10, no. 5, pp. 337–344. ISSN 1804-3119.

[4] YESIL, A., F. KACAR and H. KUNTMAN. New simple CMOS realization of voltage differencing transconductance amplifier and its RF filter application. *Radioengineering.* 2011, vol. 20, no. 3, pp. 632–637. ISSN 1805-9600.

[5] HORNG, J. W., C. C. TSAI and M. H. LEE. Novel universal voltage-mode biquad filter with three inputs and one output using only two current conveyors. *International Journal of Electronics.* 1996, vol. 80, iss. 4, pp. 543–546. ISSN 1362-3060. DOI: 10.1080/002072196137174.

[6] CHANG, C. M. Multifunction biquadratic filters using current conveyors. *IEEE Transactions on Circuits and Systems II: Analog and Digital Signal Processing.* 1997, vol. 44, iss. 11, pp. 956–958. ISSN 1057-7130. DOI: 10.1109/82.644049.

[7] HORNG, J. W., M. H. LEE, H. C. CHENG and C. W. CHANG. New CCII-based voltage-mode universal biquadratic filter. *International Journal of Electronics.* 1997, vol. 82, iss. 2, pp. 151–155. ISSN 1362-3060. DOI: 10.1080/002072197136156.

[8] HORNG, J. W. High input impedance voltage-mode universal biquadratic filters with three inputs using plus-type CCIIs. *International Journal of Electronics.* 2004, vol. 91, iss. 8, pp. 465–475. ISSN 1362-3060. DOI: 10.1080/00207210412331294612.

[9] HORNG, J. W. Voltage-mode universal biquadratic filters using CCIIs. *IEICE TRANSACTIONS on Fundamentals of Electronics, Commu-*

nications and Computer Sciences. 2004, vol. 87A, no. 2, pp. 406–409. ISSN 1745-1337.

[10] CHANG, C. M. and S. H. TU. Universal voltage-mode filter with four inputs and one output using two CCII+s. *International Journal of Electronics.* 1999, vol. 86, iss. 3, pp. 305–309. ISSN 1362-3060. DOI: 10.1080/002072199133445.

[11] HORNG, J. W. High-input impedance voltage-mode universal biquadratic filter using three plus-type CCIIs. *IEEE Transactions on Circuits and Systems II: Analog and Digital Signal Processing.* 2001, vol. 48, iss. 10, pp. 996–997. ISSN 1057-7130. DOI: 10.1109/82.974791.

[12] HOU, C. L., C. C. HUANG, Y. S. LAN, J. J. SHAW and C. M. CHANG. Current-mode and voltage-mode universal biquads using a single current-feedback amplifier. *International Journal of Electronics.* 1999, vol. 86, iss. 8, pp. 929–932. ISSN 1362-3060. DOI: 10.1080/002072199132923.

[13] HORNG, J. W., C. K. CHANG and J. M. CHU. Voltage-mode universal biquadratic filter using single current-feedback amplifier. *IEICE TRANSACTIONS on Fundamentals of Electronics, Communications and Computer Sciences.* 2002, vol. 85A, no. 8, pp. 1970–1973. ISSN 1745-1337.

[14] SHAH, N. A., M. F. RATHER and S. Z. IQBAL. A novel voltage-mode universal filter using a single CFA. *Journal of Active and Passive Electronic Devices.* 2005, vol. 1, no. 2, pp. 183–188. ISSN 1555-029X.

[15] CHANG, C. M. and H. P. CHEN. Universal capacitor-grounded voltage-mode filter with three inputs and a single output. *International Journal of Electronics.* 2003, vol. 90, iss. 6, pp. 401–406. ISSN 1362-3060. DOI: 10.1080/00207210310001612068.

[16] HORNG, J. W. High-input and low-output impedance voltage-mode universal biquadratic filter using DDCCs. *IEEE Transactions on Circuits and Systems II: Express Briefs.* 2007, vol. 54, no. 8, pp. 649–652. ISSN 1549-7747. DOI: 10.1109/TCSII.2007.899460.

[17] HORNG, J. W. Voltage-mode universal biquadratic filter using two OTAs and two capacitors. *IEICE TRANSACTIONS on Fundamentals of Electronics, Communications and Computer Sciences.* 2003, vol. 86A, no. 2, pp. 411–413. ISSN 1745-1337.

[18] KUMNGERN, M., M. SOMDUNYAKANOK and P. PROMMEE. High-input impedance voltage-mode multifunction filter with three-Input single-output based on simple CMOS OTAs. In: *International Symposium on Communications and Information Technologies, 2008. ISCIT 2008.* Lao: IEEE, 2008. pp. 426–431. E-ISBN 978-1-4244-2336-1. DOI: 10.1109/ISCIT.2008.4700228.

[19] CHANG, C. M. and H. P. CHEN. Single FDCCII-based tunable universal voltage-mode filter. *Circuits, Systems and Signal Processing.* 2005, vol. 24, no. 2, pp. 221–227. ISSN 1531-5878. DOI: 10.1007/s00034-004-0422-7.

[20] CHEN, H. P. and Y. Z. LIAO. High-input and low-output Impedance voltage-mode universal biquadratic filter using FDCCIIs. In: *9th International Conference on Solid-State and Integrated-Circuit Technology, 2008. ICSICT 2008.* Beijing: IEEE, 2008. pp. 1794–1798. E-ISBN 978-1-4244-2186-2. DOI: 10.1109/ICSICT.2008.4734904.

[21] CHEN, H. P. Voltage-mode FDCCII-based universal filters. *International Journal of Electronics and Communication (AEU).* 2009, vol. 63, iss. 4, pp. 320–323. ISSN 1434-8411. DOI: 10.1016/j.aeue.2007.05.002.

[22] KILINC, S., A. U. KESKIN and U. CAM. Cascadable voltage-mode multifunction biquad employing single OTRA. *Frequenz.* 2007, vol. 61, iss. 3-4, pp. 84–86. ISSN 2191-6349. DOI: 10.1515/FREQ.2007.61.3-4.84.

[23] HERENCSAR, N., J. KOTON and K. VRBA. Single CCTA–based universal biquadratic filters employing minimum components. *International Journal of Computer and Electrical Engineering.* 2009, vol. 1, no. 3, pp. 307–310. ISSN 1793-8163.

[24] TANGSRIRAT, W. Novel current-mode and voltage-mode universal biquad filters using single CFTA. *Indian Journal of Engineering and Materials Sciences.* 2012, vol. 17, no. 2, pp. 99–104. ISSN 0971-4588.

[25] MYDERRIZI, I., S. MINAEI and E. YUCE. DXCCII-based grounded inductance simulators and filter applications. *Microelectronics Journal.* 2011, vol. 42, iss. 9, pp. 1074–1081. ISSN 0026-2692. DOI: 10.1016/j.mejo.2011.06.008.

[26] RANJAN, A. and S. K. PAUL. Voltage mode universal biquad using CCCII. *Active and Passive Electronic Components.* 2011, vol. 2011, no. 1, pp. 1–5. ISSN 1563-5031. DOI: 10.1155/2011/439052.

[27] TANGSRIRAT, W. and O. CHANNUMSIN. Voltage-mode multifunctional biquadratic filter

using single DVCC and minimum number of passive elements. *Indian Journal of Pure & Applied Physics*. 2012, vol. 49, no. 10, pp. 703–707. ISSN 0019-5596.

[28] HORNG, J. W., C. H. HSU and C. Y. TSENG. High input impedance voltage-mode universal biquadratic filters with three inputs using three CCs and grounding capacitor. *Radioengineering*. 2012, vol. 21, no. 1, pp. 290–296. ISSN 1805-9600.

[29] PATHAK, J. K., A. K. SINGH and R. SENANI. New voltage mode universal filters using only two CDBAs. *ISRN Electronics*. 2013, vol. 2013, no. 1, pp. 1–6. ISSN 2090-8679.

[30] PUSHKAR, K. L., D. R. BHASKAR and D. PRASAD. A new MISO-type voltage-mode universal biquad using single VD-DIBA. *ISRN Electronics*. 2013, vol. 2013, no. 1, pp. 1–5. ISSN 2090-8679.

[31] SATANSUP, J. and W. TANGSRIRAT. Single VDTA-based voltage-mode electronically tunable universal filter. In: *The 27th International Technical Conference on Circuits/Systems, Computers and Communications (ITC-CSCC 2012)*. Sapporo: ITC-CSCC, 2012, pp. 1–4. ISBN 978-4-88552-272-7.

[32] PROMMEE, P., K. ANGKEAW, M. SOMDUNYAKANOK and K. DEJHAN. CMOS-based near zero-offset multiple inputs max–min circuits and its applications. *Analog Integrated Circuits and Signal Processing*. 2009, vol. 61, iss. 1, pp. 93–105. ISSN 1573-1979. DOI: 10.1007/s10470-009-9281-2.

About Authors

Witthaya MEKHUM was born in Nonthaburi, Thailand. He received Ph.D. in Technology Management from Phranakhon Rajabhat University in 2009. He has been with Department of Industrial Management, Faculty of Industrial Technology, Suan Sunandha Rajabhat University, Bangkok, Thailand since 1995. His research interests include technology management and communication system.

Winai JAIKLA was born in Buriram, Thailand. He received the B. S. I. Ed. degree in telecommunication engineering from King Mongkut's Institute of Technology Ladkrabang (KMITL), Thailand in 2002, M. Tech. Ed. in electrical technology and Ph.D. in electrical education from King Mongkut's University of Technology North Bangkok (KMUTNB) in 2004 and 2010, respectively. From 2004 to 2011 he was with Electric and Electronic Program, Faculty of Industrial Technology, Suan Sunandha Rajabhat University, Bangkok, Thailand. He has been with Department of Engineering Education, Faculty of Industrial Education, King Mongkut's Institute of Technology Ladkrabang, Bangkok, Thailand since 2012. His research interests include electronic communications, analog signal processing and analog integrated circuit. He is a member of ECTI, Thailand.

12

Power Quality Analysis in Off-Grid Power Platform

Stanislav MISAK, Jindrich STUCHLY, Jakub VRAMBA, Lukas PROKOP, Marian UHER

Department of Electrical Power Engineering, Faculty of Electrical Engineering and Computer Science, VSB–Technical university of Ostrava, 17. listopadu 15, 708 33 Ostrava-Poruba, Czech Republic

stanislav.misak@vsb.cz, jindrich.stuchly.st@vsb.cz, jakub.vramba.st@vsb.cz, lukas.prokop@vsb.cz, marian.uher@vsb.cz

Abstract. *Research projects in the field of electrical distribution systems are moving to a new philosophy of Smart Grids, where the effort is to use the maximum possible share of power from renewable energy potential. Under this philosophy the emphasis is on energy independence, reliability and safety of operation of energy distribution system. Research in this area leads for example to developing of autonomous local micro-grids with the several specific requirements. However, the problem of parameters keeping of quality of electric energy can arise together with increased penetration of distributed generation in microgrids. This problem is caused by decreased short-circuit power of local renewable energy sources, stochastic supply of electric energy from renewable energy sources and operating of active distribution grid in autonomous mode without connection to the external distribution system. General introduction to the power quality evaluation in off-grid power system is introduced in this paper. Initial results from power quality analysis from small off-grids system is presented in this text too.*

Keywords

Battery tank, off-grid system, power quality, renewable energy sources.

1. Introduction

This paper introduces the initial evaluation of power quality (PQ) of off-grid power system, which was built as research platform for renewable energy sources (RES) testing. Off-grid system was sized as a model for usual residential supply [1].

The main idea of this concept is to find links or relations between environmental (solar irradiance and wind speed) and PQ variables associated with solar and wind energy. Based on these links prediction of PQ for off-grid power system will be introduced to optimizing PQ level in off-grid power system.

In particular, this contribution concentrates on the relationships between environmental (solar irradiance and wind speed) and power quality (PQ) quantities.

The paper is organized in four main sections. Section 2 provides a brief introduction to issues of power quality in general and problems specific to RES power quality. It also brings an overview of relevant technical standards. Section 3 provides a review of work on the PQ evaluation related to off-grid power systems. Section 4 describes the overall structure of off-grid power system and also describes individual components. Section 5 describes the PQ evaluation and some of the partial results made from this data. Conclusions are presented in Section 6.

2. Background

Power quality can be explained as the fitness of consumed electrical power to user appliances and devices. Good power quality allows electrical systems to function in their intended manner without significant loss of performance or life. Without the proper power quality, an appliances (or load) may result in appliance damage or operate stoppages at all. There are many ways in which electric power can be of poor quality and many more causes of such poor quality power.

With the RES installation for off-grid power system supply, their impact on the power quality should be significant. Voltage fluctuation caused by variability of photovoltaic (PV) and wind power generation is one of the most important issues. Another significant problem is harmonic distortion, since both types of generation plants are usually connected to the power system through inverters. Wiring diagram of described off-grid power system together with individual system components are described in section 4. More detailed characterization and overall off-grid power system description is widely introduced in [1], [2].

2.1. Power Quality

It is not necessary to reach power quality levels defined by several international standards in off-grid power systems, but only good power quality allows electrical devices to function in their given manner without loss of performance or possible damage.

2.2. Standards and Rules for PQ

Usually in grid-connected power system are PQ levels defined and maintained by several international technical standard organizations. The well known are as follows IEC (Int. Electrotechnical Commission) and the IEEE (Institute of Electrical and Electronics Engineers). In off-grid applications where PQ limits are not defined by common standards we can use for the PQ evaluation limits defined for grid-connected power systems.

1) European Standards

IEC61000-4-30 (2008) "Testing and measurement techniques – Power quality measurement methods" is part of the series IEC61000 "Electromagnetic compatibility". It defines the methods of measurement and interpretation of results for power quality parameters in 50/6 0Hz AC power supply systems. The measurement methods describe, for each type of parameter, how to obtain reliable, repeatable and comparable results regardless the compliant instrument used for measurement and its environmental conditions. The standard defines three measurement classes A, S, B. Class A is used where precise measurement are necessary, e.g. for contractual application to resolve disputes. The processing requirements and instrument prices for class S are lower. Class B is obsolete. It has the weakest requirements, and it is not recommended for new designs. The standard also contains links to two other standards: IEC61000-4-7 for measurement of voltage harmonics and IEC61000-4-15 for measurement of flicker.

PQ analyzers with functionality defined in IEC61000-4-30 provide periodical data that has to be statistically processed. European standard EN 50160 describes how should be the measured data processed. It describes how to determine the main voltage characteristics of electricity supplied by public networks. It defines the "electrical product" for high, medium and low voltage systems in terms of the characteristics at the supply terminals.

2) IEEE Standards

IEEE Std 1159 (2009) "Recommended Practice for Monitoring Electric Power Quality" is maintained by the Working Group for Power Quality Monitoring, which was formed in response to the need for standardizing the rapidly expanding power quality monitoring manufacturing industry and the field use of their products by utilities and end-users. The standard introduces consistent terminology and definitions, and discusses power quality monitoring devices, application techniques, and the interpretation of monitoring results. It includes descriptions of conducted electromagnetic phenomena occurring in single and poly-phase AC power systems, and encompasses the monitoring of electrical characteristics of these systems. The standard describes nominal conditions and deviations from these conditions that may originate within the source of supply, load equipment, or from the interactions between the source and the load.

2.3. Enforcement of Standards

Both the IEEE1159 and IEC61000 series of standards cover issues of power quality. They agree on compatibility, the basic concepts and terminology. However, IEEE 1599 is not enforced; it is an informative and instructive tutorial developed by volunteers and approved by consensus. Conversely, the IEC 61000 series of standards has been developed by a group of assigned national experts through careful drafting before being approved by national voting. It is enforced at the national level by many countries.

3. Related Work

Guidelines for the measurement and assessment of power quality characteristics for the photovoltaic system are described in [10]. Harmonics, flicker, unbalance and slow voltage variations were included in these guidelines [10]. Power quality analyses for grid-connected photovoltaic plants were introduced in various papers [5], [6], [7]; also general PQ issues associated with grid-connected PV systems are discussed, for example, in [3], [10].

Power quality analyses for off-grid-connected power system were discussed only in a few papers. In [9] there were analyzed the main factors of the impact of fluctuations of wind power from the off-grid wind power systems and energy storage technology to mitigate the off-grid wind turbine power fluctuations. Smart control system for standalone and grid-connected PV systems were introduced in [4]. Also, fluctuation of total harmonics distortion (THD) in islanding mode was shortly evaluated. The THD during islanding mode of operation was around 6 % [4].

Combined photovoltaic and unified power quality controller was described in [8]. The main issue of the controller is to eliminate grid-end current harmonics caused by the distorted-unbalanced load-terminal voltages.

However, none of these papers provides overall PQ analysis of whole hybrid power system with RES. In this paper we try to find a relationship between environmental, electrical and PQ variables. The main goal is to find interconnections between environmental variables like solar irradiance or wind speed and PQ variables and arrange the basis for the PQ variables prediction for off-grid power systems.

4. System Configuration

The developed off-grid power system was designed according requirements of usual consumption of family house, whereas the mentioned system has to satisfy features of smart grid system. Basic features of developed smart grid system are:

- independence on the delivering of energy from external power grids,
- operation of the grid system with well-balanced between production and consumption of electric energy,
- energy storage possibility,
- using mainly renewable energy sources of electric energy,
- non-traditional character of load,
- new conception of protection with bi-directional power flow,
- active energy management system with supporting of methods of artificial intelligence.

The off-grid power system is created by three basic parts. The first part, part of sources, contains two types of renewable energy sources, namely wind power plant and photovoltaic system. These types of renewable energy sources were selected purposely, because mutual combination of their running ensures energy supply during all year. It is given by character of individual sources, where the electric energy is mainly supplied by wind power plant for the season of spring and winter and by photovoltaic system for the season of summer and autumn.

Only one of these sources is not applicable for specific meteorological condition of Czech Republic, where the system is installed. The block scheme of off-grid power system is introduced in Fig. 1.

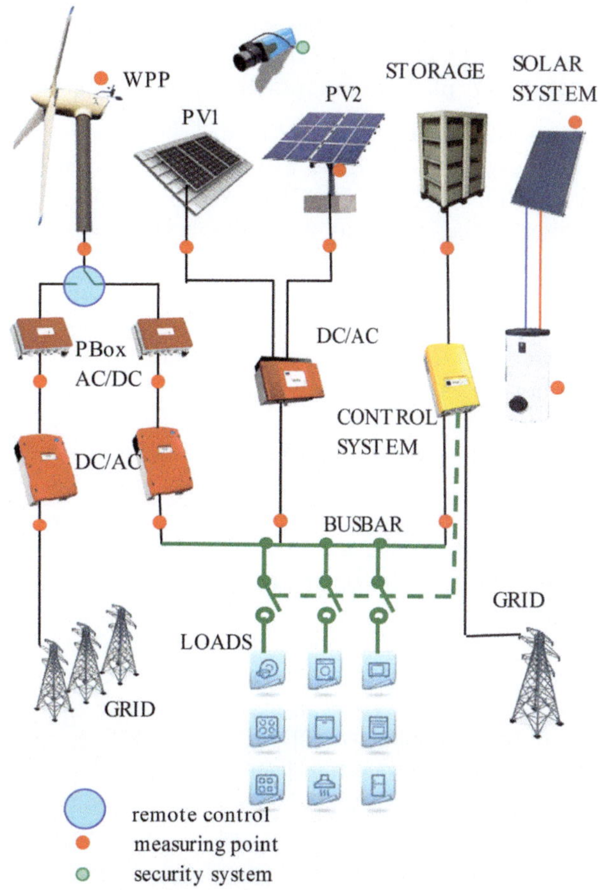

Fig. 1: The block scheme of off-grid power system.

The first source of electric energy is formed by the wind power plant (WPP) equipped with a synchronous generator with permanent magnets and the total installed capacity of 12 kVA. The synchronous generator is connected through a protection box (PBox) with controlled (AC/DC) converter and frequency converter (DC/AC) with a recuperative unit to a 1 phase AC (BUSBAR) with 230 V level and 50 Hz frequency. The photovoltaic power plant (PV1) with polycrystalline technology and the total installed output of 2 kWp is used as another renewable electric power source. The electric output from the photovoltaic system is also connected through a frequency converter (DC/AC) to the common 1 phase AC BUSBAR as a part of the transmission. The PV1 with polycrystalline technology has placed in the stable construction. The second photovoltaic system (PV2) has monocrystalline technology, identical installed output as PV1, but individual photovoltaic panels are placed on the tracker for its better using and higher efficiency of energy conversion.

Individual (LOADS) are supplied from BUSBAR, and their individual connections are controlled according to the condition of the actual state of storage device, connection priorities, character of the current surge of load and the predicted value of available power from the wind and photovoltaic power plants.

Connecting of the loads according to the requirements mentioned above is done using the smart control system (CONTROL SYSTEM), which also controls charging of the storage device (STORAGE) with the total capacity of 840 Ah and voltage 48 V DC (third part of off-grid power system). This storage unit with lead batteries balances the power and thus accumulates electric power in the case of surplus of electric power from the photovoltaic and wind power plants. This storage unit works as a backup source for synchronization of individuals power components in the case of direct supply of individual loads from the 1phase BUSBAR. The weather station is part of off-grid power system conception for definition relations between meteorological, electrical and mechanical values. Wind velocity, wind direction, temperature, relative humidity, pressure or UV radiation is possible to measure with implemented weather unit. The power flow between individual components, actual values of required parameters and total efficiency of off-grid power system is monitored by monitoring system in individual measuring points. The (SOLAR SYSTEM) is used for supply of heat energy in the off-grid power system, but detail description of heat energy power flow is not purpose of this paper.

5. Exploratory of Off-Grid PQ Data

The PQ evaluation was processed due to European standard EN 50160 and database of monitored variables (meteorological and electrical) were used. In this survey, we use positive values of variables (marked "+") for power consumption and negative values of variables (marked "-") for power production. Day with changeable solar irradiance (March 20th, 2013) was selected for presentation of the PQ evaluation in this text. We try to find the interconnection between meteorological and electrical variables, so we switch off WPP and only PV1 and PV2 were switched on. During this evaluation period there was some of the possible operation modes (on-grid and off-grid) used. Characteristic variables for selected time period are shown on Fig. 2. Selected time period should be divided on three operation modes. Time period from 0:00 to 8:00 present on-grid mode (mode 1). Since about 8:00 solar irradiance G is increasing and also PV1 and PV2 production is increasing too. For the detail see G, PV1 and PV2 on Fig. 2. At about 8:00 hybrid power system is switched to off-grid mode (mode 2), and load (P on Fig. 2) rise according to usual day consumption of typical dwelling. When total amount of PV1 and PV2 is sufficient to supply actual load, then there is used "by-pass" and load is directly supplied by PV1 and PV2. In the case that solar irradiance fluctuate and partially there is not enough power to supply load directly from PV1 and PV2 storage device is connected, and load is supplied from storage device. For details see SD on Fig. 2. After 4:00 PM load is supplied entirely from storage device (mode 3).

For power quality evaluation there were selected fundamental variables, frequency *f*, total harmonic distortion of the voltage (THDu), total harmonic distortion of current (THDi) and short (P_{ST}) and long (P_{LT}) time Severity. Soft fluctuation of frequency should be seen at Fig. 3 while hybrid power system is in on-grid mode (mode 1). Fluctuation of frequency is caused by back influence of external distribution grid in on-grid mode. Fluctuation of THDimax is more significant in this mode, and it is caused by cyclical fluctuation of load of connected appliances (refrigerator and TV and PC in stand-by mode). Load *P* is on Fig. 3 rated with total installed power 2500 W for comfortable description of discussed problems. The value of THDu is in limits set by CSN EN 50160, and it is of course related with short-circuit power of external distribution grid.

In off-grid mode, PQ variables are held by control system of SMA Sunny Island. These PQ variables are in off-grid mode more dependent on the character of load and value of short-circuit power. This situation is shown on Fig. 3 where rapid change of load cause change of frequency and THDu out of limits.

Harmonic voltage analysis is shown on Fig. 4. Harmonic voltage with frequency 150 Hz (U_{HARM3}) is 17.4 %. UHARM3 limit set by CSN EN 50160 is only 5 %, it is caused by connected device with semiconductor technology, e.g. laptop's charger, LCD screen, TV, etc.

Long and short time severity is shown on Fig. 5. When system is in mode 1 the values of long and short time severity reach values of around 0.15. Long and short time severity start to fluctuate according to increased current and reach values higher than limits set by CSN EN 50160. The value of long and short time severity is related with dynamic changes of load current.

Control system of SMA Sunny Island is not so fast to hold frequency and THDu in limits according to normative values. After finishing of rapid changes of load; control system of SMA Sunny Island is again able to stabilize frequency and THDu in off-grid mode. As it was described above, THDi is load dependent variable with direct impact of load changes. Rapid changes of THDi are usually produced by appliances with huge application of semiconductors. Details of these rapid changes of THDi are shown in Fig. 3 from about 10:00 am till 2:00 pm.

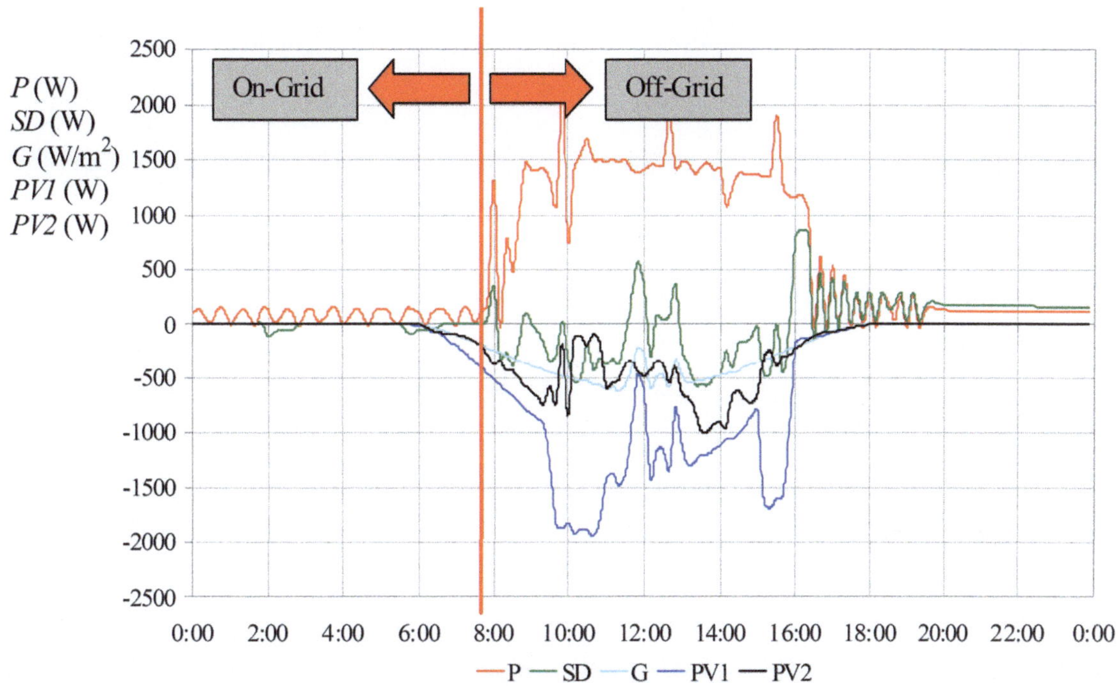

Fig. 2: Example of power variables.

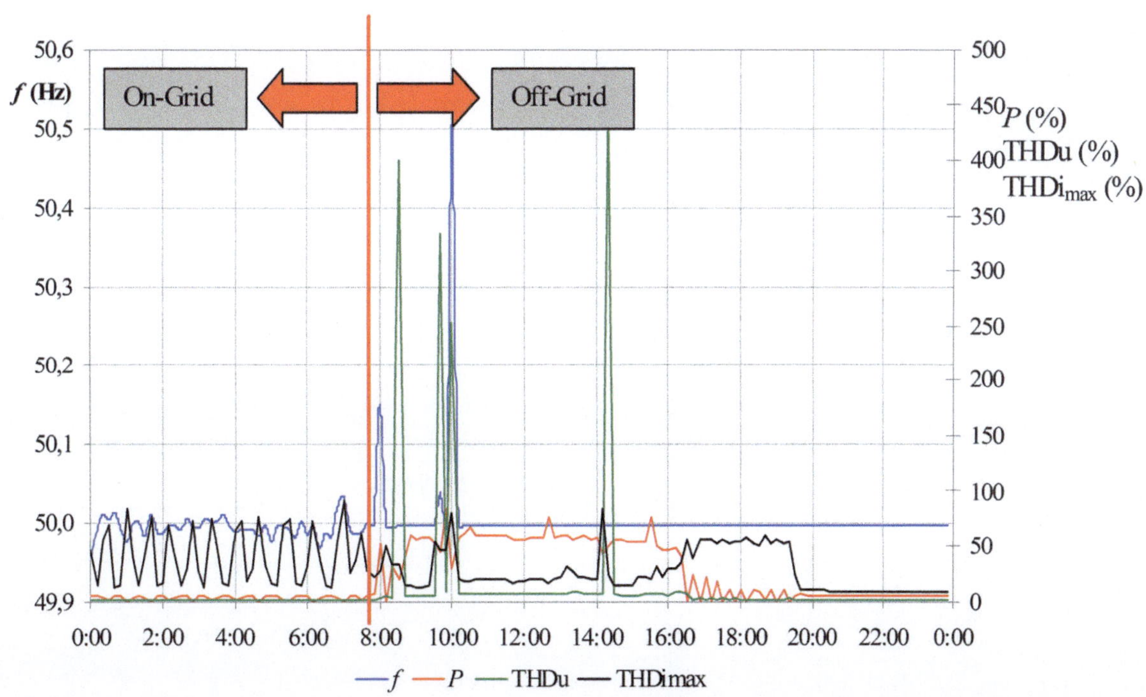

Fig. 3: Example of power quality variables.

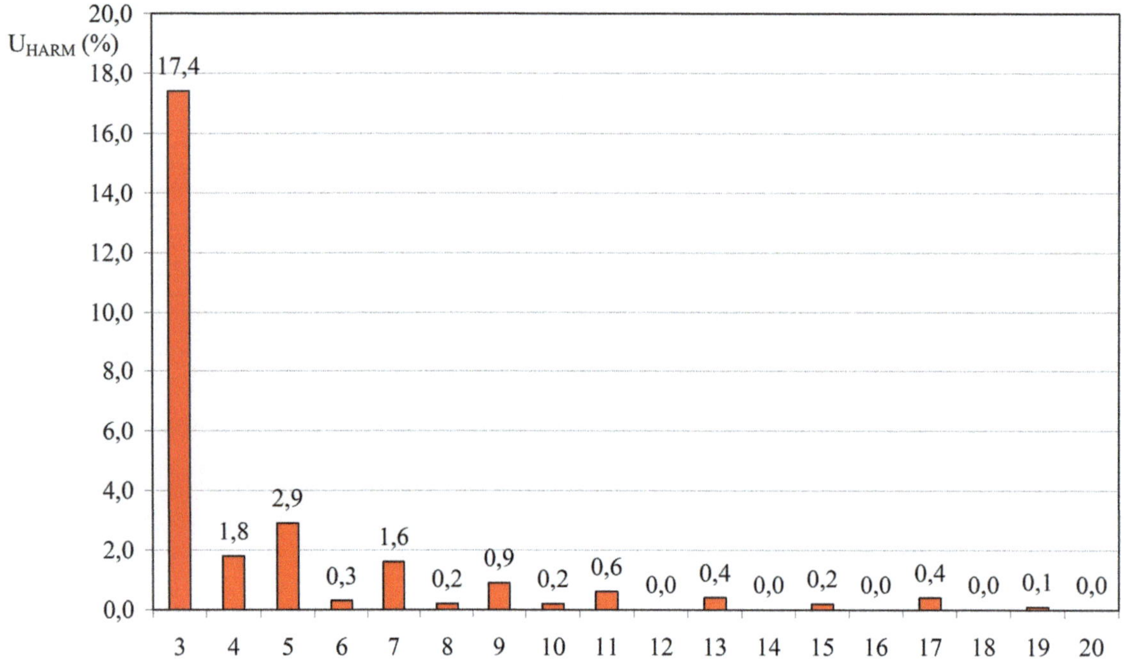

Fig. 4: Harmonic voltage analysis.

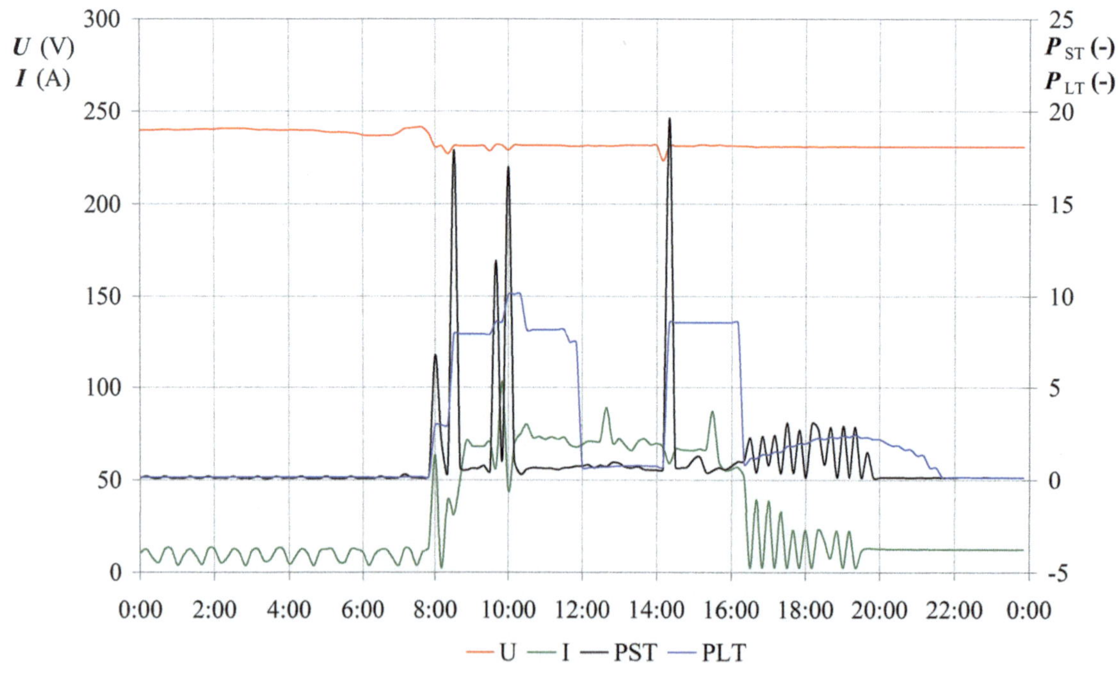

Fig. 5: Short and long time severity.

6. Conclusion and Future Work

Utilization of solar and wind power in off-grid power system for residential supply is in contradiction with the increasing PQ sensitivity of equipment and devices being connected to the off-grid power system. Definition of links and relations between the solar irradiance and wind speed and direction, it may be possible to reduce the negative effects of solar and wind power and allow their use while keeping acceptable level of PQ. In this paper, we try to introduce initial outlook on power quality evaluation in hybrid power system where on-grid and off-grid operation mode are possible.

Acknowledgment

This work was supported within the framework of the IT4Innovations Centre of Excellence project, reg. no. CZ.1.05/1.1.00/02.0070 supported by Operational Programme 'Research and Development for Innovations' funded by Structural Funds of the European Union and state budget of the Czech Republic, Additionally, this article has been elaborated in the framework of the project New creative teams in priorities of scientific research, reg. no. CZ.1.07/2.3.00/30.0055, supported by Operational Program Education for Competitiveness and co-financed by the European Social Fund and the state budget of the Czech Republic, project ENET–Energy Units for Utilization of non-Traditional Energy Sources CZ.1.05/2.1.00/03.0069, Students Grant Competition project reg. no. SP2014/49, project LE13011 Creation of a PROGRES 3 Consortium Office to Support Cross-Border Cooperation and project InterEnergy (CZ.1.07/2.3.00/20.0075).

References

[1] MISAK, S. and L. PROKOP. Technical–Economic Analysis of Hybrid Off-Grid Power System. In: *11th International Scientific Conference Electric Power Engineering 2010, EPE 2010.* Brno: Brno University of Technology, 2010. pp. 295–300. ISBN 978-80-214-4094-4

[2] MISAK, S. and L. PROKOP. Off-Grid Power Systems. In: *9th Conference on Environment and Electrical Engineering, EEEIC 2010.* Prague: IEEE, 2010. pp. 14–17. ISBN 978-1-4244-5370-2. DOI: 10.1109/EEEIC.2010.5490003.

[3] AL-HADDAD, K. Power quality issues under constant penetration rate of renewable energy into the electric network. In: *Proceedings of 14th International Power Electronics and Motion Control Conference EPE-PEMC 2010.* Ohrid: IEEE, 2010, pp. S11-39–S11-49. ISBN 978-1-4244-7856-9. DOI: 10.1109/EPEPEMC.2010.5606699.

[4] MOSTAFA, H. A. and M. M. A. SALAMA. Smart control system for standalone and grid connected PV systems. In: *2012 IEEE Power and Energy Society General Meeting.* San Diego: IEEE, 2012, pp. 1–6. ISBN 978-1-4673-2729-9. DOI: 10.1109/PESGM.2012.6344928.

[5] FARHOODNEA, M., A. MOHAMED, H. SHAREEF and H. ZAYANDEHROODI. Power Quality Analysis of Grid-Connected Photovoltaic Systems in Distribution Networks. *Przeglad Elektrotechniczny.* 2013, vol. 89, iss. 2 A, pp. 208–213. ISSN 0033-2097.

[6] ELTAWIL, M. A. and Z. ZHAO. Grid-connected photovoltaic power systems: Technical and potential problems—A review. *Renewable and Sustainable Energy Reviews.* 2010, vol. 14, iss. 1, pp. 112–129. ISSN 1364-0321. DOI: 10.1016/j.rser.2009.07.015.

[7] CAAMANO-MARTIN, E., H. LAUKAMP, M. JANTSCH, T. ERGE, J. THORNYCROFT, H. DE MOOR, S. COBBEN, D. SUNA and B. GAIDDON. Interaction between photovoltaic distributed generation and electricity networks. *Progress in Photovoltaics: Research and Applications.* 2008, vol. 16, iss. 7, pp. 629–643. ISSN 1099-159X. DOI: 10.1002/pip.845.

[8] REISI, A. R., M. H. MORADI and H. SHOWKATI. Combined photovoltaic and unified power quality controller to improve power quality. *Solar Energy.* 2013, vol. 88, pp. 154–162. ISSN 0038-092X. DOI: 10.1016/j.solener.2012.11.024.

[9] LIU, X., Y. JIA and F. ZHAO. The Method of Suppress the Output Power Fluctuations of Off-Grid Wind Power Systems. *Advanced Materials Research.* 2012, vol. 608-609, pp. 479–482. ISSN 1022-6680. DOI: 10.4028/www.scientific.net/AMR.608-609.479.

[10] ORTEGA, M. J., J. C. HERNANDEZ and O. G. GARCIA. Measurement and assessment of power quality characteristics for photovoltaic systems: Harmonics, flicker, unbalance, and slow voltage variations. *Electric Power Systems Research.* 2013, vol. 96, pp. 23–35. ISSN 0378-7796. DOI: 10.1016/j.epsr.2012.11.003.

About Authors

Stanislav MISAK received his MSc. and Ph.D. degrees in Electrical power engineering in 2003 and 2007, respectively, from VSB–Technical university of Ostrava, Czech Republic, where he is currently an Associate Professor. Dr. Misak published a number of articles in peer-reviewed journals and conference proceedings. He also holds a patent on a fault detector for high-voltage lines. In 2012 was appointed a delegate to the European Union Commission for the strategic management of renewable energy sources. He has been successful in obtaining a number of research contracts and grants from industry and government agencies for projects related to the areas of power systems and renewable energy integration. His current work includes implementation of smart grid technologies using prediction models and bio-inspired methods.

Jindrich STUCHLY received his MSc. degree in Electrical power engineering in 2011 from VSB–Technical university of Ostrava, Czech Republic. His research interests include renewable source of electric energy, implementation smart grids and artificial intelligence. He currently works towards his Ph.D. degree under the supervision of Dr. Stanislav Misak.

Jakub VRAMBA received his MSc. degree in Electrical power engineering in 2011 from VSB–Technical university of Ostrava, Czech Republic. His research interests include the operation of transmission and distribution systems, renewable energy sources, effects of renewable energy sources on distribution, and smart grids. He currently works towards his Ph.D. degree under the supervision of Dr. Stanislav Misak.

Lukas PROKOP received his MSc. degree in Electrical power engineering in 2002 from Brno University of Technology, Czech Republic, and his Ph.D. in Electrical power engineering in 2006 from VSB–Technical university of Ostrava, Czech Republic. In 2005 he joined the Department of Electrical and Computer Engineering at the VSB–Technical university of Ostrava, Czech Republic, where he is currently a research scientist. He is involved in a number of research contracts and grants from industry and government agencies, in the areas of power systems and renewable energy integration. His current work is focused on smart grid technologies, renewable energy sources, off-grid systems, power system reliability and Power Quality.

Marian UHER received his MSc. degree in Electrical power engineering in 2010 from VSB–Technical university of Ostrava, Czech Republic. His research interests include modeling of electric power grids including renewable energy sources, technical diagnostics of electrical equipment. He currently works towards his Ph.D. degree under the supervision of Dr. Stanislav Misak.

Comparison of Standard and Fast Charging Methods for Electric Vehicles

Petr CHLEBIS, Martin TVRDON, Ales HAVEL, Katerina BARESOVA

Department of Electronics, Faculty of Electrical Engineering and Computer Science, VSB–Technical University of Ostrava, 17. listopadu 15, 708 33 Ostrava-Poruba, Czech Republic

petr.chlebis@vsb.cz, martin.tvrdon@vsb.cz, ales.havel@vsb.cz, katerina.baresova.st@vsb.cz

Abstract. *This paper describes a comparison of standard and fast charging methods used in the field of electric vehicles and also comparison of their efficiency in terms of electrical energy consumption. The comparison was performed on three-phase buck converter, which was designed for EV's fast charging station. The results were obtained by both mathematical and simulation methods. The laboratory model of entire physical application, which will be further used for simulation results verification, is being built in these days.*

Keywords

Battery charging, electric vehicle, fast charging station, power losses, three-phase buck converter.

1. Introduction

The present energetic development is oriented into the considerable incorporation of alternative and renewable energy sources into the standard electrical grids in an effort to build very efficient, economic and reliable smart grids. This approach produces series of technical problems, which have to be solved. One of the most actual ones is the electric energy accumulation. The solar plant systems and especially the fast charging stations for electric vehicles and many other applications, which will be discussed further, are typical examples of energy accumulation issues.

To satisfy an efficient and economic functioning of such station, the energy surplus has to be stored into the batteries and extracted back when needed. It is also necessary to ensure sufficient battery power for fast charging of the electric vehicle's traction batteries. Voltage converters are needed to provide an efficient and reliable energetic power flow between the accumulation units and others devices. This paper discusses the efficiency of standard and fast charging in order to decide whether is more advantageous to charge an electric vehicle economically (standard charging) or user-defined (fast charging).

2. Charging Station for Electric Vehicles

Both charging methods were carefully reviewed in the sense of their implementation on the wholly charged system including models of buck converter and Li-ion batteries in order to meet initial requirements, which were defined in the abstract of this paper. The laboratory sample of three-phase buck converter for this application has been created and its wiring diagram with 3D model is described below.

The charging station unit for electric vehicles is a part of complex structure used for accumulation of electric power. Structure depicted on Fig. 1 represents an accumulation element typical for smart grids, which is eliminating time inefficient operation of alternative energy sources, for example solar power plant. High instant power for EV's fast charging will be provided from this accumulation source. Utilization of this source do not create peaks in the power grid and allows to partly regulate the energy supply system.

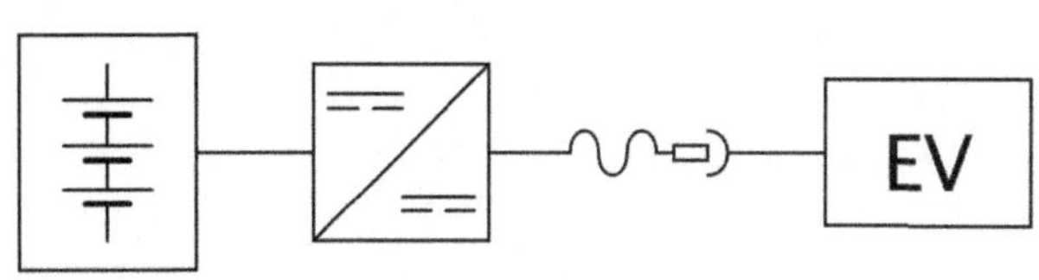

Fig. 1: Location of the converter in EV circuit topology.

The overall scheme from the point of view of charging an electrical vehicle is pictured above on the Fig. 1. The power source creates an accumulation system, which is connected directly to the power grid.

The power converter is a controllable unit between the power source and electric vehicle, which is being charged. This converter is built as a three-phase buck converter.

The inside scheme of the converter is pictured at the Fig. 2 below. The input of the converter is created by the capacitor battery used for energetic exchange caused by the fast switching of the converter. The power part contains the insulated IGBT modules with implemented freewheeling diodes. These modules are assembled on the liquid cooled heat sink, which is used for dissipation of loss heat to the coolant. On the output of the power circuit is the choke-block used for current ripple reduction, which allows the connection of individual phases to the charged battery.

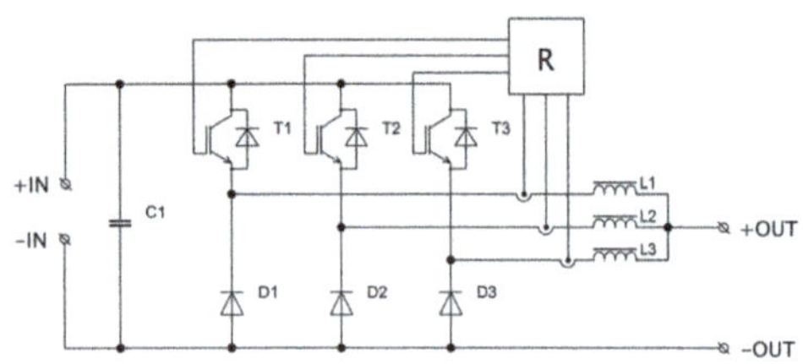

Fig. 2: Power schematic of the three-phase buck converter.

The combined AC/DC charging system according to the standard IEC 62196-3 enables connection between the power converter and the electric vehicle. This connector allows charging currents up to 125 A and voltages up to 850 V DC. Power level of this converter is preset to 70 kW. This performance is chosen to maximize the possibility of battery charging for electric vehicles.

The input voltage applied on the capacitor battery is about of 800 V DC. The voltage of charged battery must always be lower than the source voltage. The current and voltage sensors are distributed in the power circuit providing feedback for the PI controllers. The control algorithms are provided by control system equipped with a signal processor TMS320F28335 and were developed on the Department of Electronics [1].

The 3D design of three-phase buck converter is pictured at the Fig. 3 below. The converter prototype will be manufactured in the near future. The calculated and simulated results, which are presented further in this paper, will be later verified on a real system with LiFePO4 batteries, as soon as the converter will be prepared.

3. Calculation of the Charging Efficiency in Various Modes

The efficiency of fast charging is an issue that has recently been told much about. The influence of charging current size on total charging efficiency is discussed within this paper. Total charging time is indirectly proportional to the value of charging current. This ratio creates a fundamental condition for power losses assessment in particular parts of the charging system. Dependences of losses in following parts of the power circuit will be further investigated below: the power source battery; the power semiconductor converter; choke-block for ripple current reduction and the charged battery.

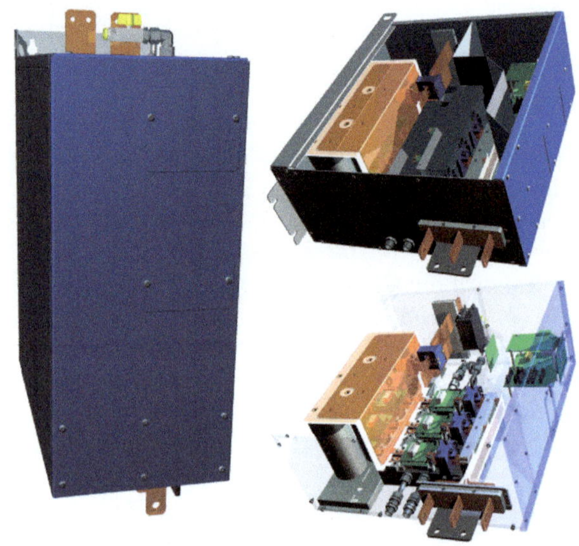

Fig. 3: Sample of design solutions of the fast charger for electric vehicles. The left image of the charger is shown in the basic position, on the other illustrations, there are several views inside of the charger.

3.1. Losses in the Battery

The electrical battery is heated during operation by both the charging and discharging process. The internal impedance of the accumulator is used for evaluation of power losses in the battery. The impedance is always defined in the datasheet of the particular accumulator. Its value is usually about 1 mΩ for one couple of LiFePO4 blocks. Couples of lithium batteries are connected in series for increasing the rated voltage of the whole battery. The energy loss in power source battery respectively in the charged battery is given by:

$$E_{Batt} = I^2 \cdot r_{int} \cdot t_{charge}, \tag{1}$$

where I is the effective value (RMS) of the taken respectively supplied current from/in the battery and t_{charge} is the time for which the energy was supplied or taken.

The energy loss is increased with the square of the current, as it can be seen from Eq. (1) and Fig. 4, which is very disadvantageous situation for fast charging.

3.2. Losses in the Buck Converter

Power losses in the semiconductor device (three-phase buck converter) are especially composed by their two main components – conduction and switching power losses. The other types of electrical power losses also emerge during converter's operation (power losses in the control circuit or capacitor battery) but these losses are negligible in comparison with the first two cases [2].

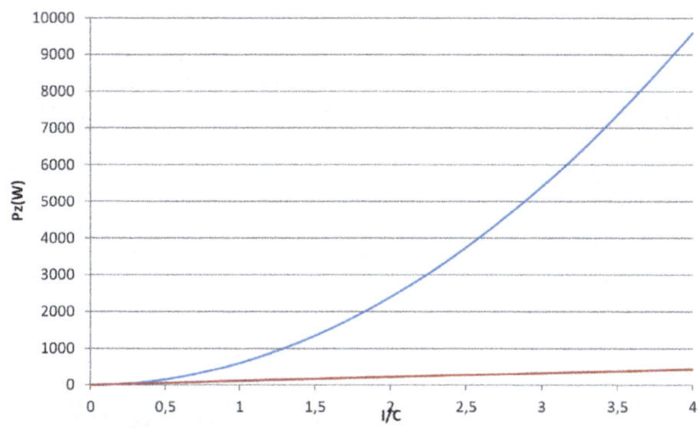

Fig. 4: The dependency of power losses on the charging current. Red line describes linear increase of the current while the blue line represents the quadratic dependency on the charging current.

1) Conducting Power Losses

Losses caused by conduction of the current are defined according to the following equation:

$$E_{Cond} = \left(I_{av} \cdot V_{CE0} \cdot I_{RMS}^2 \cdot r_d\right) \cdot t_{charge}, \qquad (2)$$

where I_{av} is the average value of the current, I_{RMS} the effective current value and V_{CE0} and r_d are threshold voltage and differential resistance of the IGBT switch.

Determination of the dependence for power losses on the current is more complicated in this case. The average current component is dependent linearly, while the effective current component increases with the square of the current. Significance of the second part is substantially lower depending on the mutual ratio between both components.

2) Switching Power Losses

Switching power losses are defined according to the following equation:

$$E_{SW} = \left[(E_{on} + E_{off}) \cdot f_{SW}\right] \cdot t_{charge}, \qquad (3)$$

where E_{on} is the switch-on energy loss of the semiconductor switch (IGBT), E_{off} is the switch-off energy loss of the IGBT and f_{SW} is the applied switching frequency.

All of these partial switching losses are linearly dependent on the current. Power semiconductor converter also contains freewheeling diodes. Losses in these bypass diodes arise equivalently as in IGBT transistors, so the current dependencies are just the same.

3.3. Losses in Chokes

The most significant component of losses in chokes is dependent on active electrical resistance of used chokes. The resulting energy loss for chokes, which are carrying the current, is given by the following equation:

$$E_L = I^2 \cdot R_L \cdot t, \qquad (4)$$

where I is the effective value of the current and R is an active electrical resistance of used chokes.

Also in this case the main part of energy losses in chokes increases with the square of the current. (There are also power losses in the core, which are usually negligible compared to the mutual ratio between electrical and magnetic losses [3].)

3.4. Evaluation of Power Losses on the Real Model of Fast Charging Station

Parameters of the solved circuit were set precisely to enable determination of total power losses on the high-fidelity computational model of the fast charging station, which is described in the first part of this paper.

The whole system is supplied from the accumulation unit with rated voltage of 800 V and rated capacitance of 100 Ah. The structural model of charged battery in the load is of the same type as the power source battery model. The charged battery has rated voltage of 400 V and rated capacitance of 100 Ah.

To compare partial and total power losses during standard and fast charging modes, the charging current was chosen to I = 0.3 C for standard charging and I = 2 C for fast charging.

The results, which were obtained for 5 % capacity increase, are listed in the Tab. 1 below. Power losses are calculated with respect for comparison with the simulation results, which will be presented further. Total charging time is 684.5 seconds for standard charging and 102.5 seconds for fast charging.

Energy losses in particular parts of the solved circuit and their total sum (E_{Total}) are written in the Tab. 1. for both standard and fast charging modes.

Tab. 1: Comparison of power losses for standard and fast charging.

Method of charging	Power losses [kJ]					
	E_{Source}	E_{Cond}	E_{SW}	E_L	E_{Load}	E_{Total}
0.3C	34.059	22.359	222.414	0.425	73.955	353.252
2C	253.750	24.602	75.024	3.655	494.107	851.138

Based on the obtained results it is possible to draw the following conclusions: Energy losses with quadratic dependence on the current (especially losses on batteries and chokes) have significantly changed and are about 7 times higher for fast charging in comparison with standard charging. On the other hand, the conducting power losses on the semiconductor devices did not almost change and switching power losses are even higher for standard charging than for fast charging. The change of these losses does not so heavily depend on the value of current. Total power losses for standard charging mode are approximately half those for the fast charging.

4. Simulation and Analysis of the Charging Efficiency in Various Modes

Simulation and analysis of the standard and fast charging was performed in the simulation software Matlab Simulink. Solving of the whole charging cycle has very high demands on the operating memory of the computer, therefore, the simulation was performed only to 5 % increase of state of charge (SoC). However, the results are not affected, because the simulation is always performed with constant currents. The charging modes were chosen the same as in the evaluation analysis presented in the previous chapter (I = 0.3 C for standard charging and I = 2 C for fast charging).

The Simulink model of charging station is depicted below. Some of the included elements are used from the Simscape library, which includes many useful models from energetics and power electronics, e.g. primary and secondary LiFePO4 battery blocks shown in Fig. 5. These blocks are configured to behave as a real lithium iron phosphate battery with nominal parameters. The high-fidelity simulation model of three-phase buck converter accurately represents the datasheet parameters of used IGBT switches [4], [5].

Tab. 2: Simulation results – power losses for standard and fast charging.

Method of charging	Power losses [kJ]			
	E_{Source}	E_{Conv}	E_{Load}	E_{Total}
0.3 C	36.664	47.681	82.42	166.765
2 C	196.358	47.952	463.62	707.93

Important parameters of the simulated circuit were recorded in the simulation output files, to enable subsequent analysis. The obtained results are making an overview about standard and fast charging processes in the solved circuit and are summarized and compared in the following Tab. 2.

Fractional values of energy losses in the three-phase buck converter structure (conducting and switching power losses) are not possible to determine positively from the simulation. However, their sum, represented as total power losses of the converter block E_{Conv} is reported in the Tab. 2. The simulation confirms the assumption about increasing power losses on batteries during fast charging, while the energy losses in the converter block are in both examined charging modes approximately the same. This is probably caused by the calculation method within the simulation model.

5. Comparison between Calculated and Simulated Results

By the comparison of calculated and simulated results it is possible to conclude, that the fast charging mode, which is very preferred in these days, is not completely advantageous because of power losses, which are several times higher than in standard charging case. This claim is confirmed by both, the calculated and simulated results obtained in Matlab Simulink.

More specific point of view on the situation is given by the comparison of both charging methods in the field of efficiency. The evaluated efficiency of standard charging is almost 95 %, while the efficiency of fast charging is only about 89 %. In other words, increasing the charging current by 6.6 times (from 0.3 C to 2 C) causes the efficiency drop by about 6 %.

The largest deviation between calculation and simulation method was observed in the three-phase buck converter block because of switching power losses, which are in the calculation method multiple times larger during standard charging compared to fast charging. The current dependencies of switching losses for calculation method were determined from power diagrams, which are stated in the datasheets of used IGBT modules, while by the simulation method are these dependencies defined inside the IGBT block and are approximately linearly proportional to the current. Therefore, power losses of the converter block E_{Conv} in the simulation method are approximately the same for both charging modes.

Verification of the simulation results will be performed on the laboratory model of fast charging station, which is currently being built.

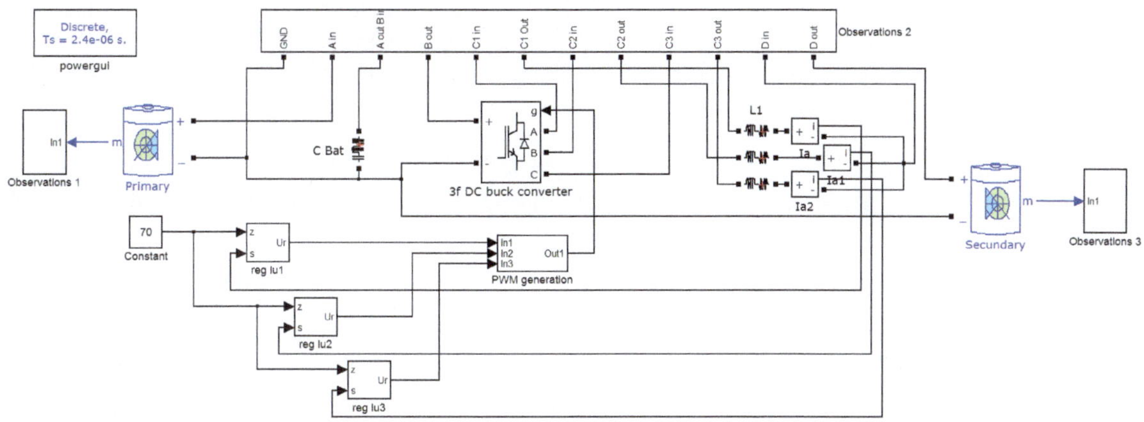

Fig. 5: Simulink model of fast charging station.

6. Conclusion

The aim of this research was to compare standard and fast charging in terms of power losses and total efficiency. Obtained results reveal a significant increase of power losses during application of fast charging. The question to ponder is to what extent is it necessary to apply fast charging on electric and hybrid electric vehicles. Recently growing infrastructure of fast charging stations can lead to inefficient management with electrical energy. According to obtained results, the process of fast charging did not create a substantial increase of power losses in the converter block with chokes (see Tab. 1), however, the situation in the power source and the charged battery is exactly the opposite (see Tab. 1 and Tab. 2).

Acknowledgment

The research described in this paper was partially supported by these projects: SGS project SP2013/80 - Utilizing of Modern SiC Components for MF - Power Semiconductor Converters and ENET CZ.1.05/2.1.00/03.0069 - Energy Units for Utilization of non Traditional Energy Sources and supervised by prof. Petr Chlebis from Department of Electronics, VSB–Technical University of Ostrava.

References

[1] VACULIK, P., A. HAVEL, M. TVRDON and M. HROMJAK. The Design and Construction of a Reversible Voltage Inverter Prototype. In: *13th International Scientific Conference Electric Power Engineering 2012*. Brno: Brno University of Technology, 2012. pp. 1059–1062. ISBN 978-80-214-4514-7.

[2] CHLEBIS, P. and J. DUDEK. Comparison of Power Losses and Quality of Output Voltage from Different Types of Inverters. In: *35th Annual IEEE Power Electronics Specialists Conference*. Aachen: IEEE, 2004. pp. 3120–3126. ISBN 0-7803-8399-0. DOI: 10.1109/PESC.2004.1355334.

[3] JUNG-SONG, M., L. JUNG-HIO, H. IN-YONG, L. TEACK-KIE and W. CHUNG-YUEN. An Efficient Battery Charging Algorithm Based on State-of-charge Estimation for Electric Vehicle. In: *International Conference on Electrical Machines and Systems (ICEMS)*. Beijing: IEEE, 2011. pp. 1–6. ISBN 978-1-4577-1044-5. DOI: 10.1109/ICEMS.2011.6073783.

[4] CHLEBIS, P., A. HAVEL, P. VACULIK and Z. PFOF. Modern Instruments for Increasing the Efficiency of the Energy Transfer in Electric Vehicles. In: *14th International Power Electronics and Motion Control Conference*. Ohrid: IEEE, 2010. pp. 89–93. ISBN 978-1-4244-7854-5. DOI: 10.1109/EPEPEMC.2010.5606586.

[5] HAVEL, A., P. VACULIK and D. SLIVKA. Efficient Propulsion Structure with an axial Flux Rotary Converter for HEV Drive Unit. *Advances in Electrical and Electronic Engineering*. 2011, vol. 9, no. 3, pp. 136–142. ISSN 1336-1376.

About Authors

Petr CHLEBIS was born in 1956. He was appointed as professor at Faculty of Electrical Engineering and Computer Science of VSB–Technical University of Ostrava in 2005. His research includes power electrical converters, their control systems and utilization of nontraditional energy sources.

Martin TVRDON was born in 1986. He obtained his Master's degree in electronics in 2011. He is currently pursuing Ph.D. study at Department of Electronics at VSB–Technical University of Ostrava on Faculty of Electrical Engineering and Computer Science. His research includes power systems, design and modeling of fast charging stations for electric vehicles.

Ales HAVEL was born in 1984. He obtained his Master's degree in the field of power electronics systems and electric machines design in 2009 and Ph.D. degree in 2013 at Department of Electronics at VSB–Technical University of Ostrava on Faculty of Electrical Engineering and Computer Science. His research includes power systems, design and modeling of electric machines.

Katerina BARESOVA was born in 1988. She obtained her Master's degree in the field of power electronics systems in 2012. She is currently pursuing Ph.D. study at Department of Electronic at VSB–Technical University of Ostrava on Faculty of Electrical Engineering and Computer Science. Her research includes power systems and modeling of electrical converters.

14

Test Frequency Selection Using Particle Swarm Optimization

Zdenek KINCL, Zdenek KOLKA

Department of Radio Electronics, Faculty of Electrical Engineering and Communication, Brno University of Technology, Technicka 12, 616 00 Brno, Czech Republic

xkincl01@stud.feec.vutbr.cz, kolka@feec.vutbr.cz

Abstract. *The paper deals with the problem of test frequency selection for multi-frequency parametric fault diagnosis of analog linear circuits. An appropriate set of test frequencies is determined by minimizing the conditionality of the sensitivity matrix based on the system of fault equations using a global stochastic optimization. A novel method based on the Particle Swarm Optimization, which provides more accurate results and improves the convergence rate, is described. The paper provides several practical examples of its application to test frequency selection for active RC filters. A comparison of the results obtained by the proposed method and by the Genetic Algorithm is also presented.*

Keywords

Fault diagnosis, frequency set selection, GA, PSO, test index measure.

1. Introduction

Nowadays, continuing miniaturization of modern electronic devices leads to more complex circuits and systems. The testing process has become more difficult and the cost of testing is still increasing. In the case of analog circuits the testing represents a much more complex problem in comparison with digital ones due to the behaviour of analog signals. To reduce test costs a development of new robust methods for Automatic Test Plan Generation (ATPG) for analog circuits is one of the main objective in this area [1].

Analog faults can be classified into several classes, e.g. manufacturing tolerances, soft, hard and catastrophic faults, depending on the deviation of network parameters from their nominal values [2]. The multi-frequency parametric fault diagnosis is a technique for estimating the actual values of some network parameters from measurements of network characteristics in the frequency domain [3]. The circuit components, whose parameters exceed the allowed tolerance intervals, are classified as faulty.

An arbitrary network function of lumped linear time-invariant analog circuit can be expressed as:

$$H(s, \mathbf{p}) = \frac{a_n(\mathbf{p})s^n + \cdots + a_0(\mathbf{p})}{b_m(\mathbf{p})s^m + \cdots + b_0(\mathbf{p})}, \qquad (1)$$

where s is the complex frequency, and the polynomial coefficients a_i and b_i are nonlinear functions of network parameters $\mathbf{p} = [p_1, \ldots, p_R]^T$.

The actual values of unknown network parameters can be estimated based on the measurements of one or more different network functions at several frequencies [2]. In the case of one excitation source, each network function corresponds to one test point. From a mathematical point of view, the parameter estimation represents solving the set of nonlinear fault equations based on the network function Eq. (1):

$$H_k(j\omega_{k,i}, \mathbf{p}) = M_{k,i}, \qquad (2)$$

where H_k represents the k-th network function with respect to the k-th test point selected, and $M_{k,i}$ are the measurements of the k-th network function on the i-th test frequency. The system can be solved for example using the Newton-Raphson iteration method.

The set of tested parameters $\mathbf{p}_t \in \mathbf{p}$, i.e. the unknowns in Eq. (2), must be selected beforehand. The remaining components are considered to have a priori known (nominal) values. A proper selection of test point(s) and test frequencies is of cardinal importance for a robust solution of Eq. (2).

As an analog circuit consists of different types of components (resistors, capacitors, inductors, etc.), the individual network parameters have different physical units and their nominal values are spread over a wide range of magnitudes. Then, the Jacobian (sensitivity) matrix associated with the system of fault equations

Eq. (2) contains mutually incomparable elements. It causes that the matrix is ill-conditioned and the fault diagnosis may return non-relevant results. To overcome the problem, the set of equations Eq. (2) should be formulated using the normalized network parameters $\tilde{p}$

$$\tilde{p} = \frac{p}{p_{\mathtt{nom}}}, \tag{3}$$

where p is the original value and $p_{\mathtt{nom}}$ is the nominal value [4].

The process of parametric fault diagnosis can be divided into two independent phases. The first phase, called testability analysis, determines which network parameters can be identified based on the selected test point(s). The maximum number of testable network parameters is referred to as the testability degree T [3]. The testability is associated with the solvability degree of the system of fault equations Eq. (2). In the case of one test point, the solvability is given as the rank of the Jacobian matrix associated with the system Eq. (2):

$$T = \mathtt{rank}\left[\frac{\partial H(j\omega_i)}{\partial \tilde{p}_j}\right]. \tag{4}$$

Theoretically, the testability degree of a circuit is independent of the fault localization method, the nominal values of network parameters and the selected set of test frequencies [5].

The set of linearly dependent columns of the Jacobian matrix determines ambiguity groups [6]. In these groups the effects of individual network parameters on the network function are indistinguishable from one another. For example, two resistors in series create such a group. Only some parameters of the ambiguity group can be estimated independently. The remaining parameters must be considered to be fixed.

When the total number of potentially faulty parameters is greater than the testability degree T or when the parameters being tested belong to the same ambiguity groups, the system of fault equations has no unique solution and the set of test points should be extended.

The second phase consists in solving the system of fault equations Eq. (2) to obtain the actual values of unknown network parameters. However, in the real fault diagnosis, the system is always perturbed by errors and for this reason the diagnosis may return non-relevant results. For example, the ill-conditioned system of equations is prone to large numerical errors of the solution, in the case of the uncertainty of fixed (untested) network parameters, the final solution is mathematically correct but the estimated values of tested network parameters may not correspond to their actual ones, the measurement of weak signals, e.g. in stop bands of analog filters, as well as the measurement at high frequencies is usually more problematic. Fortunately, all these errors can be minimized by an appropriately chosen set of test frequencies [3].

However, an optimal method for test frequency selection has not been determined yet. The methods based on maximizing the sensitivities of tested network parameters in the frequency domain [7] may lead to closely-spaced frequencies for example for band-pass filters, i.e. the system of fault equations is ill-conditioned. The method using interval analysis [8] takes into account the uncertainty of fixed network parameters, but the method is applicable only for diagnosis of catastrophic faults, i.e. soft faults can not be diagnosed. The simple heuristic solution [9] based on fixed relations between individual test frequencies is computationally very inexpensive, but the method may not be optimal in general and it can be applicable only to specific simple circuits. On the other hand, the genetic optimization method proposed in [9] discretizes the solution, the computational demands are very high and the algorithm converges slowly.

The procedure for test frequency selection proposed in this paper is based on the modified Particle Swarm Optimization (PSO). Section 2 of this paper deals with the mathematical background of the method and Section 3 provides several application examples for analog frequency filters.

2. Test Frequency Selection

The selected set of test frequencies should minimize the sensitivity of the solution of Eq. (2) to the numerical errors, uncertainty of fixed network parameters and measurements errors. The proposed method is based on the measure introduced in [9]. The error of the perturbed system of fault equations Eq. (2) as well as the convergence rate of the Newton-Raphson iteration method are taken into account in the measure formulated in the form:

$$\begin{aligned} E &= \ln\left((\mathtt{cond}(\mathbf{J}) - 1)\,\|\mathbf{J}^{-1}\|_2\right) = \\ &= \ln\left(\frac{\sigma_{\mathtt{max}} - \sigma_{\mathtt{min}}}{\sigma_{\mathtt{min}}^2}\right), \end{aligned} \tag{5}$$

where $\mathtt{cond}(\mathbf{J})$ is the condition number of the Jacobian matrix $\mathbf{J}$ associated with fault equations Eq. (2), $\sigma_{\mathtt{max}}$ and $\sigma_{\mathtt{min}}$ are the maximum and minimum singular values of $\mathbf{J}$. The minimizing of E leads to minimizing the conditionality of the Jacobian matrix and maximizing the sensitivities to unknown network parameters, which is represented by $\|\mathbf{J}^{-1}\|_2$ in Eq. (5).

The system of fault equation Eq. (2) with a small E is well-conditioned while a large E leads to an ill-conditioned problem. The computational complexity of Eq. (5) is given by the complexity of singular value decomposition. The Jacobian matrix itself can be obtained by repeatedly solving the adjoint sensitivity network [14] for each test frequency.

The minimization of E represents a complex problem with many local optimums. The use of the Genetic Algorithm (GA) was reported in [10]. The main disadvantage of the method is the discretization of the solution dependent on the bit representation of chromosomes and very slow convergence rate in the order of hundreds of steps, i.e. high computational cost. The proposed solution is based on the Particle Swarm Optimization (PSO) [11], [12]. The main advantages of the method are the high convergence rate, no discretization of the solution and a very easy implementation.

The particle swarm consists of individual agents, which move randomly in a u-dimensional space, where u is the number of test frequencies and is the same as the number of unknown parameters in Eq. (2) for only magnitude measurements. The position of each agent is associated with one set of test frequencies. The fitness function E is evaluated for each agent in each iteration step. During the optimization the agents are moving towards to the global optimum.

The agent speed $\mathbf{v}_{n+1} \in R^u$ in the next iteration step is given as:

$$\mathbf{v}_{n+1} = w_n\mathbf{v}_n + c_1r_1(\mathbf{x}_{\mathtt{pers}} - \mathbf{x}_n) + c_2r_2(\mathbf{x}_{\mathtt{glob}} - \mathbf{x}_n), \quad (6)$$

where w_n is the inertia weight in the current step, $\mathbf{v}_n$ is the actual speed, c_1 and c_2 are the weight coefficients of optimization, $\mathbf{x}_{\mathtt{pers}}$ and $\mathbf{x}_{\mathtt{glob}}$ are the coordinates of personal and global optimums found so far, $\mathbf{x}_n$ is the actual position of agent, and r_1 and r_2 are random numbers with the uniform distribution from the interval $< 0, 1 >$ generated in each iteration.

When the inertia weight w is too high, the position of each agent can oscillate around the global optimum in the final phase of optimization. Reducing w during the optimization can overcome the problem [12].

The new position of each agent $\mathbf{x}_{n+1}$ is:

$$\mathbf{x}_{n+1} = \mathbf{x}_n + \Delta t\mathbf{v}_{n+1}, \quad (7)$$

where Δt is the time step. Choosing $\Delta t < 1$ helps to damp the oscillations of agents around the global optimum.

At the beginning of optimization it is necessary to use a large number of agents to preserve the robustness of the process. During the optimization, agents move closer to the global optimum and it is unnecessary to evaluate the fitness function for all the agents used at the beginning. Therefore agents with high values of E are removed. This approach can significantly speed-up the final phase of optimization.

Since network functions are better represented in the logarithmic frequency scale and the core of PSO works well on the linear scale, it is advantageous to introduce the transformation of coordinates:

$$f_i = f_{\mathrm{L}} \exp\left(x_i \ln \frac{f_{\mathrm{H}}}{f_{\mathrm{L}}}\right), \quad (8)$$

where x_i is the agent position in the space of optimization coordinates, f_i corresponds to the actual test frequency, and f_{L} and f_{H} define the frequency interval of interest. Some areas of the search space may represent an unfeasible solution. To keep the swarm in the feasible area, use can be made of absorbing, reflecting or invisible walls [12]. The authors in [11] state that the invisible walls may produce a little better results.

The fitness function E does not depend on a particular ordering of individual test frequencies. An example for $u = 2$ is shown in Fig. 1. For this reason, the red points correspond to the same set of test frequencies. Then, the agents can be restricted to the area where $x_1 < x_2 < .... < x_u$. Restricting the search space leads to a smaller required number of agents, i.e. less computational demands.

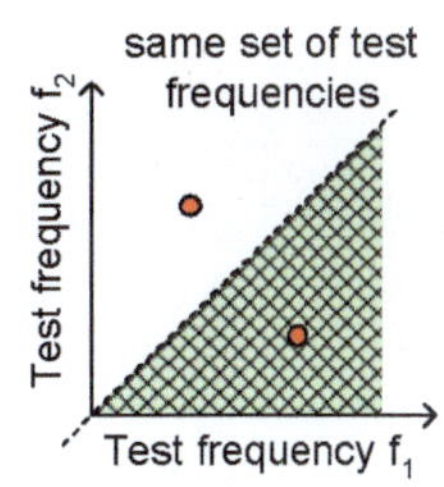

Fig. 1: Principle of test frequency set mirroring.

3. Application Examples

The method was applied to three benchmark active RC filters [10] and the results were compared with the results obtained using GA.

The following parameters of PSO according to the recommendations in [11] and [12] were used:

- Initial number of agents: 12, 26, 50.
- Final number of agents: 10.
- Time step: $\Delta t = 0,1$.
- Weight factors: $c_1 = 2$, $c_2 = 2$.

- Inertia weight: $w = 0,9 - 0,4$ (linearly decreased).
- Search space $\{\mathbf{x} \mid 0 \leq x_1 \leq x_2 \leq \cdots \leq x_u \leq 1\}$ i.e. frequency interval $\{f \mid 1\ \mathtt{Hz} \leq f_i \leq 10\ \mathtt{kHz}\}$.

The first example is the Tow-Thomas second-order band-pass filter shown in Fig. 2. The filter is tuned to the resonant frequency $f_0 = 1{,}6$ kHz with the quality factor $Q = 10$. The frequency response of the filter is shown in Fig. 3. The nominal values of components are $R_1 = 10$ kΩ, $R_2 = R_3 = R_4 = R_5 = R_6 = 1$kΩ, $C_1 = C_2 = 100$ nF, with the ideal operational amplifiers being used in the analysis. Only one test point, which corresponds to magnitude measurements of V_2, was chosen. The testability degree T is equal to 3 and the same testable group (R_1, R_4, C_1) was considered as in [10]. To determine these three unknown network parameters it is necessary to determine three test frequencies minimizing the E measure Eq. (5).

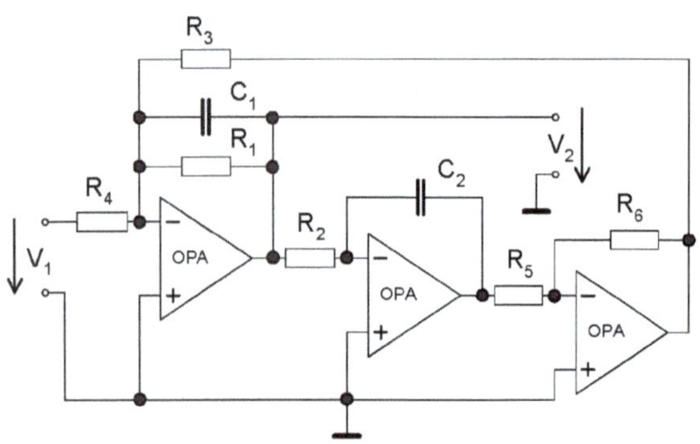

Fig. 2: Tow-Thomas band-pass filter.

As it can be seen in Fig. 4 the PSO method found, after approximately 30 iteration steps, the following solution: $f_1 = 1149$ Hz, $f_2 = 1592$ Hz and $f_3 = 4826$ Hz ($E = -1,50$). The progress of the optimization for different initial numbers of agents is also presented. In the case of 50 agents, the optimization converged quickly, but it took the longest time due to the number of agents. The results are similar to the case of 26 agents. When only 12 agents are used, the speed of convergence is reduced. The optimization took the least time, but the algorithm may not find the appropriate global optimum.

The same optimization problem was solved using the basic GA with elitism and mutations [13]. In this case, the following optimization parameters were used:

- Chromosome representation: 16 bits.
- Number of individuals: 12, 26, 50.
- Elitism: 2 individuals.
- Probability of mutation: 5 %.
- Frequency interval $\{f \mid 1\ \mathtt{Hz} \leq f_i \leq 10\ \mathtt{kHz}\}$.

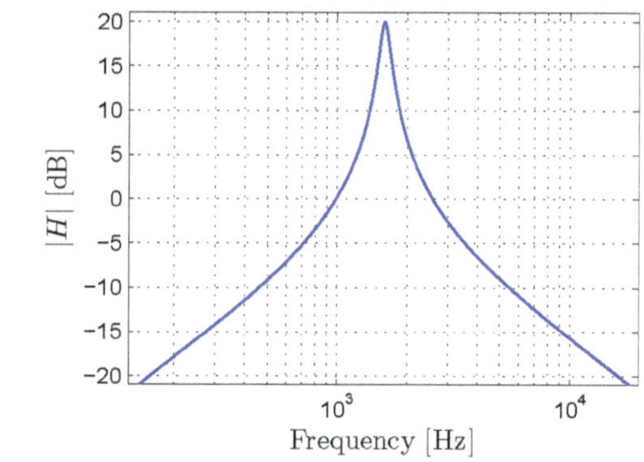

Fig. 3: Frequency response of Tow-Thomas filter.

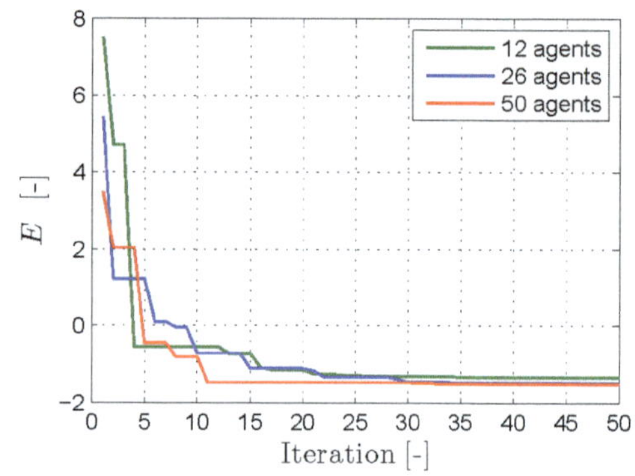

Fig. 4: Convergence of PSO optimization (Tow Thomas).

The results of GA are shown in Fig. 5. The speed of convergence is lower (approximately 3 times) than the results obtained by PSO. When a small number of bits in chromosome representation of frequency are used, the results may be inaccurate. A large number of bits reduce the speed of convergence.

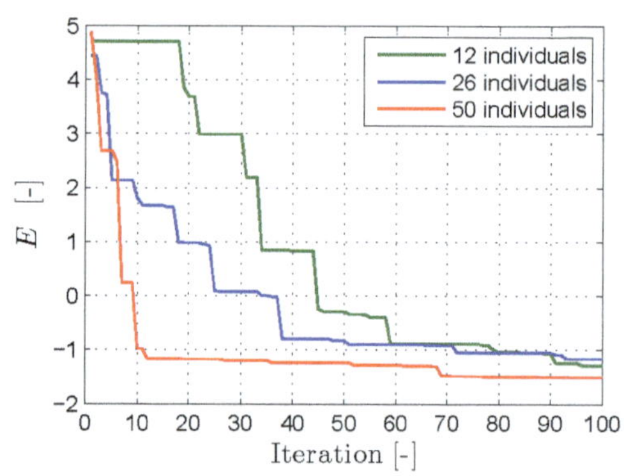

Fig. 5: Convergence of GA optimization (Tow-Thomas).

The second example is the second-order low-pass filter shown in Fig. 6. For nominal values of components $R_1 = 5{,}01$ kΩ, $R_2 = 10$ kΩ, $R_3 = 33{,}8$ kΩ, $R_4 = 200$ kΩ, $R_5 = 29{,}2$ kΩ, $R_6 = 100$ kΩ, $C_1 = 47$ nF and $C_2 = 0{,}47$ nF the filter is tuned to the center frequency $f_0 = 1{,}9$ kHz with the quality factor $Q = 1{,}4$. as shown in Fig. 7.

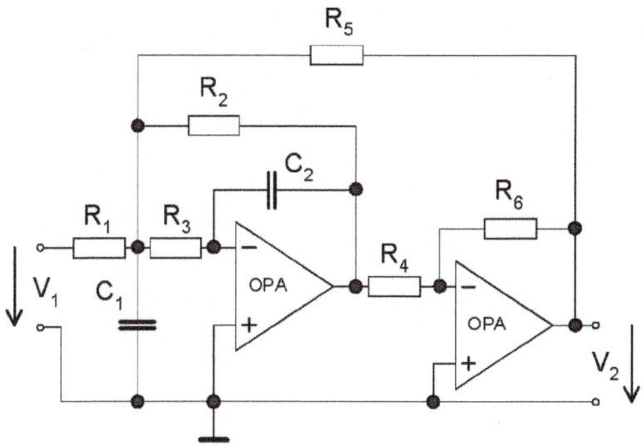

Fig. 6: Second-order low-pass filter.

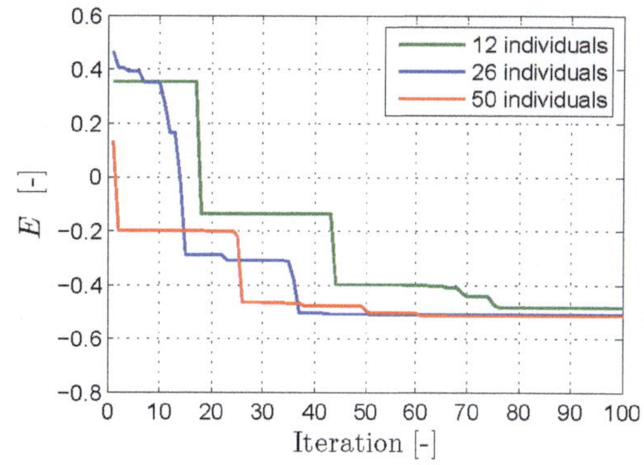

Fig. 9: Convergence of GA optimization (Low-pass).

One test point corresponding to magnitude measurements of V_2 was considered. The testability degree T is equal to 3. When proceeding as in [10] the same suitable testable group (R_6, C_1, C_2) was chosen.

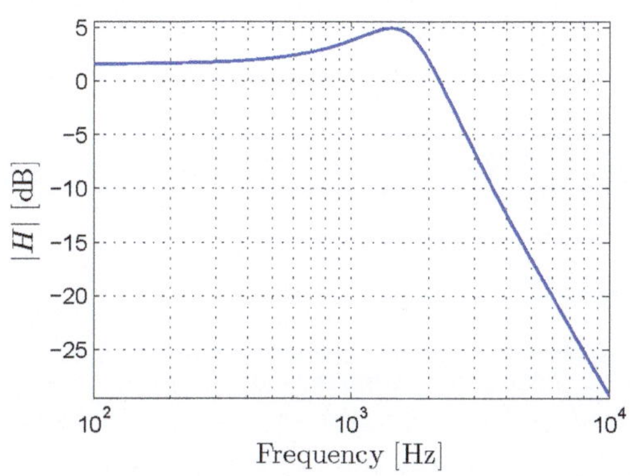

Fig. 7: Second-order low-pass filter frequency response.

After approximately 15 iteration steps, PSO found the following frequencies: $f_1 = 4{,}7$ Hz, $f_2 = 1663$ Hz and $f_3 = 8472$ Hz ($E = -0{,}51$). The results are shown in Fig. 8. As can be seen, PSO with a small number of agents did not converge to the global optimum. The results of GA are shown in Fig. 9.

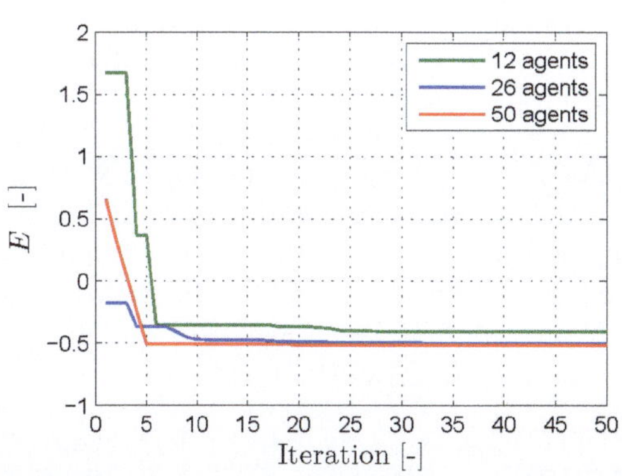

Fig. 8: Convergence of PSO optimization (Low-pass).

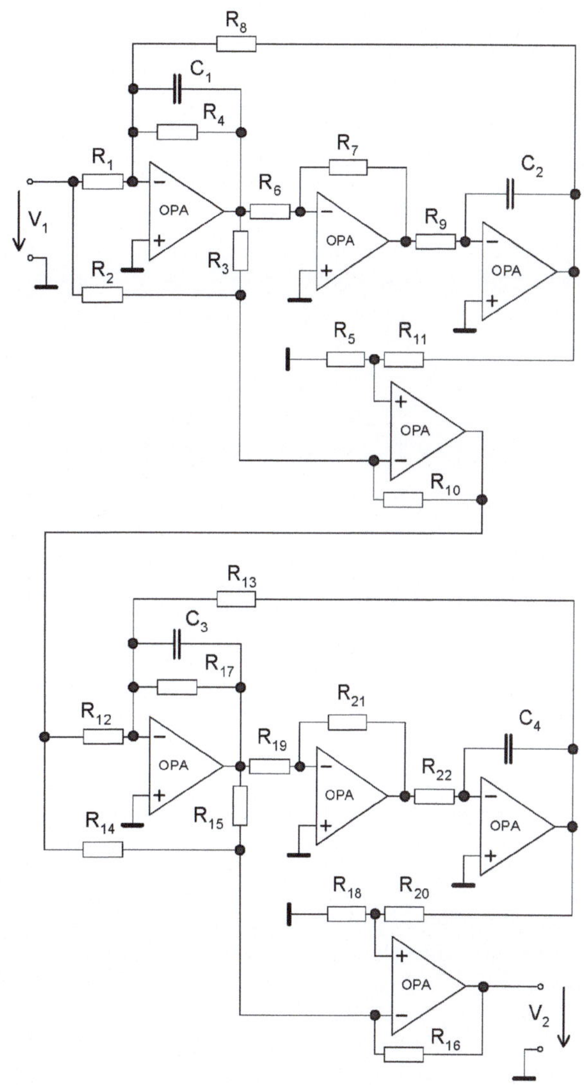

Fig. 10: Fourth-order elliptic low-pass filter.

The last example is the fourth-order elliptic low-pass filter shown in Fig. 10. For nominal values of components $R_1 = R_2 = R_5 = R_6 = R_9 = R_{10} = R_{11} = R_{12} = R_{13} = R_{14} = R_{16} = R_{18} = R_{19} = R_{20} = R_{21} = R_{22} = 1$ kΩ, $R_3 = R_4 = 2$ kΩ, $R_7 = 8$ kΩ, $R_8 = 10$ kΩ, $R_{15} = R_{17} = 500$ Ω, $C_1 = C_3 = C_4 = 100$ nF, $C_2 = 200$ nF the filter has the frequency response shown in Fig. 11.

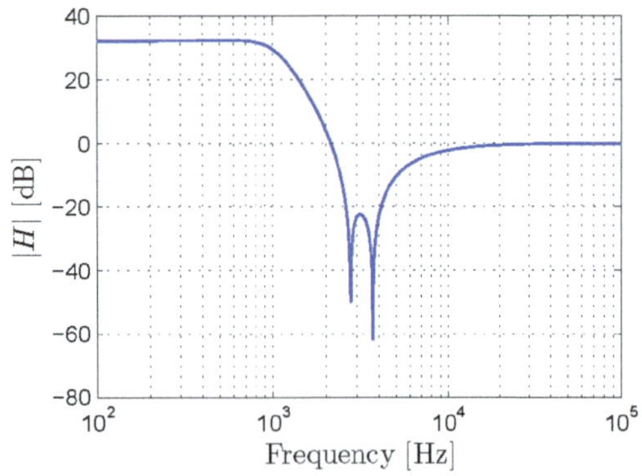

Fig. 11: Frequency response of elliptic low-pass filter.

By proceeding as in [10] the suitable testable group (R_4, R_8, R_{17}, C_1, C_3) with respect to magnitude measurements of V_2 was chosen.

After approximately 20 iteration steps, PSO found the global optimum. The set of test frequencies is $f_1 =$ 12,1 Hz, $f_2 =$ 1011 Hz, $f_3 =$ 2380 Hz, $f_4 =$ 3135 Hz and $f_5 =$ 4166 Hz ($E = 0,50$). The results are shown in Fig. 12. The speed of convergence is good and the global optimum was also found with a small number of agents.

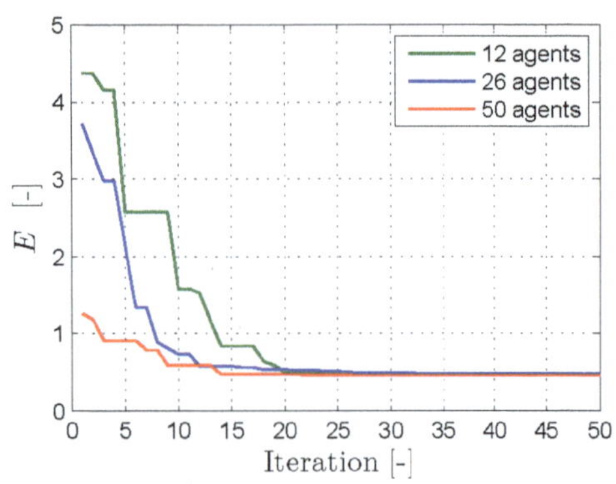

Fig. 12: Convergence of PSO optimization (Elliptic).

The results of GA are shown in Fig. 13. The speed of optimization convergence is much lower than PSO. In [10], the authors state that it may take up to thousands of iteration steps.

4. Conclusion

The paper shows a novel approach to optimum test frequency selection using the modified particle swarm optimization, which minimizes the Test Index measure E. In comparison with Genetic Algorithms the method converges faster and provides results of greater accuracy.

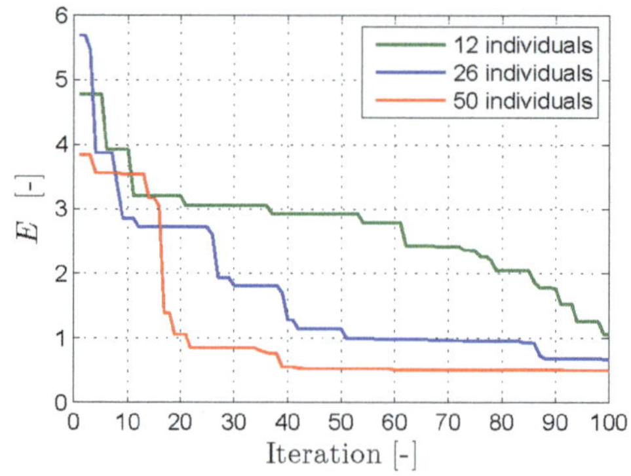

Fig. 13: Convergence of GA optimization (Elliptic).

Acknowledgment

The research was supported by the Czech Ministry of Industry and Trade under grand arrangement No. FR-TI3/431.The research was performed in laboratories supported by the SIX project no. CZ.1.05/2.1.00/03.0072, the operation program Research and Development for Innovation.

References

[1] SUN, Y. *Test and Diagnosis of Analogue, Mixed-Signal and RF Integrated Circuits: The System on Chip Approach.* London: IET Publishing, 2008. ISBN 978-0863417450.

[2] SLAMANI, M. and B. KAMINSKA. Multi-frequency analysis of faults in analog circuits. *IEEE Design Test of Computers.* 1995, vol. 12, iss. 2, pp. 70–80. ISSN 0740-7475. DOI: 10.1109/54.386008.

[3] SEN, N. and R. SAEKS. Fault diagnosis for linear systems via multifrequency measurements. *IEEE Transactions on Circuits and Systems.* 1979, vol. 26, iss. 7, pp. 457–465. ISSN 0098-4094. DOI: 10.1109/TCS.1979.1084659.

[4] KOLKA, Z., V. BIOLKOVA, Z. KINCL and D. BIOLEK. Parametric reduction of Jacobian matrix for fault analysis. In: *International Conference on Microelectronics (ICM), 2010.* Cairo: IEEE, 2011, pp. 503–506. ISBN 978-1-61284-149-6. DOI: 10.1109/ICM.2010.5696200.

[5] LIBERATORE, A., S. MANETTI, and M. C. PICCIRILLI. A new efficient method for analog circuit testability measurement. In: *Instrumentation and Measurement Technology Conference, 1994. IMTC/94.* Hamamatsu: IEEE, 1994, pp. 193–196. ISBN 0-7803-1880-3. DOI: 10.1109/IMTC.1994.352094.

[6] MANETTI, S. and M. C. PICCIRILLI. A singular-value decomposition approach for ambiguity group determination in analog circuits. *IEEE Transactions on Circuits and Systems I: Fundamental Theory and Applications.* 2003, vol. 50, iss. 4, pp. 477–487. ISSN 1057-7122. DOI: 10.1109/TCSI.2003.809811.

[7] WEI, T., M. W. T. WONG and Y. S. LEE. Efficient multifrequency analysis of fault diagnosis in analog circuits based on large change sensitivity computation. In: *Proceedings of the Fifth AsianTest Symposium, 1996.* Hsinchu: IEEE, 1996, pp. 232–237. ISBN 0-8186-7478-4. DOI: 10.1109/ATS.1996.555164.

[8] SHI, C. J. R. and M. W. TIAN. Automatic test generation of linear analog circuits under parameter variations. In: *Design Automation Conference 1998. Proceedings of the ASP-DAC '98.* Yokohama: IEEE, 1998, pp. 501–506. ISBN 0-7803-4425-1. DOI: 10.1109/ASPDAC.1998.669535.

[9] GRASSO, F., A. LUCHETTA, S. MANETTI and M. C. PICCIRILLI. Symbolic Techniques for the Selection of Test Frequencies in Analog Fault Diagnosis. *Analog Integrated Circuits Signal Processing.* 2004, vol. 40, iss. 3, pp. 205–213. ISSN 1573-1979. DOI: 10.1023/B:ALOG.0000034824.68502.99.

[10] GRASSO, F., A. LUCHETTA, S. MANETTI and M. C. PICCIRILLI. A Method for the Automatic Selection of Test Frequencies in Analog Fault Diagnosis. *IEEE Transactions on Instrumentation and Measurement.* 2007, vol. 56, iss. 6, pp. 2322–2329. ISSN 0018-9456. DOI: 10.1109/TIM.2007.907947.

[11] ROBINSON, J. and Y. RAHMAT-SAMII. Particle swarm optimization in electromagnetics. *IEEE Transactions on Antennas and Propagation.* 2004, vol. 52, iss. 2, pp. 397–407. ISSN 0018-926X. DOI: 10.1109/TAP.2004.823969.

[12] ENGELBRECHT, A. P. *Fundamentals of Computational Swarm Intelligence.* New York: John Wiley & Sons, 2006. ISBN 978-0470091913.

[13] BACK, T., U. HAMMEL, U. and H. P. SCHWEFEL. Evolutionary computation: comments on the history and current state. *IEEE Transactions on Evolutionary Computation.* 1997, vol. 1, iss. 1, pp. 3–17. ISSN 1089-778X. DOI: 10.1109/4235.585888.

[14] VLACH, J. and K. SINGHAL. *Computer Methods for Circuit Analysis and Design.* New York: John Wiley & Sons, 1983. ISBN 0442281080.

About Authors

Zdenek KINCL was born in 1985 in Brno, Czech Republic. In 2009, he graduated from the Faculty of Electrical Engineering, Brno University of Technology. At present, he is a Ph.D. student at the Department of Radio Electronics. He is interested in the fault diagnosis of analog circuits.

Zdenek KOLKA received the M.Sc. and Ph.D. degrees in 1992 and 1997 from Technical University in Brno, Czech Republic. Currently he is a professor at the Department of Radio Electronics, Faculty of Electrical Engineering and Communication, Brno University of Technology, and also in the position of deputy head of the department. Areas of his research include linear and nonlinear circuit modeling, numerical methods and circuit simulators. He was involved in the development of several methods for approximate symbolic analysis based on the simplification of circuit model.

15

Design of Two Feeder Three Phase Four Wire Distribution System Utilizing Multi Converter UPQC with Fuzzy Logic Controller

Chandra Babu PADUCHURI, Subhransu Sekhar DASH, Subramani CHINNAMUTHU

Department of Electrical and Electronics Engineering, Faculty of Engineering and Technology, SRM University, SRM Nagar, Kattankulathur, 603 203 Chennai, India

paduchuri.chandrababu@gmail.com, munu_dash_2k@yahoo.com, csmsrm@gmail.com

Abstract. *This paper proposes the instantaneous p-q theory based fuzzy logic controller (FLC) for multi-converter unified power quality conditioner (MC-UPQC) to mitigate power quality issues in two feeders three-phase four-wire distribution systems. The proposed system is extended system of the existing one feeder three-phase four-wire distribution system, which is operated with UPQC. This system is employed with three voltage source converters, which are connected commonly to two feeder distribution systems. The performance of this proposed system used to compensate voltage sag, neutral current mitigation and compensation of voltage and current harmonics under linear and nonlinear load conditions. The neutral current flowing in series transformers is zero in the implementation of the proposed system. The simulation performance analysis is carried out using MATLAB.*

Keywords

Fuzzy logic controller, instantaneous p–q theory, MC-UPQC, power quality, VSC.

1. Introduction

Modern days power systems are complicated networks with hundreds of generating stations and load centers being interconnected through power transmission lines. A few years back, the main concern of consumers of electricity was the reliability of supply. The transmission systems compound the problem further as they are exposed to the vagaries of mother nature. Power quality variations are classified as either disturbance or steady state variations. The disturbances pertained be due to abnormalities in the system such as, Voltage/currents due to a fault or some abnormal operations. A steady state variation refers to a rms deviation from the nominal quantities/harmonics.

In reality the power electronics devices are connected to the power distribution system to increase the reliability and supplied quality power to the consumers. The family of Custom Power Devices (CPD) are Distribution Statcom (DSTATCOM), Dynamic Voltage Restorer (DVR), Unified Power Quality Conditioner (UPQC). CPD can provide an integrated solution for the problems faced by the utilities and power distributors. It can improve in terms of reduced interruptions, voltage variation, load balancing, power factor correction, harmonic free voltage to consumers and voltage regulation. Among these, the UPQC alone can inject unbalanced, distorted voltages/currents; individually and also simultaneously in a dual control mode. This leads to the performance functions of load compensations and voltage control at the same time. Due to the development in recent years many power quality conditioners are proposed with the help of power electronics converters.

A new structure of three-phase four-wire (3P4W) distribution system utilizing UPQC to balance the unbalanced load currents, neutral current compensation using PI control technique is reported in [1]. Design of UPQC with minimization of DC link voltage for the improvement of power quality using FLC alone, along with PI controller and its comparison is reported in [2]. When the system is connected to 3P4W, the improvement of power quality issues like voltage, current unbalance, distortions, reactive power demanded by the load using UPQC is reported in [5], [6], [7], [8], [9], [10].

MC-UPQC consisting of three VSC's connected to the system for simultaneous compensation of voltage and current imperfection is proposed in [11], [12], [13], [8]. One feeder 3P4W distribution system can be provided the neutral conductor along with the three power lines from the generation station [8], [9], [10] to improve the unbalanced voltage and current by using PI controller. A new control strategy of 3P4W UPQC for power quality improvement is reported in [14], [15], [16], [17], to the control of shunt APF with a simple scheme based on the real component of fundamental load current ($I \cos \Phi$) to reduce the number of current sensors is applied. The above scheme also used to evaluate in terms of power factor correction, source neutral current mitigation, load balancing, mitigation of voltage and current harmonics with linear and non-linear loads.

To compensate the voltage/current unbalance there are many control techniques are available. However most of these reported controller are not compensated load and shunt neutral currents. The transformer neutral current have some distortions, it is not equal to zero while all operating conditions were not discussed in [1]. To achieve this, the proposed system to compensate the neutral current by using four leg voltage source converters in shunt.

This work proposes the instantaneous p–q theory based controller for two feeder 3P4W system fed UPQC. Here series transformer's neutral wire is connected to load as a fourth wire of 3P4W systems. The new control system has been modeled with FLC for above mentioned power quality problems of two feeders 3P4W. The system consists of two-series converters and one shunt back to back converter with common DC-link capacitor in order to maintain the voltage.

2. Proposed System Description

The MC-UPQC is connected to the distribution side to make the source and load voltage free from any distortions. The single line diagram of MC-UPQC for the proposed control scheme of two feeder 3P4W systems is shown in Fig. 1.

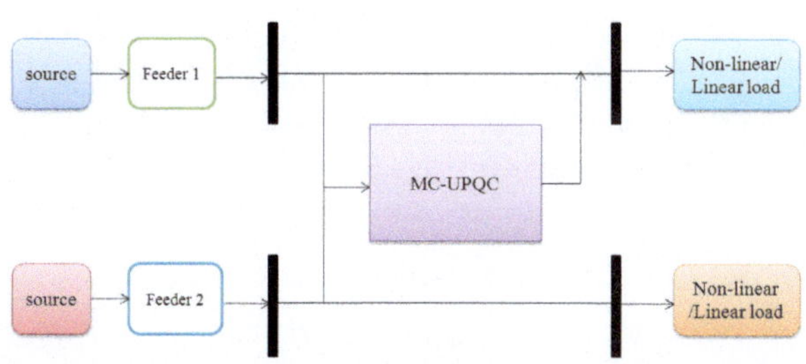

Fig. 1: Single line diagram of MC-UPQC.

At the same time, the reactive current drawn from the source should be similar to that of the currents at source side that would be in phase with utility voltages. A schematic of the proposed two feeder 3P4W distribution system is connected to MC-UPQCis shown in Fig. 2.

The schematic circuit consists of two-series converter having three-legged VSC, one shunt converter with four-legged VSC. The two feeders are connected the two different substations that are supplying the linear/non-linear loads. The active filters are connected in both shunt & series in order to eliminate the harmonic currents. However, the shunt filters are preferred more than series filters due to greater ease of protection. While considering to controlling the voltage sources in which three linear transformers connected in series along with two-series inverter.

If any neutral current is found to be present, it would flow through the fourth wire of the series transformer neutral point. The four-leg VSI based shunt active filter topology [3], [4], [10] is used to compensate the negative sequence of the source current, load balancing, power factor correction as well as to eliminate the harmonics in source currents. This result in the current at transformer neutral point will be zero. The proposed scheme can also be used to compensate the power quality problems related to different types of faults that occurs in the two feeders (L-G, L-L, L-L-G, L-L-L, L-L-L-G and without faults).

For feeder one, the source side voltages, load voltage, voltage injected by the series APF and DC link voltage between two inverters are represented by V_{s1}, V_{l1}, $V_{inj-se1}$ and V_{dc} are respectively. The current on the source side, load side, shunt current, neutral current on the load side, shunt neutral current and transformer neutral currents are represented by I_{s1}, I_{l1}, I_{sh1}, I_{ln}, I_{sh-n} and I_{sr-n} respectively. Similarly in feeder two, the source voltage, injected voltage, load voltage, source current, load current and transformer neutral current are represented by V_{s2}, V_{inj2}, V_{l2}, I_{s2}, I_{l2} and I_{sr-n2} respectively.

3. Control Strategy for the MC-UPQC

The proposed control strategy aims to generate reference signals for both shunt and series APF of the MC-UPQC. The function of series APF is to eliminate harmonics and for compensating reactive power, supply voltage in both feeders. Whereas the shunt APF functions to eliminate the supply current harmonics, distortions and also to compensate the unbalance, negative sequence current, reactive power, neutral current, thereby maintaining the transformer neutral current as zero during entire operating conditions.

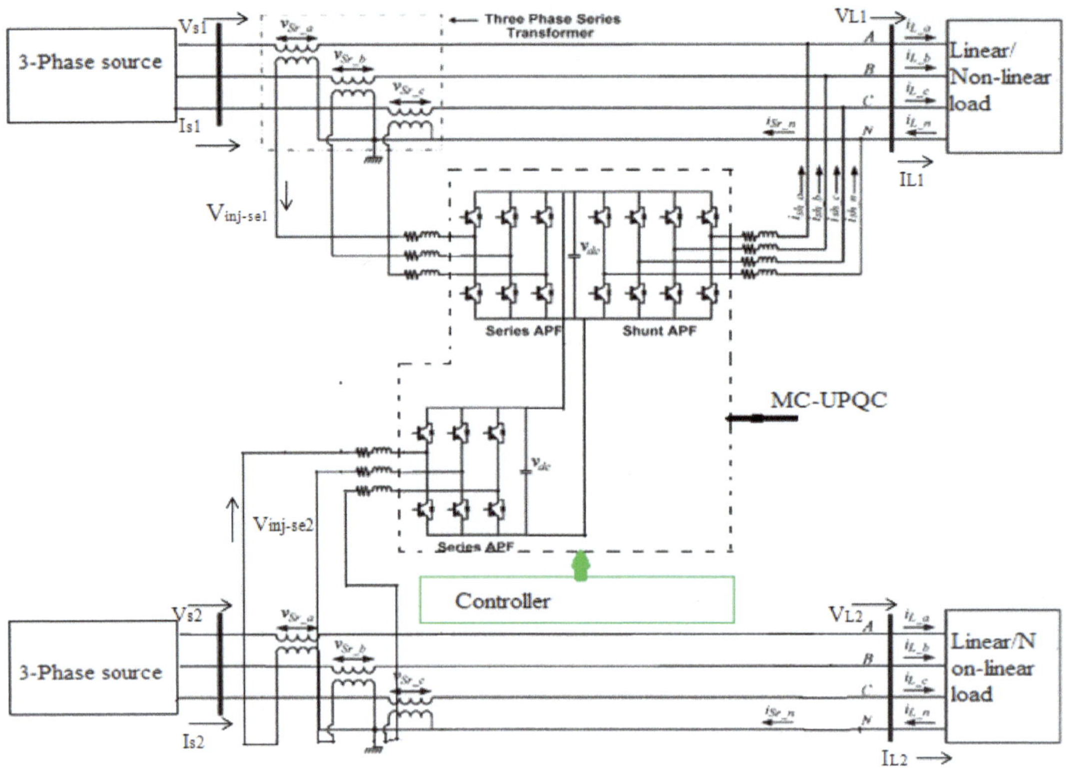

Fig. 2: The schematic of the proposed two feeder 3P4W distribution system is connected to MC-UPQC.

3.1. The Control Scheme of the Series APF's

If the supply voltage is distorted from its original value in feeder one, the phase- locked loop (PLL) is used to accomplish synchronization to bring back to its original supply voltage. The control algorithm for series APF's of two feeders is shown in Fig. 3.

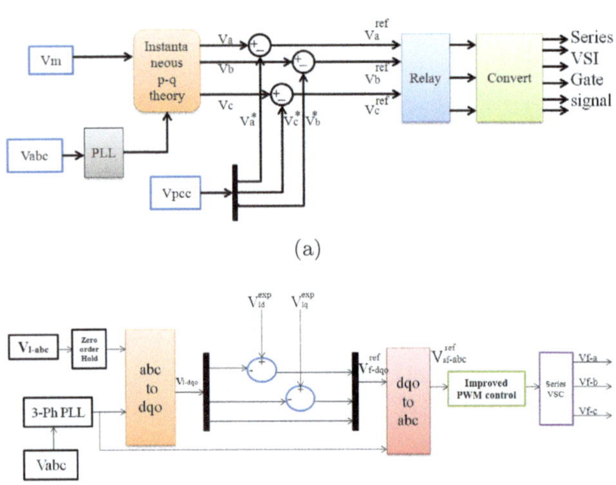

(b)

Fig. 3: Control algorithm for series APF for (a) feeder one (b) feeder two.

When any distorted supply voltages are sensed by PLL it generates two quadrature unit vectors namely, the sine and cosine outputs are used to compute the 120 ° phase displacement for each phase. The instantaneous p–q theory is expanded from the concept of single phase p-q theory [5], [6], [10], [18] and the corresponding voltage equations are mentioned below in Eq. (1), Eq. (2) and Eq. (3). TThe voltages found at the point of common coupling were added to the Eq. (1), Eq. (2) and Eq. (3) and it is given to the relay. The relay output after getting compared with the reference voltages are given to convert a block to generate a gate signal for the series inverter. According to the expanded theory, the phase angle of each phase voltage and currents can be extracted as a three independent two phase system is charitable by $\Pi/2$ lead or lag, this can be applied for three-phase balanced system as well as an unbalanced system also. The control block diagram is shown in Fig. 3(a).

$$V_a = V_m \sin(wt), \tag{1}$$

$$V_b = V_m \sin(wt - 120°), \tag{2}$$

$$V_c = V_m \sin(wt + 120°). \tag{3}$$

The series controller is connected in feeder two, leads to mitigate voltage sag, swell, voltage distortions such as harmonics and interruption. The control algorithm for the series VSC block diagram is shown in Fig. 3(b). The series VSC is designed by using modified Synchronous reference frame (MSRF) theory with improved PWM generator in the proposed scheme. The series VSC is based on the unit vector template by expanding from the concept of MSRF theory. According to this theory, this can also be applicable for both three-phase balanced and unbalanced system. The three-phase load voltages are transformed into load synchronous reference voltages using Eq. (4):

$$\begin{bmatrix} v_{l-d} \\ v_{l-q} \\ v_{l-0} \end{bmatrix} = \begin{bmatrix} v_d \\ v_q \\ v_0 \end{bmatrix} \begin{bmatrix} v_{l-a} \\ v_{l-b} \\ v_{l-c} \end{bmatrix}, \tag{4}$$

where:

$$\begin{bmatrix} v_d \\ v_q \\ v_0 \end{bmatrix} = \frac{2}{3} \begin{bmatrix} \sin wt & \sin\left(wt - \frac{2\pi}{3}\right) & \sin\left(wt + \frac{2\pi}{3}\right) \\ \cos wt & \cos\left(wt - \frac{2\pi}{3}\right) & \cos\left(wt + \frac{2\pi}{3}\right) \\ 1/2 & 1/2 & 1/2 \end{bmatrix} \begin{bmatrix} v_a \\ v_b \\ v_c \end{bmatrix}. \tag{5}$$

According to series control objective, it is that the load voltage must be kept sinusoidal with constant amplitude as far as the supply voltage is concerned. The expected load Synchronous reference dq0 voltages are subtracted with the V_{l-dq0} in Eq. (6) obtained from Eq. (4). AAgain the compensation reference feeder dq0 voltages are transformed back to the synchronous reference feeder voltages using Eq. (7):

$$\begin{bmatrix} v_{f-d}^{ref} \\ v_{f-q}^{ref} \\ v_{f-0}^{ref} \end{bmatrix} = \begin{bmatrix} v_{l-d} \\ v_{l-q} \\ v_{l-0} \end{bmatrix} \begin{bmatrix} v_{l-d}^{ref} \\ v_{l-q}^{ref} \\ v_{l-0}^{ref} \end{bmatrix}, \tag{6}$$

$$\begin{bmatrix} v_{sf-a}^{ref} \\ v_{sf-b}^{ref} \\ v_{sf-c}^{ref} \end{bmatrix} = \begin{bmatrix} v_d \\ v_q \\ v_0 \end{bmatrix}^{-1} \begin{bmatrix} v_{f-d}^{ref} \\ v_{f-q}^{ref} \\ v_{f-0}^{ref} \end{bmatrix}. \tag{7}$$

The compensated synchronous reference abc voltages were forwarded to improve the operation of PWM generator. The output of the PWM generator is directly given to control part of series VSC as shown in Fig. 3(b). However, even the source voltage is distorting the series control objective have to be maintained. The improved PWM generates switching the signal such that the voltage at PCC becomes the desired sinusoidal reference voltage. Due to the series transformer across the injected voltage which cancel out the harmonics present in the supply and load voltage with the help of ripple filter, thereby making them free-from distortion.

3.2. The Control Scheme of the Shunt APF's

It consists of the generation of three-phase reference supply currents; due to the non-linear load on the 3P4W system connected to the source side by which currents are unbalanced. The proposed control system is compensating these unbalanced source currents on the source side by expanding the concept of modified p–q theory [1]. The control algorithms for the shunt APF block diagram are shown in Fig. 4.

On the load side the three-phase load currents are different due to fundamental load active and reactive power. To overcome this unequal load power demand from the source side, the fundamental three-phase active powers can be perfectly balanced; by the following method the unequal load power should be correctly redistributed between source, MC-UPQC and load such that the three-phase load from the source would be linear and equal loads.

The power obtained from the dc-link capacitor through the series inverter and switching losses will decrease the average value of dc bus voltage. In this dc-bus voltage all other distortions like unbalance conditions, sudden changes in load current can result in oscillations, if the transient response of the dc-link capacitor voltage is very slow. In order to overcome from this error between desired capacitor voltage and measured values, at FLC is installed. The output control signal is applied to the current control system in shunt VSI, which stabilizes the dc capacitor voltage by receiving required power from the source.

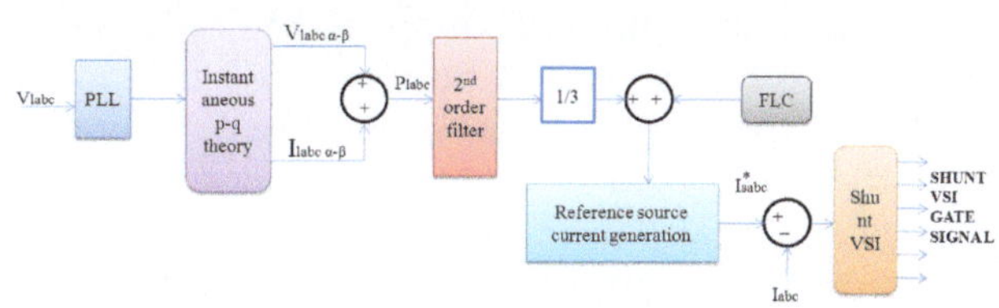

Fig. 4: Control algorithms for the shunt APF block diagram.

The three-phase load instantaneous p–q theory voltage and current can be added to get the active power estimation. The reference source current generation is extracted by the per phase fundamental active power estimation which is added to FLC output, the three-phase source currents were compared to the reference currents through which generates the appropriate shunt gate signals to be given for shunt VSI:

$$P_{labc} = V_{labc\alpha-\beta} + I_{labc\alpha-\beta}. \tag{8}$$

4. Fuzzy Logic Controller

Fuzzy Logic Control (FLC) system is composed of the following four principal components.

- Fuzzification,
- knowledge base,
- inference engine or decision making logic,

- de-fuzzification.

The output from the database and the rules of the knowledge base which were used to get the inference relation B mentioned in the Eq. (9). The input and output variables of controlled system or data bus contains a description of fuzzy sets:

$$B^{(p)} = \text{if } X_1 \text{ is } F_1 \text{ and } X_2 \text{ is } F_2, \ldots, X_n \text{ is } F_n \text{ then } Y \text{ is } C^{(p)}, \tag{9}$$

where X_1, X_2, ..., X_n is the input variables vector, Y is the output or control variable, n is the number of fuzzy variables ($N = 5$), F_1, F_2, ..., F_n is the fuzzy sets, $P = 1, 2, 3, \ldots, N$, N is the number of rules ($N = 5$).

From the given rule base the fuzzy controller has to compute necessary specific input signal conditions, which can determine its effective control action. To design an FLC, the plant control is inferred from the two input state variables, namely error dc capacitor voltage (V_{dc}) and change in reference dc capacitor voltage error (ΔV_{dc}) in Eq. (10):

$$v_{e\ fuzzy} = v_{dc} - v_{dc}^{ref}. \tag{10}$$

The proposed structure of a complete FLC is given in Fig. 5. The Fuzzy control rules are designed for a fuzzy set of the control input in each combination of fuzzy sets for V_{dc} and ΔV_{dc} through which a very small amount of real loss is required for voltage regulation taken as the output from the FLC. The per phase fundamental active power estimation is added with Eq. (10) and it is forwarded to the reference source current generation.

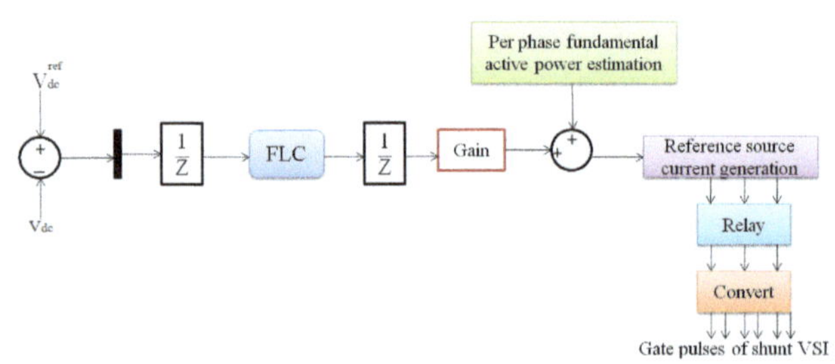

Fig. 5: Proposed structure of a complete Fuzzy logic controller.

The p–q theory based currents are directly given to relay that is sensed to control the signals of shunt VSC circuit as shown in Fig. 5.

In this paper, instead of using conventional (PI) controller mentioned in references a FLC is being used for its transient response to make MC-UPQC very fast in reducing the total harmonic distortions on source and load side voltages as well as currents on both the feeders. Here five labels of fuzzy subsets; negative large (NL), negative medium (NM), zero (ZR), positive medium (PM), positive large (PL). The control rule base table is shown in Tab. 1. In which the row and column represent the error and its changes respectively.

Tab. 1: Rule based for voltage control.

↓V	ΔV→	Input 2				
		NL	NM	ZR	PM	PL
Input 1	NL	NL	NL	NM	NM	ZR
	NM	NL	NM	NM	ZR	PM
	ZR	NM	NM	ZR	PM	PM
	PM	NM	ZR	PM	PM	PL
	PL	ZR	PM	PM	PL	PL

5. Simulation Results And Discussion

In order to justify the strategies of new control scheme discussed above, a Matlab based simulation designed as the system described in Fig. 2, and then the results are compared with a conventional system. It is observed from the results that the proposed new control scheme gives the better simulation results that the conventional system. The designed model of an MC-UPQC system connected with two feeders is realized using the MATLAB/Simulink software environment as shown in Fig. 6.

The three-phase rectifier load is turned on at time $t = 0.1$ s, to realize voltage harmonics in source voltage. The three-phase rectifier load mentioned in the design is a combination of linear and non-linear loads, where non-linear load considered being on the distribution side of feeder one, is a three-phase diode bridge rectifier with R-L load. At this time the shunt APF is connected to maintain the dc-link voltage as constant with respect to set reference value. Immediately the FLC discards load disturbance continuously without overshoot and also with a negligible steady state error.

The voltage is limited by its maximal admissible value, by a saturation function and by making the performance of a fuzzy control system in its adaptive nature. The controller is able to realize in different control scheme for each input state. The performance of the MC-UPQC connected between two feeder systems is used to voltage/current balancing, harmonic mitigation, source voltage/current compensation, power factor correction for both linear and nonlinear loads as well as maintaining the transformer neutral current as zero during all operating conditions. The control algorithm for shunt and series APF simulation diagram are shown in Fig. 7 and Fig. 8.

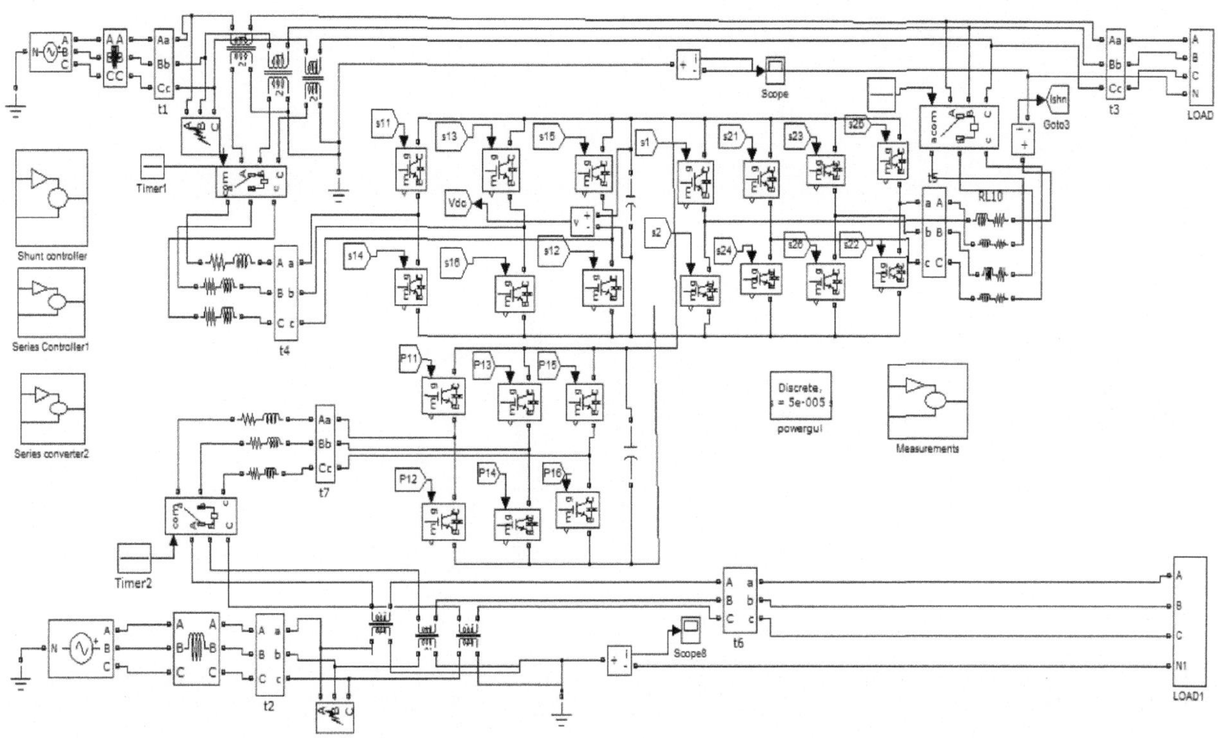

Fig. 6: Simulation circuit diagram of two feeder three phase four wire distribution system with MC-UPQC.

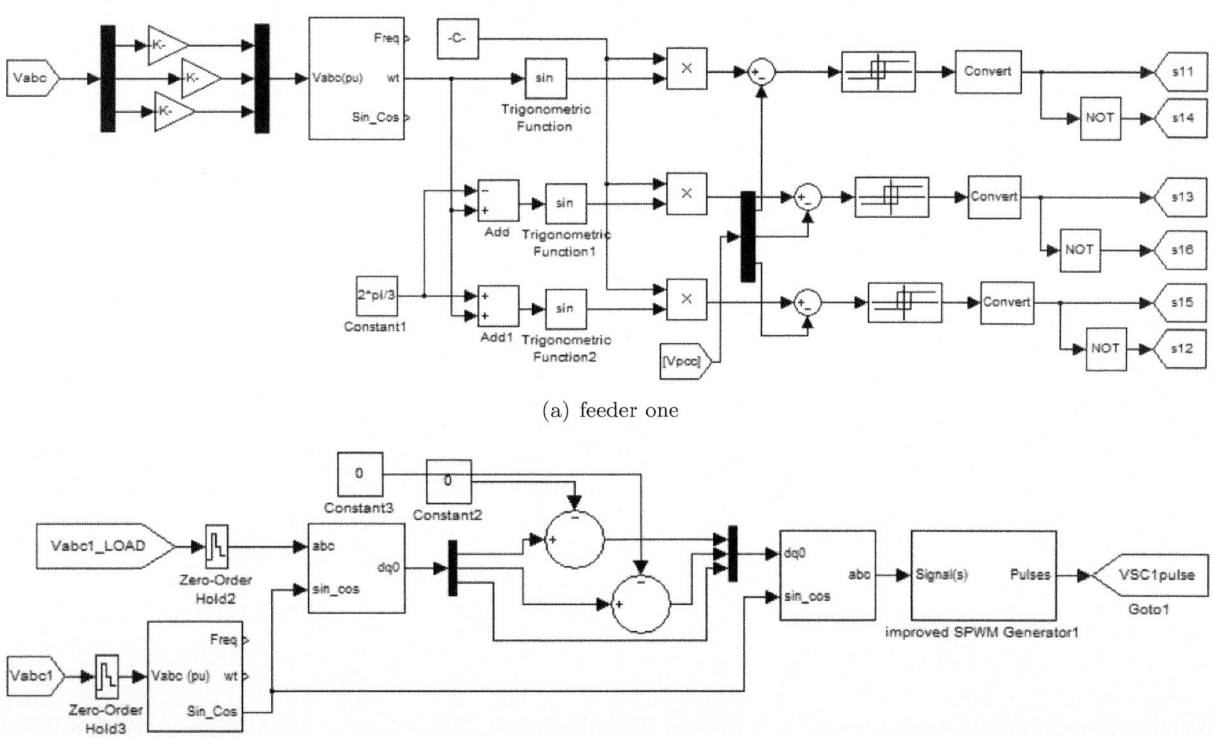

(a) feeder one

(b) feeder two

Fig. 7: Simulation circuit diagram for series APF.

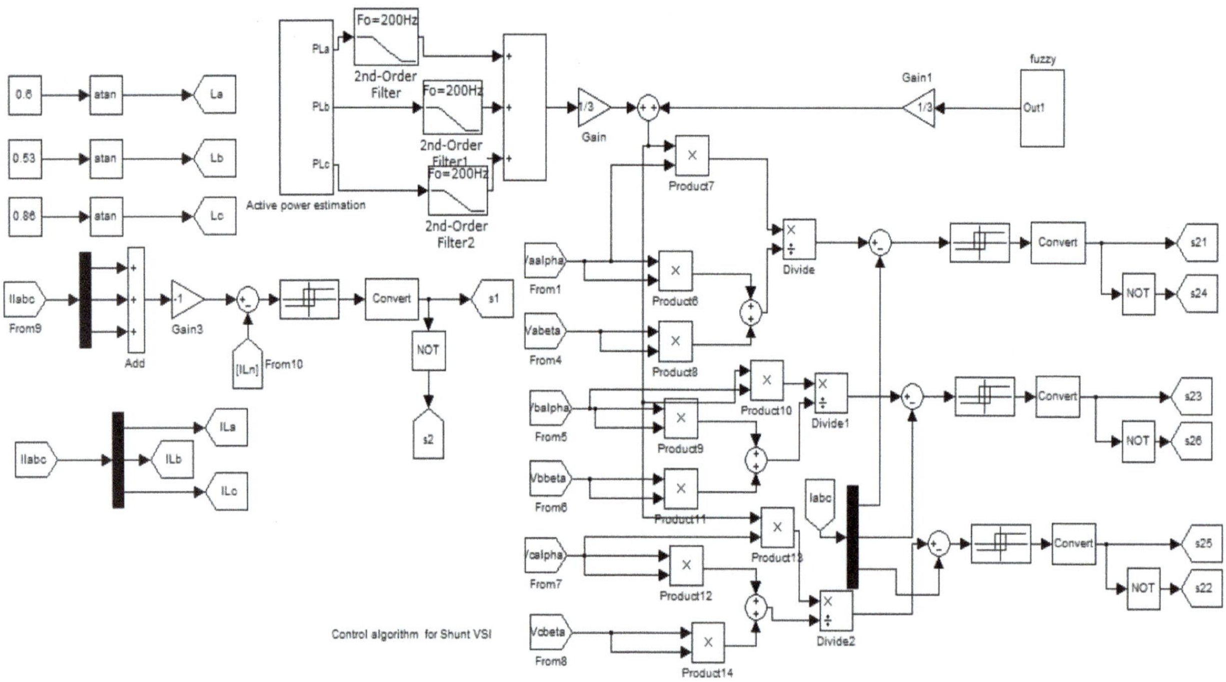

Fig. 8: Simulation circuit diagram for shunt APF.

5.1. Performance Analysis of MC-UPQC Connected to Feeder One

The performance of the MC-UPQC connected to feeder one with proposed new control techniques simulation results were presented in Fig. 9, Fig. 10, Fig. 11, Fig. 12 and Fig. 13.

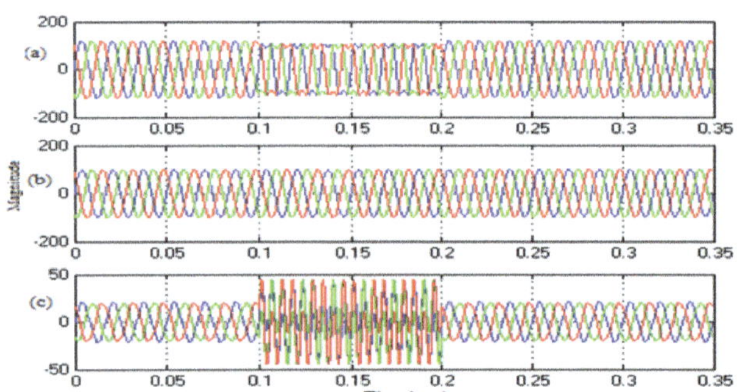

Fig. 9: Simulation results in feeder one; (a) source voltage, (b) load voltage, (c) injected voltage.

At time $t = 0.1$ s, the shunt APF is turned on which injects the compensating currents to realize the balanced source current free from distortions. The compensated source, load and injected currents are shown in Fig. 11. The THD seems to be reduced for compensated source current from 2.31 % to 2.28 %, whereas the shunt neutral current from 90.94 % to 90.90 % and load current from 12.15 % to 12.12 % respectively. The harmonic spectra of the source, load and injected currents

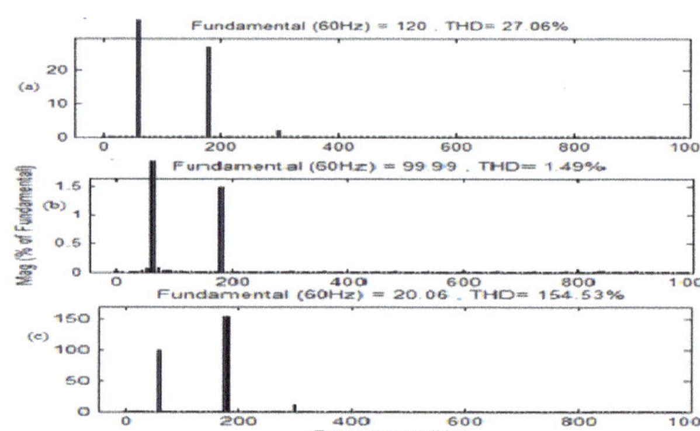

Fig. 10: Harmonic spectra of (a) source voltage, (b) load voltage and (c) injected voltage.

are shown in Fig. 12. Due to unbalanced load condition the load neutral current (25 A) that may follow towards the transformer neutral point. The load, injected shunt neutral current, dc-link capacitor voltage, transformer neutral current and harmonic spectra are shown in Fig. 13 (a)–(e) respectively. In this waveform we can see that transformer neutral current is without any distortions, with reduced THD from 0.24 % to 0.21 % and the dc-link capacitor voltage is maintained constant without any distortions.

The series APF is turned on at time $t = 0.2$ s, the injected voltage, source, and load voltage profile are shown in Fig. 9. The harmonic spectra of the source, load, and injected voltages are also shown in Fig. 10. It is observed that in the mentioned time the THD value has reduced for source voltage from 27.08 % to 27.05 %,

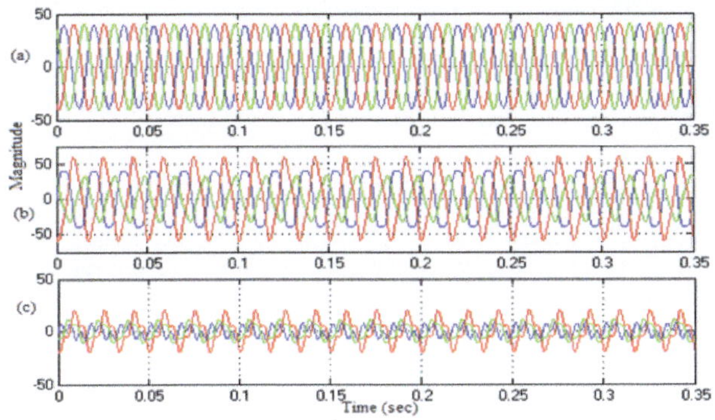

Fig. 11: Simulation results of (a) compensated source currents, (b) load currents, (c) injected currents.

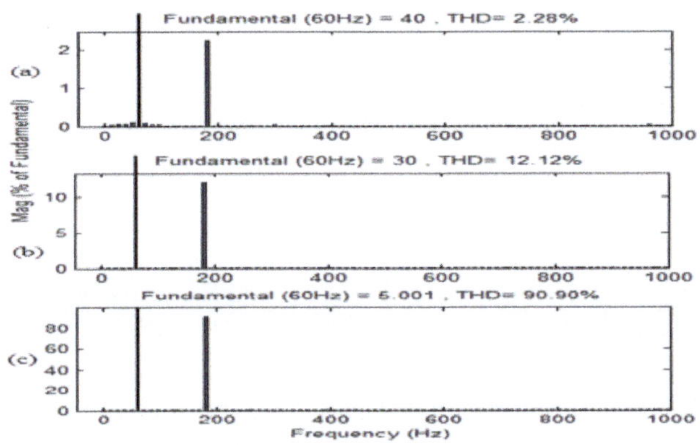

Fig. 12: Harmonic spectra of (a) compensated source currents, (b) load currents and (c) injected currents.

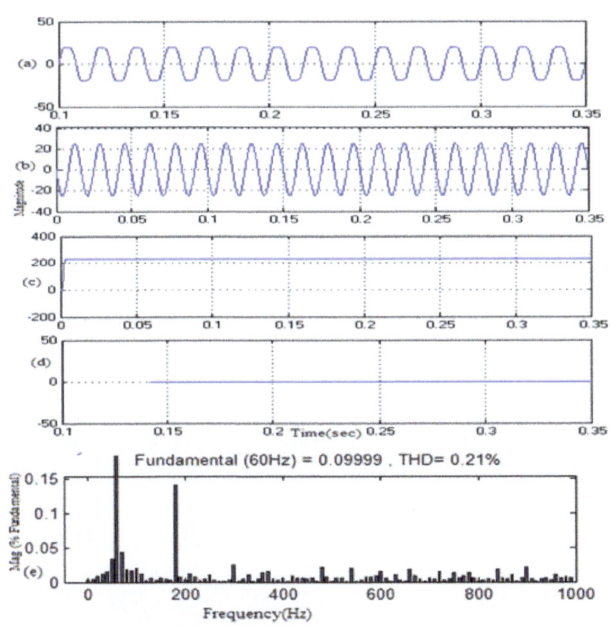

Fig. 13: Simulation results of (a) load neutral current, (b) injected shunt neutral current, (c) dc-link capacitor voltage, (d) transformer neutral current and (e) harmonic spectra for transformer neutral current.

load voltage from 1.51 % to 1.48 % and injected voltage from 154.56 % to 154.47 %.

5.2. Performance Analysis of MC-UPQC Connected to Feeder Two

The performance of MC-UPQC under new control scheme connected to feeder two is done to improve the source voltage and load current from distortion using MSRF based control techniques incorporated with FLC techniques..

The simulation results are shown in Fig. 14, Fig. 15, Fig. 16. At time $t = 0.25$ s, the series APF is turned on, the series APF gets starts compensating for the source voltages harmonics immediately by injecting out the phase harmonic voltages thus by making the source and load voltage free from distortions. The source voltage, source current, injected voltage and load voltage simulations, harmonic spectra are shown in Fig. 14, Fig. 15 respectively.

In feeder one, it is observed that source voltages are having distortions from 0.1 to 0.2 s along with injected voltage also. The source voltage in feeder one is distorted with THD value of 27.08 %. To compensate this distorted voltage in feeder one it is extended to feeder two with a new control scheme incorporated in the MC-UPQC results in reduced THD value of 0.19 %. The feeder two load current and transformer load neutral simulations, harmonic spectra are shown in Fig. 16.

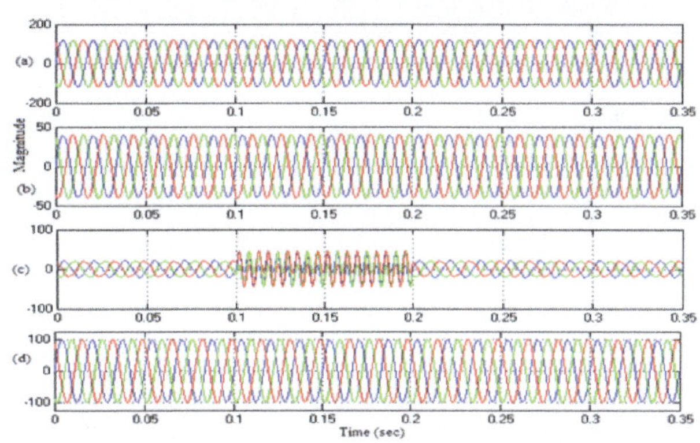

Fig. 14: Results of feeder two (a) source voltages, (b) source current, (c) injected voltage and (d) load voltage.

The comparison between conventional and new control scheme with reduced THD values are mentioned in Tab. 2. From the results it is observed that the proposed new control scheme gives better results even under the system is subjected to different types of faults like L-G, L-L, L-L-G, L -L-L and without faults. The harmonic spectra of different types of faults are listed in Tab. 3. The system with and without controller, the load neutral and shunt neutral having 25 A is verified. Not only these currents were compensated but also load neutral current flowing in the fourth wire of transformer neutral point is also maintained zero during all the operating conditions.

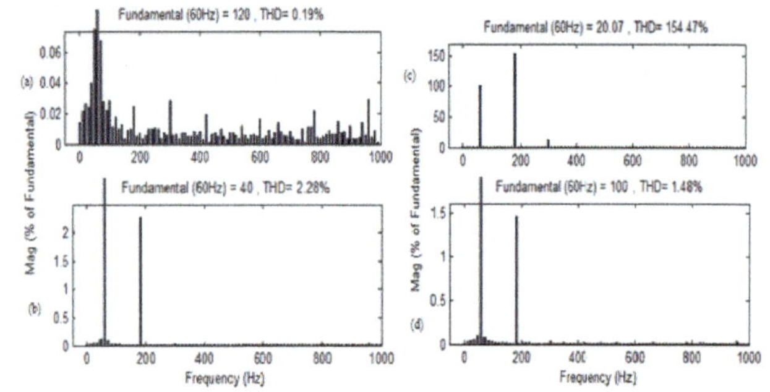

Fig. 15: THD Results of Feeder two (a) source voltage, (b) source current, (c) injected voltage and (d) load voltage by using Fuzzy Logic Controller.

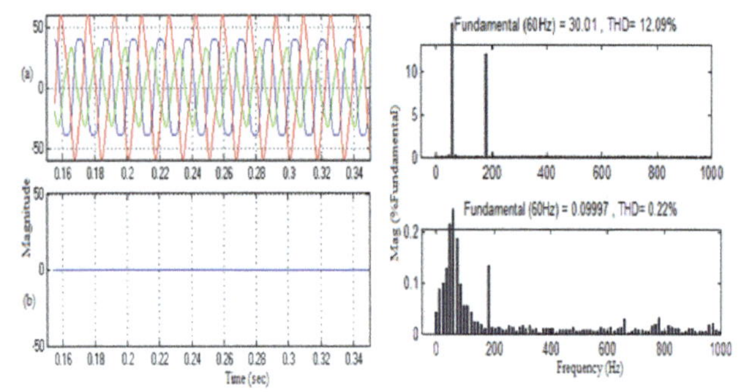

Fig. 16: Simulation results and THD value of (a) feeder two load current, (b) transformer load neutral current.

Tab. 2: % THD values comparison between existing system and proposed control strategy.

Voltage/ Current	Existing System [1]	Proposed Controller
Feeder 1	Considered	Considered
Source Voltage	27.08	**27.05**
Load Voltage	1.51	**1.48**
Injected Voltage	154.56	**154.47**
Source Current	2.31	**2.28**
Load Current	12.15	**12.12**
Shunt Current	90.94	**90.90**
Neutral Current	0.24	**0.21**
Feeder 2	Not Considered	Considered
Source Voltage	Not Considered	**0.19**
Load Voltage	Not Considered	1.48
Injected Voltage	Not Considered	154.47
Source Current	Not Considered	2.28
Load Current	Not Considered	**12.09**
Neutral Current	Not Considered	0.21

6. Conclusion

In this paper, the proposed control topology for two feeders three-phase four-wire system utilizing MC-UPQC is validated extensively under various operating conditions and compared with a conventional controller. The main advantages found from the proposed system compared to the conventional system are to compensate the power quality issues, such as the load voltage, load currents balancing, voltage and current harmonics mitigation. The source, load voltage and source, neutral currents with its harmonic levels are maintained according to IEEE 1159-1995 std., under all the operating conditions. Moreover, the transformer load neutral current is compensated, and it is found to be zero without any distortions in the load side. The effect of series converter is to eliminate the distortions on the supply side voltages due to unbalanced load conditions; the effect of shunt converter is not only to compensate the neutral current it also to make the current balanced on the source side, when they act individually. The proposed transient response of DC-link capacitor voltage is very fast when compare with the PI controller.

Tab. 3: % THD values comparison between different types of loads.

Voltage/ Current	L-G	L-L	L-L-G	L-L-L	Without faults
Feeder 1	**Considered**				
Source Voltage	7.16	34.54	11.02	683.34	18.04
Load Voltage	1.51	1.51	1.52	1.51	1.50
Injected Voltage	1.53	32.00	1.60	31.02	38.70
Source Current	2.31	2.31	2.32	2.31	2.30
Load Current	12.15	12.15	12.16	12.15	12.14
Shunt Current	90.93	90.93	90.96	90.93	90.92
Neutral Current	0.24	0.24	0.25	0.26	0.23
Feeder 2	**Considered**				
Source Voltage	35.42	54.19	54.00	28492.8	27.08
Load Voltage	1.51	1.51	1.52	1.51	1.50
Injected Voltage	1.53	32.00	1.60	31.02	38.70
Source Current	2.31	2.31	2.32	2.31	2.30

References

[1] KHADKIKAR, V. and A. CHANDRA. A Novel Structure for Three-Phase Four-Wire Distribution System Utilizing Unified Power Quality Conditioner (UPQC). *IEEE Transactions on Industry Applications*. 2009, vol. 45, iss. 5, pp. 1897–1902. ISSN 0093–9994. DOI: 10.1109/TIA.2009.2027147.

[2] RAMA RAO, R. V. D. and S. S. DASH. Design of UPQC with minimization of DC link voltage for the improvement of power quality by Fuzzy logic controller. *International Journal on Electrical & Power Engineering*. 2011, vol. 2, iss. 1, pp. 36–42. ISSN 2158-7574.

[3] SINGH, B., K. AL-HADDAD and A. CHANDRA. A review of active power filters for power quality improvement. *IEEE Transactions on Industrial Electronics.* 1999, vol. 46, iss. 5, pp. 960–971. ISSN 0278-0046. DOI: 10.1109/41.793345.

[4] QUINN, C. A. and N. MOHAN. Active filtering of harmonic currents in three-phase, four-wire systems with three-phase and single-phase nonlinear loads. In: *Applied Power Electronics Conference and Exposition, 1992. APEC '92. Conference Proceedings 1992., 7th Annual.* Boston: IEEE, 1992, pp. 829–836. DOI: 10.1109/APEC.1992.228328.

[5] CORREA, J. M., S. CHAKRABORTY, M. G. SIMOES and F. A. FARRET. A single phase high frequency AC microgrid with an unified power quality conditioner. In: *38th IAS Annual Meeting. Conference Record of the Industry Applications Conference, 2003.* Salt Lake City: IEEE, 2003, pp. 956–962. ISBN 0-7803-7883-0. DOI: 10.1109/IAS.2003.1257652.

[6] KHADKIKAR, V., A. CHANDRA, A. O. BARRY and T. D. NGUYEN. Application of UPQC to protect a sensitive load on a polluted distribution network. In: *IEEE Power Engineering Society General Meeting.* Montreal: IEEE, 2006, pp. 6–12. ISBN 978-1-4244-0493-2. DOI: 10.1109/PES.2006.1709522.

[7] KHADKIKAR, V., A. CHANDRA, A. O. BARRY and T. D. NGUYEN. Conceptual Study of Unified Power Quality Conditioner (UPQC). In: *IEEE International Symposium on Industrial Electronics.* Montreal: IEEE, 2006, pp. 1088–1093. ISBN 1-4244-0496-7. DOI: 10.1109/ISIE.2006.295788.

[8] AREDES, M., K. HEUMANN and E. H. WATANABE. An universal active power line conditioner. *IEEE Transactions on Power Delivery.* 1998, vol. 13, iss. 2, pp. 545–551. ISSN 0885-8977. DOI: 10.1109/61.660927.

[9] FARANDA, R. and I. VALADE. UPQC compensation strategy and design aimed at reducing losses. In: *Proceedings of the IEEE International Symposium on Industrial Electronics, (ISIE 2002).* L'Aquila: IEEE, 2002, pp. 1264–1270. ISBN 0-7803-7369-3. DOI: 10.1109/ISIE.2002.1025971.

[10] CHEN, G., Y. CHEN and K. M. SMEDLEY. Three-phase four-leg active power quality conditioner without references calculation. In: *Nineteenth Annual IEEE Applied Power Electronics Conference and Exposition (APEC '04).* Anaheim: IEEE, 2004, pp. 587–593. ISBN 0-7803-8269-2. DOI: 10.1109/APEC.2004.1295866.

[11] MOHAMMADI, H. R., A. Y. VARJANI and H. MOKHTARI. Multiconverter Unified Power-Quality Conditioning System: MC-UPQC. *IEEE Transactions on Power Delivery.* 2009, vol. 24, iss. 3, pp. 1679–1686. ISSN 0885-8977. DOI: 10.1109/TPWRD.2009.2016822.

[12] FUJITA, H. and H. AKAGI. The unified power quality conditioner: The integration of series and shunt active filters. In: *27th Annual IEEE Power Electronics Specialists Conference (PESC '96).* Baveno: IEEE, 1996, pp. 494–501. ISBN 0-7803-3500-7. DOI: 10.1109/PESC.1996.548626.

[13] GHOSH, A. and G. LEDWICH. A unified power quality conditioner (UPQC) for simultaneous voltage and current compensation. *Electric Power Systems Research.* 2001, vol. 59, iss. 1, pp. 55–63. ISSN 0378-7796. DOI: 10.1016/S0378-7796(01)00141-9.

[14] PAL, Y., A. SWARUP and B. SINGH. A novel control strategy of three-phase, Four-wire UPQC for power quality improvement. *Journal of Electrical Engineering & Technology.* 2012, vol. 7, no. 1, pp. 1–8. ISSN 1975-0102. DOI: 10.5370/JEET.2012.7.1.1.

[15] GUNTHER, E. W. and H. MEHTA. A survey of distribution system power quality. *IEEE Transactions on Power Delivery.* 1995, vol. 10, iss. 1, pp. 322–329. ISSN 0885-8977. DOI: 10.1109/61.368382.

[16] DUGAN, R. C., M. F. MCGRANAGHAN and H.W. BEATY. *Electric Power System Quality.* 3rd edition. New York: McGraw-Hill Companies, 2012. ISBN 978-0071761550.

[17] FUCHS, E. and M. A. S. MAUSOUM. *Power Quality in Power Systems and Electrical Machines.* London: Elsevier Academic Press, 2008. ISBN 978-0123695369.

[18] HAQUE, M. T. Single-phase PQ theory. In: *IEEE 33rd Annual Power Electronics Specialists Conference (PESC 02).* Cairns: IEEE, 2002, pp. 1815–1820. ISBN 0-7803-7262-X. DOI: 10.1109/PSEC.2002.1023074.

About Authors

Chandra Babu PADUCHURI was born in 1987. He received B.Tech from JNTU-K and M.Tech from SRM UNIVERSITY Chennai in 2006 and 2012 respectively. He is currently pursuing his research at SRM University. His area of Interests includes Power

System operation, Control & Power Electronic Converters, Drives and FACTS, DSP. He has published 8 international conference papers.

Subhransu Sekhar DASH was born in 1972. He received A.M.I.E graduation from I.E (India). The M.E Degree from U.C.E, Burla and Ph.D. degree in Electrical Engineering from Anna University, Chennai in 1994, 1996 and 2006 respectively. He has published more number of Papers in National and International reputed Journals. He is presently working as Professor and HOD (EEE) in SRM Engineering College, SRM University, Chennai, India and his areas of interests include FACTS, Power System Operation, Control & Stability, Power Electronics Drives and Intelligent Controlling Techniques.

Subramani CHINNAMUTHU was born in 1976. was born in 1976. He received B.E graduation from Bharathiar University, M.E degree from Anna University, Chennai in 2004 and 2006 respectively. He was completed his Ph.D. degree in Electrical and Electronics Engineering from SRM University in 2012. He is presently working as Assistant professor in the Department of EEE at SRM University. He has published 20 Papers in referred International Journals and more than 40 papers in International/National Conference. He is having 15 years of teaching and industrial experience. His areas of interest are power system stability, power system control and FACTS, electrical machine design.

Fuzzy Multi-Objective Optimal Power Flow Using Genetic Algorithms Applied to Algerian Electrical Network

Ahmed SALHI[1], *Djemai NAIMI*[1], *Tarek BOUKTIR*[2]

[1]Laboratory of Electrical Engineering (LGEB), Department of Electrical Engineering, University Mohamed Khider of Biskra, BP 145, 070 00 Biskra, Algeria
[2]Department of Electrical Engineering, Faculty of Technology, University Ferhat Abbas of Setif, Cite Mabouda, 190 00 Setif, Algeria

a_salhi_m@yahoo.fr, d.naimi@univ-biskra.dz, tarek.bouktir@esrgroups.org

Abstract. *This paper presents a mathematical model for solving Multi-Objective Optimal Power Flow problem considering uncertainties modeled by fuzzy numbers affecting three objective functions given by total generation cost, total gas emission and voltage profile index. The presented resolution approach is based on Genetic Algorithm (GA), where the parameters of this algorithm are determined and optimized after many tests of execution. A model for analyzing trade-off between profit and security constraint is developed. The probabilities of crossover and mutation optimized for GA parameters, dedicated to the presented approach are used to demonstrate a performance and effectiveness of the algorithm compared to other approaches mentioned in this paper. The mathematic model is applied in the Algerian electrical network for 59-bus test system.*

Keywords

Fuzzy logic, gas emission, generation cost, genetic algorithms, optimal power flow, voltage profile index.

1. Introduction

Modern power grids are facing many paradigms that directly affect their state of planning and operations, while being affected by several uncertain variables. Under the influence of market processes and operating constraints, the decision-making for power utilities is influenced mainly by the volatility of fuel prices, fluctuating pollutant gas emission and instability of the voltage caused by load changes from time to another. For such reasons, the Optimal Power Flow (OPF) problem must deal with three problems mentioned above as an optimization problem under uncertainties, where objective functions are handled as fuzzy numbers. We are faced a Multi-Objective Optimal Power Flow (MOOPF) issue for three objectives, the fuel cost minimization (economic generation), less polluted environment with minimization of total emission of environmental pollutants (can be caused by nitrogen oxides NOx and sulphur oxides emitted from generation units based on fuel fossil) and best security of power system described by flatter voltage profile (safety operation of equipments in power system is considered for preferred voltage magnitude at buses closer to 1 p.u).

The ordinary applications of the traditional OPF optimize only a single objective function and the rest of objectives functions must be dealt as constraints [1]. Nevertheless, when such objective functions describe trade-off statements with each other, the development of an efficient algorithm of MOOPF becomes a necessity to solve the problem [2].

Many deterministic optimization techniques in the literature have been applied for solving OPF problems [3], [4], [5], [6], they depend to the convex form of the objective function to find the global optimal solution. Nevertheless, the most OPF problems are nonlinear, non-convex and non-differential, where the previous techniques do not ensure a global optimal solution. Furthermore, those traditional methods used to optimize specific aspects of power system operations are not efficient, because they are not appropriate to deal with many practical aspects encountered in recent electrical networks, including the uncertainties of the objective functions and operational constraints [7]. According to these issues, it is necessary to extend the problem to Fuzzy MOOPF (FMOOPF) problem using Artificial Intelligence (AI) [8]. The application of AI in the recent years has undergone a significant evolution

[9], they include for example Differential Evolution [10], Particle Swarm Optimization (PSO) [11] and Genetic Algorithms (GA) [12].

In this paper, the study is based on the optimization of three objective functions previously cited, while taking into the account the adjustment devices in the electrical network. We describe three cases for the optimization problem. In the first case, each objective function is optimized independently to another one, with determination of maximal and minimal values in order to define the membership function for each one. In the second case, FMOOPF problem is solved to get the best compromise solution for all membership functions using maximization of Degree of Satisfaction (DS). In the third case, the optimization problem is extended for various scenarios by increasing the load to 20 % and 30 % compared to the base case. The problem resolution is based on the optimal choice of GA parameters (crossover and mutation probability), considered as key factors for efficiency and accuracy of the algorithm. The GA parameters adjustment is determined with respect to our own experimental tests (crossover probability with two points). The mathematical optimization model proves its efficiency to overcome other metaheuristic methods. The mathematical algorithm application is interested in the Algerian electrical network for 59 bus test system. For this aim, our paper is structured as follows: After the introduction, section 2 presents the problem formulation that focuses on the FMOOPF Problem describing the approach and the objective functions. Section 3 is devoted to GA dedicated to the resolution of the FMOOPF problem. Section 4 illustrates the optimization strategy and its application on the mathematical model. Simulation, results and discussions are reserved to the Section 5 with comparison of the algorithm to some other methods. Finally, we finished with conclusion.

2. Problem Formulation

The problem formulation is based on fuzzy quantification of objective functions in the aim to define the Multi-Objective Problem.

2.1. Fuzzy Multi-Objective Problem

Regarding human ambiguity, it is more convenient for the Decision Maker (DM) to have such fuzzy targets as, "it is desirable to make each objective function below a certain value f_o" and "it is necessary for each objective function to have a desirable limit value f_m" Fig. 1.

Then the quantitative implementation can be made by defining a membership function for each objective function. For the multi-objective problem formulated

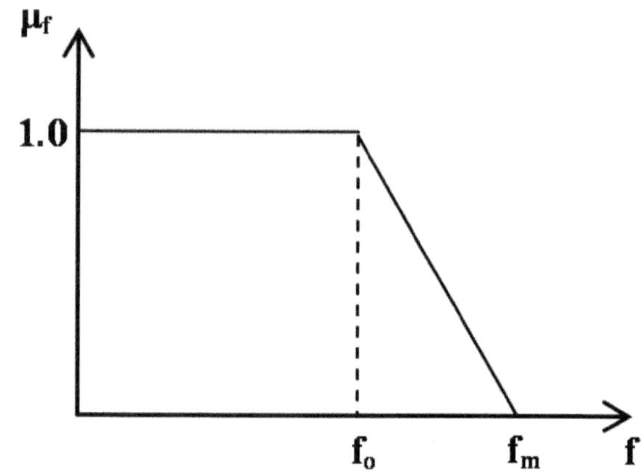

Fig. 1: Membership function for one objective function.

as a vector minimization, it becomes a vector of maximization of all membership functions and evaluation of membership function can be expressed as formulating the DS of DM by changing the problem to conventional OPF with the aim to maximize the DS [13]. The work in this paper is based on the above formulation using GA. We adopt a fuzzy linear distribution for each objective function as shown in the following expression.

$$\mu_{fi}(x) = \begin{cases} 1 & 0 \le f_i(x) \le f_{i(o)} \\ \dfrac{f_i(x) - f_{i(m)}}{f_{i(o)} - f_{i(m)}} & f_{i(o)} \le f_i(x) \le f_{i(m)} \\ 0 & f_i(x) \ge f_{i(m)} \end{cases} . \quad (1)$$

2.2. Objective Functions

The three most important problems considered in the power system operation are economy, environment, and security conflicts selected as the evaluation of three functions described as below:

1) Total Cost Generation Function

The most commonly used objective in the OPF problem formulation is the minimization of the total operating cost of the fuel consumed in producing electric power. The generation cost related to each thermal unit is considered as an economic function with respect to real power, describing a curve in second order and expressing a quadratic function. The total cost of power generation can be calculated as:

$$f_c = \sum_{i=1}^{n_g} (a_i + b_i P_{gi} + c_i P_{gi}^2) \; [\$/h], \quad (2)$$

P_{gi}: active power generation at unit i. n_g: number of total units for active power generation. a_i, b_i and c_i : are the cost coefficients of the ith generator.

2) Total Gas Emission Function

The Kyoto Protocol, adopted in 1997, has forced electricity power companies to reduce their pollutant gases emitted by power plants to keep a clean environment. The objective function of total emission gases is expressed as a function of active power generation and can be summarized as follows:

$$f_E = \sum_{i=1}^{n_g} (a_i + \beta_i P_{gi} + \gamma_i P_{gi}^2 + \delta \exp(\varepsilon P_{gi}))\ [Ton/h], \quad (3)$$

α_i, β_i, γ_i, δ_i and ε_i are coefficients of the ith generator emission characteristics.

3) Function of Voltage Profile Index

The voltage magnitude of load buses must be in proximity of 1 p.u (nominal values), which means a flattened voltage profile for a security issue. The function of Voltage Profile Index (VPI) is as below:

$$f_{VPI} = \sqrt{\sum_{i=1}^{N_{pg}} (V_i - 1)^2}, \quad (4)$$

V_i - Voltage magnitude in p.u at PQ bus i, N_{pg} - Number of PQ buses.

2.3. Fuzzy Multi-Objective Optimization

The generalized problem optimization can be formulated in the following manner:

$$\begin{aligned} & min(f_C, f_E, f_{VPI}) \\ \text{Subject to} \quad & g(x) = 0 \qquad , \\ & h(x) \leq 0 \end{aligned} \quad (5)$$

where *g(x)* and *h(x)* are respectively the set of equality and inequality constraints described respectively in Eq. (6) and Eq. (7).

We have two types of equality constraints; the first is the necessity to keep at each moment, the total generation power equal to the total load demand plus transmission line losses. The second and third equality constraints are described by the power flow equations reflected by the injection of active and reactive powers at each bus.

$$\begin{cases} \sum_{i=1}^{ng} P_{gi} - \sum_{k=1}^{nld} P_{dk} - P_{loss} = 0 \\ Q_i = Q_{gi} - Q_{di} = \sum_{j=1}^{nb} V_i V_j (G_{ij} \cos\theta_{ij} + B_{ij} \sin\theta_{ij}) \\ P_i = P_{gi} - P_{di} = \sum_{j=1}^{nb} V_i V_j (G_{ij} \sin\theta_{ij} - B_{ij} \cos\theta_{ij}) \end{cases} \quad (6)$$

P_{gi} and P_{di}, are the active powers of generation and demand respectively of the *i-th* bus, Q_{gi} and Q_{di} are the reactive powers of generation and demand respectively of the *i-th* bus. The voltage magnitudes of the *i-th* and *j-th* buses are V_i and V_j respectively. The angle θ_{ij} signify the difference of voltage angles for *i-th* and *j-th* buses, while the transfer admittance is modeled by real and imaginary parts G_{ij} and B_{ij} for the same buses. The parameters *nb* is the number of buses and *nld* is the number of loads in power system, Ploss is the total active losses.

The inequality constraints are reflected by the operating limits of electrical network components which are considered for secure power system operation. The most adopted inequality constraints in OPF problem are the limited active and reactive powers at generation buses, voltage limits corresponding to voltage generators at PV-buses and the limited interval of transformer tap setting.

$$\begin{cases} P_{gi}^{min} \leq P_{gi} \leq P_{gi}^{max} & i = 1, 2, ...ng \\ Q_{gi}^{min} \leq Q_{gi} \leq Q_{gi}^{max} & i = 1, 2, ...ng \\ V_{gi}^{min} \leq V_{gi} \leq V_{gi}^{max} & i = 1, 2, ...r \\ T_i^{min} \leq T_i \leq T_i^{max} & i = 1, 2, ...p \end{cases}, \quad (7)$$

where: p and r are the number of transformer tap settings and the number of generators with controlled voltages respectively. $\vec{x}$ is the vector of control and state variables. The control variables are:

- Unit active power output.
- Generation bus voltage magnitude.
- Transformer-tap settings.

The state variables are voltage and angle of load buses.

For the optimization in fuzzy set theory, objectives and constraints are modeled as membership functions. The decision making of Bellman and Zadeh [14] in fuzzy environment is adopted to get the overall degree of membership using the minimum operator with conjunction of all membership functions. The intersection of all fuzzy sets must lead to the optimal solution (optimal decision for the DM) [13]. If $\mu_{f(x)}$ is the membership function of all fuzzy objective functions and $\mu_{h(x)}$

is the membership function of all fuzzy constraints, the overall degree of membership is given as follows :

$$\mu_{DS} = \min(\mu_{f(x)}, \mu_{h(x)})$$
$$\mu_{f(x)} = \min(\mu_{FC(x)}, \mu_{FE(x)}, \mu_{VPI(x)}) \ , \quad (8)$$
$$\mu_{h(x)} = \min(\mu_{h1(x)}, \mu_{h2(x)}, ..., \mu_{hm(x)})$$

m: number of inequality constraints.

The maximization of the overall degree of membership μ_{DS} leads to the optimal solution defined with the highest degree of membership, subject to the crisp and fuzzy constraints. The multi-objective optimization problem Eq. (5) can be transformed as:

$$\max_{x \in X}(z = \mu_{DS}(x))$$
$$z \leq \mu_{fi}(x)$$
$$z \leq \mu_{hk}(x) \ , \quad (9)$$
$$0 \leq z \leq 1$$

with $i = 1,...,N_{obj}$; $k = 1,...,m$ and X denotes the feasible region satisfying all goals and constraints of the problem. N_{obj} is the number of objective functions.

3. Genetic Algorithms

The GA is proposed as a computational model inspired from the natural evolution of survivals by Holand [15] in 1970. It is an optimization procedure to find the globally optimal solution using natural selection process, in which stronger individuals of the population have the highest probability to reproduce based on their level of goodness. Each individual represents a potential solution reflecting a chromosome structure.

In the beginning of GA algorithm, an initial population (encoded in chromosome) satisfying all constraints of the problem is selected. The evaluation of fitness function based on the objective function value is accomplished to rank all members of the population in the goodness order and to perform the selection process of chromosomes that must be reproduced. The combination of the current solutions of the population (pairs from subsets) is achieved based on a crossover process to create a new population of children (called offspring) and mutation mechanism describing occasional interchanges on the genes of chromosomes (particular variables of the problem). The process continues with the same reasoning described previously in every generation until the number of generations is reached. The genetic operators; number of generations, crossover and mutation forms the key factors to find the optimum solution efficiently and accurately [16]. GA increases the probability to achieve an absolute optimum point of the optimization problem without trapping in local optimum points [17]. The parameters of GA for this paper are given as follows:

3.1. Chromosome Type

Each variable of the optimization problem plays as a gene and a set of genes form a chromosome [18]. In the presented paper approach, the chromosome is an answer to the optimization problem as it is represented in Fig. 2.

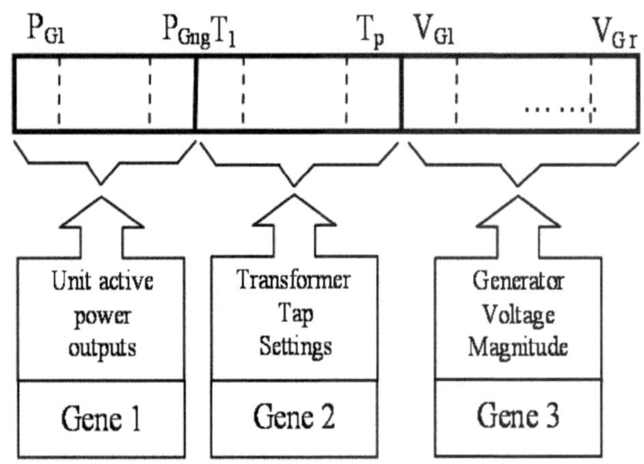

Fig. 2: Chromosome structure described by genes.

3.2. Crossover Type

We must specify a crossover type to produce children from one pair of parents. Among crossover type used in GAs, we choose crossover two points cleared as below: for example, choose randomly two integers *pi* and *qi* from 1 to variables number. The process must select:

- Sub-vector from the parent number 1, for entries numbered less than or equal to *pi*.
- Sub-vector from the parent number 2, for entries numbered from *pi*+1 to *qi*.
- Sub-vector from the parent number 1, numbered greater than *qi*.

The Fig. 3 illustrates two point crossover process with $pi = 2$ and $qi = 5$.

3.3. Mutation Type

The mutation is an essential process of the genetic algorithm, by applying random changes to a single in-

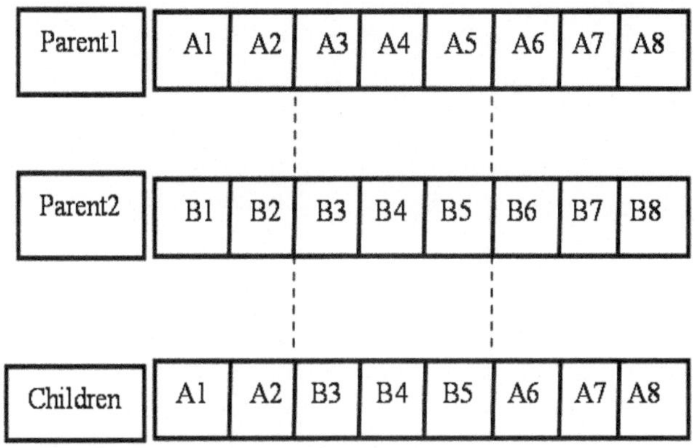

Fig. 3: Two points crossover process illustration.

dividual in the current generation to create a child. It is a mechanism which ensures the diversity of the population and thereby increases the likelihood that the algorithm will generate individuals with better fitness function values. In this paper, a uniform mutation is used and defined in a two-step process. First, the algorithm selects a fraction of the vector entries of an individual for mutation, where each entry has a probability rate of being mutated. In the second step, the algorithm replaces each selected entry by a random number selected uniformly from the range for that entry [19].

3.4. Fitness Function

GAs are usually designed so as to maximize the Fitness Function (FF), which is a measure of the quality of each candidate solution. In this paper the fitness function is defined as follows:

$$FF = \frac{1}{F_{obj} + WP_{fun}}, \tag{10}$$

F_{obj}: Objective Function. P_{fun}: Penalty function for violating constraint in the first equation of system (6). W: Penalty factor.

4. Optimization Strategy

The strategy of optimization aims at the maximization of the DS for the problem formulated in Eq. (9) by GA while being based on the couple (f_o, f_m) for each objective function f, resulting from DM. For this objective, three cases are presented:

- Case1: Each objective function is optimized independently to another one and the couple (f_o, f_m) is extracted. f_o: optimal value of the objective function f given independently to another one. f_m: maximum value of the objective function f imposed during the optimization of another objective functions. The membership function for each objective function is defined based on the couple (f_o, f_m). The objective function in Eq. (10) becomes f_C, f_E or f_{VPI}.
- Case 2: The couple (f_o, f_m) is introduced for each objective function in problem Eq. (9), the DS is maximized by GA and the objective function in Eq. (10) becomes

$$F_{obj} = \frac{1}{DS}. \tag{11}$$

- Case 3: The optimization is focused on various load levels with different scenarios given by:

$$\begin{aligned} P_D &= \lambda P_{DO} \\ Q_D &= \lambda Q_{DO} \end{aligned}, \tag{12}$$

where, P_{DO} and Q_{DO} are active and reactive power demands at base case. λ: is the load factor. (λ=1 for the base case).

5. Simulation and Results

Simulation is applied on the Algerian electrical network for 59-bus test system represented in Fig. 4. The model has 9 generators (except 13th generator bus, no generation power at output), 83 branches (lines and transformers) and 50 load buses. Number of control variables is 22 with $n_g = 9$, $p = 5$ and $r = 8$. The first type of control variables is reserved to the power generation output at buses 1, 2, 3, 4, 27, 37, 41, 42 and 53 the second type is devoted to tap setting transformers at branches 20–22, 29–30, 33–34, 52–53 and 58–59, while the third type is addressed to voltage of generating buses (except the slack bus). Upper and lower active power generating limits, reactive power limits and unit cost coefficients are given in Tab. 1. Emission characteristics of generators for the test system are presented in [20]. Lower and upper limits of voltage magnitude for generator buses are 0,95 and 1,05 p.u respectively. Permissible range related to tap setting transformers is considered from 0,9 to 1,1 p.u.

The proposed approach is developed by the use of MATLAB 7.9 tested with Pentium (R) - Dual-Core CPU 3 GHz, 1 GHz DDR RAM, consistently acceptable results were observed. For base case λ=1, the total active and reactive load are P_{dt}=684,1 [MW] and Q_{dt}=311,6 [MVAR].

Tab. 1: Lower and upper limits of parameters related to generators with cost coefficients.

Bus Gen	Pg min [MW]	Pg max [MW]	Qg min [MVAR]	Qg max [MVAR]	a [$/h]	b [$/MWh]	c [$/MW^{2}h]
1	8	72	−10	15	0	1,5	0,0085
2	10	70	−35	45	0	2,5	0,0170
3	30	510	−35	55	0	1,5	0,0085
4	20	400	−60	90	0	1,5	0,0085
27	10	100	−20	35	0	2,5	0,0170
37	10	100	−20	35	0	2	0,0030
41	15	140	−35	45	0	2	0,0030
42	18	175	−35	55	0	2	0,0030
53	30	450	−100	160	0	1,5	0,0085

Tab. 2: Optimization of one objective function independently to another one at base case.

Control Variables & Objectives	Load level λ=1 (base case)		
	Min FC	*Min FE*	*Min VPI*
Pg(1)[MW]	57, 4753	70, 1355	34, 6141
Pg(2)[MW]	24, 0218	66, 2597	65, 6863
Pg(3)[MW]	101, 2526	97, 1995	73, 1261
Pg(4)[MW]	109, 7631	91, 7974	100, 0925
Pg(27)[MW]	26, 6854	92, 1636	68, 4728
Pg(37)[MW]	51, 8554	58, 2120	49, 8656
Pg(41)[MW]	95, 8106	68, 7312	15, 2855
Pg(42)[MW]	142, 5365	74, 2913	60, 6954
Pg(53)[MW]	103, 8911	88, 6956	250, 0076
T_{20-22}	0, 9677	0, 9812	0, 9510
T_{29-30}	0, 9938	0, 9949	0, 9512
T_{33-34}	1, 0499	1, 0496	0, 9503
T_{52-53}	0, 9501	0, 9597	0, 9506
T_{58-59}	1, 0008	0, 9895	0, 9512
Vg_2[p.u]	1, 0490	1, 0470	1, 0136
Vg_3[p.u]	1, 0499	1, 0496	1, 0006
Vg_4[p.u]	1, 0179	1, 0102	1, 0246
Vg_{27}[p.u]	1, 0169	1, 0115	1, 0260
Vg_{37}[p.u]	1, 0500	1, 0469	1, 0319
Vg_{41}[p.u]	1, 0500	1, 0453	1, 0247
Vg_{42}[p.u]	1, 0500	1, 0485	1, 0479
Vg_{53}[p.u]	1, 0498	1, 0422	1, 0260
F_c[$/h]	1693, 69	1840, 8	2118, 2
F_e[Ton/h]	0, 4918	0, 3970	0, 8824
F_{VPI}[p.u]	0, 2921	0, 2811	0, 1718
P_{loss}[MW]	29, 1918	23, 3859	33, 7459

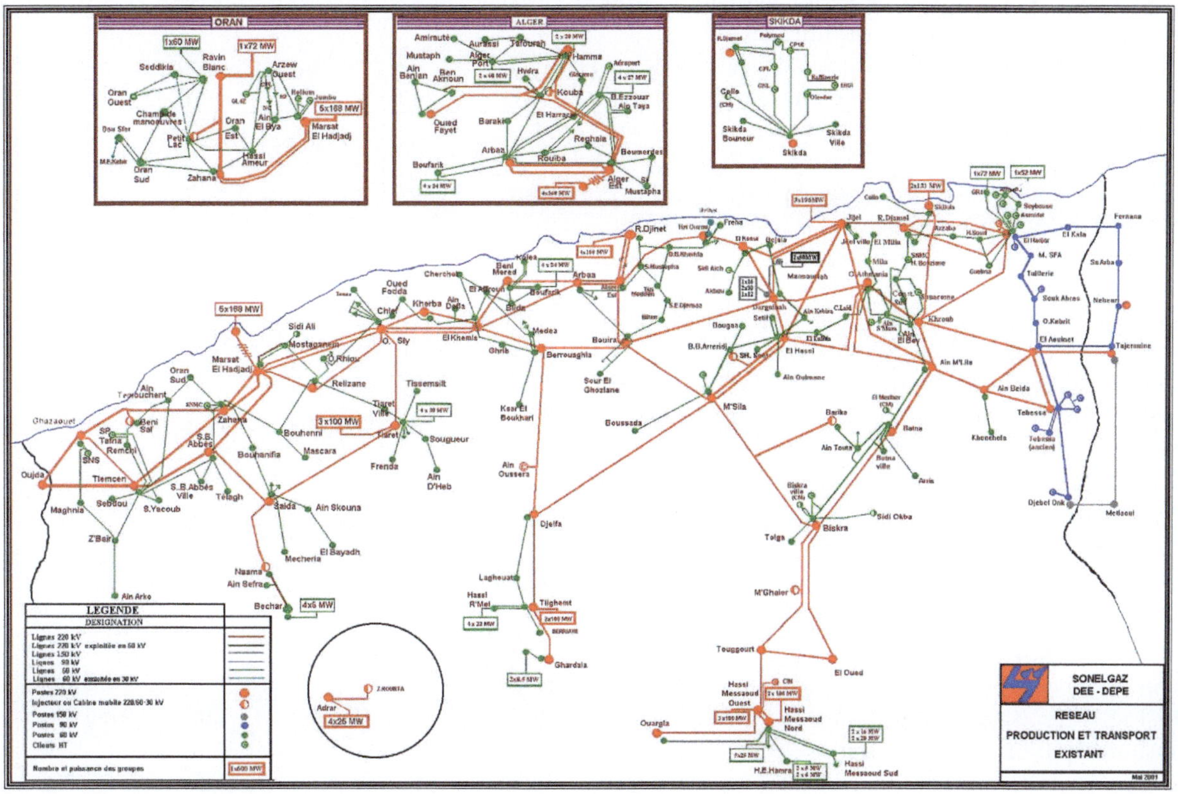

Fig. 4: Topology of the Algerian production and transmission network.

5.1. First Case

The optimal results of several objective functions independently of the one to another are illustrated in Tab. 2 in the base case, specifying the optimal values of the parameters of the adjusting devices. Generation's number, population's size, crossover probability (two points) and mutation probability are respectively 250, 200, 0,7 and 0,061. These parameters are optimized based on our own experimental results (after several tests of execution).

For the fuel cost minimization, our approach gives a more important profit of 1693,6 \$/h compared to three other approaches Genetic Algorithm/Fuzzy Rules (FGA) [21], Decomposed Parallel GA (PGA) [22] and Ant Colony Optimization (ACO) [23], given in Tab. 3. The good simulation results show the best choice of the probabilities of crossover and mutation for the presented approach (compared to the choice of the same parameters indicated in [21] and [22]). Ploss have an acceptable value of 29,19 MW and acceptable voltage profile with minimal value of voltage magnitude 0,9166 p.u referred to Fig. 5. The previous figure describes the voltage profile for each case of optimization.

Best situation of electrical power system security is assigned to the voltage profile index minimization with value of 0,1718 and relatively higher active total losses 33,74 MW, but maximal values of the total cost generation 2118,2 \$/h and emission level 0,882 ton/h referred to other functions minimization. Such remark shows the trade-off between security and cost-benefits problems.

In the case of emission level minimization, 0,3970 ton/h is the better results assigned to our approach compared to optimal emission values for PGA method in [22] with 0,4213 ton/h and for Fast Successive Linear Programming method (FSLP) with 0,4329 ton/h in [24] based on Tab. 4. Medium values of generation cost 1840,8 \$/h and Ploss of 23,3859 MW with violation of voltage magnitude (0,8891 p.u at 35th bus noticed in Fig. 5). All control variables are in their allowable limits referred to declared values in the beginning of this section. It is clearly remarked the trade-off between market benefits, gas emission and voltage security issues.

5.2. Second Case

The couple (f_o, f_m) to define each membership function related to each objective function is shown in Tab. 5 referred to Tab. 2.

The DS is optimized using GA (maximizing the fitness function i.e optimization of three functions simultaneously). The crossover probability pc chosen from four values 0,6; 0,7; 0,8 and 0,9 based on simulation results taken at base case with fixed mutation probability at 0,061. The best value of F_{obj} to maximize the DS is observed for simulation results in Fig. 6 and for pc equal to 0,7 (which proves the best choice previously indicated), where it is established in following simulations. The previous figure exposes the change of the objective function at each generation, with a multitude of crossover probability values and mutation probability of 0,061. Control variables are extracted after simulation for the optimization of three functions simultaneously, DS is maximized and optimal values of objective functions are obtained in the base case and results are arranged in Tab. 6. The convergence has been achieved after 226,116 seconds with 250 iterations.

Voltage magnitudes are closer to 1 p.u for the optimization of three functions simultaneously than the optimization of the cost function and emission function Fig. 5 and Fig. 7 (voltage profile index has the best value in optimization of f_{VPI}). Based on Fig. 7, voltage profile is assigned to the DS maximization without voltage violation (minimal voltage magnitude 0,9005 at 35th bus).

In the base case and for the optimization of three functions simultaneously, we remark based on simulation results in Tab. 6 intermediary's values of three conflicting functions. The tap setting transformers and generation bus voltage magnitudes have contributed to the improvement of the objective functions. As seen in Tab. 7, the presented approach is significantly better than PGA [22] and FSLP [24] methods (for simultaneous minimization of fuel cost and gas emission). The optimized choice of GA parameters confirms the efficiency of the proposed approach which makes it promising to solve the FMOOPF.

5.3. Third Case

For this case, the load level λ=1,2 and λ=1,3 (load is increased by 20 % and 30 %), we extend the code of simulation to obtain the state of the test system in such cases. For the same reasoning as described in the base case, results are detailed in Tab. 8, Tab. 9 and Tab. 10. Increased values are assigned to our objective functions due to expansion of demand (active and reactive total load). Results appear in Tab. 8 for each couple (f_o, f_m). Tab. 9 and Tab. 10 show simulation results at such cases obtained by the proposed approach for DS maximization (control variables are between lower and upper limits).

Tab. 3: Comparison of the proposed approach with other techniques for the total cost minimization case.

Generation [MW]	FGA [21]	PGA [22]	ACO [23]	Our Approach
Pg(1)	11, 193	41, 272	64, 01	57, 4753
Pg(2)	24, 000	37, 319	22, 75	24, 0218
Pg(3)	101, 70	133, 83	82, 37	101, 2526
Pg(4)	84, 160	142, 32	46, 21	109, 7631
Pg(27)	35, 220	24, 80	47, 05	26, 6854
Pg(37)	56, 800	39, 70	65, 56	51, 8554
Pg(41)	121, 38	39, 54	39, 55	95, 8106
Pg(42)	165, 520	119, 78	154, 23	142, 5365
Pg(53)	117, 32	123, 46	202, 36	103, 8911
P_{dt}[MW]	684, 1	684, 1	684, 1	684, 1
P_{loss}[MW]	33, 1930	17, 921	39, 98	29, 19
Min Cost [$/h]	1768, 5	1769, 7	1815, 7	**1693, 6**

Tab. 4: Comparison of the proposed approach with other techniques for total gas emission minimization case.

Generation [MW]	PGA [22]	FSLP [24]	Our Approach
Pg(1)	30, 5995	28, 558	70, 1355
Pg(2)	70, 00	70, 000	66, 2597
Pg(3)	109, 40	114, 200	97, 1995
Pg(4)	79, 80	77, 056	91, 7974
Pg(27)	80, 58	87, 575	92, 1636
Pg(37)	34, 86	32, 278	58, 2120
Pg(41)	70, 04	63, 176	68, 7312
Pg(42)	100, 62	95, 645	74, 2913
Pg(53)	128, 02	135, 540	88, 6956
P_{dt}[MW]	684, 1	684, 1	684, 1
P_{loss}[MW]	19, 8195	19, 93	23, 3859
Min Emission [Ton/h]	0, 4213	0, 4329	**0, 3970**

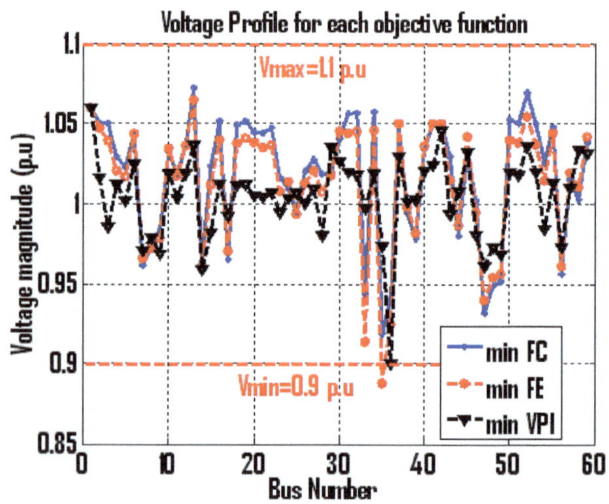

Fig. 5: Voltage profile for the optimization of one objective function at base case.

We obtain the Pareto solutions in wide case, they will be increased by percentage ΔF (%) (for each objective function) related to the base case (for DS maximiza-

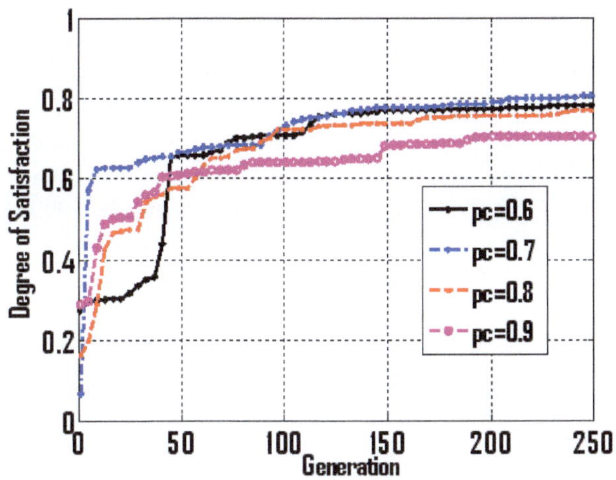

Fig. 6: Evolution of DS with respect to crossover probability at base case λ=1.

Tab. 5: f_o and f_m values for each objective function at base case.

Load level	Base case λ=1	
Function	f_o	f_m
f_C[\$/h]	1693.6	2118.2
f_E[Ton/h]	0.3970	0.8824
f_{VPI}[p.u]	0.1718	0.29214

Tab. 6: Optimization of three functions simultaneously (best compromise solution) at base case.

Control Var. & Objectives	Load level λ=1 (base case)	Control Var. & Objectives	Load level λ=1 (base case)
	Max DS		*Max DS*
Pg(1)[MW]	57, 1366	V_{g2}[p.u]	1, 0131
Pg(2)[MW]	47, 3324	V_{g3}[p.u]	1, 0179
Pg(3)[MW]	120, 6109	V_{g4}[p.u]	1, 0146
Pg(4)[MW]	79, 1477	V_{g27}[p.u]	1, 0159
Pg(27)[MW]	41, 3378	V_{g37}[p.u]	1, 0324
Pg(37)[MW]	52, 7766	V_{g41}[p.u]	1, 0438
Pg(41)[MW]	68, 9375	V_{g42}[p.u]	1, 0500
Pg(42)[MW]	103, 4737	V_{g53}[p.u]	1, 0247
Pg(53)[MW]	139, 8473	F_c[\$/h]	1759, 9
T_{20-22}	0, 9509	F_e [Ton/h]	0, 4669
T_{29-30}	0, 9579	F_{VPI} [p.u]	0, 1905
T_{33-34}	0, 9519	P_{loss}[MW]	26, 50
T_{52-53}	0, 9510	DS	0, 8438
T_{58-59}	0, 9537	P_{dt}[MW]	684, 1

tion) described in Tab. 11. Simulation model allows us to have a database of several best compromise solutions for different load scenarios, without recourse to a Pareto front.

6. Conclusion

An efficient algorithm of Multi-Objective Optimal Power Flow with Fuzzy Logic using Genetic Algorithms is applied on Algerian electrical network with accurate results and best solutions compromise to deal with different types of objective functions.

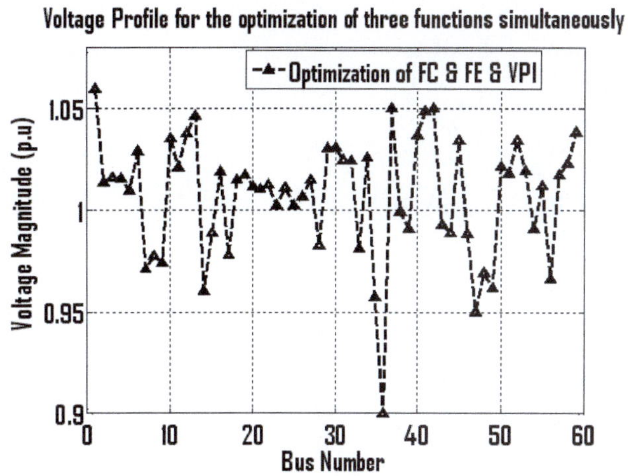

Fig. 7: Voltage profile for the optimization of three functions simultaneously at base case.

Tab. 7: Comparison of the proposed approach with other techniques for DS maximization case.

Generation [MW]	PGA [22]	FSLP [24]	Max DS Approach
Pg(1)	36, 8311	37, 464	57, 1366
Pg(2)	53, 170	52, 675	47, 3324
Pg(3)	119, 06	116, 080	120, 6109
Pg(4)	138, 32	141, 490	79, 1477
Pg(27)	22, 860	28, 286	41, 3378
Pg(37)	39, 800	34, 565	52, 7766
Pg(41)	59, 900	56, 644	68, 9375
Pg(42)	109, 52	101, 800	103, 4737
Pg(53)	122, 92	133, 920	139, 8473
F_C[\$/h]	1765, 7	1786, 000	**1759, 9**
F_E[Ton/h]	0, 4723	0, 4746	**0, 4669**
P_{loss}[MW]	18, 2811	18, 83	26, 50

Tab. 8: f_o and f_m values for each objective function at wide cases λ=1,2 and λ=1,3.

Load level	Wide case λ=1,2		Wide case λ=1,3	
Function	Value of f_o	Value of f_m	Value of f_o	Value of f_m
f_C[\$/h]	2161, 8	2571, 8	2421, 9	3020, 1
f_E[Ton/h]	0, 4964	1, 1347	0, 5709	1, 8284
f_{VPI}[p.u]	0, 2244	0, 3471	0, 2737	0, 3393

The optimized GA parameters of the proposed approach show its effectiveness to enhance the solution quality in comparison to other Algorithms PGA, FSLP and ACO presented in this paper. The security constraints are satisfied for voltage magnitudes ($0,9 < V < 1,1$ p.u) for the optimization of three functions simultaneously at the base and wide cases. Results

Tab. 9: Optimization of three functions simultaneously in wide case λ=1,2 (maximization of DS).

Control Var. & Objectives	Load level λ=1,2 (base case)	Control Var. & Objectives	Load level λ=1,2 (base case)
	Max DS		*Max DS*
Pg(1)[MW]	64,0361	Vg_2[p.u]	1,0213
Pg(2)[MW]	44,5210	Vg_3[p.u]	1,0395
Pg(3)[MW]	157,2404	Vg_4[p.u]	1,0194
Pg(4)[MW]	151,6268	Vg_{27}[p.u]	1,0175
Pg(27)[MW]	51,2728	Vg_{37}[p.u]	1,0500
Pg(37)[MW]	59,2180	Vg_{41}[p.u]	1,0500
Pg(41)[MW]	66,2862	Vg_{42}[p.u]	1,0500
Pg(42)[MW]	113,6315	Vg_{53}[p.u])	1,0307
Pg(53)[MW]	141,6863	F_c[$/h]	2241,5
T_{20-22}	0,9697	F_e [Ton/h]	0,6213
T_{29-30}	0,9502	F_{VPI} [p.u]	0,2477
T_{33-34}	0,9589	P_{loss}[MW]	28.599
T_{52-53}	0,9500	DS	0,8043
T_{58-59}	0,9517	P_{dt}[MW]	820,92

Tab. 10: Optimization of three functions simultaneously in wide case λ=1,3 (maximization of DS).

Control Var. & Objectives	Load level λ=1,3 (base case)	Control Var. & Objectives	Load level λ=1,3 (base case)
	Max DS		*Max DS*
Pg(1)[MW]	65,5808	Vg_2[p.u]	1,0281
Pg(2)[MW]	41,0067	Vg_3[p.u]	1,0433
Pg(3)[MW]	154,0971	Vg_4[p.u]	1,0361
Pg(4)[MW]	165,9475	Vg_{27}[p.u]	1,0372
Pg(27)[MW]	63,0611	Vg_{37}[p.u]	1,0500
Pg(37)[MW]	62,5830	Vg_{41}[p.u]	1,0500
Pg(41)[MW]	69,3981	Vg_{42}[p.u]	1,0500
Pg(42)[MW]	140,0954	Vg_{53}[p.u]	1,0451
Pg(53)[MW]	159,1458	F_c[$/h]	2490,5
T_{20-22}	0,9592	F_e [Ton/h]	0,7152
T_{29-30}	0,9512	F_{VPI} [p.u]	0,2812
T_{33-34}	0,9625	P_{loss}[MW]	31,585
T_{52-53}	0,9500	DS	0,8853
T_{58-59}	0,9633	P_{dt}[MW]	889,33

Tab. 11: Rate of increase of each objective function (in wide cases) referred to base case.

Load level	λ=1,2	λ=1,3
Function	ΔF (%)	ΔF (%)
f_C[$/h]	27,36	41,51
f_E[Ton/h]	34,55	53,18
f_{VPI}[p.u]	30,02	47,61

signal the trade-off between economic environmental dispatch and voltage security constraints.

For the case of DS maximization, a best compromise solution can be obtained without resort to Pareto front. Uncertainties about constraints, ambiguities about practical data description language provide an interactive algorithm with decision maker. The proposed approach creates a platform of relationship analysis between objective functions denoted by the trade-off to get a set of Pareto solutions (best compromise solutions).

References

[1] MOMOH, J. A., M. E. EL-HAWARY and R. ADAPA, Ramababu. A review of selected optimal power flow literature to 1993. II. Newton, linear programming and interior point methods. *IEEE Transactions on Power Systems.* 1999, vol. 14, iss. 1, pp. 105–111. ISSN 0885-8950. DOI: 10.1109/59.744492.

[2] XUEBIN L. Study of multi-objective optimization and multi-attribute decision-making for economic and environmental power dispatch. *Electric Power Systems Research.* 2009, vol. 79, iss. 5, pp. 789–795. ISSN 0378-7796. DOI: 10.1016/j.epsr.2008.10.016.

[3] YANG, Xin-She. *Engineering optimization: an introduction with metaheuristic applications.* Hoboken, N.J.: John Wiley, 2010. ISBN 04-705-8246-4.

[4] RAO, S. *Engineering optimization: theory and practice. 4th ed.* Hoboken, N.J.: John Wiley, 2009. ISBN 978-0-470-18352-6.

[5] ZHU, Jizhong. *Optimization of power system operation: Theory and practice.* 4th ed. Chichester: John Wiley, 2009. IEEE. ISBN 04-702-9888-X.

[6] CAPITANESCU F., M. GLAVIC, D. ERNST and L. WEHENKEL. Interior-Point based Algorithms for the Solution of Optimal Power Flow Problems. *Electric Power Systems Research.* 2007, vol. 77, iss. 5–6, pp. 508–517. ISSN 0378-7796. DOI: 10.1016/j.epsr.2006.05.003.

[7] BEDRINANA, M. F., J. A. BOSCO, C. A. F. MURARI and C. A. CASTRO. Decisions in Power System Operation based on Fuzzy Economic and Environmental Dispatch. In: *IEEE Power Tech 2007.* Lausanne: IEEE: 2007, pp. 1296–1301. ISBN 978-1-4244-2189-3. DOI: 10.1109/PCT.2007.4538503.

[8] HAIDAR, Ahmed M. A., A. MOHAMED and F. MILANO. A computational intelligence-based suite for vulnerability assessment of electrical power systems. *Simulation Modelling Practice and Theory.* 2010, vol. 18, iss. 5, pp. 533–546. ISSN 1569-190X. DOI: 10.1016/j.simpat.2009.12.009.

[9] ABIDO M. A. *Multiobjective Evolutionary Algorithms for Electric Power Dispatch Problem.* Berlin: Springer Berlin Heidelberg, 2009. ISBN 978-3-642-01798-8. DOI: 10.1007/978-3-642-01799-5_3.

[10] ABOU EL ELA, A. A., M. A. ABIDO and S. R. SPEA. Optimal power flow using differential evolution algorithm. *Electric Power Systems Research.* 2010, vol. 80, iss. 7, pp. 878–885. ISSN 0378-7796. DOI: 10.1016/j.epsr.2009.12.018.

[11] SAFARI, A. and H. SHAYEGHI. Iteration particle swarm optimization procedure for economic load dispatch with generator constraints. *Expert Systems with Applications.* 2011, vol. 38, iss. 5, pp. 6043–6048. ISSN 0957-4174. DOI: 10.1016/j.eswa.2010.11.015.

[12] RAHUL, J., Y. SHARMA and D. BIRLA. A New Attempt to Optimize Optimal Power Flow Based Transmission Losses Using Genetic Algorithm. In: *2012 Fourth International Conference on Computational Intelligence and Communication Networks.* Mathura: IEEE, 2012, pp. 566–570. ISBN 978-1-4673-2981-1. DOI: 10.1109/CICN.2012.212.

[13] LIANG, Ruey-Hsun, Sheng-Ren TSAI, Yie-Tone CHEN and Wan-Tsun TSENG. Optimal power flow by a fuzzy based hybrid particle swarm optimization approach. *Electric Power Systems Research.* 2011, vol. 81, iss. 7, pp. 1466–1474. ISSN 0378-7796. DOI: 10.1016/j.epsr.2011.02.011.

[14] ABOU EL-ELAA A. A., M. BISHRA, S. AL-LAMA and R. EL-SEHIEMY. Optimal Preventive Control Actions using Multi-Objective Fuzzy Linear Programming Technique. *Electric Power Systems Research.* 2005, vol. 74, iss. 1, pp. 147–155. ISSN 0378-7796. DOI: 10.1016/j.epsr.2004.08.014.

[15] MAN, K. F., K. S. TANG and S. KWONG. Genetic algorithms: concepts and applications. *IEEE Transactions on Industrial Electronics.* 1996, vol. 43, iss. 5, pp. 519–534. ISSN 0278-0046. DOI: 10.1109/41.538609.

[16] SIVANANDAM, S. *Introduction to genetic algorithms.* Berlin: Springer, 2008. ISBN 978-3-540-73189-4.

[17] KUMARI, M. S. and S. MAHESWARAPU. Enhanced Genetic Algorithm based computation technique for multi-objective Optimal Power Flow solution. *International Journal of Electrical Power.* 2010, vol. 32, iss. 6, pp. 736–742. ISSN 0142-0615. DOI: 10.1016/j.ijepes.2010.01.010.

[18] LAKSHMI, G. Venkata and K. AMARESH. Optimal Power Flow with TCSC using Genetic Algorithm. In: *2012 IEEE International Conference on Power Electronics, Drives and Energy Systems (PEDES).* Bengaluru: IEEE, 2012, pp. 1–6. ISBN 978-1-4673-4506-4. DOI: 10.1109/PEDES.2012.6484394.

[19] User's Guide. *Global Optimization Toolbox.* Mathworks Inc, 2011.

[20] MAHDAD B., T. BOUKTIR and K. SRAIRI. OPF with environmental constraints with SVC controller using decomposed parallel GA: Application to the Algerian network. In: *Power & Energy Society General Meeting (PES'09).* Calgary, AB: IEEE, 2009, pp. 1–8. ISSN 1944-9925. ISBN 978-1-4244-4241-6. DOI: 10.1109/PES.2009.5275817.

[21] MAHDAD B., T. BOUKTIR and K. SRAIRI. Optimal power Flow of the Algerian Network using Genetic Algorithm/Fuzzy Rules. In: *Power and Energy Society General Meeting - Conversion and Delivery of Electrical Energy in the 21st Century, IEEE General Meeting 2008.* Pittsburgh: IEEE, 2008, pp. 1–8. ISSN 1932-5517. ISBN 978-1-4244-1905-0. DOI: 10.1109/PES.2008.4596656.

[22] MAHDAD B., T. BOUKTIR and K. SRAIRI. OPF with Environmental Constraints with Multi Shunt Dynamic Controllers using Decomposed Parallel GA: Application to the Algerian Network. *Journal of Electrical Engineering & Technology.* 2009, vol. 4, iss. 1, pp. 55–65. ISSN 1975-0102.

[23] BOUKTIR T. and L. SLIMANI. Optimal Power Flow of the Algerian Electrical Network using an Ant Colony Optimization Method. *Leoanardo Journal of Sciences.* 2005, vol. 4, iss. 7, pp. 43–57. ISSN 1583-0233.

[24] ZEHAR K. and S. SAYAH. Optimal Power Flow with Envirenmental Constraint using a Fast Successive Linear Programming Algorithm: Application to the Algerian Power System. *Energy Conversion and Management.* 2008, vol. 49, iss. 11, pp. 3362–3366. ISSN 0196-8904. DOI: 10.1016/j.enconman.2007.10.033.

About Authors

Ahmed SALHI was born in Biskra, Algeria in 1967. He received the B.Sc. degree in Electrical Engineering Power system from Biskra University Algeria in 1992, his M.Sc. degree from Batna University in 2003. He is spirit to prepare his Ph.D. degree in power system. His areas of interest are the application of the meta-heuristic methods in optimal power flow, FACTS control and improvement in electric power systems, Multi-Objective Optimization for power systems, and Voltage Stability and Security Analysis.

Djemai NAIMI was born in Batna, Algeria in 1967. He is a professor in the department of electrical engineering in Biskra University, Algeria. He received his M.Sc. degree in electrical engineering on 2003 from Constantine University, Algeria. He is also a member of LGEB laboratory. His research activities include integration of renewable energy in power system grid, optimization, and power flow and power system stability. His teaching includes modeling and optimization in power system and power system stability.

Tarek BOUKTIR was born in Ras El-Oued, Algeria in 1971. He received the B.Sc. degree in Electrical Engineering Power system from Setif University (Algeria) in 1994, his M.Sc. degree from Annaba University in 1998, his Ph.D. degree in power system from Batna University (Algeria) in 2004. He is with the Department of Electrical Engineering in Ferhat Abbes University (Setif), ALGERIA. His areas of interest are the application of the meta-heuristic methods in linebreak optimal power flow, FACTS control and improvement in electric power systems, Multi-Objective Optimization for power systems, and Voltage Stability and Security Analysis. He is the Editor-In-Chief of Journal of Electrical Systems (Algeria), He currently serves on the editorial boards of TELKOMNIKA Journal, Indonesia. He serves as reviewer with the Journals: IEEE Transactions on SYSTEMS, MAN, AND CYBERNETICS, IEEE Transactions on Power Systems (USA), ETEP-European Transactions on Electrical Power Engineering. He is also the head of the research team "Algerian Smart Grid".

17

Direct Thrust Force and Flux Control of a PM-Linear Synchronous Motor Using Fuzzy Sliding-Mode Observer

Saeed MASOUMI KAZRAJI, Mohammad Bagher BANNAE SHARIFIAN

Electrical Power Department, Faculty of Electrical and Computer Engineering, University of Tabriz, 29 Bahman Blvd, 5166616471 Tabriz, Iran

masoumikazraji@tabrizu.ac.ir, sharifian@tabrizu.ac.ir

Abstract. *In this paper, a fuzzy sliding-mode observer, which uses sigmoid function for speed sensorless control of permanent-magnet linear synchronous motor (PMLSM) is proposed. Most of the observers use sign or saturation functions that need low pass filter in order to detecting back electromotive force. In this paper, the sigmoid function is used instead of discontinuous sign function to decrease undesirable chattering phenomenon. By reducing the chattering, detecting back EMF can be done directly from switching signal without any low pass filter. So the delay time because of low pass filter, in the proposed observer is eliminated. Furthermore, there is no need to compensate phase fault in position estimating. The simulation results show advantages of proposed observer over conventional ones.*

Keywords

Direct Thrust Force and Flux Control (DTFC), Fuzzy Logic Control (FLC), Permanent Magnet Linear Synchronous Motor (PMLSM), Sliding Mode Observer (SMO).

1. Introduction

Permanent-magnet linear synchronous motors (PMLSMs) are used widely in systems that need linear drives. They are good alternatives for rotary motors that work with mechanical converters to producing linear movement. The main reasons of absorbing many attentions by PMLSMs are larger power density, high efficiency and faster response due to the presence of the magnetic field that linear induction motors do not have [1], [2], [3], [4] and [5].

Direct thrust force and flux control (DTFC) approach that is accepted as a form of vector control methods beside vector current control. This control approach carries out a strong operation in both of the transient and steady state and has a structure that responses fast and robust to variation of parameters. It does not have the complexity of field oriented control and can be feasible [6], [7]. In [8] and [9] DTFC approach for rotary synchronous motor has been turned into a PMLSM successfully. In this control method, the estimation of the flux, the position and the speed of the motor are paid much attention. Sensorless control is divided into two main types:

- Estimate with use of observers.
- Signal injection method.

Comparative methods, Kalman filter, and sliding mode are some of estimating methods that are based on observers. In general, the first type methods depend on the precision of the motor's model, especially the comparative methods. In other words, whatever the model of the motor is accurate; efficacy of this method will be improved. Kalman filter method requires a large amount of computing. In order to reducing the time of the computations, the system must be improved by the hardware, but it is not affordable. Between mentioned methods for estimating based on observers, sliding mode with sign function has a simple algorithm and is firm against disturbances, fault of parameters and noise. In conventional sliding mode observers, a low-pass filter is used to operation of filtering, but using the low-pass filter causes a lag of phase that depends on the cut frequency and the angular frequency of the input signal. In order to compensating the phase lag completely, the received information from the real angular velocity must be used, and the information, which is given by the estimated angular velocity, is not enough for compensating [10], [11], [12]. As it mentioned above, in conventional sliding mode observer method there are the chattering phenomenon and the phase lag. Thus, in order to avoid of using the low pass filter and the phase compensator based on back EMF, in this paper a fuzzy sliding mode observer with sigmoid function for detecting the back EMF in a PMLSM is designed to estimate the speed and the position of the rotor.

Overview of the mathematical model of PMLSM is given in Section 2. In Section 3, the proposed fuzzy sliding mode observer design is presented. Section 4 presents estimation rotor position and velocity by using proposed method. The simulation result is presented in Section 5. Finally, section 6 presents summary and conclusions.

2. Mathematical Model of PMLSM

In order to obtain the PMLSM model in d-q reference frame, first the stator voltage equation should be introduced [13]:

$$u_d(t) = R_s i_d + \frac{\mathrm{d}\psi_d}{\mathrm{d}t} - \omega\psi_q, \tag{1}$$

$$u_q(t) = R_s i_q + \frac{\mathrm{d}\psi_q}{\mathrm{d}t} - \omega\psi_d, \tag{2}$$

$$\omega = \frac{\pi}{\tau} v_{lin}, \tag{3}$$

where u_d and u_q are the d and q axis components of voltage space vector, i_d and i_q are the d and q axis components of current space vector and R_s is the phase resistance of the armature.

The linkage fluxes ψ_d and ψ_q in above equations are yielded from these:

$$\psi_d = L_d i_d + \psi_{PM}, \tag{4}$$

$$\psi_q = L_q i_q, \tag{5}$$

where L_d and L_q are the inductance of armature and ψ_{PM} is the linkage flux of permanent magnet.

The input three phase instant power of armature p is obtained by follows:

$$P = u_A i_A + u_B i_B + u_C i_C = \frac{3}{2}(u_d i_d + u_q i_q), \tag{6}$$

where u_A, u_B and u_C are the instant phase voltages, i_A, i_B and i_C are the instant phase currents, u_d and u_q are the d and q axis voltages, i_d and i_q are the d and q axis currents.

By using the stator voltage equations, the power equation is expressed as:

$$\begin{aligned} u_d i_d + u_q i_q &= R_s i_d^2 + \tfrac{\mathrm{d}\psi_d}{\mathrm{d}t} i_d + R_s i_q^2 + \tfrac{\mathrm{d}\psi_q}{\mathrm{d}t} i_q + \\ &+ \omega(\psi_d i_d - \psi_q i_q), \end{aligned} \tag{7}$$

where the last term is the electromagnetic power of two poles synchronous machine in each phase. In three phase machines, the power equation is:

$$\begin{aligned} P_{elm} &= \tfrac{3}{2}\omega(\psi_d i_q - \psi_q i_d) = \\ &= \tfrac{3}{2}(\psi_{PM} + (L_d - L_q) i_d) i_q. \end{aligned} \tag{8}$$

The electromagnetic thrust of PMLSM F_{thrust} with p pair poles is concluded from Eq. (3) and Eq. (8) as follows:

$$\begin{aligned} F_{thrust} &= \tfrac{3}{2} p \tfrac{\pi}{\tau}(\psi_d i_q - \psi_q i_d) = \\ &= \tfrac{3}{2} p \tfrac{\pi}{\tau}(\psi_{PM} + (L_d - L_q) i_d) i_q. \end{aligned} \tag{9}$$

The Eq. (1), Eq. (2), Eq. (3), Eq. (4), Eq. (5) and Eq. (9) equations are the bases of PMLSM model. using transformation dq reference frame to $\alpha\beta$, its equations are as follows:

$$\frac{\mathrm{d}i_\alpha}{\mathrm{d}t} = -\frac{R_s}{L_s} i_\alpha - \frac{1}{L_s} e_\alpha + \frac{1}{L_s} u_\alpha, \tag{10}$$

$$\frac{\mathrm{d}i_\beta}{\mathrm{d}t} = -\frac{R_s}{L_s} i_\beta - \frac{1}{L_s} e_\beta + \frac{1}{L_s} u_\beta, \tag{11}$$

$$e_\alpha = -\psi_{PM}\omega \sin\theta, \tag{12}$$

$$e_\beta = -\psi_{PM}\omega \cos\theta. \tag{13}$$

In motion control application, nonlinear elements in a linear motor cause insignificant errors in tracking or increase settling time. So, they must be considered carefully to prevent their negative effects. The dynamic behavior of the linear motor can be expressed as:

$$\begin{aligned} F_{thrust} &= m_{tot} \tfrac{\mathrm{d}v_{lin}}{\mathrm{d}t} + F_{load}(t) + \\ &+ F_{friction}(v_{lin}) + F_{disturb}(x), \end{aligned} \tag{14}$$

where m_{tot} is the total mass of the mover and the load, v_{lin} is the linear velocity of the mover, $F_{friction}$ is the friction force which is caused by viscosity, coulomb and static effects, F_{load} is the extra force which is produced by the load and $F_{disturb}$ is the force which includes cogging and end effect.

3. The Proposed Fuzzy Sliding Mode Observer Design

3.1. The Proposed Sliding Mode Observer

The equations, which are used in a conventional sliding mode observer, are written as Eq. 10. By using a mathematical model of PMLSM and defining slip level as $S = \hat{i}_s - i_s = 0$ the equations are written as follows [14], [15]:

$$L_s \frac{di_\alpha}{dt} = -R_s \hat{i}_\alpha + u_\alpha - k \cdot \text{sign}\left(\hat{i}_\alpha - i_\alpha\right), \quad (15)$$

$$L_s \frac{di_\beta}{dt} = -R_s \hat{i}_\beta + u_\beta - k \cdot \text{sign}\left(\hat{i}_\beta - i_\beta\right). \quad (16)$$

Due to reducing the chattering phenomenon, the sign function must be exchanged for a continuous function which is defined as:

$$F(x) = \left[\frac{2}{(1+e^{-\alpha x})}\right] - 1, \quad (17)$$

where α is the adjustable parameter. By defining the continuous function as above, the sliding mode observer equation is rewritten as:

$$L_s \frac{d\hat{i}_\alpha}{dt} = -R_s \hat{i}_\alpha + u_\alpha - k \cdot F\left(\hat{i}_\alpha - i_\alpha\right), \quad (18)$$

$$L_s \frac{d\hat{i}_\beta}{dt} = -R_s \hat{i}_\beta + u_\beta - k \cdot F\left(\hat{i}_\beta - i_\beta\right). \quad (19)$$

For stability analysis of the above sliding mode observer, Lyapunov function is chosen as:

$$V = \frac{1}{2} \mathbf{S}(X)^T \mathbf{S}(X). \quad (20)$$

Requisite condition for stability of sliding mode observer is obtained as follows:

$$\dot{V} = \frac{1}{2} \mathbf{S}(X)^T \dot{\mathbf{S}}(X) \le 0. \quad (21)$$

By subtracting Eq. (10), Eq. (11), Eq. (12) and Eq. (13) from Eq. (18) and Eq. (19), the error equation is concluded:

$$L_s \left[\frac{dS_\alpha(X)}{dt}\right] = \\ = -R_s S_\alpha(X) + e_\alpha - k \cdot F\left(\hat{i}_\alpha - i_\alpha\right), \quad (22)$$

$$L_s \left[\frac{dS_\beta(X)}{dt}\right] = \\ = -R_s S_\beta(X) + e_\beta - k \cdot F\left(\hat{i}_\beta - i_\beta\right). \quad (23)$$

$\mathbf{S}(X)$ is defined as:

$$\mathbf{S}(X) = \begin{bmatrix} S_\alpha(X) \\ S_\beta(X) \end{bmatrix} = \begin{bmatrix} \hat{i}_\alpha - i_\alpha \\ \hat{i}_\beta - i_\beta \end{bmatrix}. \quad (24)$$

By derivative from the above equation, the stability condition becomes as Eq. (29), thereupon:

$$k > \max\left(|e_\alpha|\,|e_\beta|\right), \quad (25)$$

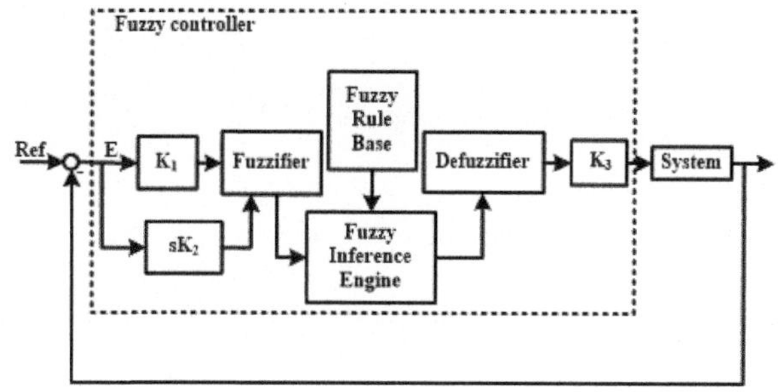

Fig. 1: The structure of the fuzzy controller.

because k is enough big, the asymptotic stability and slip movement seems certain. When the system gets to the slip level:

$$\dot{\mathbf{S}}(X) = \mathbf{S}(X) = 0. \quad (26)$$

By putting above equation in Eq. (22) and Eq. (23), which is based on equivalent control way, follows are given:

$$e_\alpha = k \cdot F\left(\hat{i}_\alpha - i_\alpha\right), \quad (27)$$

$$e_\beta = k \cdot F\left(\hat{i}_\beta - i_\beta\right). \quad (28)$$

1) Fuzzy Logic Control (FLC)

Fuzzy logic controller is suitable in controller designing for complex or nonlinear systems [16]. Fig. 1 shows the structure of the fuzzy controller which is used in this paper.

As it is obvious in Fig. 1, the inputs of the fuzzy controller are the error signal and its derivative, the output is the reference value which is applied to the system. In this fuzzy controller, the input and output membership functions are normalized values. The gains K_1, K_2, and K_3 are scaling factor to adapt the variables to the normalized scales. So, determining correct values for K_1, K_2, and K_3 is very important.

Table 1 shows fuzzy roles collection, where e and ce denote error signal and its derivative respectively. The other abbreviated forms which are used in this table are: NB (Negative Big), NM (Negative Medium), NS (Negative Small), Z (Zero), PS (Positive Small), PM (Positive Medium), PB (Positive Big).

Figure 2(a) and Fig. 2(b) are the inputs and output normalized membership functions and three dimensions input-output diagram respectively. In this paper, triangular membership function is used due to simply and good control work.

In practice, the fuzzy rule sets usually have several antecedents that are combined using fuzzy operators, such as AND, OR, and NOT, though again the definitions tend to vary: AND, in one popular definition,

$$\dot{V} = \frac{1}{2}\mathbf{S}(X)^T\dot{\mathbf{S}}(X) = S_\alpha\dot{S}_\alpha + S_\beta\dot{S}_\beta = \frac{1}{L_S}\left[\left(\hat{i}_\alpha - i_\alpha\right)e_\alpha - k\left(\hat{i}_\alpha - i_\alpha\right)F\left(\hat{i}_\alpha - i_\alpha\right)\right] + \\ +\frac{1}{L_S}\left[\left(\hat{i}_\beta - i_\beta\right)e_\beta - k\left(\hat{i}_\beta - i_\beta\right)F\left(\hat{i}_\beta - i_\beta\right)\right] - \frac{R_s}{L_s}\left[\left(\hat{i}_\alpha - i_\alpha\right)^2 + \left(\hat{i}_\beta - i_\beta\right)^2\right] \le 0. \tag{29}$$

Tab. 1: Fuzzy roles.

ce	e						
	NB	**NM**	**NS**	**Z**	**PS**	**PM**	**PB**
NB	NB	NB	NB	NB	NM	NS	Z
NM	NB	NB	NB	NM	NS	Z	PS
NS	NB	NB	NM	NS	Z	PS	PM
Z	NB	NM	NS	Z	PS	PM	PB
PS	NM	NS	Z	PS	PM	PB	PB
PM	NS	Z	PS	PM	PB	PB	PB
PB	Z	PS	PM	PB	PB	PB	PB

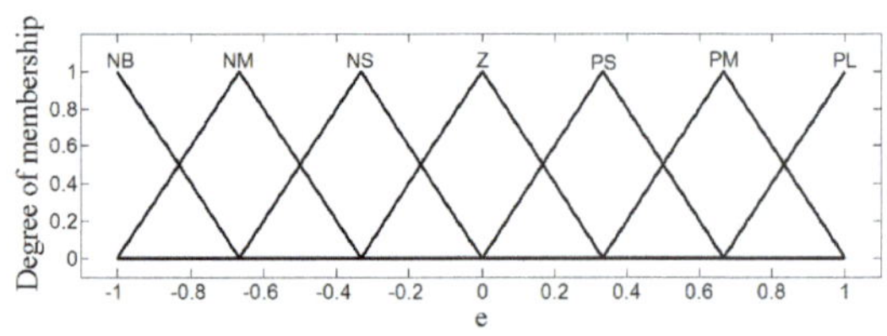

(a)

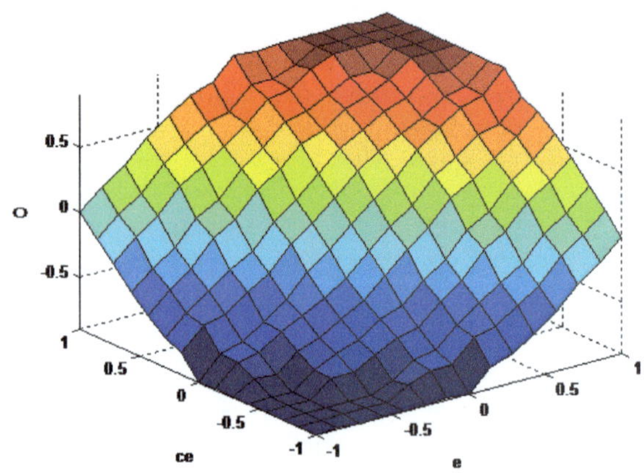

(b)

Fig. 2: (a) Inputs and output membership functions, (b) Three dimensions input-output diagram.

simply uses the minimum weight of all the antecedents, while OR uses the maximum value. There is also a NOT operator that subtracts a membership function from 1 to give the "complementary" function. There are several ways to define the result of a rule, but one of the most common and simplest is the "max-min" inference method, in which the output membership function is given the truth value generated by the premise. Rules can be solved in parallel in hardware, or sequentially in software. The results of all the rules that have fired are "defuzzified" to a crisp value by one of the several methods. There are dozens, in theory, each with various advantages or drawbacks. The "centroid" method is very popular, in which the "center of mass" of the result provides the crisp value. Another approach is the "height" method, which takes the value of the biggest contributor. The centroid method favors the rule with the output of the greatest area, while the height method obviously favors the rule with the greatest output value [16] and [17].

4. Estimation Rotor Position and Speed by Using Proposed Method

Back EMF can be obtained by described sliding mode observer, but the signals still include high-frequency components. Therefore, they cannot be used directly for an estimate the rotor position and speed. In a conventional sliding mode observer, a low pass filter is used for filtering operation, which causes the phase lag that depends on the cut frequency and the input signal angular one. In order to compensating the phase lag completely, the acquired information from the real linear speed must be used, and the information, which is given by the estimated linear speed, is not enough for compensating. Thus, due to avoid of using the low pass filter and the phase compensator based on back EMF, in this paper a fuzzy sliding mode observer for detecting the back EMF in a PMLSM is designed to estimate the speed and the position of the rotor. Because the changes of the motor linear speed are smaller than the changes of the stator's current, it is assumed that $\dot{\omega} = 0$. So, back EMF of PMLSM can be expressed as:

$$\frac{de_\alpha}{dt} = -\omega e_\beta, \tag{30}$$

$$\frac{de_\beta}{dt} = -\omega e_\alpha. \tag{31}$$

In consideration of the above equation, the back EMF observer is designed which is based on follows:

$$\frac{d\hat{e}_\alpha}{dt} = -\hat{\omega}\hat{e}_\beta - l\left(\hat{e}_\alpha - \hat{e}_\beta\right), \tag{32}$$

$$\frac{d\hat{e}_\beta}{dt} = -\hat{\omega}\hat{e}_\beta - l\left(\hat{e}_\beta - \hat{e}_\alpha\right), \tag{33}$$

$$\frac{d\hat{\omega}}{dt} = \left(\hat{e}_\alpha - e_\alpha\right)\hat{e}_\beta - \left(\hat{e}_\beta - e_\beta\right)\hat{e}_\alpha, \tag{34}$$

where l is the gain of observer which is more than 1.

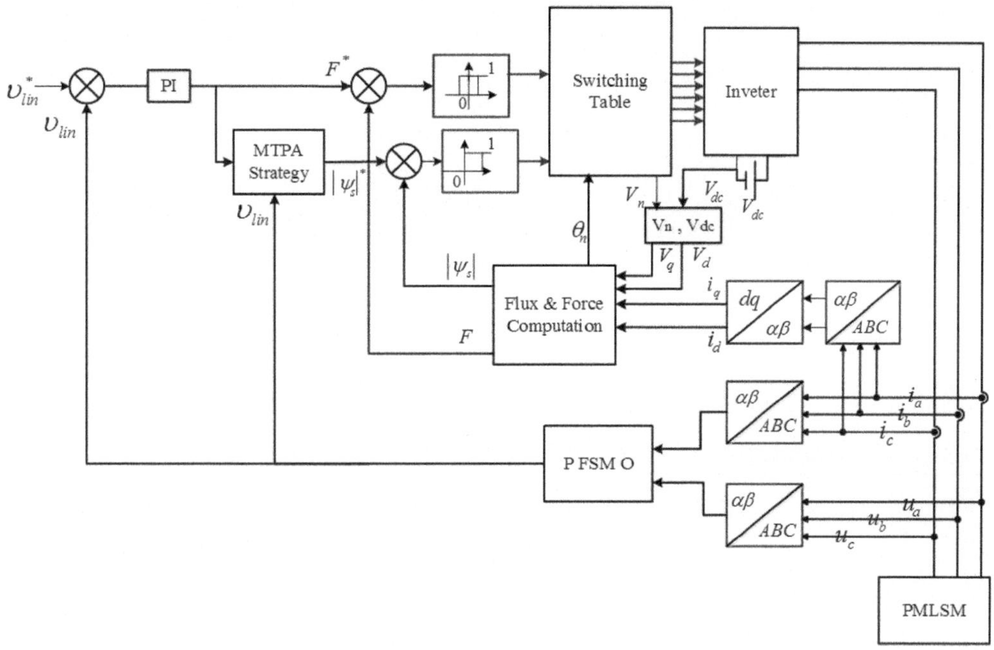

Fig. 3: Block diagram of the PMLSM sensorless control.

By subtracting Eq. (30), Eq. (31) from Eq. (32), Eq. (33) and Eq. (34), the error equation of the observer is obtained from:

$$\frac{\mathrm{d}\hat{e}_\alpha}{\mathrm{d}t} = -\tilde{\omega}\hat{e}_\beta - \omega\tilde{e}_\beta - l\tilde{e}_\alpha, \tag{35}$$

$$\frac{\mathrm{d}\hat{e}_\beta}{\mathrm{d}t} = -\tilde{\omega}\hat{e}_\alpha - \omega\tilde{e}_\alpha - l\tilde{e}_\beta, \tag{36}$$

$$\frac{\mathrm{d}\tilde{\omega}}{\mathrm{d}t} = \tilde{e}_\alpha\hat{e}_\beta - \tilde{e}_\beta\hat{e}_\alpha, \tag{37}$$

where $\tilde{e}_\alpha = \tilde{e}_\alpha - e_\alpha$, $\tilde{e}_\beta = \tilde{e}_\beta - e_\beta$ and $\tilde{\omega} = \tilde{\omega} - \omega$.

Due to show that Eq. (32), Eq. (33) and Eq. (34) is stable, Lyapunov function is defined as follows:

$$V = \frac{\left(\tilde{e}_\alpha^2 + \tilde{e}_\beta^2 + \tilde{\omega}^2\right)}{2}, \tag{38}$$

derivative from the above equation is led to:

$$\dot{V} = \tilde{e}_\alpha\dot{\tilde{e}}_\alpha + \tilde{e}_\beta\dot{\tilde{e}}_\beta + \tilde{\omega}\dot{\tilde{\omega}}. \tag{39}$$

By putting Eq. (35), Eq. (36) and Eq. (37) in Eq. (39), it is given that:

$$\dot{V} = -l\left(\tilde{e}_\alpha^2 + \tilde{e}_\beta^2\right) \leq 0. \tag{40}$$

From mentioned equation, it results that the proposed back EMF observer is asymptotically stable. Thus by use of back EMF that is obtained from the observer and the relationship between back EMF and the situation of the rotor, the position signal is estimated from follows:

$$\hat{\theta} = -\arctan\left(\frac{\hat{e}_\alpha}{\hat{e}_\beta}\right). \tag{41}$$

The speed is obtained easily there to by using an integrator in the observer. The overall scheme of the PMLSM sensorless control and the block diagram of the proposed fuzzy sliding mode observer are shown in sequence in Fig. 3 and Fig. 4. In first one, it must be noticed that the input of the sliding mode observer comes from the motor's voltage and the output voltages of the current loop u_α^* and u_β^* are not used. Usage of u_α and u_β reduces relatively the dead time effect in the inverter. So, the motor's voltage is obtained more precise and thus estimates accuracy of the motor position and speed are improved.

5. Experiment

The parameters given by the manufacturer of the motor and the inverter parameters are presented in the Tab. 2, [18]. The voltage source inverter (VSI) drives are most widely used. There are mainly two types of VSI drives available:

- two level devices,
- three level devices.

The typical structure of the two-level VSI drive is used in this paper. The step function is considered as a reference signal. The speed, which is estimated by a sliding mode observer, has been shown in Fig. 5. Chattering phenomenon in these types of observers which use discontinuous sign function is seen obviously.

The speed, which is estimated by the proposed fuzzy sliding mode observer, has been shown in Fig 6. As it is

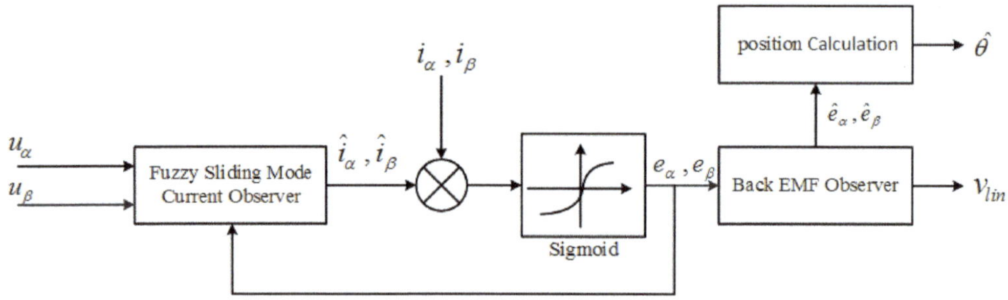

Fig. 4: Block diagram of the proposed fuzzy sliding mode observer.

Tab. 2: Parameters of the motor and inverter model.

Symbol	Value	Parameter
R	1.6 Ω	Phase resistance
L_d	0.013 H	d-axis inductance
L_q	0.013 H	q-axis inductance
p	1	Number of pole pairs
τ	0.012 m	PM pole pitch
ψ_{PM}	0.237 Wb	PM flux linkage
U_N	560 V	Rated voltage inverter
P_N	3.5 kW	Rated power
F_N	1800 N	Rated force
V_{max}	4.5 m·s^{-1}	Peak speed

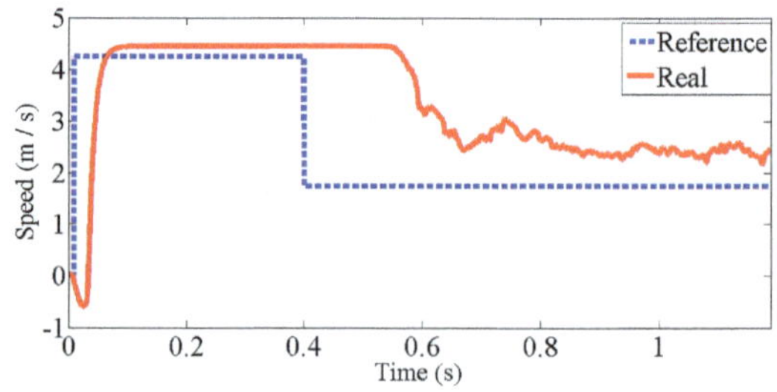

Fig. 5: Speed reference signal and real speed of the motor by sliding mode observer.

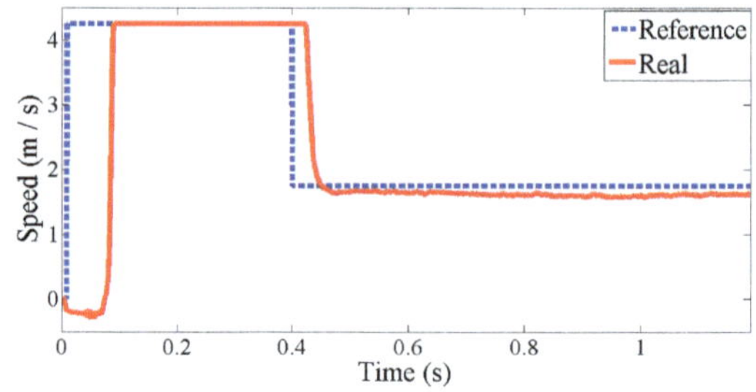

Fig. 6: Speed reference signal and speed actual of the motor by the proposed sliding mode observer.

seen, comparison with conventional observers, chattering phenomenon is reduced effectively. Moreover, low pass filter and phase lag compensator are not required.

Figure 7(a) and Fig. 7(b) show acceleration of the mover in order for a conventional observer and the proposed fuzzy sliding mode. Figure 8(b) and Fig. 8(b) show the estimated force and Fig. 9(a) and Fig. 9(b) show the force which has been made by the mover by considering end effects cogging and friction forces. Comparison Fig. 8(a) with Fig. 8(a) shows that the reference and estimated forces in system that uses the proposed fuzzy sliding mode observer is more accurate than others that use conventional ones, because choosing continuous function instead of discontinuous sign function leads to eliminate chattering phenomenon. The mentioned explanations are also true for Fig. 7.

Figure 10 show the current of the system that uses the proposed sliding mode observer during accelerating time of the mover at the stability speed and when acceleration of the mover is reduced.

6. Conclusion

In this paper for PMLSM sensorless control, the improved fuzzy sliding mode observer, which uses sigmoid function, was proposed. It was illustrated that using a sigmoid function instead of sign function that is usual in sliding mode observers leads to reducing undesirable chattering phenomenon effectively. Reducing chattering phenomenon results that back EMF can be detected directly from switching signal without any need to low pass filter. So the delay time which is caused by presence of low pass filter, in the proposed observer is omitted. Moreover, there is no need to compensate phase fault in estimated position. Advantages of the proposed observer over conventional ones have been shown by results, which are obtained from simulation by MATLAB/Simulink software.

References

[1] BATELAAN, J. A Linear Motor Design Provides Close and Secure Vehicle Separation of Many Transit Vehicles on a Guideway. *IEEE Transactions on Industrial Electronics*. 2007, vol. 54, iss. 3, pp. 1778–1782. ISSN 0278-0046. DOI: 10.1109/tie.2007.895079.

[2] TUTELEA, L. N., M. C. KIM, M. TOPOR, J. LEE and I. BOLDEA. Linear Permanent Magnet

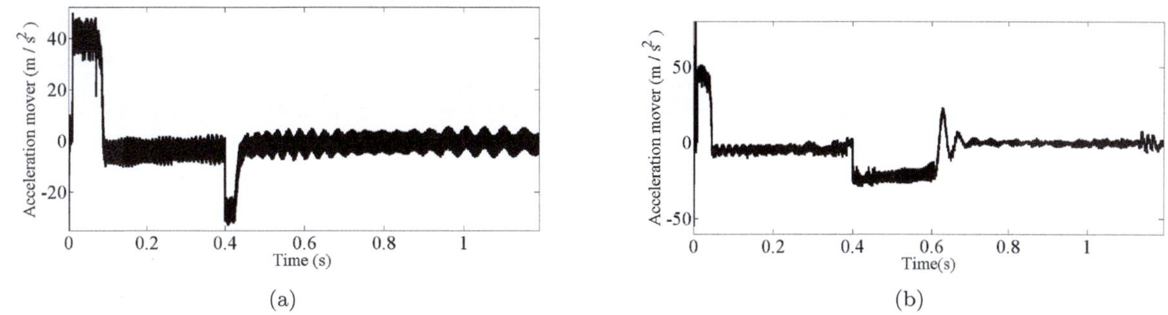

Fig. 7: Acceleration of the mover (a) by using a conventional sliding mode observer (b) by using the proposed fuzzy sliding mode observer.

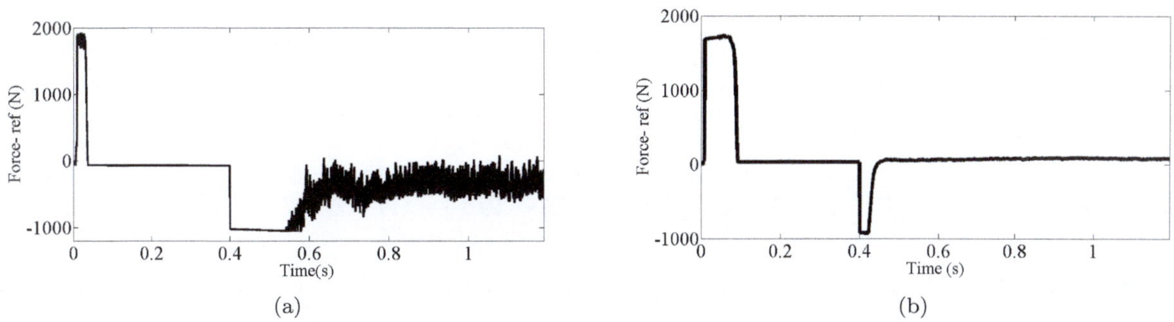

Fig. 8: Force reference (a) by using a conventional sliding mode observer (b) by using the proposed sliding mode observer.

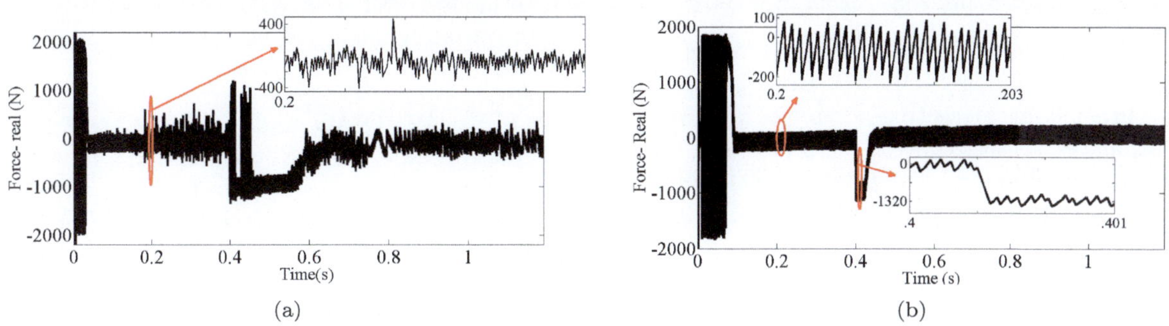

Fig. 9: The produced force of the mover (a) by using a conventional sliding mode observer (b) by using the proposed sliding mode observer.

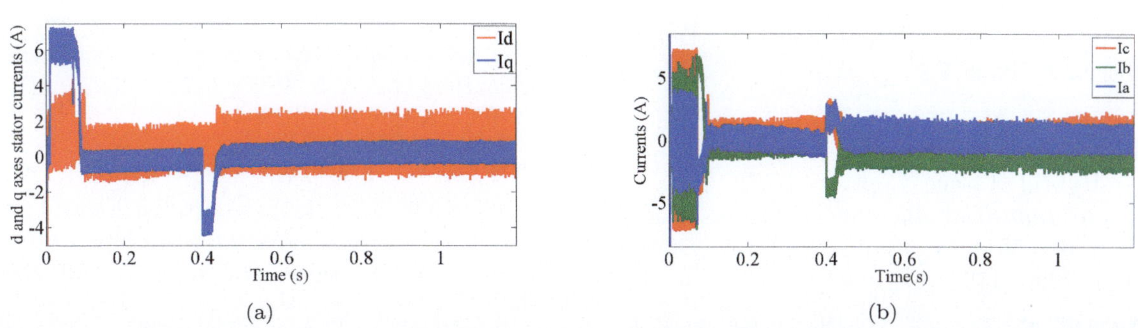

Fig. 10: The currents of PMLSM, which uses the proposed sliding mode observer in d and q axes.

Oscillatory Machine: Comprehensive Modeling for Transients With Validation by Experiments. *IEEE Transactions on Industrial Electronics.* 2008, vol. 55, iss. 2, pp. 492–500. ISSN 0278-0046. DOI: 10.1109/tie.2007.911936.

[3] BOUSSAK, M. Implementation and Experimental Investigation of Sensorless Speed Control With Initial Rotor Position Estimation for Interior Permanent Magnet Synchronous Motor Drive. *IEEE Transactions on Power Electronics.* 2005, vol. 20, iss. 6, pp. 1413–1422. ISSN 0885-8993. DOI: 10.1109/tpel.2005.854014.

[4] KUNG, Y.-S.. Design and Implementation of a High-Performance PMLSM Drives Using DSP Chip. *IEEE Transactions on Industrial Electronics.* 2008, vol. 55, iss. 3, pp. 1341–1351. ISSN 0278-0046. DOI: 10.1109/tie.2007.909736.

[5] LIU, T.-H., Y.-Ch. LEE and Y.-H. CHANG. Adaptive controller design for a linear motor control system. *IEEE Transactions on Aerospace and Electronic Systems.* 2004, vol. 40, iss. 2, pp. 601–616. ISSN 0018-9251. DOI: 10.1109/taes.2004.1310008.

[6] ZHU, Y.-W. and Y.-H. CHO. Thrust Ripples Suppression of Permanent Magnet Linear Synchronous Motor. *IEEE Transactions on Magnetics.* 2007, vol. 43, iss. 6, pp. 2537–2539. ISSN 0018-9464. DOI: 10.1109/tmag.2007.893308.

[7] TAKAHASHI, I. and Y. IDE. Decoupling control of thrust and attractive force of a LIM using a space vector control inverter. *IEEE Transactions on Industry Applications.* 1993, vol. 29, iss. 1, pp. 161–167. ISSN 0093-9994. DOI: 10.1109/28.195902.

[8] GARCIA DEL TORO, X., A. ARIAS, M. G. JAYNE and P. A. WITTING. Direct Torque Control of Induction Motors Utilizing Three-Level Voltage Source Inverters. *IEEE Transactions on Industrial Electronics.* 2008, vol. 55, iss. 2, pp. 956–958. ISSN 0278-0046. DOI: 10.1109/tie.2007.896527.

[9] MOREL, F., J.-M. RETIF, X. LIN-SHI and C. VALENTIN. Permanent Magnet Synchronous Machine Hybrid Torque Control. *IEEE Transactions on Industrial Electronics.* 2008, vol. 55, iss. 2, pp. 501–511. ISSN 0278-0046. DOI: 10.1109/tie.2007.911938.

[10] VELUVOLU, K. C. and Y. Ch. SOH. High-Gain Observers With Sliding Mode for State and Unknown Input Estimations. *IEEE Transactions on Industrial Electronics.* 2009, vol. 56, iss. 9, pp. 3386–3393. ISSN 0278-0046. DOI: 10.1109/tie.2009.2023636.

[11] NGUYEN, Q. D. and S. UENO. Improvement of sensorless speed control for nonsalient type axial gap self-bearing motor using sliding mode observer. In: *IEEE International Conference on Industrial Technology.* Via del Mar: IEEE, 2010, pp. 373–378. ISBN 978-1-4244-5696-3. DOI: 10.1109/icit.2010.5472709.

[12] JAAFAR, A., E. GODOY, P. LEFRANC, X. L. SHI, A. FAYAZ and N. LI. Nonlinear sliding mode observer and control of high order DC-DC converters. In: *36th Annual Conference on IEEE Industrial Electronics Society.* Glendale: IEEE, 2010, pp. 180–186. ISBN 978-1-4244-5226-2. DOI: 10.1109/iecon.2010.5675208.

[13] ABROSHAN, M., K. MALEKIAN, J. MILIMONFARED and B. A. VARMIAB. An optimal direct thrust force control for interior Permanent Magnet Linear Synchronous Motors incorporating field weakening. In: *International Symposium on Power Electronics, Electrical Drives, Automation and Motion.* Ischia: IEEE, 2008, pp. 130–135. ISBN 978-1-4244-1664-6. DOI: 10.1109/speedham.2008.4581088.

[14] QIAO, Z., T. SHI, Y. WANG, Y. YAN, Ch. XIA and X. HE. New Sliding-Mode Observer for Position Sensorless Control of Permanent-Magnet Synchronous Motor. *IEEE Transactions on Industrial Electronics.* 2013, vol. 60, iss. 2, pp. 710.-719. ISSN 0278-0046. DOI: 10.1109/tie.2012.2206359.

[15] ZHAO, Y., W. QIAO and L. WU. An Adaptive Quasi-Sliding-Mode Rotor Position Observer-Based Sensorless Control for Interior Permanent Magnet Synchronous Machines. *IEEE Transactions on Power Electronics.* 2013, vol. 28, iss. 12, pp. 5618–5629. ISSN 0885-8993. DOI: 10.1109/tpel.2013.2246871.

[16] SILVA NETO, J. L. and H. L. HUY. An improved fuzzy learning algorithm for motion control applications. In: *24th Annual Conference of the IEEE Industrial Electronics Society.* Aachen: IEEE, 1998, pp. 1–5. ISBN 0-7803-4503-7. DOI: 10.1109/IECON.1998.723934.

[17] CHOPRA, S., R. MITRA and V. KUMAR. Auto Tuning of Fuzzy PI Type Controller Using Fuzzy Logic. *International Journal of Computational Cognition.* 2008, vol. 4, iss. 4, pp. 12–18. ISSN 1542-5908.

[18] TECNOTION. *A primer of Tecnotion Linear Motors.* Version 2.1. Amelo, 2008. Available in: `http://www.electromate.com/db_support/downloads/LinearMotorPrimer_V21_20080912.pdf`.

About Authors

Saeed MASOUMI KAZRAJI was born in Tehran, Iran in 1989. He received the B.Sc. degree from the Sbzevar University of Tarbiat Moallem, Mashhad, Iran in 2011 and the M.Sc. degree from University of Tabriz, Tabriz, Iran in 2013 both in Electrical Power Engineering. He is currently pursuing the Ph.D. degree in electrical engineering (machine and electric drives) University of Tabriz. His research interests include drive and motion control of electric machines, machine design, linear actuator and power electronic converters.

Mohammad Bagher BANNAE SHARIFIAN studied Electrical Power Engineering at the University of Tabriz, Tabriz, Iran. He received the B.Sc. and M.Sc. degrees in 1989 and 1992 respectively from University of Tabriz. In 1992 he joined the Electrical Engineering Department of the University of Tabriz as a lecturer. He received the Ph.D. degree in Electrical Engineering from the same University in 2000. In 2000 he rejoined the Electrical Power Department of Faculty of Electrical and Computer Engineering of the same university as Assistant Professor. He is currently Professor of the mentioned Department. His research interests are in the areas of design, modeling and analysis of electrical machines, transformers, linear electric motors, and electric and hybrid electric vehicle drives.

18

LOW-VOLTAGE POWER SUPPLY SUBSYSTEM FOR A SUB-ORBITAL PARTICLE PHYSIC INSTRUMENT

Hector Hugo SILVA LOPEZ[1], *Gustavo MEDINA-TANCO*[1], *Lauro SANTIAGO CRUZ*[2]

[1]Laboratorio de Detectores, Instituto de Ciencias Nucleares, Universidad Nacional Autonoma de Mexico, Circuito exterior s/n Ciudad Universitaria AP, 70-543 Coyoacan CP. 04510, Distrito Federal, Mexico
[2]Instrumentacion, Instituto de Ingenieria, Universidad Nacional Autonoma de Mexico, Circuito exterior s/n Ciudad Universitaria AP, 70-543 Coyoacan CP. 04510, Distrito Federal, Mexico

hhsilvalopez@gmail.com, gmtanco@nucleares.unam.mx, lsc@pumas.iingen.unam.mx

Abstract. *The Japanese Experiment Module–Extreme Universe Space Observatory (JEM-EUSO) is a wide-field (+/− 30° of aperture) 2.5 m refractor telescope to be installed in the International Space Station (ISS). The instrument looks downward from its orbit, into Earth's atmosphere, with the main objective of observing ultra-violet (UV) fluorescence light generated by Ultra-High Energy Cosmic Rays (UHECR) extensive air showers (EAS). It is a frontier particle-physics experiment, the first of its kind. The validation of the technical readiness level of such a complex and unique instrument requires prototypes at several levels of integration. At the highest level, the EUSO-Balloon instrument has been conceived, through French space agency (CNES). At a smaller scale and in suborbital flight, EUSO-Balloon integrates all the sub-systems of the full space JEM-EUSO telescope, allowing end-to-end testing of hardware and interfaces, and to probing the global detection chain and strategy, while improving at the same time our knowledge of atmospheric and terrestrial UV background. EUSO-Balloon will be flown by CNES for the first time from Timmins, Canada; on spring 2014. This article presents the low-voltage power supply (LVPS) subsystem development for the EUSO-Balloon instrument. This LVPS is the fully operational prototype for the space instrument JEM-EUSO. Besides design and construction, all performance tests and integration results with the other involved subsystems are shown.*

Keywords

DC-DC converter, power supply, stratospheric balloon, thermal-vacuum test.

1. Introduction

JEM-EUSO consists of a large telescope with a wide field-of-view, to be mounted on the ISS. It is the first space mission to explore the universe at extreme energies through the detection of UHECR impinging the Earth's atmosphere. Such interactions produce an EAS, a huge numbers of charged particles, mainly electrons and positrons, travelling through the atmosphere along the original direction of the incoming UHECR. They excite the atmospheric N_2, which immediately decays by fluorescence producing UV photons with wavelengths between 300 and 400 nm. Furthermore, the electrons and positrons are super-luminal inside the atmosphere and, therefore, also produce UV Cherenkov radiation. Both radiations are observed by JEM-EUSO from orbit at 400 km of height. Since electrons and positrons travel all at roughly the same speed, essentially that of light the net effect is that, from the point of view of the telescope, an EAS is observed as a light spot moving at the speed of light, thus defining a track on its focal surface. It is worth noticing that the Cherenkov photons are emitted along the propagation direction of the particles in the EAS, i.e., downwards. Therefore, those photons are only observed because of scattering in the atmosphere or after diffuse reflection on the top of optically thick clouds or on the ground. The latter produces a strong Cherenkov mark signaling the impact point of the EAS and it is extremely important for data reconstruction [1].

Although theoretically sound, the observational principle has never been tested from space. There are also considerable uncertainties with regard to the intensity of the Earth up-going UV background produced by the atmosphere and the ground due to both, the Moon and terrestrial sources, as well as the UV diffuse reflectivity of different types of ground (sand, snow, ice, clouds, vegetation, oceans, etc.) which can affect the reconstruction efficiency and the effective duty cycle of the instrument. In addition to instrument complexity, an EAS has never been observed yet from space. Thus, a large scale comprehensive prototype is mandatory: EUSO-Balloon is such prototype.

EUSO-Balloon had being design and built by a consortium which, under the umbrella of CNES, includes laboratories from France, Germany, Italy, Japan, Mexico, Poland, South Korea and Spain. The instrument is a reduced version of JEM-EUSO which still comprises all the subsystems of the JEM-EUSO main mission. The instrument will operate at an altitude of 40 km during the night and measure photons in the UV range with a time resolution of 2.5 µs with a field of view (FOV) of the optics 12°× 12° (squared) and a large dynamic range. EUSO-Balloon also has the potential to detect for the first time a cosmic ray atmospheric shower from above, marking a key milestone in the development of UHECR science, and paving the way for any future large scale, space-based UHECR observatory, including of course JEM-EUSO [2].

Regardless of the payload, the power supply system is a key component. The Electrical Power System (EPS) of a spacecraft comprises power conversion, energy storage, regulation, power conditioning, overvoltage (OVP) and over-current (OCP) protections, and power distribution to the various users via the on-board (OB) LVPS [3]. These stages are generally divided into three sections: *the primary energy source*, which converts a fuel into electricity power. The *secondary energy source* for storage and regulation of energy and subsequently delivers electrical power to spacecraft and its payload. And finally, the *power control and distribution* network that is required to deliver appropriate voltage-current levels to all spacecraft loads when required [4].

The present work describes the LVPS developed for the EUSO-Balloon instrument. The paper is organized as follows: Section 2 gives a general description of the instrument; Section 3 presents the power distribution architecture; Section 4 details the LVPS requirements; Section 5 gives the specification subsystems; Section 6 describes the several modules that conform the subsystem; Section 7 summarizes several design considerations and Section 8 presents the various tests that were applied for the characterization of the system and validation of the compliance to its requirements.

2. EUSO-Balloon Instrument

EUSO-Balloon instrument consists, firstly, of a Fresnel optic array comprised by tree polymethyl-methacrylate (PMMA) square lenses for UV transmitting. After that, a focal plane detector composed of 36 multi-anode photomultiplier tubes (MAPMTs) with 64 pixels each representing 2.304 pixels, called photo-detector module (PDM). A 2°× 2° MAPMTs array, called elementary cell (EC), has its associated application-specific integrated circuit (ASIC), high voltage supply (HV) and HV switches. The 15°×15° cm PDM and the 100 cm × 100 cm Fresnel lenses provide a field of view of ±6 deg and shall observe in "nadir mode" that can be varied, in principle, during later flights in a range between 0° to 30°. A schematic description of the instrument is shown in Fig. 1.

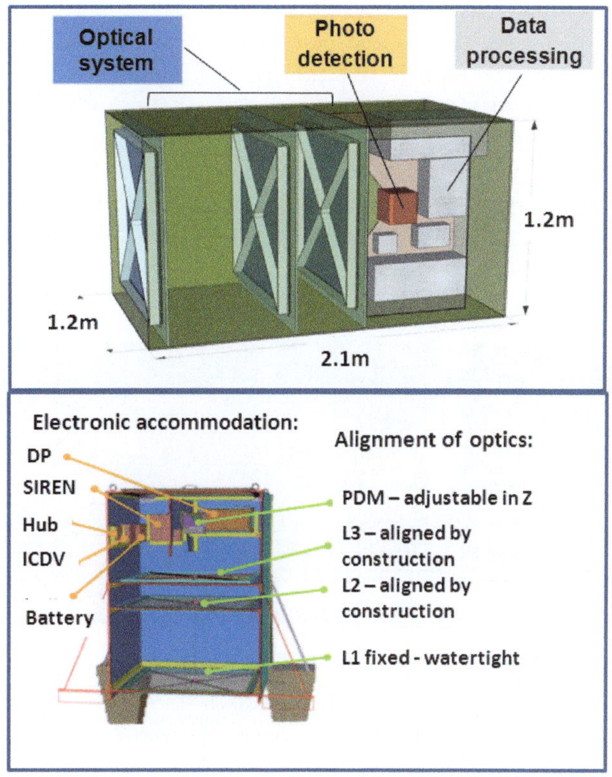

Fig. 1: EUSO-Balloon instrument description.

Therefore, the UV light entering in optics system is imaged on the PDM. The later normally works in single-photoelectron mode; however it can also extend its dynamic range in order to observe intense light signals thanks to the implementation of gain switches. The signal from each MAPMT is processed by the Front-End Electronics (based on the ASIC) and preliminary processed in the so-called PDM board, which is within the PDM structure where the first-level trigger is performed. Data are then transferred to the Digital Processor (DP) component where the second level trigger is first applied by the cluster control board (CCB), after that data are prepared to be sent as a part of telemetry information. Data management and storage, instrument control and commanding are managed by components within the DP structure. As shown in Fig. 2, the main subsystems of the EUSO-Balloon instrument are: telescope (TES), electronic (ELS), power (PWS), and the camera (CAM), [2].

As can be observed in the block diagram of the instrument in Fig. 2, the tree main components of PDM structure are: EC, high-voltage power supply (HVPS), and PDM board (PDMB). The cores of these components are based on ASICs and field programmable gate arrays (FPGAs) in order to perform signal acquisition and digitalization as a part of first trigger level detection. The DP structure comprises the CPU, data storage (DST) hard disks, cluster control board (CCB), clock board (CLKB) which dictates the system master clock, GPS receiver (GPSR), and the housekeeping (HK). The latter monitors the instrument; generate alarms, reset signals and powers on/off every other subsystem at the request of either the CPU or ground. As in the PDM structure, the cores of all these components are FPGAs and microprocessors unit (μPU) which have tight constraints about voltages levels requirements. As a first approximation, one LVPS module has been considered for each structure: LVPS-PDM and LVPS-DP.

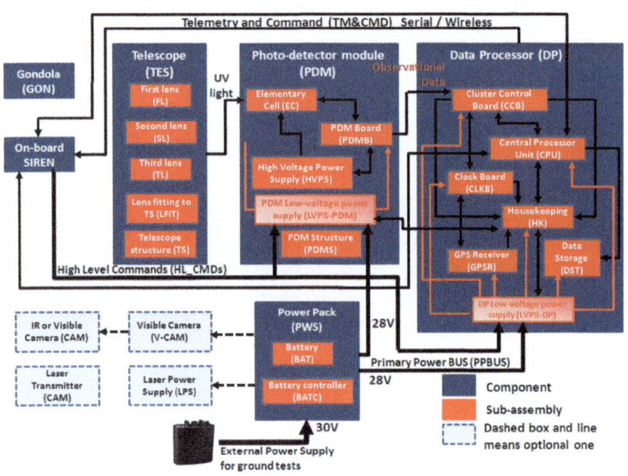

Fig. 2: EUSO-Balloon block diagram.

3. Power Distribution Architecture

The latest high-performance FPGAs, ASIC's and μPU require increased high performance on voltage levels from power supply. Typical requirements include low voltages (sometimes less than 1 V), high currents (up to 3 A), tight regulation (less than 1 %), fast transient response, and even supply voltage sequencing.

In space and near-space environment, it is recommended [3] that buses voltage should not be selected below 20 Vdc and not to exceed 125 Vdc. Bus voltages below 20 Vdc might lead to unwieldy current densities, while bus voltage above 125 Vdc might easily generate dangerous potential gradients, leading to ionization of air molecules (plasma), causing coronal and electric arc discharges, in particular during tests in partially pressurized environments. A 28 V power bus is often used as internal power bus in satellites, aircrafts and vehicles under specific military standards or program requirements [5]. It also has to be considered that the biggest selection of components of-the-shelf (COTS) is available for operating voltages between 28 Vdc and 50 Vdc [3].

Many issues play an important role in determining the best power distribution architecture, e.g., the kind of loads, the circuitry dimensions, the amount of current and low output voltage values, thermal management, transient response, etc. The latter produces lack of scalability and puts small configurations at cost disadvantage [6]. Therefore, a detailed analysis and a trade-off study must be performed in to ensure the most appropriate selection of power distribution architecture to meet the specific mission requirements [7]. For that purpose, a system design criteria must be stated considering factors like cost, low profile, system efficiency, dynamic response, reliability, availability of components, heat distribution, redundancy and development schedule. These factors help not only in selecting the most appropriate power distribution architecture, but will also constitute a basis for selecting the most appropriate power technology.

The DC-DC converter unit (DDCU) is a power device that typically receives power from a main power bus and regulates their outputs to a narrow voltage range. As already mentioned, the satellite power bus voltage can range from 20 Vdc to 120 Vdc. Meanwhile, payload electronic and bus system typically require 3.3 Vdc to 28 Vdc, although at present the voltage requirements are decreasing towards 1 Vdc and less for digital loads [8]. Taking into account the high cost of these type of missions and the impossibility of repairs, the DDCU must rigorously comply with general requirements as reliability tests, good efficiency (above 80 %), good load and line regulation, low output ripple, electromagnetic compatibility (EMC) standards and input/output (I/O) protections.

Due to the instrument hierarchical disposition and number of components, see Fig. 2, four LVPS modules were proposed, developed and housed inside the DP structure. The LVPS represent a transformation and isolation stage between PWS and payload. Meanwhile, final transformation and tight regulation are met by implementing non-isolated Points of Load (niPOL) converters or linear regulators (LRgs) OB of each individual subsystem, as shown in Fig. 3. The increased number of LVPS modules has the undeniable disadvantage of pushing up costs; nevertheless, it improves heat dissipation, system reliability, while decreasing electromagnetic interference (EMI) contamination by ground loops generated because of ground planes sharing of different voltage levels subsystems.

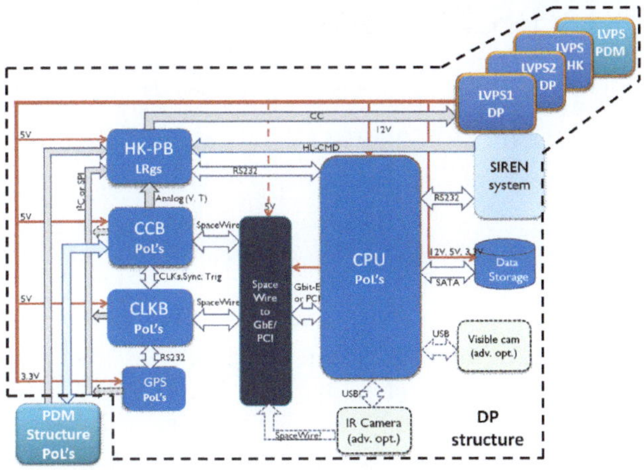

Fig. 3: EUSO-Balloon power distribution architecture inside of DP structure.

Additionally, in order to avoid radiated and conducted EMI, generated by DDCU within the PDM detector structure, which in principle has the most EMI-susceptible components, it was decided to install the LVPS module of PDM subsystem (LVPS-PDM) inside of the DP rack, see Fig. 3. Additionally, the part of LVPS devoted to the DP components was distributed into three modules, according to importance and criticality of the respective components to instrument performance. This strategy also facilitates heat dissipation and increases the overall reliability of the instrument. The three modules are: LVPS-HK, which powers the monitoring subsystem, HK; LVPS1-DP powering CCB, responsible for the second level trigger process, CLKB, the clock board, and the GPS receiver (GPSR); and LVPS2-DP which powers both the CPU and the data storage subsystem (DST) comprised by two 1 TB solid state disks.

4. LVPS Requirements

The LVPS subsystem must distribute power in a reliable way to each one of the other subsystems of the EUSO-Balloon instrument, while complying with the various specification requirements of those subsystems. Those requirements are, in fact, the input data of the design stage. We group the requirements in 3 broad categories listed below: functional, environmental and of communication.

4.1. Functional Requirements

- The LVPS shall supply power in a reliable way to PDM and DP structures.
- The LVPS subsystem shall consist, at least, of two different power distribution modules: LVPS-PDM and LVPS-DP.
- The LVPS modules shall provide isolation interface between 28 V bus of PWS and all payload subsystems.
- The isolation stage will comprise isolated DDCU with efficiencies equal or higher than 80 %.
- The LVPS modules shall withstand a lower voltage level, as input, from battery power pack (PWP).
- The regulation stage shall be performed directly at load (as close as possible) for best performance.
- Tight regulation stage will comprise niPOL converters with efficiencies higher than 90 %.
- Propagation failures inside of LVPS modules should be controlled. If any failure event occurs, it cannot be propagated to either the primary or secondary sides of the isolation stage.
- The LVPS modules shall provide remote on/off functionality, in order to be controlled from OB-SIREN.
- The DDCU shall provide wide input voltage range in order to withstand input voltage oscillations.
- The DDCU shall provide input circuit protection function in order to protect circuitry attached to main bus by some abnormalities.
- The DDCU shall provide output over current protection function.
- Input and Output EMI filters should be considered in the design of the LVPS modules.
- The maximum PCB dimensions shall conform to the 3U Eurocard size board standard [19].
- The weight of LVPS subsystem must not exceed the 5 kg.

4.2. Environmental Requirements

- The LVPS elements must withstand temperature variations in the range −20 °C to +50 °C during operation.
- The LVPS elements must accept a storage temperature > −40 °C.
- The LVPS elements must withstand shock acceleration levels of at least 3 G.
- The LVPS elements must be able to operate at very low atmospheric density, 3 mbar of pressure.

4.3. Communication Requirements

- Tele-command (TC) shall be used for turning on/off remotely LVPS modules.
- Status on/off by contact closure (CC) signal shall be provided from LVPS modules.
- Telemetry (TM) data, about voltage and current (V&I) levels, shall be generated and transmitted from LVPS modules to the HK subsystem.

5. LVPS Specification Subsystems

The LVPS requirements stated in the previous section define, to some extent, the interaction level with other subsystems. On other hand, the specifications describe the technical subsystem features. This explicit information provides the procedure to be followed for determining whether the requirements are met [9]. Table 1 lists the subsystem specifications regarding the EUSO-Balloon block diagram shown in Fig. 2.

The PWP in Tab. 1 is the instrument power source during flight, and it provides a non regulated bus (NRB) nominally of 28 V with a power capability of 100 W. However, this NRB may drift down up to 18 V, due to the normal discharge of its battery cell array, during the flight. Those variations must be accommodated by the LVPS subsystem by a DDCU proper selection.

Moreover, as can be seen in the voltage column of Tab. 1, it is necessary to create an intermediate bus semi-regulated (SRB) between the isolation stage and tight regulation, in order not to decrease the overall efficiency of the first stage. However, a SRB value must be selected based on two aspects: the minimum and maximum output voltage of first stage (tolerance) and input voltage range of next stage (final transformation and tight regulation).

Commonly, supply power for digital logic and IC processing devices is done through linear regulators and niPOL converters, depending of load, which nominal input voltage is 5 V. At this value the output voltage is maintained, with efficiencies higher than 90 %, within a tight regulation band to ensure proper functionality of the devices [10]. Output voltage tolerances of 10 % in isolation stage are expected in order to comply with the input voltage range of payload.

6. LVPS Modules Description

As stated in section 3, Four LVPS modules were proposed: LVPS1-DP, LVPS2-DP, LVPS-PDM and LVPS-HK. As an example, Fig. 4 shows LVPS2-DP diagram description.

Each LVPS module has, at least, three header connections represented in Fig. 4 as DB connectors: input power, input/output (I/O) TM&TC connector and output power. An important goal of the present work was to propose the use of industrial isolated DDCU (denoted with DDCU1 and DDCU2) in order to keep costs as low as possible. However, this proposal carries a certain level of risk, since most DDCU are tested by the manufacturer only under operation condition at 2 km of height. For example, Tab. 2 shows the maximum operational conditions tests reported by TDK Lambda switching power supplies manufacturer.

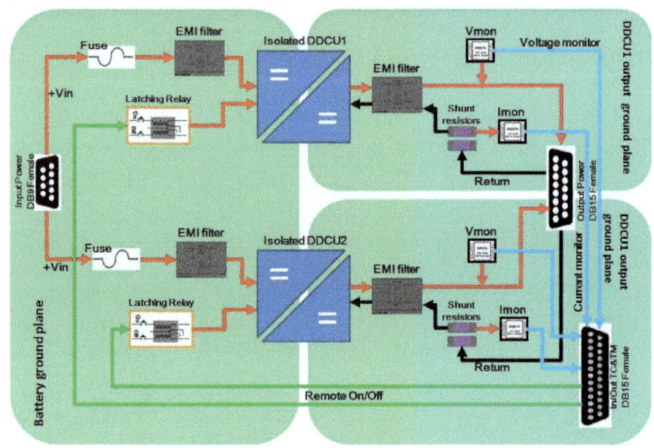

Fig. 4: LVPS2-DP diagram description.

In order to provide instrument flexibility and reliability, a remote On/Off function was implemented (green lines in Fig. 4) based on a power switch circuit and dual pole dual throw latching relay (DPDT-LR), the last showed in Fig. 4 orange framed. On other hand, LC and pi filters at I/O DDCU were implemented (denoted as EMI filters) in order to offer protection to conducted emissions interference sources of other equipments coming from cables and wires.

Also, from the point of view of flexibility and regarding future flights scheduled for EUSO-Balloon, fuses at each input converter were located for inrush energy and maximum dc input current protection. Furthermore, V&I monitoring circuits (labeled as Vmon and Imon on Fig. 4) were implemented in order to provide TM of LVPS performance subsystem. Finally, in order to keep isolation between main power bus and payload secondary side three different ground planes were planed during PCB design. The last helps avoiding ground loops and less contamination between subsystems.

Tab. 1: Subsystem specifications.

Subsystem	**Component**	**Element Subsystem**	**Power (W)**	**Voltage (V)**	**Electronic**
PWS	PWP	Battery	100	28	N/A
	PDM	EC	10	1.5	4 ASICs
				3.3	
		PDMB	10	1	1 FPGA
				1.5	
				1.8	
				2.5	
				3.3	
	DP	CCB	5	1.2	1 FPGA
				2.5	
				3.3	
		CLKB	4	1.8	1 FPGA
				2.5	
				3.3	
		DST	15	5	2 Hard Disks
		HKB	5.5	±12	1 MCU
				5	
			0.5	3.3	
		CPU	12	12	Intel Atom MPU
				5	
		GPSR	1.5	5	GPS module receiver

Tab. 2: TDK Lambda DDCU test operational conditions (based on MIL-STD-810F).

Test condition	Operating combined temperature/altitude		Non-operating combined temperature/altitude	
Temperature (°C)	0	40	−40	70
Altitude (km)	12.2	12.2	15.2	15.2
Test duration (hr)	2	2	6	6

7. LVPS Design Considerations

Is important to note that a power supply subsystem assumes a very important role within any system, it gives power in a consistent and reliable way to circuits system. Besides protects the system against electrical abnormalities like a transients and short circuits events. In a few words, if the supply experiences a failure within itself, it must fail gracefully and not allow the failure to reach the subsystems [11].

Following these considerations, and in order to produce a good design, many aspects must be considered prior to the beginning of the design process. In particular, the EUSO-Balloon instrument environmental operation condition at 40 km of altitude implies some constraints related to components selection, heat dissipation, EMC, grounding architecture, etc.

7.1. DDCU Selection

As new generations of programmable digital devices (FPGA's, DSP's, ASIC's, etc.) enter the market to meet the growing demand for increased bandwidth, and data processing speed OB subsystem instruments, these devices place unprecedented constraints in power supplies (PS). Low voltage supply and high operating current are the main features required from new digital devices.

Correspondingly, a detailed analysis and a trade-off study must be performed to ensure that the specific mission requirements are met [7]. For that purpose, design criteria factors, such as cost, low-profile, electrical performance (input voltage range, load and line regulation, efficiency, tolerances, etc.), availability (procurement times), heat distribution and EMC were considered.

Based on the above considerations, isolated DDCU from three different manufacturers (TDK Lambda, Tracopower-TPW and XPPower-XPP) were selected. The DDCU main features are summarized in Tab. 3 according to each LVPS module. As can be observed in Tab. 3, LVPS1-DP which is devoted to CCB, CLKB and GPSR subsystems creates a 5 V secondary bus voltage, as LVPS-PDM does, with the main aim of providing a better regulation for FPGA by implementing niPOL converter OB in each subsystem.

Meanwhile LVPS2-DP and LVPS-HK, create a 12 V secondary bus voltage in order to power linear regulators, which supply 5 V level for the microprocessor unit (MPU) and microcontroller unit (MCU), respectively. 5 V and 3.3 V were also generated separately for not critical components (gates, drivers, etc.).

7.2. Electromagnetic Compatibility

EMC is the ability of electrical or electronic equipment to function in the intended operating electromagnetic environment without causing or experiencing performance degradation due to intentional EMI [12]. In order to achieve EMC, the influence of electrical noise in a system must be minimized. Therefore, it is necessary to establish an EMC design flow diagram based on three parallel branches: mechanical design, electrical design and system design as is showed in Fig. 5, [13].

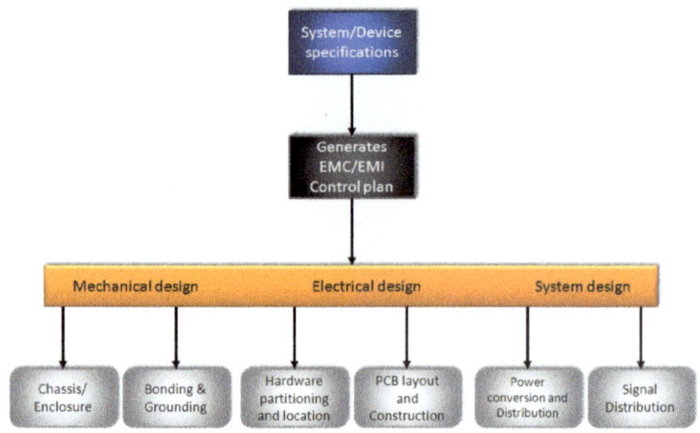

Fig. 5: EMC design flow diagram.

Since switching mode power supplies (SMPS) have high voltage and high current switching waveforms that generate EMI, in both conducted and radiated emissions, enclosure and grounding become a very important issue [14].

Therefore, in order to minimize radiated emissions, LVPS modules were housed inside 19" plug-in units which serve as accommodation up to two PCB. The mechanical dimensions comply with the IEC-60297-3-102/IEEE 1101.1/11 standard. Figure 6, shows the LVPS modules frame type plug-in unit enclosure. This plug-in units match 3U eurocard standard [19], and also includes a conductive foam EMI gasket consisting of a polyester textile yarn uniformly platted which provides high shielding performance (60–100 dB), high level of flame resistance (UL94-V0), mounted using pressure sensitive adhesive (PSA) and is able to operate at higher frequencies due to no gaps on their construction.

Additionally, these plug-in cabinets are fitted into metal chassis DP rack box that serves as a shield for all the subsystems housed inside it. Figure 7 shows the DP box (rack) with most of the sub-systems installed and labeled.

Therefore, the fields from all the radiating surfaces are shielded, with noise coming only from cables or wires that enters/exit holes and slots made in the box. In order to achieve this, the shield must be a good conduction plate with the least possible number of me-

Tab. 3: Main specifications of DDCU.

<table>
<tr><th>Parametr</th><th colspan="2">LVPS1-DP</th><th>LVPS_PDM</th><th colspan="2">LVPS2-DP</th><th colspan="2">LVPS-HK</th></tr>
<tr><td>Mfgr/DDCUs qty</td><td>TDK (1)</td><td>TPW (2)</td><td>TPW (2)</td><td>TPW (1)</td><td>XPP (1)</td><td colspan="2">XPP (2)</td></tr>
<tr><td>Subsystem</td><td>CCB</td><td>CLKB and GPRS</td><td>FPGA and EC-ASIC</td><td>CPU</td><td>DST</td><td colspan="2">HK-MCU and HK_LVDS</td></tr>
<tr><td>Input voltage range (V)</td><td colspan="3">18 to 36</td><td colspan="4">9 to 36</td></tr>
<tr><td>Output power (W) with safety factor</td><td>6</td><td>5 and 2 (respec)</td><td>15 (each one)</td><td>15</td><td>30</td><td>8</td><td>2</td></tr>
<tr><td>Output voltage (V) secondary bus voltage</td><td colspan="2">5</td><td>5.5</td><td>12</td><td>5</td><td>±12</td><td>3.3</td></tr>
<tr><td>Line and load regulation (%)</td><td colspan="2">0.2–1</td><td>0.2–0.3</td><td colspan="2">0.2–0.5</td><td colspan="2">0.2–0.5</td></tr>
<tr><td>Efficiency (%) at full load</td><td colspan="2">80–82</td><td>84</td><td colspan="2">87–91</td><td colspan="2">80–86</td></tr>
<tr><td>Operating temperature ranges (°C)</td><td colspan="5">−40 to 85</td><td colspan="2">−40 to 100</td></tr>
<tr><td>EMC</td><td colspan="3">Safety stds: UL60950-1
IEC60950-1
EN60950-1
Conducted noise:
EN55022 Class A nad B,
FCC part 15</td><td colspan="4">Safety stds: UL60950-1
IEC60950-1
EN60950-1
Conducted noise:
EN55022 Class A and B,
FCC part 15 ESD
Radiated and conducted immunity: EN61000-4-2, 4-3, 4-6 respectively</td></tr>
<tr><td>Vibration</td><td colspan="3">MIL-STD-810F
10–55 Hz, 2G, 30 min along X, Y</td><td colspan="2">MIL-STD-810F</td><td colspan="2">—</td></tr>
</table>

Tab. 4: Load regulation of LVPS modules.

Module	DDCU subsystem	Secondary bus voltage (V)	%LdReg @Vin_{nom}	%LdReg @Vin_{low}
LVPS1-DP	CCB	5	1.4	1
	CLKB	5	1.2	1
	GPRS	5	2.8	2.6
LVPS2-DP	CPU	12	0.17	0.25
	DST	5	0.74	0.96
LVPS-PDM	FPGA	5	1.36	1.31
	EC-ASIC	5	1.35	1.32
LVPS-HK	Arduino MCU	±12	7.3	7.4
	LVDS	3.3	10	7.96

chanical breaks and lower impedances between the current sources and the ground reference point (GRP).

The GRP at the EUSO-Balloon instrument is based on a single point grounding topology, with star connection tying all subsystem together, throughout the LVPS modules and up to the negative borne of the battery, which is called mechanical ground (GND_M), see Fig. 8.

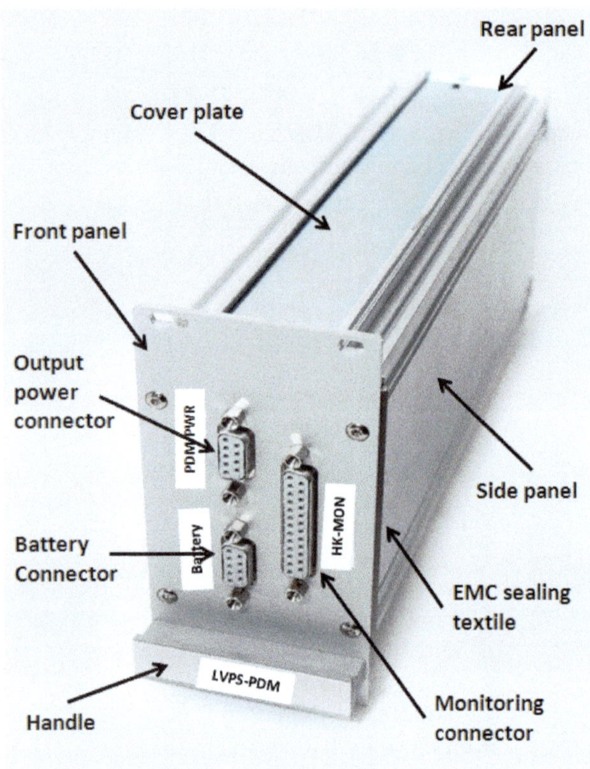

Fig. 6: 3U plug-in unit of LVPS-PDM module (100 mm × 227 mm × 61 mm).

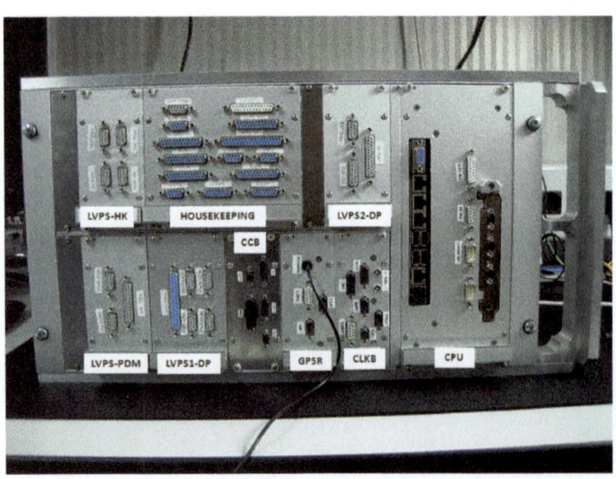

Fig. 7: DP box with all subsystems fitted (220 mm × 235 mm × 498 mm).

This topology requires a considerable number of conductors, which is generally not practical in large distributed subsystems [15]. On other hand, this architecture accomplishes the functions of signal return, while helping to control common-impedance interference coupled between subsystems. Additionally, to avoid closed-loop paths for noise current contamination, separately ground planes were implemented for each LVPS module (see Fig. 8) at secondary side output (right winding transformer) of magnetic isolation transformer. These ground planes are denoted in Fig. 8 with small squares per each winding output transformer. On other hand, most of DDCU were selected built-in metal enclosure in order to decrease radiated interference, as well as, be less susceptible to EMI. Cables were implemented as a twisted pair and connectors provided with back-shielding in order to minimize EMI radiation.

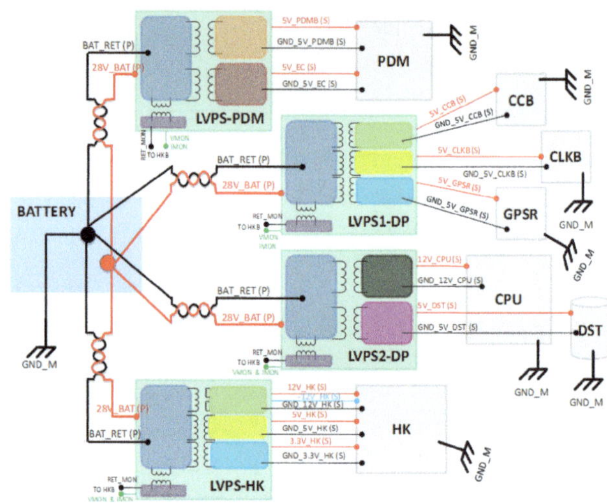

Fig. 8: EUSO-Balloon grounding architecture.

7.3. Thermal Management

Due to the lack of natural convection and reduced air heat conductivity in heat dissipation issue, some specific techniques must be implemented to conduct the heat generated by DDCU on different subsystems to the external structure for dissipation.

LVPS subsystem contains semiconductor devices, capacitors and other components that are vulnerable to thermal stress. In this particular case, the power devices inside of DDCU modules represent the major heat source. Therefore, an adequate thermal architecture becomes vital to limit the DDCU temperature to a lower value for extremely high reliability performance. As an example, Fig. 9 shows in a schematic description the thermal parameters involved on a DDCU mounted in two layers PCB. Parameters denoted with θ are described as thermal impedance of some heat power dissipation path. In that way, θ_{JA} represent thermal impedance from the junction, internally of DDCU under test, to the ambient environment and can be expressed as Eq. (1) shows:

$$\theta_{JA} = \frac{T_J - T_A}{P_D}. \quad (1)$$

There are two primary heat paths for the DDCU in Fig. 9, represented by their associated thermal resistance. The first path travels from the junction of DDCU to the plastic molding at the top of the case, θ_{JC}, and then to the ambient air, θ_{CA}. The second path is from the junction of the DDCU to an exposed pad, θ_{JEP}, when it exists. Therefore, the exposed pad is connected to the PCB, where the heat travels through the surface and finally to the ambient air [16]. Following the same path, θ_{Cu} represent the thermal resistance of our board's cooper to lateral heat transfer. Meanwhile, θ_{FR4} is the thermal resistance between the cooper planes provided by the vertical resistance of FR4 laminate, and θ_{SA}, represent the thermal resistance from the surface of the PCB to the ambient air.

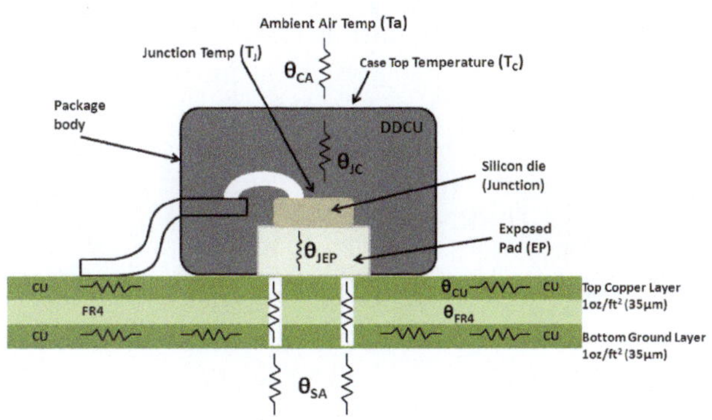

Fig. 9: DDCU thermal resistance model for a two layers PCB.

Therefore and concerning in our case. The most critical module, because of power consumption and transient response conditions, is the DDCU parallel arrangement for DST section of LVPS2-DP. For this arrangement the heat transfer path is through the case to ambient air.

Furthermore, special attention is needed in order not to exceed $\sim 105\,°C$ case temperature for a reliable operation, otherwise thermal protection is activated. Taking into account this issue, a preliminary setup under ambient temperature and normal pressure conditions, was proposed in order to get a better knowledge about how long does it take reach this critical temperature. As a first result, 20 min (at 5 A) were necessary for reaching $\sim 85\,°C$ with a positive slope in the curve obtained. Therefore, it is necessary to implement heat dissipation architecture, according to our case, to maintain the temperature bellow critical level. This thermal architecture is based on conduction method using a heatsink (HS) as interface material between DDCU of LVPS2-DP and DP rack structure.

Regarding this issue and the thermal background above described, now θ_{JA} is actually made up of at least two separate thermal resistances in series. One is the thermal resistance inside the device package, θ_{JC}, and the other one is the resistance between the case and the ambient, θ_{CA}. The last term can be splitted into θ_{CS} and θ_{SA}, thermal resistance from case to HS and thermal resistance from HS to ambient respectively [17] resulting in Eq. (3):

$$\theta_{JA} = \theta_{JC} + \theta_{CS} + \theta_{SA}. \quad (2)$$

However, in order to select a HS the term θ_{SA} has to be calculated. Therefore, assuming the next safety conditions and specification from DDCU datasheet: $T_{Jmax} = 100\,°C$, $T_{JAmax} = 50\,°C$, $T_{Cmax} = 90\,°C$, $P_D = 7$ W and, using Eq. (1), Eq. (3) and Eq. (4), θ_{SA}term can be obtained by Eq. (5):

$$\theta_{JC} = \frac{T_{Jmax} - T_{Cmax}}{P_D}, \quad (3)$$

$$\theta_{CS} = \frac{t}{k \cdot A_C}, \quad (4)$$

where t is the thickness material (9.6×10^{-3} m^2), k is the thermal conductivity of aluminum 3003-O (190 W/m·°C), and Ac is the contact area between DDCU and the HS (1.3×10^{-3} m^2):

$$\theta_{SA} = \theta_{JA} - (\theta_{JC} + \theta_{SC}). \quad (5)$$

Consequently, a $\theta_{SA} = 5.68\,°C/W$ was obtained which implies a selection of HS rated at $5.68\,°C/W$, or less. Therefore, and regarding the plug-in unit box in Fig. 6, the thermal architecture and its components showed in Fig. 10 were developed.

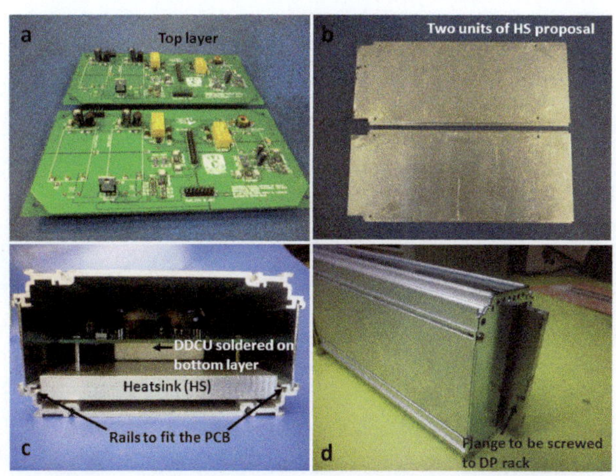

Fig. 10: (a) LVPS2-DP PCB, (b) Heatsink proposed, (c) PCB, heatsink and unit box assembly, (d) Heatsink rear side to be assembly with cooling plate of DP structure.

As can be seen in Fig. 10(a), most of elements are in the top layer while the converters are in the bottom layer, unlike the rest of the power modules PCB in which all the components are in the top layer. Consequently, the converters make direct contact with HS surface (as is showed in Fig. 10(b)) conducting the heat to HS edge coming out across the rear side panel of plug-in box, as can be seen in Fig. 10(d). Finally,the HS edge (rear side) has holes in order to be screwed to DP rack cooling metal backplane, if it is necessary. This metal backplane is connected to heat metal conductive bars, and then connected to external structure of balloon gondola improving in that way the thermal management of all instrument.

Fig. 11: Electrical performance set-up: main power (1), LVPS module (2), DC electronic load (3), HK subsystem (4), PC software (5), andauxiliary power supply (6).

8. Tests

In order to qualify the LVPS subsystem performance under the balloon environmental conditions, two sorts of tests were done. The first one was devoted to evaluate the electrical performance of LVPS modules under different load ranges (10 % to 100 % of full load) considering worst case (WC) scenario. The second one was focused in evaluating the critical electrical performance levels (at full load) under approximate vacuum environment (~3 mbar) applying temperatures variations, during the test, of $-20\,^{\circ}$C to $55\,^{\circ}$C.

8.1. Electrical Performance

The aim of this test is to characterize the four LVPS modules according to each subsystem specifications. A dc electronic load (dcEL) simulator, specifically developed at our group for this purpose, was used in order to automate the test process. The main idea of this dcEL is based on the use of a simple feedback loop to allow the transistor work as a current drain in current regulation mode, or as well as a voltage source in voltage regulation mode [18]. At this particular case, current regulation mode is used for characterizing voltage sources, in which the power source must deliver a specific current level setting by serial software interface to dcEL hardware. Figure 11 shows the electrical performance setup used to perform this test.

Basically, the setup above consists of a main laboratory power supply (LPS) labeled as 1, which provides the 28 V primary power bus voltage to the LVPS modules at number 2. Inside of LVPS modules, DDCU isolate and transform to a secondary power bus voltage depending on each subsystem. On other hand, dcEL at 3 receives commands from HK subsystem labeled as 4, powered with an auxiliary power source at number 6. HK communicates with a PC software interface, showed at 5 via, serial port. The software fixes the current level that should provided LVPS module to dcEL. At same time LVPS module provide V&I levels as TM information about DDCU status consumption to HK subsystem.

Therefore, based on setup explanation above, parameters like load and line regulation (LdReg & LnReg) were obtained at each LVPS module output connector using Eq. (6) and Eq. (7), [11]. Table 4 shows load regulation values for each LVPS module sections. Performance curves related to efficiency vs output current and input voltage (LnReg) also were obtained. Figure 12, Fig. 13, Fig. 14 and Fig. 15 show the line regulation according to voltage level provided by LVPS modules to each subsystem.

$$\% Load_{\mathrm{REg}} = \\ = \left(\frac{V_{OUT_{Full\_load}} - V_{OUT_{10\%\_load}}}{V_{OUT_{50\%\_load}}} \right) \cdot 100, \tag{6}$$

$$\% Line_{\mathrm{REg}} = \\ = \left(\frac{V_{OUT_{High\_Vin}} - V_{OUT_{Low\_Vin}}}{V_{OUT_{Nom\_Vin}}} \right) \cdot 100. \tag{7}$$

The percentage of LdReg indicates the variation in the output voltage in response to a change in load current, and is a reference to LVPS subsystem in order to comply with the output voltage tolerances. Load regulation percentage measurements were made varying the load from 10 % to full load in both cases, with nominal input voltage (28 V) and low input voltage (18 V) which is WC condition level.

Figure 12, Fig. 13, Fig. 14 and Fig. 15 shows LnReg plots of subsystems powered from 5 V, 5.5 V, 12 V and 3.3 V secondary SRB respectively. Those curves shows output voltage variations less than 10 %, at 90 % of full load, which comply with the subsystem specification requirement. The behavior of curves is almost linear, a slightly increment (less than 10 mV) at the

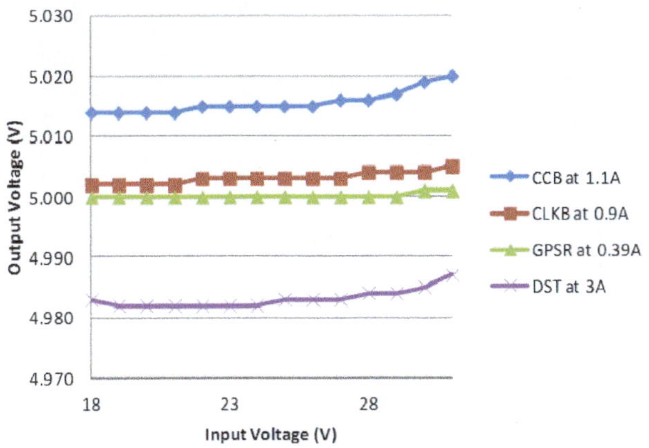

Fig. 12: Line regulation of subsystems powered with 5 V.

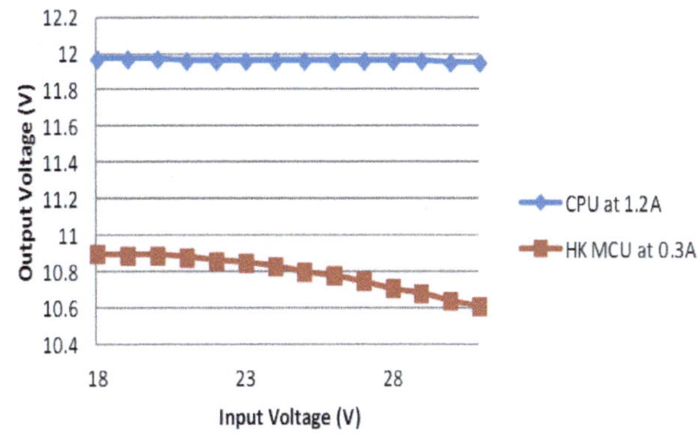

Fig. 14: Line regulation of subsystems powered with 12 V.

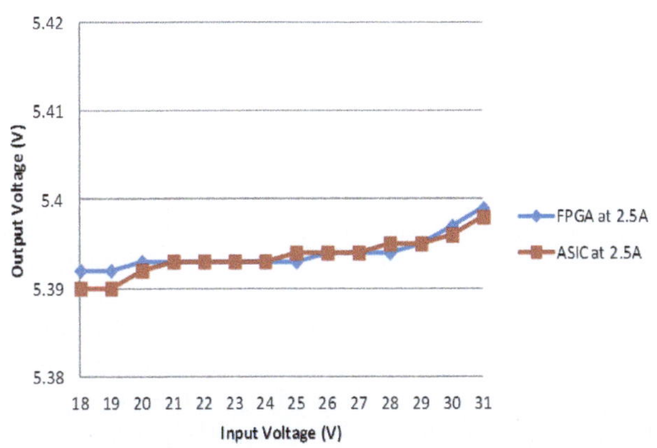

Fig. 13: Line regulation of LVPS-PDM subsystems powered with 5.5 V.

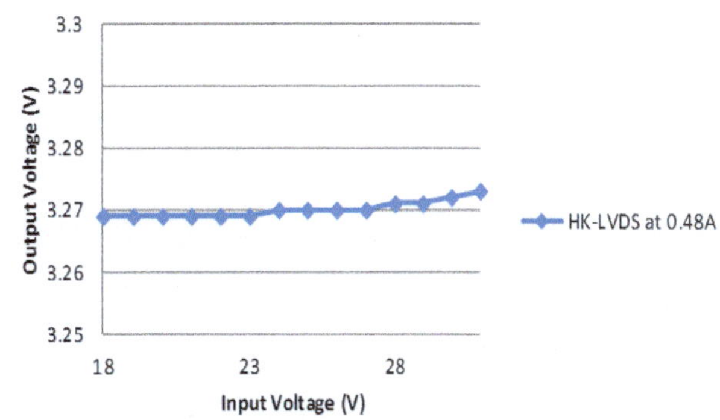

Fig. 15: Line regulation of subsystems powered with 3.3 V

end of curves ($Vin > 28$ V) is due to the temperature increase in resistors loads which cause increasing resistance value and consequently decreasing in current level demand which is compensated increasing the output voltage in order to keep a constant output power.

The LnReg term represent the change in output voltage in response to a change in the input voltage; these tests were planned considering the possible voltage range provided by PWP (battery), which goes from 18 V to 30 V. Besides, it is important to note that the DDCU input voltage range fits perfectly with variations range from battery.

On other hand, in Fig. 14 it can be observed a voltage drop in HK curve from 12 V to less than 11 V at 18 V input voltage on "x" axe. Figure 14 shows, at 24 V, a voltage drop that exceeds the 10 % specified. This abnormality is due to bad LdReg DDCU value obtained in Tab. 4. However, this abnormality is acceptable and compensated by the input voltage range of linear regulator in HK arduino board, from 7 V to 15 V. Voltage drop due to resistance cables represents a critical issue in any power distribution system. For this particular case, 0.5 m cables length with resistance of 0.15 Ω including connectors at the end were used. The latter do not represent a significant voltage drop at most cases. Moreover, efficiency curves of each LVPS modules were obtained, according to each subsystem demand. The test consisted in varying output current demanded with increment of 10 % steps of full load.

The last was done under WC condition ($Vin = 18$ V) in order to observe the behavior of each LVPS subsystem at maximum current levels. Therefore, in order to have a better appreciation of efficiency behavior on each case, Fig. 16 combines efficiency curves of those subsystems which current consumption is about 3 A, Fig. 17 groups subsystems which current consumption is about 1.2 A; meanwhile Fig. 18 shows those subsystem which demand is about 0.5 A.

The curves describe how efficiency improves meanwhile full load consumption is reached. However, is important to note that the maximum efficiency obtained in curves (from 60 % to 87 %) do not match with the converters efficiency, at full load, which are higher. The last is because of test measurements includes PCB parameters as impedance traces, planes, components, connectors, etc.

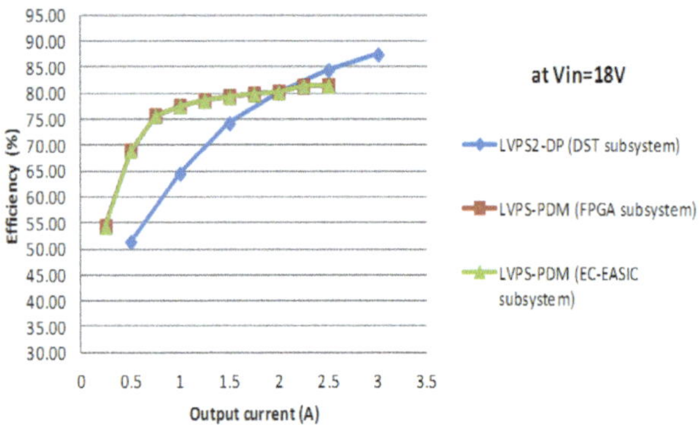

Fig. 16: Efficiency curves subsystems of approximately 3 A current consumption.

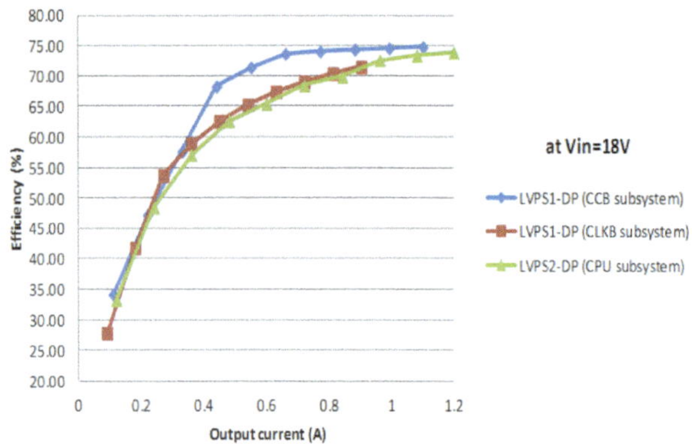

Fig. 17: Efficiency curves subsystems of approximately 1.2 A current consumption.

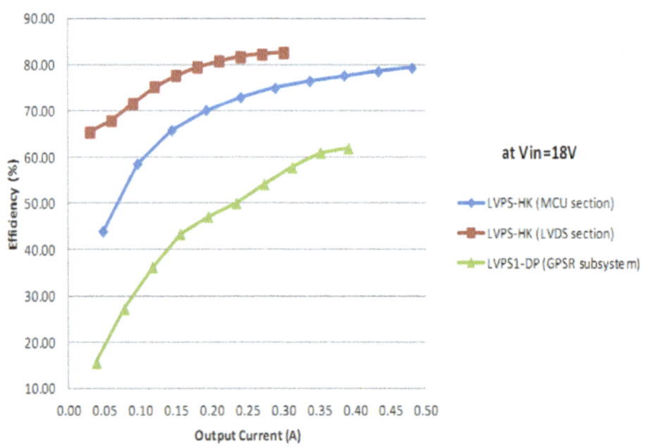

Fig. 18: Efficiency curves subsystems of approximately 0.5 A current consumption.

1) Turning On/Off LVPS Modules

In order to test the reliability of each module of power distribution system, and at the same time the interaction with HK (also designed and provided by our group) the setup showed in Fig. 19 was implemented.

Commands provided by an external PC were sent to HK subsystem. The commands were interpreted and

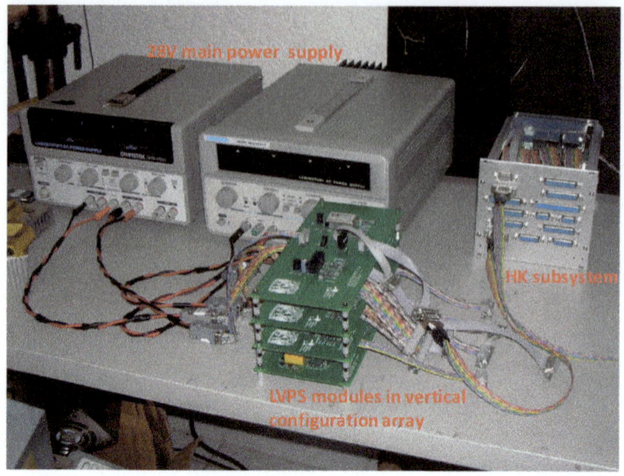

Fig. 19: Set-up for turning on/off LVPS modules.

executed by HK; which sends two high level commands (HL_CMD's) to LR's located OB LVPS modules. A HL_CMD is represented by a 9 V to 15 V amplitude of 10 ms pulse width. These commands are used in order to bias the internal coils of LR's, and diodes were included in design for feedback current protection.

In order to minimize the test space required, LVPS modules were arranged in a vertical assembled configuration as can be observed in Fig. 19. Finally, two dual laboratory power sources were added to the test setup in order to provide the 28 V primary bus voltage. This more complex setup was used to verify the four LVPS modules power consumption (without load) at the moment of turning-on, and the TM information of current consumption levels from current monitoring circuits at the LVPS modules. Also, a mechanical assembly was done in order to improve reliability during the testing process.

Table 5 shows the power consumption of LVPS modules (without load) after a successful turning on/off process, and monitored voltage levels obtained from voltage monitoring circuit as TM information. The voltage monitoring circuit is based on a non-inverter amplifier, with a constant gain factor of 0.326. In that way, a 3.3 V level correspond to 1.07 V, 5 V level correspond to 1.63 V, meanwhile 12 V level correspond 3.9 V output voltage.

Related to current monitoring and due to maximum current consumption subsystem (~4 A), it was decided to use a shunt resistor as input of circuit based on OpAmp which relationship of output voltage level and current consumption is 1:1. Therefore, TM information levels obtained are presented in section 8.3.

Tab. 5: Load regulation of LVPS modules.

Subsystem	Secondary voltage by LVPS (V)	Consumption without load (A)	Monitored voltage (V)
CCB	5	0.03	1.613
CLKB	5	0.005	1.688
GPRS	5	0.02	1.678
PDM FPGA	5.5	0.025	1.716
PMD EC-ASIC	5.5	0.025	1.707
CPU	12	0.07	3.677
DST	5	0.1	1.6
HK-MCU	± 12	0.015	3.735
HK-LVDS	3.3	0.01	0.914

2) Mechanical Assembly of LVPS Modules

As mentioned in section 7.2, plug-in unit boxes were used for enclosing each LVPS module. Figure 20 shows the final mechanical assembly of LVPS modules.

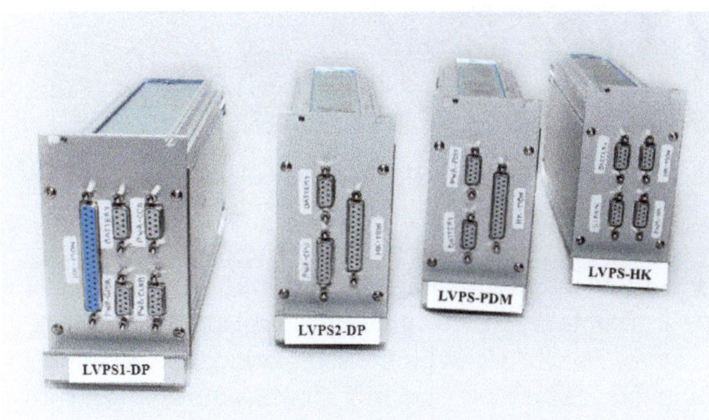

Fig. 20: Mechanical assembly of LVPS modules.

These mechanical boxes were housed into a provisional euro rack structure, together with the HK subsystem, in order to interconnect them in easier way. Then, we were ready to perform the thermo-vacuum tests.

8.2. Heat Dissipation

The suborbital environmental conditions limit heat dissipation to conduction and radiation. This, coupled with the very localized injection of heat at the electronic boards, makes of heat dissipation an important issue. In particular, in the case of the LVPS, the heat generated by the DDCU must be piped to the external mechanical structure. LVPS2-DP PCB, in particular, represents a critical case due to the power demand of its clients, which may reach up to ~4 A for DST and 1.25 A for CPU.

Furthermore, this steady state consumption due to different CPU operation modes and to several hardisk accessing processes (read/write), many transient events were generated. These transients cause not only high peaks of current, but also demand a faster response from the DDCU. Such transients are difficult to characterize and, after several tests during the first integration phase of the LVPS with the DP unit at INFN-Naples, Italy, a PCB design based on two DDCU in parallel was proposed to deal with them.

In order to manage the increase heat production on the latter design, a HS was designed in order to improve the dissipation from three DDCU of LVPS2-DP module. The Fig. 21 shows the mechanical assembly between HS and LVPS2-DP PCB inside of module.

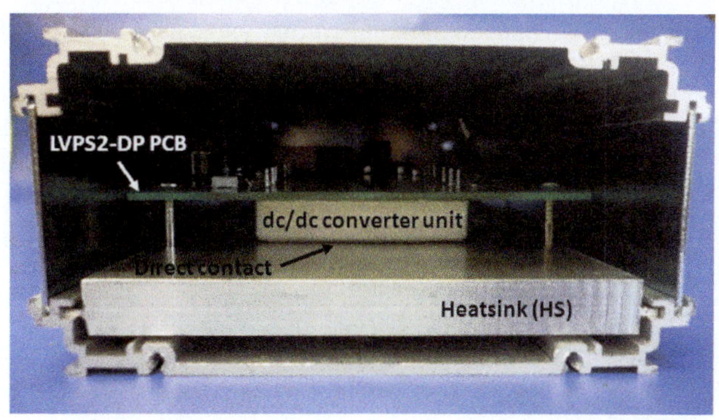

Fig. 21: Mechanical assembly between LVPS2-PCB and heatsink.

8.3. Thermo-Vacuum

In order to ensure the physical integrity of the LVPS subsystem in sub-orbital flight conditions (upper atmosphere), a series of thermo-vacuum tests have been conducted in a custom-made vacuum chamber with thermal control. The chamber uses a mechanical pump in order to reach the minimum pressure conditions.

The experimental instrument has a volume of roughly 60L, inside which an aluminum plate (30 cm × 30 cm × 1 cm) is resting on 4 stainless steel rods. The aluminum plate is connected to a refrigerator, which allows cooling down to −37 °C. A heater, made of NiCr, attached to the surface of the plate, is used to regulate the aluminum plate temperature in the range from −37 °C to +80 °C. In addition to the mechanical pump, if required, a second turbo pump stage can be used to reach 10^{-6} Torr. However, the most demanding test for the system is at the higher pressure of 3 mbar, where thermal conduction is highly reduced and convection is negligible, but there is still enough air in the rarefied atmosphere to produce unhindered sparks.

The test results will warrant the operation of LVPS modules in vacuum environment (3 mbar), for temperatures ranging from −20 °C to +50 °C, validating the overall thermal requirements imposed to the EUSO balloon experiment. The baseline duration time of the first flight of the EUSO-balloon is approximately 12 hours. Furthermore, at the moment all the phases of balloon ascension cannot be simulated with the current experimental setup owing conditions of humidity and radiation exposure are not implemented for performing inside vacuum jar.

Therefore, a series of tests comprising module on/off activation, constant full load operation, V&I monitoring information, status relay and temperature monitoring were conducted over a period beyond 12 hours in vacuum conditions in compliance with [20], [21]. Figure 22 shows the setup for testing LVPS modules, already mounted in the provisional euro rack structure, at constant full load.

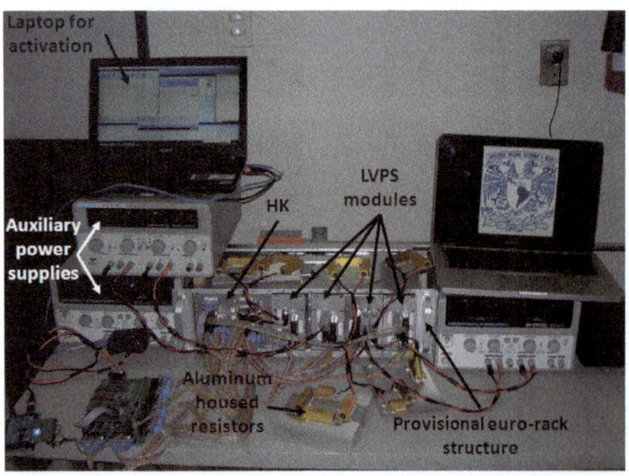

Fig. 22: Set-up for testing LVPS modules outside vacuum chamber.

On other hand, Fig. 23 shows an electronic setup with the four LVPS modules and HK system housed in the same rack. The HK was used both, through its regular interface with the LVPS subsystem, and as a probe inside vacuum for testing temperatures at several critical components of the LVPS modules. Input power twisted supply cables from two main laboratory power supplies were used to bias LVPS modules through 28 V bus voltage. Meanwhile, differential cables from monitoring LVPS connectors (twisted color cables in figure above) come out for transmitting to HK unit. Furthermore, twisted output power cables from 14 AWG to 20 AWG gauge were connected to fixed value aluminum housed power resistors which were calculated in order to demand full load operation. Finally, three thermistors at most critical and power demanded DDCU were located: one at CPU, and 2 at DST section.

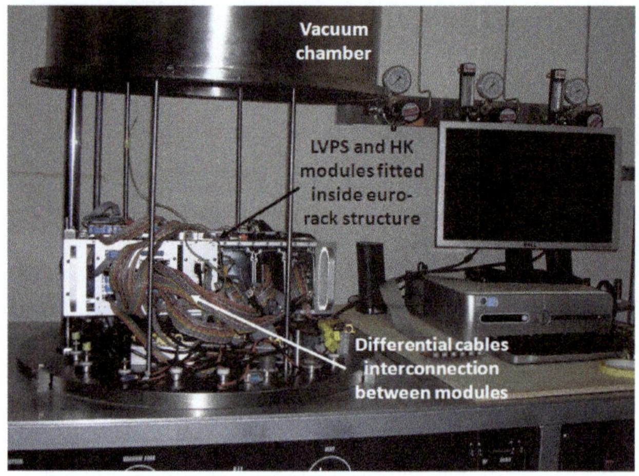

Fig. 23: Set-up for testing LVPS modules inside vacuum chamber.

As was mentioned above in thermal management section, as a first test, was implemented a setup (at room temperature ∼ 25 °C) which helps basically to know three aspects. The first one was the measurements (at full load) from current monitoring circuits. Table 6 shows the measurements according to each subsystem. These values were acceptable regarding cable length losses.

Tab. 6: DDCU monitored current measurements at full load.

Subsystem	Experimental current load (A)	Current monitoring (A)
CCB	1	0.982
CLKB	0.793	0.971
GPRS	0.357	0.347
PDM FPGA	2	1.95
PDM EC-ASIC	3	1.97
CPU	1	0.988
DST	3	2.89
HK-MCU	N/A	0.107
HL-LVDS	N/A	0.183

The second aspect was in order to be sure about similar current consumption of parallel configuration comprised by two DDCU of DST section (see Fig. 10(a)). Consequently, temperature rise of both converters was monitored. As can be seen in Fig. 24, the temperature difference between both converters (DC/DC1 and DC/DC2) before 15 min was less than 2 °C), which is acceptable. After that, the difference increase which probably indicates an unbalanced load operation.

The third reason was in order to figure out how long it takes to reach ~ 85 °C (maximum case temperature). Approximately in 20 min both converters reach the temperature barrier. Therefore, in order to decrease the temperature on converters, the HS manufactured and implemented in LVPS2-DP module (see Fig. 10(c)) was used as a part of the setup showed in Fig. 25. This setup comprise HK subsystem, using through its regular interface with LVPS subsystem and as a probe for testing temperatures of LVPS components; and two LVPS modules, LVPS2-DP and LVPS-HK, the last powering HK subsystem. The LVPS2-DP module was fixed in vertical position, in that way, the rear side of HS (coming out from LVPS2-DP module) was in direct contact with aluminum plate of vacuum jar and the heat transfer between them was improved by using thermal grease. A set of thermistors were implemented as was described previously. Furthermore, two more thermistors were added, one on HS over the edge (HS and aluminum plate boundary contact) and other on aluminum plate. This test was done at room temperature inside of vacuum jar (not functioning) with the main aim to compare the thermal performance over 20 min.

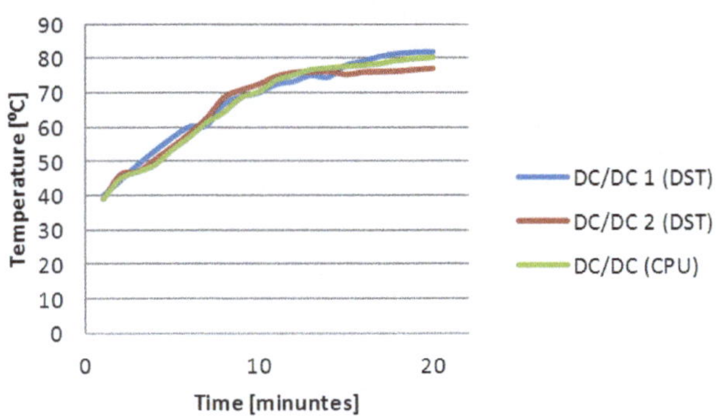

Fig. 24: Rise temperature of DST section DDCU on LVPS2-DP module.

The result was that temperature decreases at 35 °C in 20 min which represent an improvement on LVPS module thermal performance. However, it is necessary to perform a longer test under environmental conditions of Balloon project. Therefore, in order to continue the module test, vacuum jar was closed and pressure ambient was set at 3 mbar. The module performance continued working during 90 min in which temperature was reaching ~ 55 °C as can be observed at Fig. 26.

Finally, keeping the same pressure conditions and the setup of Fig. 25, the LVPS2-DP module was subjected to two thermal increments at the plate (the

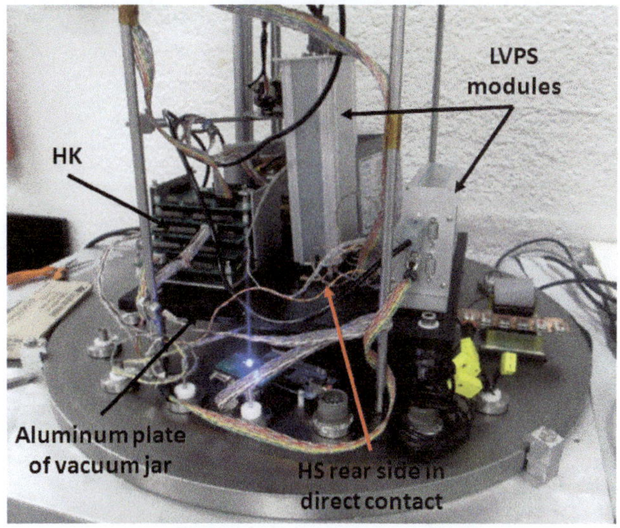

Fig. 25: Thermal performance test of DDCU of LVPS2-DP module at room temperature.

amount of thermal steps is shown in Fig. 27, Fig. 28 and Fig. 29, in blue line) through heater located under aluminum plate.

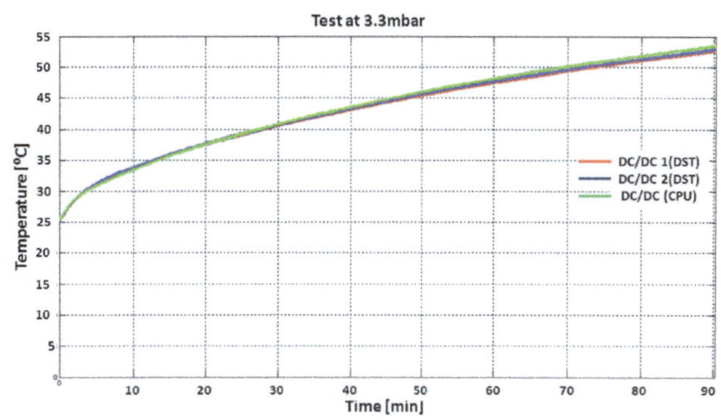

Fig. 26: Thermal performance of LVPS2-DP converters at 3 mbar conditions.

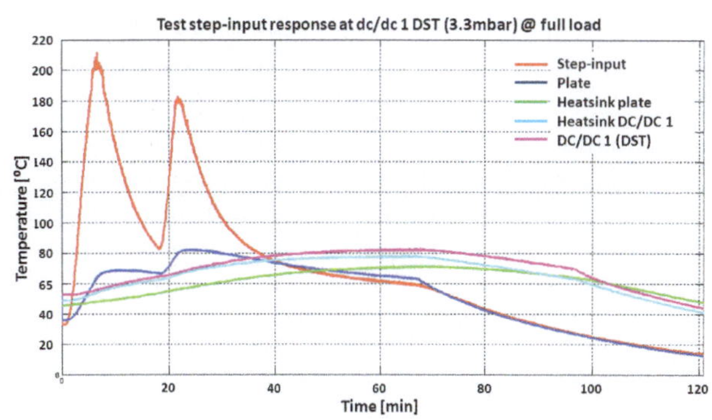

Fig. 27: Thermal performance of LVPS2-DP DDCU1 of DST section.

The last was done with the main aim to figure out the converters behavior when reach the maximum case temperature considering those environmental conditions provided by vacuum jar. The red curve describes

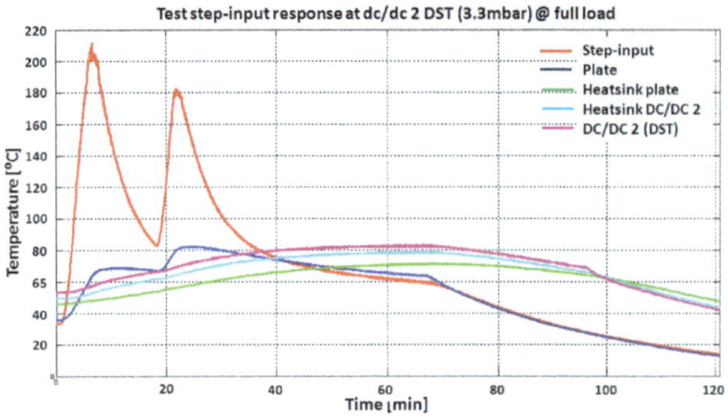

Fig. 28: Thermal performance of LVPS2-DP DDCU2 of DST section.

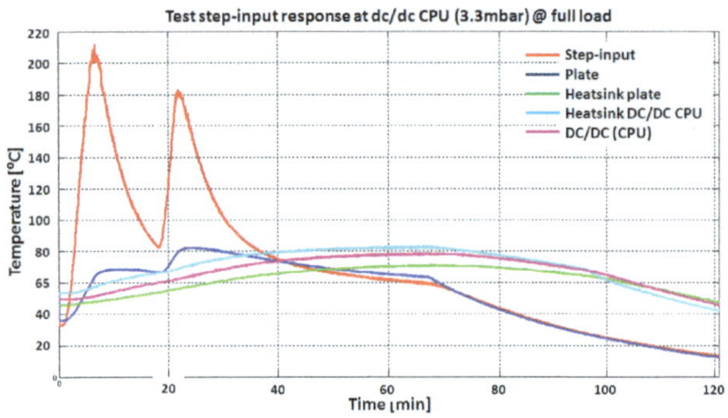

Fig. 29: Thermal performance of LVPS2-DP DDCU of CPU section.

the thermal steps in heater (attached to plate) providing those two increments in aluminum plate; meanwhile green line describes the temperature at the edge of HS and plate. The remaining two lines, light-green and purple, describe the rise temperature at heat sink (on DDCU) and lateral side case of DDCU under test, respectively. Therefore, Fig. 27, Fig. 28 and Fig. 29 describe, in separately way, the thermal performance over 120 min of three DDCU of LVPS2-DP module.

Consequently, in figures above can be observed that maximum case temperature after 60 min of continuous performance (at full load approximately) is reached. Also, is important to mention that load conditions test are not the same of real performance of subsystems balloon. However, the tests here described and performed have the main aim to characterize the LVPS modules under WC conditions.

Finally and, getting advantage of final integration phase of EUSO-Balloon project at INFN-Naples, Italy, LVPS modules were tested under real load and environmental balloon conditions with all subsystems. However, DST section power consumption was higher than specified causing an extra heating on DDCU of DST section on LVPS2-DP module. This inconvenient was resolved attaching the heat sink (of power module) to cooling plate of DP rack in order to increase surface heat dissipation. Figure 30 shows how temperature is stabilized, at 51 °C, after reach 10 hrs of continues working. This temperature does not represent danger on DDCU performance.

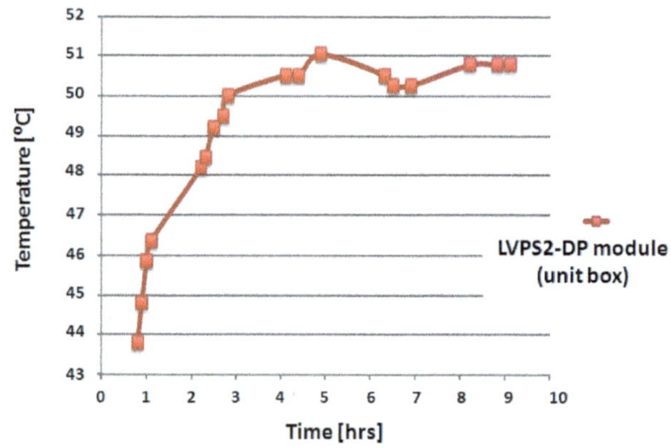

Fig. 30: Thermal performance of LVPS2-DP module during final integration test in Naples, Italy.

8.4. Weight

Weight is often a critical issue in every payload of space and sub-orbital project. Therefore, Tab. 7 shows how each module of LVPS subsystem meet the requirements regarding weight constrain. Furthermore, weight of provisional rack structure is also included in Tab. 7.

Tab. 7: Weight of each LVPS subsystem modules.

Subsystem	Weight (kg)
LVPS-PDM	0.9
LVPS1-DP	1
LVPS2-DP	0.9
LVPS-HK	0.9
Sub-total	**3.7**
Rack	1.3
Total	**5**

9. Conclusion

A fully operational low-voltage power supply subsystem has been developed for a sub-orbital EUSO-Balloon instrument. The LVPS subsystem has the capability to operate at 3 mbar pressure environment under maximum load current consumption in a temperature range from −20 °C to +50 °C. The subsystem comprises four power modules which can be remotely activated, in reliable way, through 12 V HL_CMD pulses provided by external subsystems, HK and SIREN. The four modules developed provide isolation between 28 V main power bus and secondary power bus to payload users, avoiding failures propagation. At same time, the modules provide voltage transformation to low-voltage levels ±12 V, 5 V, 5.5 V and 3.3 V, allowing to implement tight regulation stage ($\leq \pm 1$ %) on-board subsystems by the use of linear regulators or niPoL. The modules are able to withstand low excursion voltage level from PWP (from 18 V to 30 V) and provide 10 % tolerance on output voltage levels (secondary power bus).

Regarding environmental conditions, importance level of some subsystems, and considering that EUSO-Balloon is a recoverable instrument (future flights planned) the LVPS modules provide I/O short circuit protection by the use of fuse at input, over-temperature protection, and since the switching power supplies are the major EMI noise contributors on instrument, I/O EMI filters were implemented (according to manufacturer specifications) in order to minimize the instrument contamination by conducting interference from tracks, connectors, cables, etc.

Additionally, PCB power modules were enclosed into individually unit boxes with the main aim to minimize the radiated interference and, at the same time, be less susceptible to external EMI. Another capability of LVPS subsystem consists of providing voltage and current levels information (proportionally to 0 V to 5 V output range monitoring circuits) as TM data to HK. Furthermore, the power modules allow verification status of LR's clamps in order to control reliable power supply to each subsystem. The power modules also comply with the Eurocard size dimensions, and consequently, meet the weight restriction imposed at requirements section, ≤ 5 kg.

Thermal management was a complicated issue. A heat sink of $\theta_{SA} = 5.68\,^{\circ}\mathrm{C/W}$ was manufactured in order to provide an extra path conduction dissipation for LVPS2-DP, which did represent the most critical module related to thermal performance. Stable thermal performance of 51 °C was obtained attaching HS designed to DP cooling plate.

Acknowledgment

The authors were greatly benefitted from working on the collaborative atmosphere of the JEM-EUSO consortium and through the discussion with several colleagues and personnel from CNES. In particular, we would wish to thank Michel Dupieux, Pierre Pratt, Phillippe Gorodetzky, Giuseppe Osteria, Guillaume Prevot, Jorg Bayer, Aera Jung, Pascale Danto and to the colleagues from UNAM: Cesar Lopez, Javier Rojas, Adrian de la Cruz, Socrates Guerrero, Greta Leon and Leonardo Munoz. Without their advice and cooperation this work would not have been possible. This work was supported by CONACyT through CB and RedCyTE projects, and by PAPIIT-UNAM projects.

References

[1] CABALLERO, R., J. C. D'OLIVO, G. MEDINATANCO, L. NELLEN and F. A. SANCHEZ. Instrument Owerview of the JEM-EUSO mission. In: *International Cosmic Ray Conference.* Merida: Universidad Nacional Autonoma de Mexico, 2008, pp. 1077–1080. ISBN 978-970-32-5155-1.

[2] PARIZOT, E. and S. DAGORET-CAMPAGNE. *EUSO BALLOON: Organization Note of the EUSO-Balloon Instrument.* v 1.0. 2012. Available at: `http://euso-balloon.lal.in2p3.fr/IMG/pdf/EUSO-ON-INST-201-LAL_V1_0.pdf`.

[3] LEY, W., K. WITTMANN and W. HALLMANN. *Handbook of space technology.* Chichester: Wiley, 2009. ISBN 04-706-9739-3.

[4] FORTESCUE, P., J. STARK and G. SWINERD. *Spacecraft systems engineering.* New York: Wiley, 2003. ISBN 978-0-470-75012-4.

[5] BUTLER, S. Point-of-load power solutions for low-voltage military applications. In: *Defense Electronics* [online]. 2005, pp. 14–17. ISSN 0278-3479. Available at: `http://defenseelectronicsmag.com/site-files/defenseelectronicsmag.com/files/archive/rfdesign.com/mag/512RFDSF2.pdf`.

[6] MARCHETTI, R. Power Systems Architectures What's in? What's out?. In: *Vicor Power* [online]. 2003. Available at: `http://cdn.vicorpower.com/documents/industry_articles/power_systems_article.pdf`.

[7] BUSSARAKONS, T. High Efficiency Space Grade DC-DC Converter is a Prime Solution for Modern Day Digital Loads. In: *International Rectifier* [online]. 2011. Available at: `http://www.irf.com/pressroom/articles/000eet1107.pdf`.

[8] BUSSARAKONS, T. Thick film hybrid dc-dc converters are standard "bricks" for satellite power systems. In: *International Rectifier* [online]. 2004. Available at: `http://www.irf.com/product-info/hi-rel/tp-hybridbrick.pdf`.

[9] GRAF, R. F. *Modern dictionary of electronics.* Boston: Newnes, 1999. ISBN 07-506-9866-7.

[10] SODERBERG, B. and T. BUSSARAKONS. Compatibility Analysis of Space Qualified Intermediate Bus Converter and Point of Load Regulators for Digital Loads. In: *International Rectifier* [online]. 2011. Available at: `http://www.irf.com/technical-info/whitepaper/sbbzacompatibility.pdf`.

[11] BROWN, M. *Power supply cookbook.* Boston: Newnes, 2001. ISBN 07-506-7329-X.

[12] SINPRO CO, LTD. Introduction: EMI, EMC. In: *SINPRO CO, LTD.* Available at: http://www.sinpro.com.tw/file/Technical_Knowledge/IntroductionEMIEMC%20_2_.pdf.

[13] COLOTTI, J. EMC Design Fundamentals. In: *IEEE Long Island Setion.* 2005. Available at: http://www.ieee.li/pdf/viewgraphs/emc_design_fundamentals.pdf.

[14] KAISER, K. L. *Electromagnetic compatibility handbook.* Boca Raton: CRC Press, 2005. ISBN 08-493-2087-9.

[15] JOFFE, E. B. and K.-S. LOCK. *Grounds for grounding: a circuit-to-system handbook.* Hoboken, N.J.: Wiley, 2010. ISBN 04-716-6008-6.

[16] TEXAS INSTRUMENTS. AN-2020 Thermal Design By Insight, Not Hindsight. In: *Texas Instruments.* 2013. Available at: http://www.ti.com/lit/an/snva419c/snva419c.pdf.

[17] TEXAS INSTRUMENTS. Understanding Thermal Dissipation and Design of a Heatsink. In: *Texas Instruments.* 2011. Available at: http://www.ti.com/lit/an/slva462/slva462.pdf.

[18] ROCHE, S. and S. ROSA. Power up a microcontroller with pre-power-down data. 2005. In: *m.eet.com* [online]. Available at: http://m.eet.com/media/1130553/13338-31705di.pdf.

[19] IEEE 1101.1. *Standard for Mechanical Core Specifications for Microcomputers Using IEC 60603-2 Connectors.* 1998. Available at: https://ph-dep-ese.web.cern.ch/ph-dep-ese/crates/standards/1101_1.pdf.

[20] MEDINA-TANCO, G. and F. TRILLAUD. *EUSO-BALLOON: Thermal analysis.* V1.0. 2012. Available at: http://euso-balloon.lal.in2p3.fr/IMG/pdf/EUSO-TA-INST-409-IRAP_V1-0.pdf.

[21] MEDINA-TANCO, G. EUSO-BALLOON. *Thermal policy and thermal test procedure.* 2013.

About Authors

Hector Hugo SILVA LOPEZ was born in Mexico City 1979. He received his M.Sc. in the field of Electronic Systems from National Autonomous University of Mexico (UNAM) in 2008. Currently, he is in the process to obtain his Ph.D. in Electronic Instrumentation as active member of JEM-EUSO collaboration. His research interests include power electronics and electronic analog design for harsh environment.

Gustavo MEDINA-TANCO was born in Jujuy, Argentina 1962. He got his Ph.D. in the field of Physics from Instituto Astronomico e Geofisico, Universidade de Sao Paulo, Brasil in 1992. Currently, he is a main researcher "C" in the Nuclear Science Institute at UNAM, and he is involved in Auger and JEM-EUSO projects PI leader. His work field lies in Theory and experimentation in Ultra-high energy cosmic rays; with deep interest in space technology development.

Lauro SANTIAGO CRUZ was born in Puebla, Mexico 1957. He received his M.Sc. in the field of Electronic Digital Systems Design from National Autonomous University of Mexico (UNAM). Currently, he is Main Tachnicia type "B" at Engineering Institute in UNAM; and also is professor of Electronic Instrumentation at UNAM. His work is focused in the field of Electronic Instrumentation design for Energy projects development.

19

Optimal Placement of Non-Site Specific DG for Voltage Profile Improvement and Energy Savings in Radial Distribution Networks

Mudathir Funsho AKOREDE[1], *Edris POURESMAEIL*[2], *Hashim HIZAM*[3], *Ishak ARIS*[3], *Mohd Zainal AB-KADIR*[3]

[1]Department of Electrical & Electronics Engineering, Faculty of Engineering and Technology, University of Ilorin, P.M.B. 1515 Ilorin, Nigeria
[2]Centre for Smart Energy Solutions, University of Southern Denmark, Campusvej 55, DK-5230 Odense, Denmark
[3]Electrical & Electronic Engineering Department, Faculty of Engineering, Universiti Putra Malaysia, 43400 UPM Serdang, Selangor Darul Ehsan, Malaysia

makorede@ieee.org, edris_pooresmaeil@yahoo.com, hashim@eng.upm.edu.my, ishak@eng.upm.edu.my, mzk@upm.edu.my

Abstract. *This paper proposes a model based on Fuzzy Genetic Algorithm (FGA) to determine the optimal capacity and location of a DG unit in a radial distribution network. In the FGA, a fuzzy controller is integrated into GA to adjust the crossover and mutation rates dynamically to maintain the proper population diversity during GA's operation. This effectively overcomes the premature convergence problem of the simple genetic algorithm (SGA). The main objective functions considered in this study are maximisation of cost savings arising from energy loss, minimisation of voltage drops across all lines, and maximisation of the transfer capability of the system. The model takes into account the peculiarities of radial distribution networks, such as high R/X ratio, voltage dependency and composite nature of loads. The proposed model is evaluated on three radial test distribution systems, and the results obtained are very impressive, with high computational efficiency, when compared with those of the existing approaches cited in the literature.*

Keywords

Distributed generation, fuzzy genetic algorithm, radial distribution systems, voltage profile.

1. Introduction

Generally speaking, the term distributed generation (DG), also known as embedded generation (EG) or decentralised generation (DG) in some quarters is defined as any source of electric energy of limited capacity that is directly connected to the existing power grid distribution network, where it is required by the end-users. Presently, there is no consensus on what exactly the capacity limit of DG as well as the voltage level at the point of common coupling (PCC) to the grid should be. The definitions of these two parameters somewhat vary from country to country [1]. The potential benefits, the drivers responsible for the recent global surge in the interest for DG in the realm of power generation, as well as the issues in connecting DG to the existing grid, have been well reviewed and reported by the authors [2].

To maximise the benefits of DG and/or avoid putting the system operation in jeopardy, we clearly need to optimise DG size and location on the system. Previous research has shown that improper location of DG units on the system could result in more total line loss. To this end, a number of researchers have proposed different approaches in determining the best capacity and placement of DG sources in power distribution networks. A critical review of these approaches has been carried out and reported in [3], and so only a few of them would be referenced in this paper. For example, Acharya, et al. [4] used exact loss formula which involves the calculation of the bus admittance from its impedance to determine the optimal size and location of DG. However, the large size and complexity of distribution networks would render this approach to fall short of the robustness requirements.

Elnashar [5] categorised the impacts of DG on the distribution systems as positive, e.g. voltage profile improvement, and negative, such as increased system loss and short circuit level, and used these to formulate a multi-objective function. This problem formulation is flawed because DG would lead to increase in loss only when it is sited in improper locations, and so its impact cannot be considered negative in all cases. Hedayati et al [6] used continuation power flow to determine the most sensitive buses to voltage collapse. In [7], Tabu search approach was employed, and Kumar et al [8], [9] adopted genetic algorithm (GA) approach, while Haghifam [10] used the concept of Pareto optimality based on non-dominant sorting genetic algorithm (NSGA-II) for the optimisation problem. Two analytical approaches for meshed and radial networks were proposed by Wang and Nehrir [11], even though DG size was not optimised, and Celli et al [12] combined GA with ϵ-constrained technique for the multi-objective problem.

However, a number of these early works [4], [6], [7], [8], [9], [12], [13] did not consider the effect of load modelling in their studies. It is pertinent to mention here that constant power load model is commonly used in the power flow studies of most conventional load flow techniques. Adopting this model alone for distribution systems will not however, give accurate results in view of the voltage dependency nature of most load buses. Therefore, to present a real life scenario, distribution systems require a composite of constant power, constant current and constant impedance load models as adopted in this work.

Similarly, the popular load flow algorithms such as Gauss Siedel, Newton Raphson, and Fast decoupled methods that are usually employed for power flow analysis of transmission networks were adopted by many of these early works [4], [6], [7], [9]. These methods would not work efficiently for radial distribution systems due to the special peculiarities of the latter. Details of this are discussed in Section 3. On the strength of the foregoing, a new power flow technique proposed for radial distribution systems by Eminoglu and Hocaoglu [14] is adopted for the power flow studies carried out in this paper.

The focus of this study is, therefore, to formulate a multi-objective model to determine the optimal capacity and location of a single DG unit in a radial distribution network. A single DG unit is considered on the distribution feeder since connection of a higher number may not be economically viable and could lead to violation of certain system constraints. The objectives considered in the present work are maximisation of energy loss cost savings, minimisation of lines voltage drops, and maximisation of the power transfer capability of the network. These are formulated into a single multi-objective function, and solved using fuzzy genetic algorithm (FGA).

In the proposed approach, weighting factors are allocated to each objective according to the preference of the system planner at any particular time. Also, the composite voltage-dependent load modelling carried out in our previous work, to reflect a real life scenario in the distribution systems, is adopted in this study. The results obtained from the proposed model are compared with those obtained from the existing methods, and they are found to be more accurate (optimal) and computationally faster. The proposed model in this work is suitable for non-site specific DG such as microturbine, diesel generators, fuel cell units, etc.

The present work is suitable for an Independent Power Producer (IPP) in an unbundled deregulated energy market environment, where a generating company's main target is to produce and sell electricity to attain a maximum profit without accurately satisfying system demand/reserve. In this study, a decentralised scheduling, such as that obtainable in Norway, where scheduling is the responsibility of individual generating companies, is assumed. Since the objective of an IPP is to maximise profit, the scheduling is done as if all generation is sold in the spot market. Therefore, generation is assumed at full capacity of the DG unit.

2. Problem Formulation

The aim of the multi-objective function formulated in this study is to maximise the energy loss cost savings, minimise voltage drops across all lines, and to maximise the power transfer capability of the system. Each of these is treated next in the literature.

2.1. Energy Loss Cost Savings (ELCS) Maximisation

The energy loss cost savings for the distribution company (DISCO) could be maximised by installing a DG unit in the system. The mathematical expression given in Eq. (1) is formulated for this objective function. It is the percentage decrease in total energy loss cost when a DG is installed in the network and run for T_d hours in a year. It is desired to have a DG size that maximises this objective function when located at a particular bus.

$$\max(ELCS) = \frac{C_E \sum_{m=1}^{N-1} T_d \left(P_{Lm}^{B} - P_{Lm}^{DG}\right)}{C_E \sum_{m=1}^{N-1} T_d P_{Lm}^{B}}. \quad (1)$$

This expression is subject to Eq. (2).

$$0 \leq \sum_{m=1}^{N-1} P_{Lm}^{DG} < \sum_{m=1}^{N-1} P_{Lm}^{B}, \tag{2}$$

where C_E is unit energy cost [$/MWh]; P_{Lm}^{B} and P_{Lm}^{DG} are line m active power loss without and with DG installation respectively; N is the number of buses.

2.2. Line Voltage Drop (LVD) Minimisation

To improve the voltage profile of all buses, it is essential to minimise the voltage drop on all lines of the network. Voltage drops are a consequence of increased line loading, as well as line losses. The higher the line loading, the higher the losses incurred and consequently the higher the voltage drop on the line. Given the ever-increasing demand for power, power systems are expected to be more heavily loaded in future. On the strength of this, there is need to minimise voltage drops in anticipation.

The active power loss prior to DG connection to the network on line m is presented in Eq. (3) [15], where line m is between buses i and j. ΔV_m, R_m, and X_m are voltage drop, resistance, and reactance of line m respectively.

$$P_{Lm} = \Delta V_m^2 \frac{R_m}{R_m^2 + X_m^2}. \tag{3}$$

The voltage drop on line m, from the expression in Eq. (3), is as given in Eq. (4).

$$\Delta V_m = \sqrt{\frac{P_{Lm}\left(R_m^2 + X_m^2\right)}{R_m}}. \tag{4}$$

Now the total voltage drops on the system is a sum of the voltage drop on all the lines of the power system, which is formulated as in Eq. (5).

$$\sum_{m}^{N-1} \Delta V_m = \sum_{m}^{N-1} \sqrt{\frac{P_{Lm}\left(R_m^2 + X_m^2\right)}{R_m}}. \tag{5}$$

Minimising the voltage drop on a line is synonymous to maximising the difference between the voltage drop on the line before and after DG connection to the network. This can be formulated in percentage form as in Eq. (6).

$$\min(VLD) = \max\left(\frac{\sum_{m=1}^{N-1} \Delta V_m^B - \sum_{m=1}^{N-1} \Delta V_m^{DG}}{\sum_{m=1}^{N-1} \Delta V_m^B}\right). \tag{6}$$

In other words, the highest value of the expression on the RHS of Eq. (6) minimises the LVD.

2.3. Power Transfer Capability (PTC) Maximisation

The power transfer capability PTC, of a power system is the maximum power that can be transported via a power line from one point to another, without compromising the system security. DG is a viable alternative to distribution system expansion in response to the ever-increasing population growth since environmental and economic constraints prohibit expansion of the existing system. The overall PTC of a distribution system could be computed from Eq. (7), which is the sum of all the power transported on a line m, when P_{DG} size is connected to bus B_k.

$$PTC = \frac{\left(\left(P_{DG}^{B_k}+P_{sw}\right) - \sum_{m=1}^{N-1} P_{Lm}^{DG}\right) - \left(P_{sw} - \sum_{m=1}^{N-1} P_{Lm}^{B}\right)}{P_{sw} - \sum_{m=1}^{N-1} P_{Lm}^{B}}. \tag{7}$$

Simplifying the expression in Eq. (7) leads to Eq. (8).

$$(PTC) = \frac{P_{DG}^{B_k} - \sum_{m=1}^{N-1} P_{Lm}^{DG} + \sum_{m=1}^{N-1} P_{Lm}^{B}}{P_{sw} - \sum_{m=1}^{N-1} P_{Lm}^{B}}. \tag{8}$$

Equation (8) is subject to bus voltage, line thermal limit and DG capacity constraints presented in Eq. (9).

$$|V_i|_{\min} \leq |V_i| \leq |V_i|_{\max} \; S_m \leq S_m^{\max} \quad \text{and } P_{DG} \leq 0.4 \sum_{i=1}^{N} P_{Di}. \tag{9}$$

In fixing the maximum DG capacity constraint, [7] assumed 35 % of the total connected loads. On the other hand, [11] varies DG capacity from 47 to 100 % of the total load demand while [4] and [13] obtained an optimal DG size equal to 47.63 % of the load demand on the test system. However, DG capacity is constrained to 40 % of the total system load demand in this study to avoid any provocation of power quality and protection issues, particularly during the contingency period. This is based on the outcome of our earlier studies on the impact of different penetration levels of DG on the overall network during fault periods.

2.4. The Multi-Objective Function (MOBF)

Combining the three individual objective functions formulated in the previous subsections results in the expression presented in Eq. (10). The bus B_k that maximises MOBF when P_{DG} is sited there is regarded the optimal location in this study. The essence of the percentage in each objective function is to scale them individually so as to have the same magnitude range.

$$MOBF_{Bk} = \max\left(W_1 ELCS + W_3 PTC\right) + \min\left(W_2 LVD\right), \tag{10}$$

where the weighting factors $W_1 + W_2 + W_3 = 1$; (0.4, 0.3 and 0.3 respectively as used in this work). These weights are allocated by the system planner to indicate the relative importance of each objective.

3. Power Flow Method and Static Load Modelling

The popular conventional power flow techniques such as Gauss-Seidel, Newton Raphson, and Fast decoupled methods, usually employed for solving transmission system load flow problems, have been reported to be inefficiently suitable for solving radial distribution systems [13], [14], [16]. In spite of the robustness of Gauss Seidel method, it requires many iterations that consequently slows it down. Newton Raphson (NR) method requires a large computer memory, and worse still the high R/X ratio of distribution systems makes its convergence poor. Actually, fast decoupled approach has the advantage of computational speed and small storage facility when compared to NR. However, the assumptions usually made to simplify the analysis such as: the bus voltage magnitudes are close to 1 p.u, the phase angles are not large in magnitude, and R is far less than X, and therefore ignored [17]; are often invalid in distribution system applications. These are due to the special features of distribution systems which include their radial configuration, wide-ranging resistance and reactance values, unbalanced distributed loads, extremely large numbers of branches and nodes, etc.

On the strength of the foregoing, a power flow technique that would sufficiently take into cognisance, all these special features of distribution systems, and yet with a good convergence performance is required. As a result, the new power flow technique proposed by Eminoglu and Hocaoglu [14] known to offer a better solution for radial distribution networks with voltage-dependent loads, is employed in this study. This method is based on polynomial equations on the forward process and backward ladder equations for each branch of radial distribution systems. Line shunt capacitance and exponents of static load are incorporated in the said power flow method.

In carrying out power flow studies, most conventional techniques assume constant power at load buses, regardless of the voltage magnitude of such buses. This approach is usually employed in transmission networks. However, this assumption would not give accurate results when applied to distribution networks. The reason being that different types of loads such as residential, commercial, as well as industrial loads are commonly found in distribution systems. By the nature of these types of loads, their active and reactive powers are determined by the values of the voltage of the system; hence, they are referred to as voltage-dependent loads.

To reflect this scenario, the steady-state active and reactive load power demand at bus j can be expressed in an exponential form as given in Eq. (11) and Eq. (12) respectively.

$$p_{Dj} = p_{Dj}^{n}\left(\frac{V_j}{V_j^n}\right)^{n_p}, \tag{11}$$

$$Q_{Dj} = Q_{Dj}^{n}\left(\frac{V_j}{V_j^n}\right)^{n_q}, \tag{12}$$

where n_p and n_q are the load power exponents. Values of 0, 1, and 2 are respectively allocated to np and nq for constant power, constant current, and constant impedance loads. Other common exponential values for different static load models could be found in [14]. P_{Dj}^n and Q_{Dj}^n stand for the active and reactive power demand at bus j at the nominal voltage V_j^n; while the j load bus voltage is represented by V_j.

As it is generally known that distribution loads are composite loads that comprise of constant impedance (Z), constant current (I), and constant power (P). This load model in Eq. (11) and Eq. (12) is called ZIP model, which could be modified to reflect the composite nature of the voltage-dependent loads, as presented in Eq. (13) and Eq. (14).

$$P_{Dj} = P_{Dj}^{n}\left(\alpha\left(\frac{V_j}{V_j^n}\right)^2 + \beta\left(\frac{V_j}{V_j^n}\right) + \gamma\right). \tag{13}$$

$$Q_{Dj} = Q_{Dj}^{n}\left(\alpha\left(\frac{V_j}{V_j^n}\right)^2 + \beta\left(\frac{V_j}{V_j^n}\right) + \gamma\right). \tag{14}$$

4. A Brief Overview of Fuzzy Genetic Algorithm

4.1. Genetic Algorithm (GA)

Genetic algorithm (GA) is a stochastic search and optimisation technique based on the mechanism of natural selection and natural genetics [18]. It begins with an initial set of random solutions known as *population*. The population is composed of individual *chromosomes*, each of which represents a possible solution

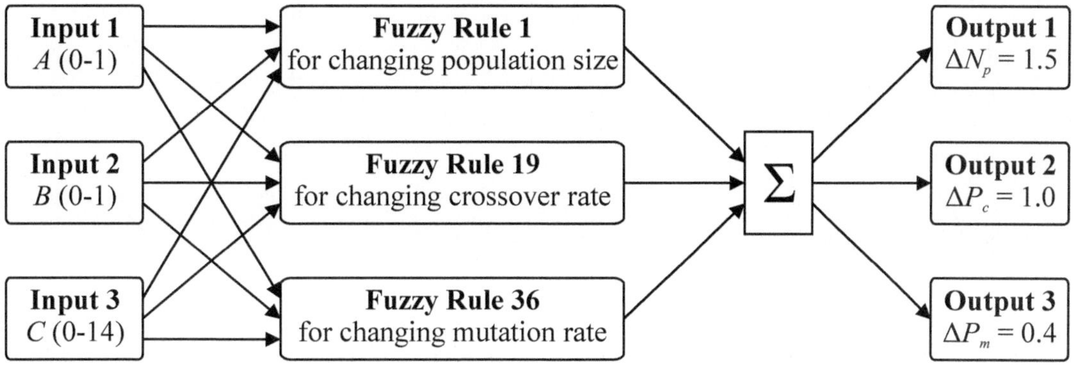

Fig. 1: Fuzzy inference process.

to the optimisation problem at hand. These chromosomes, which are usually binary bit strings, evolve through successive iterations, otherwise known as *generations.*

A common method of encoding GAs is binary encoding; other types include *permutation*, *value*, and *tree* encodings. In binary encoding every chromosome is a string of bits, 0 or 1. The function to be optimised, called the objective function, provides the mechanism to evaluate each string. To maintain uniformity over various problem domains, *fitness function* is used to scale the objective function to a convenient range of 0 to 1. Similar to other optimisation techniques, GAs have certain control parameters which include population size, crossover rate, and mutation rate. The choice of the control parameters, often left to the GA user, itself can be a complex nonlinear optimisation problem [19], [20].

In this study, each DG is represented by a string of 8 binary bits, which depicts the size of DG, P_{DG}. In order to represent the location in the system, a new string B_k, also of 8 binary bits is created. Therefore the solution to the optimisation problem is a concatenation of the two strings. In other words, $P_{DG}B_k$ that is a possible solution to the optimisation problem now contains 16 bits. With an initial population size, N_p of 30, it has a total of $30 \times 16 = 480$ bits.

4.2. Fuzzy Set Theory

Unlike a *classical* set that either wholly includes or wholly excludes any given element, a fuzzy set is a set without a clearly defined boundary. It can contain elements with only a partial degree of membership. In fuzzy logic, the truth of any statement is a matter of degree [21]. Fuzziness measures the degree to which an event occurs, and not whether it occurs or not. The use of fuzzy methods by rule-based systems is applied to solve many types of real life problems. This application becomes necessary most especially when a system is difficult to model using the conventional Boolean models; controlled by a human operator, or when ambiguity or vagueness is common [22].

Fuzzy Inference Systems is the process of formulating the mapping from a given input to an output using fuzzy logic. This mapping then provides a basis from which decisions could be made. The most popular type of fuzzy inference is the Mamdani-type proposed in 1975 [23], and the other commonly used is the Sugeno-type. While the two are similar in many respects, the main difference between them is that the latter's output membership functions are either linear or constant, but the former has continuous value output. The process of fuzzy inference system shown in Fig. 1 consists of membership functions, logical operations, and IF-THEN rules.

1) Membership Functions (MFs)

Membership function (MF) is defined as a curve that describes the way each point in the input space is mapped to a membership value of between 0 and 1. A "0" value signifies none membership while 1 means full membership. Other possibilities are in-between these two values. The simplest MFs are built using straight lines to form *triangular* or *trapezoidal* MFs. Other types of MFs include *bell*, *Gaussian*, and *sigmoidal* MFs, etc; details of which could be found in [21]. Generally speaking, MFs are usually given the designation μ. Figure 2 illustrates a triangular type of membership functions for both input and output variables, as adopted in this study. Figure 2(a), Fig. 2(b) and Fig. 2(c) are respectively inputs of outputs in Fig. 2(d), Fig. 2(e) and Fig. 2(f). For our FGA, adjacent MFs are allowed to overlap fully.

2) Logical Operations

The most crucial thing to remember about logical reasoning is the fact that it is a superset of standard Boolean logic. In this context, keeping the fuzzy values at their extremes (0 and 1) would hold true, just as

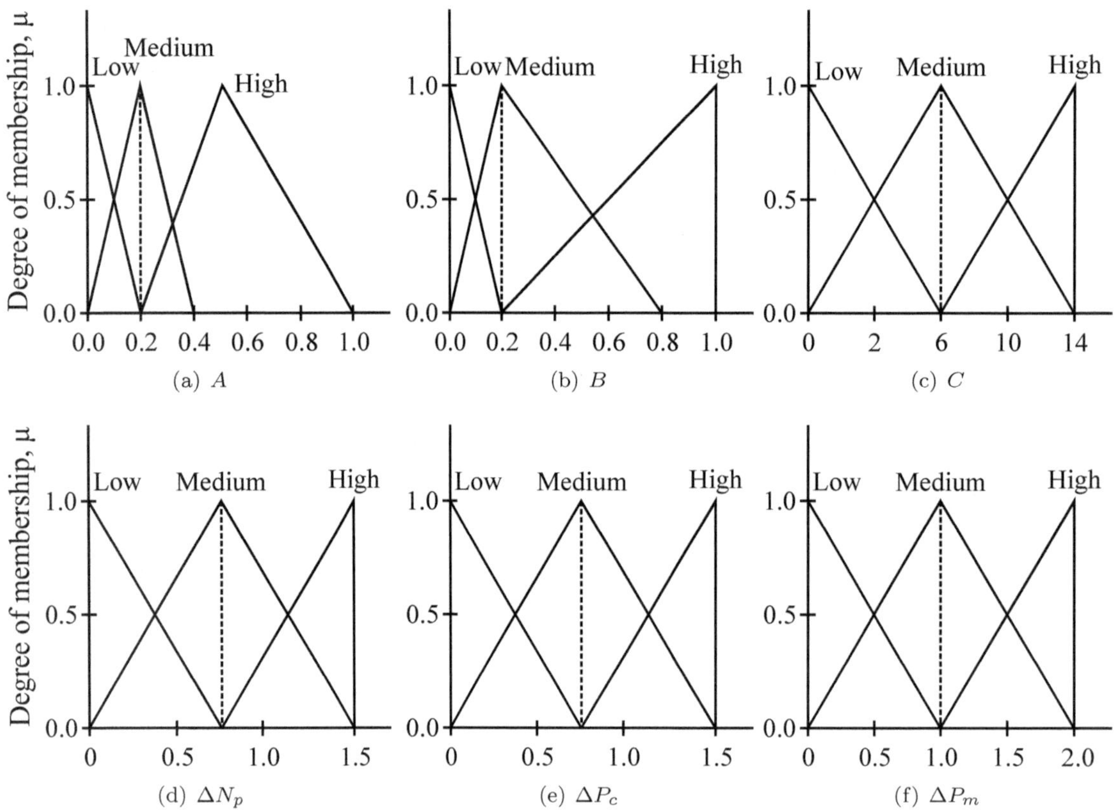

Fig. 2: Fuzzy inference process.

in the standard logical operations. In standard logical operations, logics AND, OR, and NOT are popularly used. To preserve the results of AND truth table, *min* operation could be used in fuzzy logic. In other words, A AND B could be written as *min*(A,B) in fuzzy logic, where A and B are constrained to a range of 0 and 1. Similarly, using the same reasoning, an OR operation can be replaced with the max function, so that A OR B now becomes *max*(A,B). As for the operation NOT A, the equivalent logical reasoning would be operation 1 - A. With these three functions, it is possible to resolve any construction using fuzzy sets and fuzzy logical operation AND, OR and NOT [21].

3) IF-THEN Rules

By using a linguistic approach, fuzzy theory can be integrated into control theory [24]. This could be achieved by using rules of the form IF condition THEN action. The IF part of the rule is called the *antecedent* or premise, while the THEN part is known as the *consequent* or conclusion. From these rules, a functional controller could be created. The maximum number of rules in a system is determined by the number of possible combinations of input sets. In this study for example, three inputs which are composed of three fuzzy sets and three output variables are used. Tab. 3 in the Appendix presents the 51 rules used in this study.

In control applications, the output of a fuzzy knowledge-based system should be a single crisp value. This is achieved by the process of *defuzzification*. In this study, defuzzification of the outputs was performed using the fuzzy centroid [25], which calculates the centre of gravity of the area under the membership function.

5. The Proposed Algorithm

In this study, a fuzzy genetic algorithm (FGA) is created by systematically integrating fuzzy expert systems into GA, to dynamically control the GA parameters during operation. Experiments show that FGA can search faster and more effectively than the simple GA in solving optimisation problems [26]. In achieving this goal, three input variables - $A = (\textit{average fitness})/(\textit{best fitness})$, $B = (\textit{worst fitness})/(\textit{average fitness})$, and C, which represents the change in the best fitness since the last control action; and three output variables - variations in the current crossover rate P_c, mutation rate P_m, and population size N_p are used in the fuzzy controller. A is defined as a measure of the current population diversity's usefulness. If the value is close to 1 (i.e. high), then convergence has been reached, on the other hand, a low value of close to 0 means the population shows a high level of diversity. These input and output quantities are defined as follows $A \in [0,1]$, $B \in [0,1]$, $C \in [0,14]$, $N_p \in [30,45]$, $P_c \in [0.6,0.9]$, and $P_m \in [0.01,0.015]$ in this study, as implemented in [27].

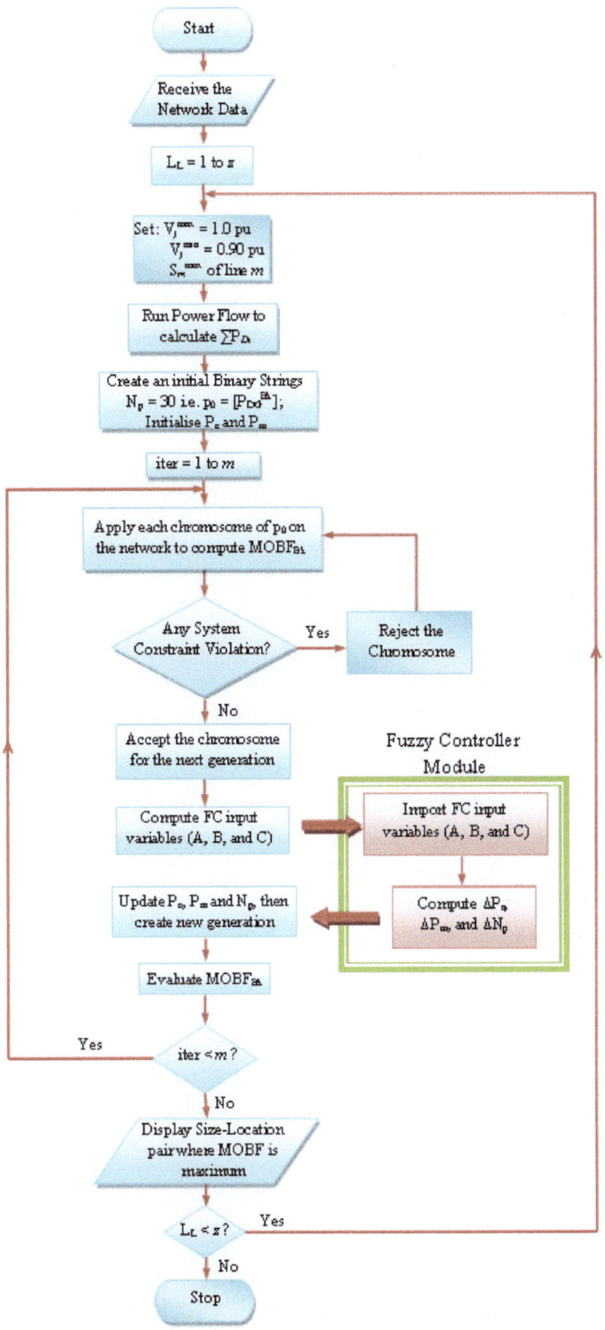

Fig. 3: The proposed algorithm flowchart.

The linguistic value sets of the inputs and outputs are all Low, Medium, High with respect to the lower and upper limit values respectively. The fuzzy IF-THEN rules previously described [22], [28] are adopted to combine fuzzy sets in this paper. The FGA has been executed in this study with an initial population size of 30, and 0.6 and 0.01 as the initial crossover and mutation rates respectively. The maximum generation of 100 is used, using Roulette Wheel selection. The output change ΔP_c, ΔP_m, and ΔN_p are multiplied by the current crossover rate, the mutation rate, and the population size respectively. The results obtained by implementing the 51 fuzzy control rules are presented in Tab. 3 in the Appendix.

The proposed DG placement algorithm for radial distribution systems is summarised in the flowchart presented in Fig. 3. The algorithm is successively implemented in MATLAB. In this work, the best location is determined by selecting the bus with the highest value of MOBF. This is the bus that maximises energy loss cost savings and power transfer capability of the system, and also minimises the sum of voltage drops on all the lines of the network simultaneously.

6. Application to Test Networks

To test the accuracy of the proposed method described in Section 5, three test radial distribution systems are used. A natural gas fired combustion turbine generator is assumed for the DG unit in this study. The results obtained with the approach presented in the literature on the two test systems are very accurate and reliable. This study also assumes there is no forced outage on any of the lines and the substation generator and transformer throughout the period under consideration. In calculating the power loss for the year, three loading levels are adopted; light, normal and heavy loadings.

6.1. Case 1 - 30-Bus Radial Distribution System

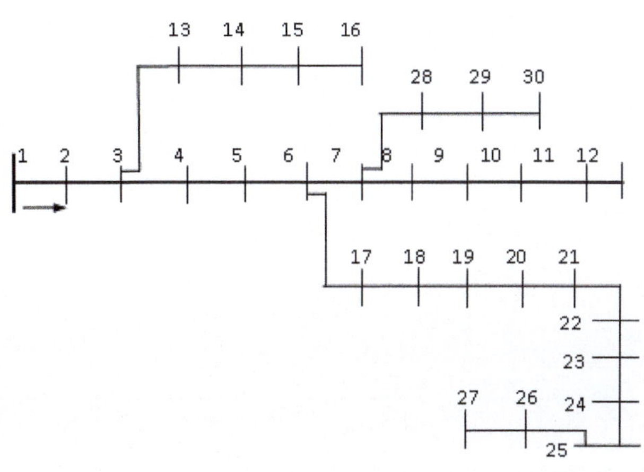

Fig. 4: 30-bus radial distribution system.

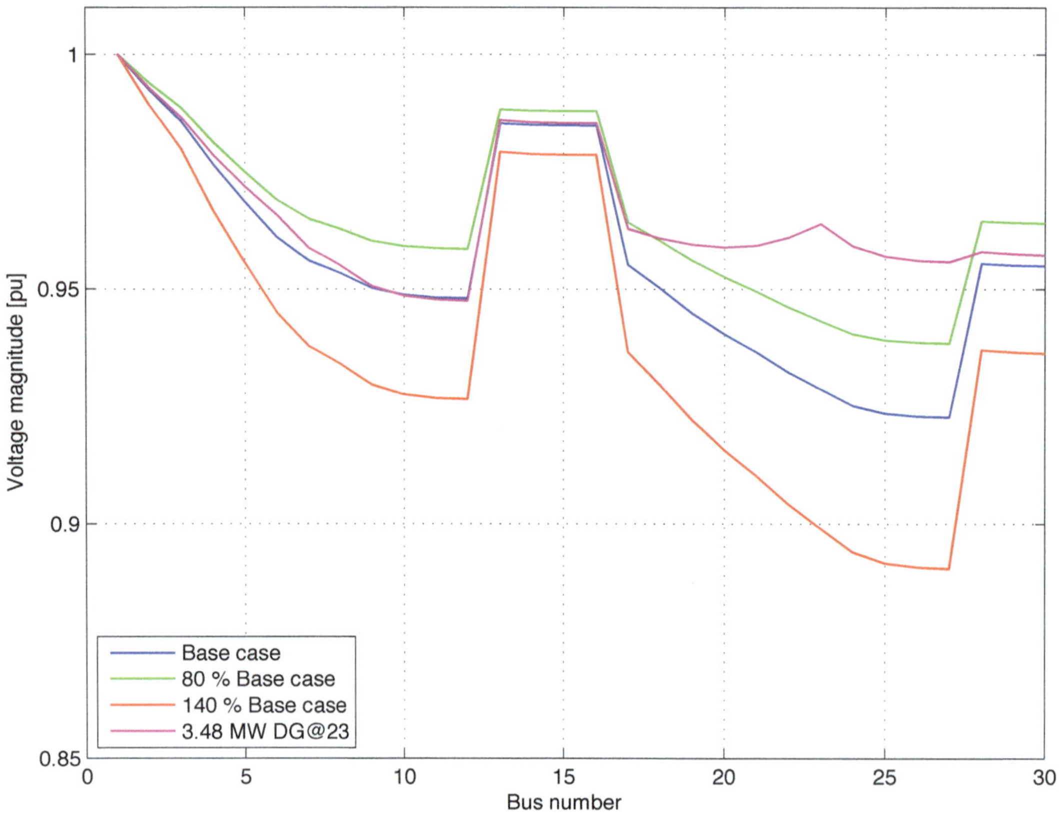

Fig. 5: Voltage profile of the 30-bus system.

The system obtained from [14] is an 11 kV distribution system having 29 load buses and 1 source bus. It has a main feeder and three laterals as shown in Fig. 4. Although the load composition of the distribution system is not stated, the loads have been modelled in this work as composite loads consisting of 46 % constant power, 31 % constant current and 23 % constant impedance. This is done to reflect the real life scenario of distribution systems load models.

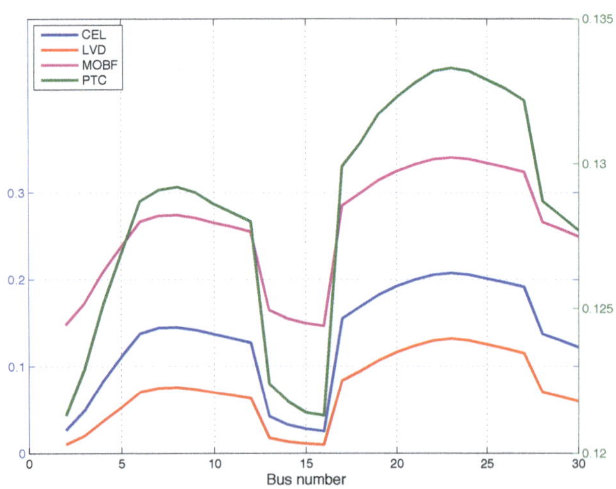

Fig. 6: Objective functions values for the 30-bus system with $W_1 = 0.4$, $W_2 = W_3 = 0.3$.

This test system is slightly modified to have a base case active and reactive powers capacity of 6.22 MW and 3.83 MVar respectively. However, increasing the loading of the system under study linearly at a constant power factor across the buses by 40 % would force some buses to violate the specified voltage limits. For example, buses 23–27 fall below the lower bound of 0.9 V as shown in Fig. 5.

Applying the proposed algorithm to the over loaded system (i.e. 8.7163 MW, 5.4034 MVars); an optimum DG size of 3.48 MW is obtained. The optimum bus is determined from Fig. 6 as the load bus that maximises the multi-objective function, MOBF. This is bus 23; hence declared the best location to site the DG unit. The three objective functions are also plotted in the figure, where the secondary axis is exclusively for the power transfer capability (PTC) objective function.

The effect on the voltage profile of locating the optimal DG size in bus 23 could be seen in Fig. 5. The voltage plots of other scenarios are also shown in the figure. Similarly with a DG unit of 3.48 MW, the power loss is plotted against load buses as shown in Fig. 7. This graph further confirms the results of the proposed model that bus 23 is the best location for the DG since it has the least total system power loss.

Tab. 1: Annual energy loss calculation based on three loading levels.

Load Level	**80 % (light)**		**100 % (normal)**		**140 % (heavy)**	
Duration/yr [hrs]	1 700		4 200		1 300	
30-Bus System						
	With DG	Without DG	With DG	Without DG	With DG	Without DG
Power Loss [MW]	0.1198	0.2445	0.1880	0.3851	0.3711	0.7697
Energy Loss /yr [MWhr]	203.66	415.65	789.60	1 617.42	482.43	1 000.61
Energy Loss Cost/yr [$]	22 198.94	49 878.00	86 066.40	194 090.40	52 584.87	120 073.20
Total Energy Loss Cost per yr [$]	With DG 160 850.21 Without DG 364 041.60					
Savings per year [$]	203 191.39					
69-Bus System						
	With DG	Without DG	With DG	Without DG	With DG	Without DG
Power Loss [MW]	0.0538	0.1329	0.0846	0.2247	0.1677	0.4387
Energy Loss /yr [MWhr]	91.46	225.93	355.32	943.74	218.01	570.31
Energy Loss Cost/ yr [$]	9 969.14	27 111.60	38 729.88	113 248.80	23 763.09	68 437.20
Total Energy Loss Cost per yr [$]	With DG 72 462.11 Without DG 208 797.60					
Savings per year [$]	136 335.49					
With 2.13 MW DG sited at Bus 60 in the 69-Bus System						
	With DG	Without DG	With DG	Without DG	With DG	Without DG
Power Loss [MW]	0.05914	0.1329	0.09318	0.2247	0.18533	0.4387
Energy Loss /yr [MWhr]	100.54	225.93	391.36	943.74	240.93	570.31
Energy Loss Cost/yr [$]	10 958.64	27 111.60	42 657.80	113 248.80	26 361.26	68 437.20
Total Energy Loss Cost per yr [$]	With DG 79 877.71 Without DG 208 797.60					
Savings per year [$]	128 919.89					

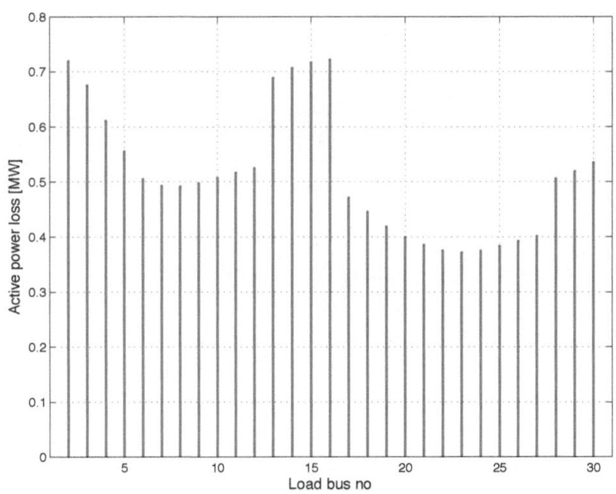

Fig. 7: Power loss with DG at various locations.

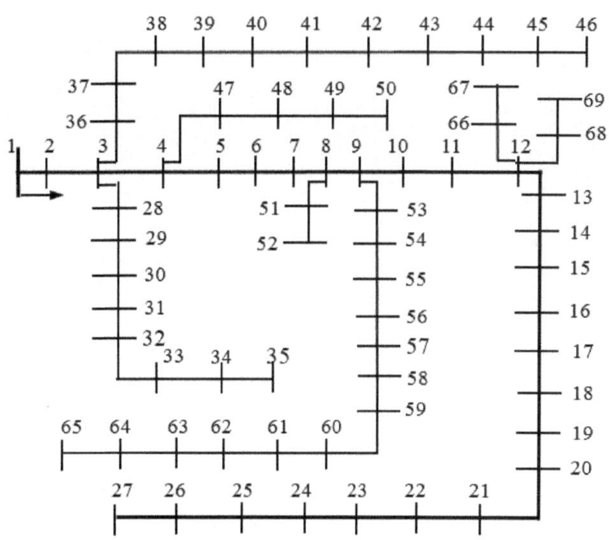

Fig. 8: 69-bus radial distribution system.

Locating a DG unit of 3.48 MW at bus 23 in the test system reduces the total power loss from 0.7697 MW (without DG) to 0.3711 MW. This translates to a savings of $203 191.39 for the DISCO for a period of one year if the DG unit is assumed to run for 300 days in a year. The annual power loss using three loading levels is calculated as shown in Tab. 1. The unit costs of electric energy used in this study are $109 and $120 for DG and grid respectively.

6.2. Case 2 - 69-Bus Radial Distribution System

The 69-bus distribution network is as shown in the single line diagram in Fig. 8. The system obtained from [29], has a base case capacity of 3.8021 MW and 2.6945 MVar active and reactive power respectively at 12.66 kV. The load data could be found in the cited reference.

The optimum DG size as determined by the proposed model is 2.13 MW for 140 % base case (5.32 MW, 3.77 MVar) capacity. Looking at Fig. 9, it is found that bus 61 has the maximum value of MOBF at 0.5118; hence, declared the optimal location for the DG. The effect of the weighting factors on each objective function is illustrated in Fig. 10, where MOBF now has a maximum value of 0.4720, though at the same bus.

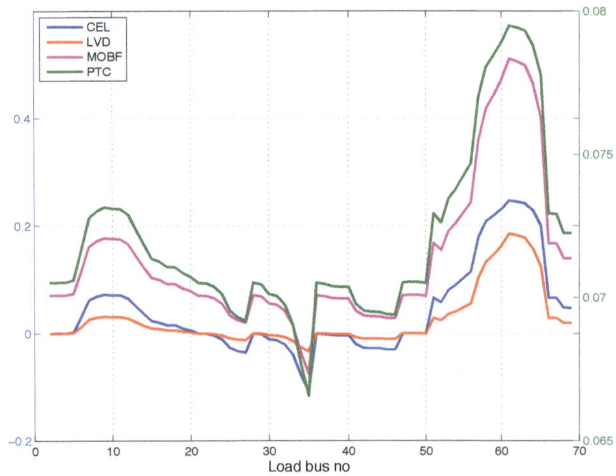

Fig. 9: Objective functions values for the 69-bus system with $W_1 = 0.4$, $W_2 = W_3 = 0.3$.

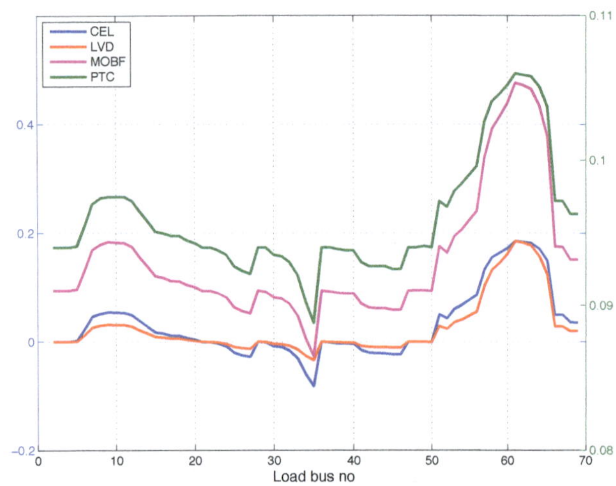

Fig. 10: Objective functions values for the 69-bus system with $W_1 = W_2 = 0.3$, $W_3 = 0.4$.

Interestingly, this is corroborated by the plot in Fig. 11, since the minimum power loss is when the DG is located in bus 61. The power loss without DG is 0.4387 MW; while 0.1677 MW is the recorded loss when 2.13 MW DG is installed on the system. This is equivalent to a savings of $136 335.49 for the DISCO for the period of one year. If for any reason, the IPP decides to install the DG unit at bus 60 instead; his savings reduces to $128 919.89 for the same period as shown in Tab. 1.

Figure 12 shows the voltage profiles of the system at base case, 80 % base case, 140 % base case before DG, and when 2.13 MW DG is installed on bus 61. It is evident from the figure that there is a significant improvement in the voltage profile due to DG connection.

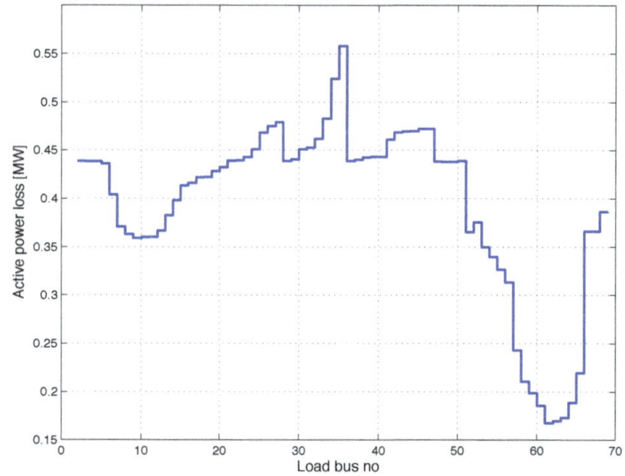

Fig. 11: Power loss with DG at various locations.

Nevertheless, it is worthy of note that siting the DG size in buses 21–27, 29–35, 37–46, and 50 would amount to a great disservice to the distribution network, as the power loss when DG is placed in any of these buses is greater than when no DG is sited in the system at all. This is shown in Fig. 9 with objective function values less than 0. It is therefore recommended to the IPP not to attempt to site DG in any of these buses.

7. Results Comparison with Others

The results obtained with the proposed approach were evaluated against the existing methods on the base case of a number of test systems, and were found to be highly impressive, as presented in Tab. 2. For example, while the proposed method obtained the optimal size to be 1.5208 MW which results in 0.0877 MW total system power loss on the 69-bus system, Gozel [13] and Vinothkumar [30], whose optimal sizes are 1.8078 MW and 2.0 MW are only able to reduce the total power loss to 0.0920 MW and 0.091 MW respectively. It is obvious and 2.0 MW are only able to reduce the total power loss from this that the proposed method produces a much better result. Even though the Acharya's methods [4] recorded the least power loss of 0.0814 MW and 0.08133 MW, a further critical analysis would however reveal that the results have gone beyond the most optimal level. For instance, the extra loss reduction of 0.0063 MW (0.0877–0.0814 MW) vis-a-vis the proposed method is attained with a 0.2892 MW DG size extra (1.81–1.5208 MW) in the table. This would certainly not motivate the DISCO to invest more on the

DG capacity since there is no commensurate economic justification from the output.

This simply underscores the fact that focusing attention on power loss alone, without a fair consideration for other factors, would not yield the most optimal result.

Similar analysis is applicable to the 33-bus test system when compared with the proposed method. These results obviously show that the proposed method produces more optimal results (DG size and location) than other approaches as demonstrated in the table. While refs [4], [13], [31], [32]] obtained bus 6 as the optimal location for 33-bus, the proposed method obtained bus 8, instead. On the other hand, [30] and [4] obtained buses 7 and 10 as the optimal locations respectively for the same system. The optimality of the results obtained in this study is demonstrated in the DG size and consequently power loss shown in Tab. 2. Where the energy savings are higher than that of the proposed method, the investment cost on the extra DG capacity may not encourage the DISCO to implement it.

8. Conclusion

A multi-objective model for optimal allocation of a single DG unit in a radial distribution network has been proposed in this paper. The Fuzzy Genetic Algorithm based approach was premised on maximising loss cost savings, minimisation of lines voltage drops, and maximisation of the power transfer capability of the system. In the study, fuzzy controller was implemented to dynamically control the GA control parameters. The essence of this is to overcome the premature convergence of the simple GA. The proposed model takes into cognisance, the peculiarities of radial distribution systems, such as high R/X ratio, voltage dependency and composite nature of the loads, which most of the early works did not consider. In this study, an appropriate weighting factor is allocated to each of the objective functions by the network planner, based on his preference for each objective at any time instant.

The accuracy and reliability of the proposed approach was evaluated using three radial distribution networks. The results obtained appeared to be more optimal than those of the earlier existing methods cited in the literature, and would help the DISCO save energy cost. One other advantage of the model proposed in this paper is the exceptionally high processing speed. For example, while it took more than 40 seconds on a Pentium Dual-Core CPU E6500 at 2.93 GHz and 2 GB RAM personal computer to produce results for the 69-bus system using the approach of [31], it only took the proposed method just less than 3 seconds for the same test system.

Appendix

Table 3 presents the fuzzy control rules as used in this work.

Tab. 3: Fuzzy control rules.

Rule no.	Input			Output
	A	B	C	ΔN_p
1	Low	Low	High	High
2	Low	Medium	Low	High
3	Low	Medium	Medium	Low
4	Low	Medium	High	High
5	Low	High	Low	Low
6	Low	High	High	Medium
7	Medium	Low	Low	Low
8	Medium	Low	Medium	Medium
9	Medium	Low	High	Low
10	Medium	Medium	Low	Low
11	Medium	Medium	Medium	Medium
12	Medium	Medium	High	High
13	High	Low	Low	Medium
14	High	Low	High	Low
15	High	Medium	Low	High
16	High	Medium	Medium	High
17	High	Medium	High	High
	A	B	C	ΔP_c
18	Low	Low	Low	Low
19	Low	Low	Medium	Medium
20	Low	Low	High	Low
21	Low	Medium	Medium	High
22	Low	High	Low	Medium
23	Low	High	Medium	Low
24	Low	High	High	Low
25	Medium	Low	Low	Medium
26	Medium	Low	Medium	Medium
27	Medium	Medium	Medium	High
28	Medium	Medium	High	Low
29	Medium	High	Low	Low
30	High	Low	Medium	Low
31	High	Low	High	High
32	High	Medium	Medium	Low
33	High	Medium	High	Low
34	High	High	Low	Medium
35	High	High	High	Medium
	A	B	C	ΔP_m
36	Low	Low	Low	Low
37	Low	Low	High	Medium
38	Low	Medium	Medium	Medium
39	Low	Medium	High	High
40	Low	High	Low	Medium
41	Medium	Low	Low	Medium
42	Medium	Low	Medium	High
43	Medium	Low	High	Medium
44	Medium	Medium	Low	Medium
45	Medium	High	Low	High
46	Medium	High	Medium	Medium
47	Medium	High	High	Medium
48	High	Low	Low	Low
49	High	Low	High	High
50	High	Medium	High	High
51	High	High	High	Low

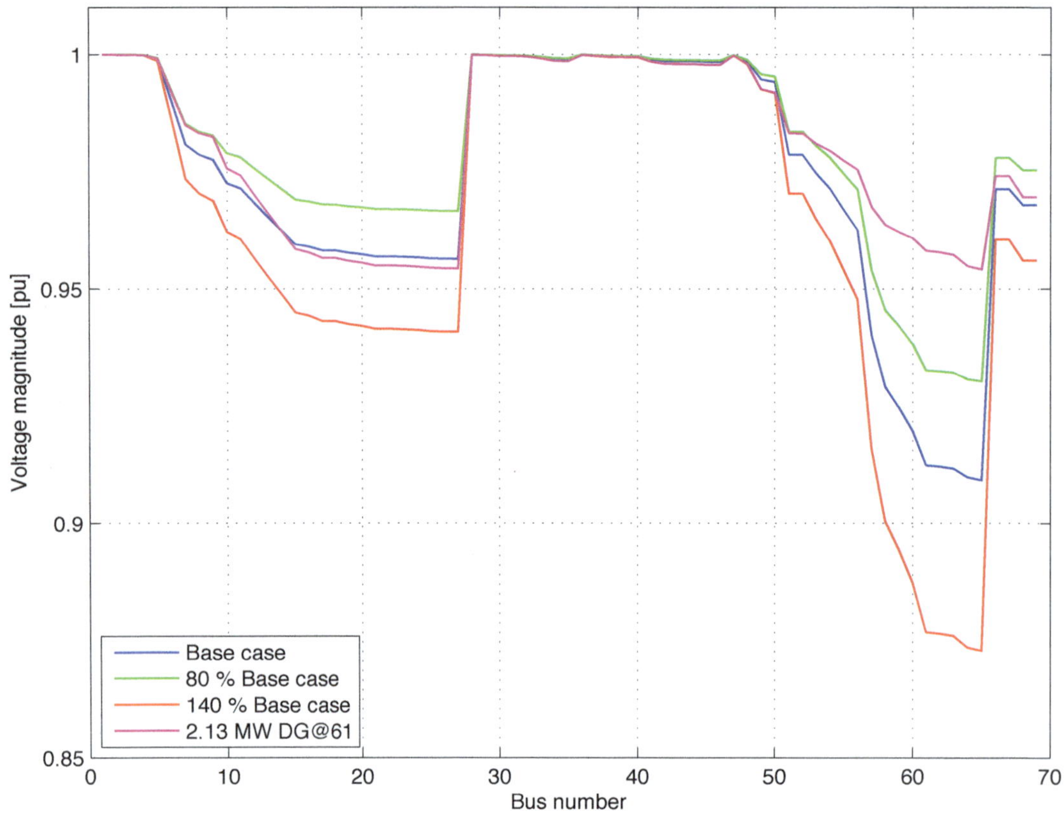

Fig. 12: Voltage profile on 69-bus system.

Tab. 2: Results comparison.

Author	Method	Test system	PL without DG [MW]	Opt loc	Opt size [MW]	PL with DG [MW]	Energy savings [$/hr]	DG Invest Cost [$ million]
Acharya [4]	Analytical	69-bus	0.2193	61	1.81	0.0814	16.55	2.61
Acharya [4]	Loss Sensitivity	69-bus	0.2193	61	1.9	0.08133	16.56	2.74
Gozel [13]	Analytical	69-bus	0.2249	61	1.8078	0.092	15.95	2.60
Gozel [31]	Analytical	69-bus	0.225	61	1.8727	0.0832	17.02	2.70
Shukla [32]	GA	69-bus	0.225	61	1.872	0.0832	17.02	2.70
Vinothkumar [30]	FGA	69-bus	0.22	61	2.00	0.091	15.48	2.88
Akorede	Proposed	69-bus	0.225	61	1.5208	0.0877	16.48	2.19
Acharya [4]	Analytical	33-bus	0.2112	6	2.49	0.1112	12.00	3.59
Acharya [4]	Loss Sensitivity	33-bus	0.2112	10	1.4	0.12382	10.49	2.02
Gozel [13]	Analytical	33-bus	-	-	-	-	-	-
Gozel [31]	Analytical	33-bus	0.211	6	2.5902	0.111	12.00	3.73
Shukla [32]	GA	33-bus	0.216	6	2.49	0.1328	9.98	3.59
Vinothkumar [30]	FGA	33-bus	0.2	7	2.00	0.11	10.80	2.88
Akorede	Proposed	33-bus	0.211	8	1.486	0.1206	10.85	2.14

References

[1] CIRED, WORKING GROUP WG04. *Dispersed Generation: Preliminary Report of CIRED*. Nice, 1999.

[2] AKOREDE, M., H. HIZAM, I. ARIS and M. Z. AB-KADIR. Re-emergence of Distributed Generation in Electric Power Systems: Incentives, Values and Issues. *Energy & Environment*. 2010, vol. 21, no. 2, pp. 75–92. ISSN 0958-305X. DOI: 10.1260/0958-305X.21.2.75.

[3] AKOREDE, M., H. HIZAM, I. ARIS and M. Z. AB-KADIR. A Critical Review of Strategies for Optimal Allocation of Distributed Generation Units in Electric Power Systems. *International Review of Electrical Engineering*. 2010, vol. 5, no. 2, pp. 593–600. ISSN 1827-6660.

[4] ACHARVA, N., P. MAHAT and N. MITHULANANTHAN. An analytical approach for DG allocation in primary distribution Net-

work. *International Journal of Electrical Power and Energy Systems.* 2006, vol. 28, iss. 10, pp. 669–678. ISSN 0142-0615. DOI: 10.1016/j.ijepes.2006.02.013.

[5] ELNASHAR, M. M., R. EL SHATSHAT and M. M. A. SALAMA. Optimum siting and sizing of a large distributed generator in a mesh connected system. *Electric Power Systems Research.* 2010, vol. 80, iss. 6, pp. 690–697. ISSN 0378-7796. DOI: 10.1016/j.epsr.2009.10.034.

[6] HEDAYATI, H., S. A. NABAVINIAKI and A. AKBARIMAJD. A Method for Placement of DG Units in Distribution Networks. *IEEE Transactions on Power Delivery.* 2008, vol. 23, iss. 3, pp. 1620–1628. ISSN 0885-8977. DOI: 10.1109/TPWRD.2007.916106.

[7] GOLSHAN, M. E. H. and S. A. AREFIFAR. Optimal allocation of distributed generation and reactive sources considering tap positions of voltage regulators as control variables. *European Transactions on Electrical Power.* 2007, vol. 17, iss. 3, pp. 219–239. ISSN 1546-3109. DOI: 10.1002/etep.130.

[8] KUMAR, V., H. C. R. KUMAR, I. GUPTA and H. O. GUPTA. DG Integrated Approach for Service Restoration Under Cold Load Pickup. *IEEE Transactions on Power Delivery.* 2010 vol. 25, iss. 1, pp. 398–406. ISSN 0885-8977. DOI: 10.1109/TPWRD.2009.2033969.

[9] SEDIGHIZADEH, M. and A. REZAZADEH. Using Genetic Algorithm for Distributed Generation Allocation to Reduce Losses and Improve Voltage Profile. *World Academy of Science, Engineering and Technology.* 2008, vol. 2, no. 1, pp. 251-256. ISSN 2010-3778.

[10] HAGHIFAM, M. R., H. FALAGHI and O. P. MALIK. Risk-based distributed generation placement. *IET Generation, Transmission & Distribution.* 2008, vol. 2, iss. 2, pp. 252–260. ISSN 1751-8687. DOI: 10.1049/iet-gtd:20070046.

[11] WANG, C. and M. H. NEHRIR. Analytical approaches for optimal placement of distributed generation sources in power systems. *IEEE Transactions on Power Systems.* 2004, vol. 19, iss. 4, pp. 2068–2076. ISSN 0885-8950. DOI: 10.1109/TPWRS.2004.836189.

[12] CELLI, G., E. GHIANI, S. MOCCI and F. PILO. A multiobjective evolutionary algorithm for the sizing and siting of distributed generation. *IEEE Transactions on Power Systems.* 2005, vol. 20, iss. 2, pp. 750–757. ISSN 0885-8950. DOI: 10.1109/TPWRS.2005.846219.

[13] GOEZEL, T. and M. H. HOCAOGLU. An analytical method for the sizing and siting of distributed generators in radial systems. *Electric Power Systems Research.* 2009, vol. 79, iss. 6, pp. 912–918. ISSN 0378-7796. DOI: 10.1016/j.epsr.2008.12.007.

[14] EMINOGLU, U. and M. U. HOCAOGLU. A new power flow method for radial distribution systems including voltage dependent load models. *Electric Power Systems Research.* 2005, vol. 76, iss. 1–3, pp. 106–114. ISSN 0378-7796. DOI: 10.1016/j.epsr.2005.05.008.

[15] VERBIC, G., M. PANTOS and F. GUBINA. On voltage collapse and apparent-power losses. *Electric Power Systems Research.* 2006, vol. 76, iss. 9–10, pp. 760–767. ISSN 0378-7796. DOI: 10.1016/j.epsr.2005.10.007.

[16] DEVI, A. L. and B. SUBRAMANYAM. Optimal DG Unit Placement for Loss Reduction in Radial Distribution System-A Case Study. *ARPN Journal of Engineering and Applied Sciences.* 2007, vol. 2, no. 6, pp. 57–61. ISSN 1819-6608

[17] ARAVINDHABABU, P. A new fast decoupled power flow method for distribution systems. *Electric Power Components and Systems.* 2003, vol. 31, iss. 9, pp. 869–878. ISSN 1532-5008. DOI: 10.1080/15325000390227209.

[18] GEN, M., R. CHENG and L. LIN. *Network models and optimization: Multiobjective genetic algorithm approach.* Berlin: Springer Verlag, 2008. ISBN 978-1-84800-180-0.

[19] SRINIVAS, M. and L. M. PATNAIK. Genetic algorithms: A survey. *Computer.* 1994, vol. 27, iss. 6, pp. 17–26. ISSN 0018-9162. DOI: 10.1109/2.294849.

[20] MOMOH, J. A. *Electric power system applications of optimization.* New York: Marcel Dekker Inc., 2001. ISBN 978-1420065862.

[21] Fuzzy Logic Toolbox™: User's Guide. In: *Mathworks* [online]. 2014. Available at: `http://www.mathworks.com/help/pdf_doc/fuzzy/fuzzy.pdf`.

[22] LAI, W.-H. and C.-T. TSAI. Fuzzy rule-based analysis of firm's technology transfer in Taiwan's machinery industry. *Expert Systems with Applications.* 2009, vol. 36, iss. 10, pp. 12012–12022. ISSN 0957-4174. DOI: 10.1016/j.eswa.2009.03.054.

[23] MAMDANI, E. H. and S. ASSILIAN. An experiment in linguistic synthesis with a fuzzy logic controller. *International Journal of Man-Machine Studies.* 1975, vol. 7, iss. 1, pp. 1–13. ISSN 1071-5819.

[24] HOMAIFAR, A. and E. McCORMICK. Simultaneous design of membership functions and rule sets for fuzzy controllers using genetic algorithms. *IEEE Transactions on Fuzzy Systems.* 1995, vol. 3, iss. 2, pp. 129–139. ISSN 1063-6706. DOI: 10.1109/91.388168.

[25] RUNKLER, T. A. Selection of appropriate defuzzification methods using application specific properties. *IEEE Transactions on Fuzzy Systems.* 1997, vol. 5, iss. 1, pp. 72–79. ISSN 1063-6706. DOI: 10.1109/91.554449.

[26] XU, H. Y. and G. VUKOVICH. A fuzzy genetic algorithm with effective search and optimization. In: *International Joint Conference on Neural Networks.* Nagoya: IEEE, 1993, pp. 2967–2970. ISBN 0-7803-1421-2. DOI: 10.1109/IJCNN.1993.714345.

[27] AKOREDE, M., H. HIZAM, I. ARIS and M. Z. AB-KADIR. Effective method for optimal allocation of distributed generation units in meshed electric power systems. *IET Generation, Transmission & Distribution.* 2011, vol. 5, iss. 2, pp. 276–287. ISSN 1751-8687. DOI: 10.1049/iet-gtd.2010.0199.

[28] LEE, M. A. and H. TAKAGI. Dynamic control of genetic algorithms using fuzzy logic techniques. In: *5th International Conference on Genetic Algorithms.* San Francisco: Morgan Kaufmann Publishers Inc., 1993, pp. 76–83. ISBN 1-55860-299-2.

[29] CHAKRAVORTY, M. and D. DAS. Voltage stability analysis of radial distribution networks. *International Journal of Electrical Power & Energy Systems.* 2001, vol. 23, iss. 2, pp. 129–135. ISSN 0142-0615. DOI: 10.1016/S0142-0615(00)00040-5.

[30] VINOTHKUMAR, K. and M. P. SELVAN. Fuzzy Embedded Genetic Algorithm Method for Distributed Generation Planning. *Electric Power Components and Systems.* 2011, vol. 39, iss. 4, pp. 346–366. ISSN 1532-5008. DOI: 10.1080/15325008.2010.528533.

[31] GOEZEL, T., U. EMINOGLU and M. HOCAOGLU. A tool for voltage stability and optimization (VS&OP) in radial distribution systems using matlab graphical user interface (GUI). *Simulation Modelling Practice and Theory.* 2008 vol. 16, iss. 5, pp. 505–518. ISSN 1569-190X. DOi: 10.1016/j.simpat.2008.02.003.

[32] SHUKLA, T., S. SINGH, V. SRINIVASARAO and K. NAIK. Optimal Sizing of Distributed Generation Placed on Radial Distribution Systems. *Electric Power Components and Systems.* 2010, vol. 38, iss. 3, pp. 260–274. ISSN 1532-5008. DOI: 10.1080/15325000903273403.

About Authors

Mudathir Funsho AKOREDE was born in Nigeria. He received his B.Eng (Hons) in Electrical Engineering from University of Ilorin, Ilorin in 2000, and M.Eng in 2006 from Bayero University Kano; both in Nigeria. He bagged his Ph.D. in Electrical Power Engineering from Universiti Putra Malaysia in 2011. Dr. Akorede is currently a senior lecturer in the Department of Electrical & Electronics Engineering, University of Ilorin, Ilorin, Nigeria. His research interests include optimization of distributed energy resources in power systems, smart grid, power system analysis, etc. Presently, Dr. Akorede is the Managing Editor, Nigerian Journal of Technological Development, published by Faculty of Engineering and Technology of his university.

Edris POURESMAEIL was born in Iran. He received the B.Sc. and M.Sc. degrees in Electrical Engineering from University of Mazandaran, Babol, in 2003 and 2005, respectively. He received the Ph.D. degree in Electrical Engineering from Technical University of Catalonia, BarcelonaTech (UPC), Barcelona, Spain, in 2012. After his Ph.D. he joined the Department of Elec. & Comp. Eng., at University of Waterloo, ON, Canada, as a Postdoctoral Research Fellow. He is currently an Associate Professor with the Centre for Smart Energy Solutions at University of Southern Denmark (SDU). His research interests include power electronics converters for distributed generation, control and management of microgrids, integration of renewable energies in smart grids, and energy hub management system.

Hashim HIZAM was born in Malaysia. He graduated with a B.Sc. and M.Sc. in Electrical and Electronic Engineering from Polytechnic University, Brooklyn, New York in 1993 and 1994, respectively. In 2004, he obtained his Ph.D. from Queen University Belfast, Northern Ireland in the United Kingdom. Dr. Hizam is currently an Associate Professor in the Department of Electrical and Electronic Engineering, Faculty of Engineering, Universiti Putra Malaysia. His areas of research are in Power System Protection and Power System Analysis.

Ishak ARIS Ishak ARIS was born in Malaysia. He graduated from the George Washington University, USA, in Electrical Engineering in 1988. He obtained M.Sc. in Power Electronics Engineering from University of Bradford/ Loughborough University of Technology, United Kingdom in 1991 and Ph.D.

from University of Bradford, United Kingdom in 1995. He is currently a Professor in the Department of Electrical and Electronic Engineering, Universiti Putra Malaysia. His area of research interest is in Power Electronics & Drives System (Electric & Hibrid Vehicles), Artificial Intelligence and Robotic System, and Automative Power Train Engine Management System.

Mohd Zainal AB-KADIR was born in Malaysia. He received B.Eng. Electrical and Electronics Engineering from Universiti Putra Malaysia in 2001 and Ph.D. in High Voltage Engineering in 2006 from University of Manchester, United Kingdom. He is currently the Head and a Professor in the Department of Electrical and Electronic Engineering, Universiti Putra Malaysia. Professor Ab-Kadir's research interests include High Voltage Engineering, Lightning Protection and EMC/EMI.

20

A New Control Strategy Based Multi Converter UPQC Using Fuzzy Logic Controller to Improve the Power Quality Issues

Chandra Babu PADUCHURI, Subhransu Sekhar DASH, Subramani CHINNAMUTHU

Department of Electrical and Electronics Engineering, Faculty of Engineering and Technology, SRM University, SRM Nagar, Kattankulathur, 603 203 Chennai, India

paduchuri.chandrababu@gmail.com, munu_dash_2k@yahoo.com, csmsrm@gmail.com

Abstract. *A design of multiconverter unified power quality conditioner to improve the power quality issues is presents in this paper. Modified SRF theory and fuzzy logic controller technique are incorporated in this modelling. This newly designed controller is connected to a source in order to compensate voltage and current in the two feeders. The expanded concept of UPQC is multi converter-UPQC; this system has two series voltage source converter (VSC) and one shunt VSC connected back to back. In the proposed system, the power can be conveyed from one feeder to another in order to mitigate the voltage sag, swell, interruption and transient response of the system. The control strategies of multi converter-UPQC are designed based on the modified synchronous reference frame theory with fuzzy logic controller. The transient response of the fuzzy logic controller in dc-link voltage controller will be very fast. The relevant simulation and compensation performance analysis of multi converter-UPQC with fuzzy logic controller is performed using MATLAB/Simulink software.*

Keywords

Fuzzy logic controller, MC-UPQC, modified synchronous reference frame theory, power quality, voltage source converter.

1. Introduction

The Flexible AC Transmission System (FACTS) is generally being used for faster damping of power swing, to control the power flow and to load the transmission line within their thermal limits securely. In the distribution system the power electronic devices were used to increase the reliability and quality of the power supplied to consumers. A new technology with the application of power electronics devices in power distribution system for the benefit of consumer or a group of consumers is called "Custom Power Devices (CPD)". The use of power electronics device such as UPQC is connected combination of both in shunt and series side; to mitigate problems such as frequency of rare power interruption, over/under voltages of magnitude and duration within limits; low harmonic distortion, low phase unbalance, low voltage flicker in the supply side and frequency of the supply voltage (According to IEEE 1159-1995 standard).

A custom power device provides an integrated solution to the present problems that are faced by the utilities and power distribution. The various compensating devices are DSTATCOM, DVR and UPQC; among these the UPQC provides good solutions when compared to the other devices. Because it is a very versatile device that can inject current in shunt side and voltage in series side simultaneously in a dual control mode. Hence it can perform both the functions in load compensations. Power quality variations are classified as disturbances or steady state variations. Disturbances pertain to be abnormalities in the system voltage /currents due to a fault or some abnormal operations [1].

The Multi-Converter UPQC (MC-UPQC) system has three Voltage Source Converter (VSC's) connected to two feeder lines to compensate the voltage and current imperfection in both feeders [2]. The control parts of the shunt and series Active Power Filters (APF) are proposed based on Synchronous Reference Frame (SRF) theory with Proportional Integral (PI) controller. SRF based control for a dynamic model in three phase system under different load consideration is used to improve the Power Quality (PQ) by using multiconverter with power conditioner [3]. The Interline Unified Power Quality Conditioner (IUPQC) consists of series VSC and shunt of VSC both joined together by a common dc bus. It can also be used to demonstrate how it is connected between two independent feeders in regulating the voltage across a sensitive load from the other feeder [4]. The Generalized Unified Power Quality Conditioner (GUPQC) is a combination of one shunt and two series VSC to compensate current imperfections in one feeder and voltage imperfection in the other two feeders [5].

In the study of reference current generation techniques using VSC based DSTATCOM for reactive power compensation, source current balancing and harmonic mitigation in delta connected for different control techniques such as Instantaneous Reactive Power (IRP), SRF and Symmetrical component (SC) theory have been used [6]. The Convertible Unified Power Quality conditioner (CUPQC) is connected to multi-bus/multifeeder distribution system to mitigate current and voltage interruptions [7]. The control strategy of the UPQC is to contribute on the flow of instantaneous active and reactive powers inside the UPQC [8]. The comparison of Unified Power Flow Controller (UPFC) and UPQC are studied and its advantages are found to be over conventional APF's [9]. A four wire capacitor midpoint shunt APF with a predictive control technique is used to mitigate the current harmonics and the neutral current. A different type of controller methods exists to improve the power quality problems based on the PI control also many theories are available [11], [12], [13], [14], [15], [16], [17], [18], [19].

In all the above mentioned techniques PI controller is used for designed UPQC. In order to regulate the DC-link capacitor voltage, a conventional PI controller is used to maintain the DC-link voltage at the reference value. The transient response of the PI controller in DC-link voltage will be very slow. To overcome this problem a better controller is proposed to improve the transient response of the dc-link voltage. The conventional UPQC is also modified; with the new control techniques based on Modified Synchronous Reference Frame theory (MSRF) to overcome the power quality problems such as voltage/current unbalance, harmonics, reactive power compensation, voltage sag, swell and interruptions.

2. Proposed System Description

The MC-UPQC is connected before the load to make load voltage and current free from any distortions. A two feeder system is designed to interconnect with MC-UPQC as shown in Fig. 1.

The reactive current drawn from the source will be in such a way that it is in-phase with the feeder voltages. The schematic structure of the proposed MC-UPQC is given in Fig. 2. In this configuration, feeder one is connected to a non-linear load and feeder two is connected to a linear load; with MC-UPQC the two feeders are connected. The MC-UPQC operation is the combination of two series voltage converter and one shunt voltage converter which are connected back-to-back with a common DC-link capacitor, by which it can be Controlled independently to compensate the power quality problems. Due to non-linear load in feeder one, system is affected with unbalanced voltage/currents, harmonic distortions in the source as well as in load sides of the both feeders.

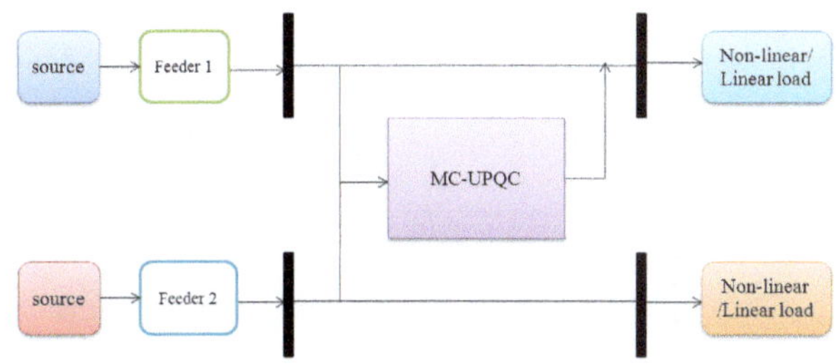

Fig. 1: Single line diagram of MC-UPQC.

As per feeder two the load is a linear load hence it has no harmonic distortion, sag, swell, interruption, voltage/current unbalance, hence it has no effect in both the feeders. The two series VSC's are connected in the series between the two feeders through a series transformer, the shunt VSC is connected to feeder one through a shunt transformer in load side. To avoid the flow of switching harmonics in MC-UPQC the power RC high pass filter with commutation reactor (L) is connected to all VSC's. To achieve this, we examine one suitable structure of the MC-UPQC with Fuzzy Logic Controller (FLC).

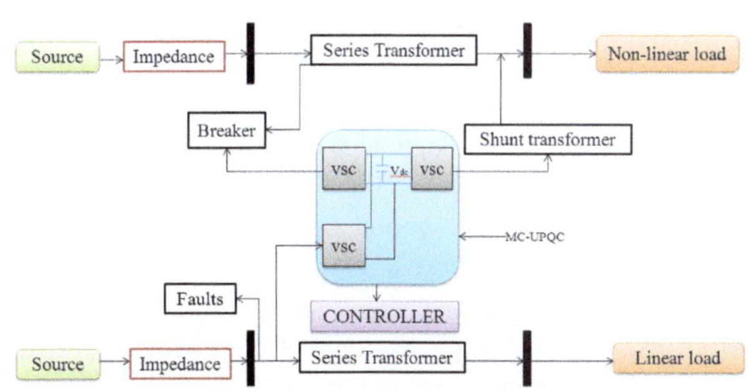

Fig. 2: Single line diagram of MC-UPQC.

3. Control Strategy for the MC-UPQC

The proposed control strategy aims to generate reference signals for both shunt and series voltage source converter of the MC-UPQC. The control technique is capable of extracting most of the load currents, source

$$\begin{bmatrix} i_{l-d} \\ i_{l-q} \\ i_{l-0} \end{bmatrix} = \begin{bmatrix} I_d \\ I_q \\ I_0 \end{bmatrix} \begin{bmatrix} i_{l-a} \\ i_{l-b} \\ i_{l-c} \end{bmatrix} \tag{1}$$

$$\begin{bmatrix} I_d \\ I_q \\ I_0 \end{bmatrix} = \frac{2}{3} \begin{bmatrix} \sin wt & \sin\left(wt - \frac{2\pi}{3}\right) & \sin\left(wt + \frac{2\pi}{3}\right) \\ \cos wt & \cos\left(wt - \frac{2\pi}{3}\right) & \cos\left(wt + \frac{2\pi}{3}\right) \\ \frac{1}{2} & \frac{1}{2} & \frac{1}{2} \end{bmatrix} \begin{bmatrix} I_a \\ I_b \\ I_c \end{bmatrix} \tag{2}$$

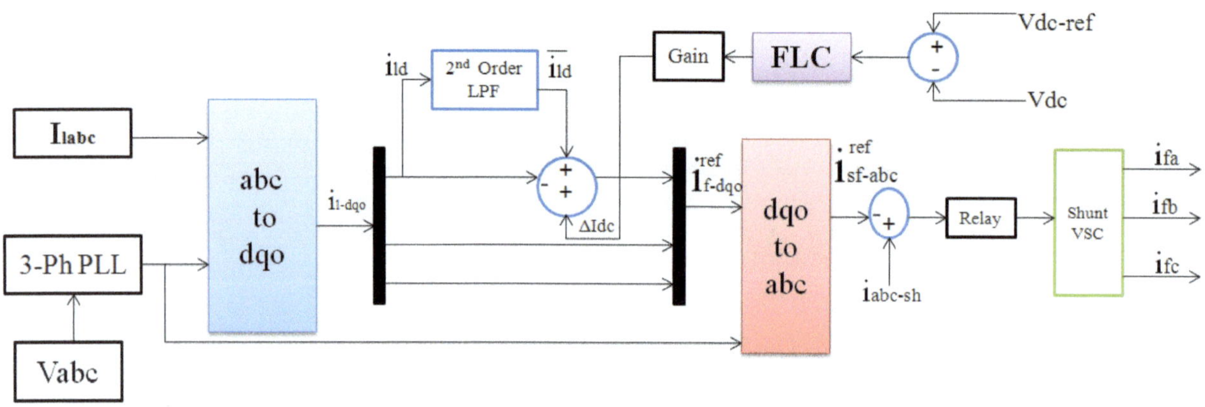

Fig. 3: SRF based control strategy of the shunt VSC.

voltage distortions, voltage sags, swell and harmonics, voltage and current unbalance, reactive and harmonic component of the both feeders.

3.1. The Control Scheme of the Shunt VSC

The control algorithm for the shunt VSC block is shown in Fig. 3. The shunt VSC used in the simulation is designed by using MSRF theory with FLC technique. When compared to the conventional method [2], the designed system of shunt VSC gives the better compensating of harmonics, reactive components of feeder one load current as well as to regulate the common DC-link capacitor voltage. When the supply voltage is got distorted, a phase-locked loop (PLL) is used to achieve synchronization with the supply voltage. The distorted supply voltage is sensed and given to PLL to generate two quadrature unit vectors, namely sine and cosine outputs from the PLL in order to compute the 120° phase displacement for each phase. The shunt VSC is based on the unit vector template, based on the concept of MSRF theory. According to this theory, the phase angle of each phase voltage and currents can be extracted as a three independent two-phase system usually given by $\Pi/2$ lead or lag. This theory can be applied for three-phase balanced system as well as unbalanced of each phase system independently.

The three phase load currents for feeder one is transformed into load synchronous reference currents using Eq. (1) and Eq. (2).

The fundamental direct axis component current is transferred into DC quantities using 2nd order low-pass-filter and it is added to the fuzzy logic output to generate a new reference shunt feeder currents in Eq. (3) and Eq. (4):

$$i_{f-d}^{ref} = \overline{i_{ld}} + \Delta I_{dc}, \tag{3}$$

$$i_{f-q}^{ref} = i_{l-q}, \tag{4}$$

$$\begin{bmatrix} i_{f-a}^{ref} \\ i_{f-b}^{ref} \\ i_{f-c}^{ref} \end{bmatrix} = \begin{bmatrix} I_a \\ I_b \\ I_c \end{bmatrix} \begin{bmatrix} i_{f-d}^{ref} \\ i_{f-q}^{ref} \\ i_{f-0}^{ref} \end{bmatrix}. \tag{5}$$

The power received from the DC-link capacitor through the series inverter and switching losses can be used to decrease the average value of DC bus voltage. All other distortions like unbalance conditions and sudden change in load current can also result in oscillations in DC bus voltage. In order to overcome the

$$\begin{bmatrix} v_{l-d} \\ v_{l-q} \\ v_{l-0} \end{bmatrix} = \begin{bmatrix} v_d \\ v_q \\ v_0 \end{bmatrix} \begin{bmatrix} v_{l-a} \\ v_{l-b} \\ v_{l-c} \end{bmatrix} \tag{6}$$

$$\begin{bmatrix} v_d \\ v_q \\ v_0 \end{bmatrix} = \frac{2}{3} \begin{bmatrix} \sin wt & \sin\left(wt - \frac{2\pi}{3}\right) & \sin\left(wt + \frac{2\pi}{3}\right) \\ \cos wt & \cos\left(wt - \frac{2\pi}{3}\right) & \cos\left(wt + \frac{2\pi}{3}\right) \\ \frac{1}{2} & \frac{1}{2} & \frac{1}{2} \end{bmatrix} \begin{bmatrix} v_a \\ v_b \\ v_c \end{bmatrix} \tag{7}$$

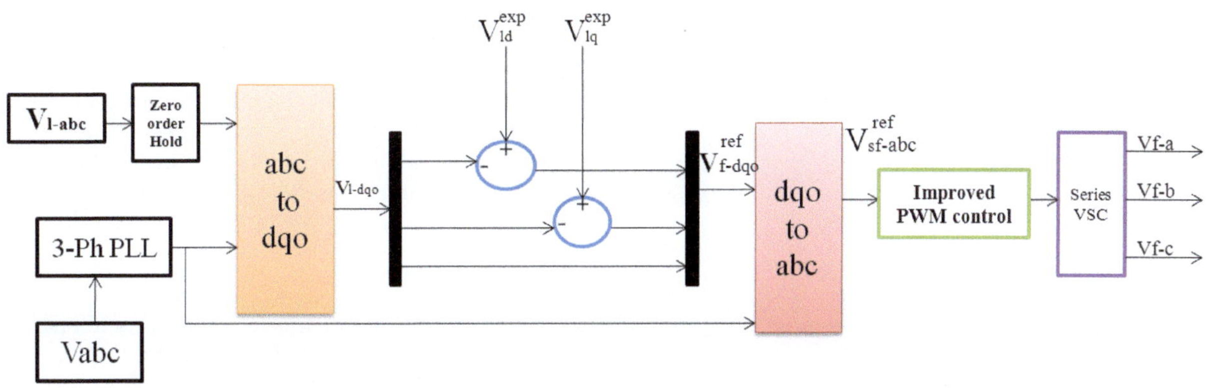

Fig. 4: SRF based control strategy of the series VSC.

error between the desired capacitor voltage and measured values, both are applied to fuzzy logic controllers. The output controlling signal is applied to the current control system of shunt VSC, which stabilizes the DC capacitor voltage by receiving required power from the source. The direct component of the feeder current is subjected to load direct components and quadrature components of the feeder current is subjected to zero. Hence, there is no harmonic current and reactive component in feeder one. The new reference shunt feeder currents in Eq. (3) and Eq. (4) are transformed back to the abc reference currents in Eq. (5).

The shunt currents are added to the abc reference frame currents and it is sensed by the relay to control the currents. The compensation currents of shunt VSC are directly given to controller part is shown in Fig. 3. In section 4 are discussed details about fuzzy logic control operation.

3.2. The Control Scheme of the Series VSC

The control algorithm for the series VSC block is shown in Fig. 4. The series VSC is proposed using MSRF theory with the improved PWM generator are proposed.

Compared to the conventional method [2], the proposed system of series VSC's gives the better compensation of voltage sag, swell and interruptions in feeder two alone, voltage distortions, harmonic distortions and load voltage unbalance in both feeders. The series VSC block is based on the unit vector template by the new MSRF theory. The distorted three-phase supply voltages were sensed by PLL to generate two quadrate unit vectors. The three phase load voltages are transformed into load synchronous reference voltages using Eq. (6), Eq. (7).

According to series control objective, even if the supply voltage is concerned the load voltage must be kept sinusoidal with constant amplitude. So, the expected load Synchronous reference *dqo* voltages are subtracted from the V_{l-dqo} in Eq. (8) and its compensation reference feeder dqo voltages are transformed back to the synchronous reference feeder voltages using Eq. (9):

$$\begin{bmatrix} v_{f-d}^{ref} \\ v_{f-q}^{ref} \\ v_{f-0}^{ref} \end{bmatrix} = \begin{bmatrix} v_{l-d} \\ v_{l-q} \\ v_{l-0} \end{bmatrix} \begin{bmatrix} v_{f-d}^{exp} \\ v_{f-q}^{exp} \\ v_{f-0}^{exp} \end{bmatrix}, \tag{8}$$

$$\begin{bmatrix} v_{sf-a}^{ref} \\ v_{sf-b}^{ref} \\ v_{sf-c}^{ref} \end{bmatrix} = \begin{bmatrix} v_d \\ v_q \\ v_0 \end{bmatrix}^{-1} \begin{bmatrix} v_{f-d}^{ref} \\ v_{f-q}^{ref} \\ v_{f-0}^{ref} \end{bmatrix}. \tag{9}$$

The compensation synchronous reference *abc* voltages are forwarded to the improved PWM generator.

The output of the PWM generator compensation voltage is directly given to control part of series VSC as shown in Fig. 4.

3.3. Design of New Source Controller

The design of the new source controller which gives the supply voltage to the system block diagram is shown in Fig. 5. In this controller six sine waveforms (SW) are considered in continuous mode operation. It is modified by the change of amplitude, angular frequency and phase sequence to get the discrete sine waveform. The sine wave determines the computational technique used the parameters in the two types are related Eq. (10) and Eq. (11):

$$S_p = \frac{2\Pi}{(ft+\Phi)} + B, \tag{10}$$

$$N_{os} = \frac{P \times S_p}{2\Pi}, \tag{11}$$

$$S_0 = V_{sin}.(ft+\Phi) + B, \tag{12}$$

where Sp = samples per period, f = frequency, t = time period, B = bias, Φ = phase angle, N_{os} = number of offset samples, P = phase sequence, S_0 = signal output, V = sine wave amplitude.

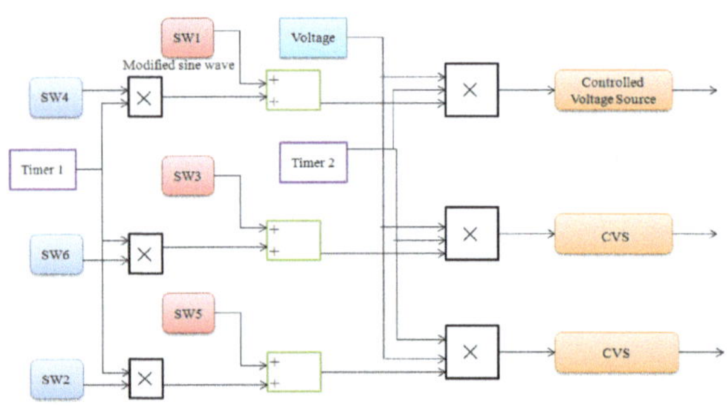

Fig. 5: Block diagram of Source controller.

Here, it is considered the bias value is zero, with phase degree of 120° phase shift. The signal output equation of the sine wave is given in Eq. (12). The modified SW1, SW3, SW5 are connected with timer 2 (t_2). Similarly the SW4, SW6, SW2 are connected with timer 1 (t_1). The SW1, SW4 are added and the output of the signals is multiplied by phase voltage and it is sensed signal to CVS. The three phase source voltage is generated by using below Eq. (13), Eq. (14) and Eq. (15).

$$S_a = V_p[SW1 + (SW4 \times t_1)] \times t_2, \tag{13}$$

$$S_b = V_p[SW3 + (SW6 \times t_1)] \times t_2, \tag{14}$$

$$S_c = V_p[SW5 + (SW2 \times t_1)] \times t_2. \tag{15}$$

Where, V_p is the phase-phase rms voltages, t_1 and t_2 are the timers to generate a signal changing at a specified time. Assume SW1, SW3, SW5 are upper switches and SW4, SW6, SW2 are lower switches. Generally, an ideal voltage source is a mathematical abstraction that simplifies the analysis of electrical circuits. If the voltage across an ideal voltage source can be specified independent voltage source. Conversely, if the voltage across an ideal voltage source is determined by some other voltage or current in a circuit then it is called a "dependent or controlled voltage Source (CVS)".

The use of controlled voltage source is converting the simulink input signal into an equivalent voltage source. The generated voltage is driven by the input signal of the block, and then it initializes the circuit with a specific AC to DC voltage [20]. To start the simulation in steady state, the block input must be connected to a signal starting as a sinusoidal or DC waveform corresponding to the initial values. The design parameters of source controller are shown in Tab. 1.

Tab. 1: Design parameters of source controller.

Parameters	**Upper switches (SW1, SW3, SW5)**	**Lower switches (SW4, SW6, SW2)**
Amplitude	0.22	1
Bias	0	0
Φ [deg]	120°	120°
ω [rad/s]	$2 \cdot \Pi \cdot 50 \cdot 7$	$2 \cdot \Pi \cdot 50$
Sample time	$50e^{-6}$	$50e^{-6}$

4. Fuzzy Logic Controller

Fuzzy logic control (FLC) system is composed of the following four principal components:

- fuzzification,
- knowledge base,
- inference engine or decision making logic,
- de-fuzzification.

The output from the database and the rules of the knowledge base which were used to get the inference relation B mentioned in Eq. (16). The input and output variables of the controlled system or data bus contain a description of fuzzy sets. The paper deals with fuzzy memberships that are designed based on the Eq. (16).

$$B^{(p)} = IF\ X_1\ is\ F_1\ AND\ X_2\ is\ F_2 \ldots X_n\ is\ F_n \\ then\ Y\ is\ C^{(p)}, \tag{16}$$

where $X_1, X_2 ... X_n$ is the input variables vector, Y is the output or control variable, n is the nomber of fuzzy variables (N=5), $F_1, F_2 ... F_n$ is the fuzzy sets, $P = 1, 2, 3 ... N$, N is the number of rules (N=5).

From the given rule base, the fuzzy controller has to compute necessary specific input signal conditions that can determine its effective control action.

To design an FLC, the plant control is inferred from the two input state variables, namely error DC capacitor voltage (V_{dc}) and change in reference DC capacitor voltage error (ΔV_{dc}) in Eq. (17):

$$v_{e\ fuzzy} = v_{dc} - v_{dc}^{ref}. \tag{17}$$

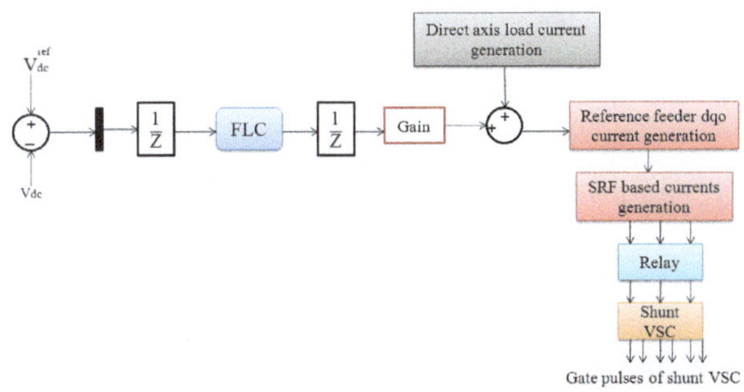

Fig. 6: Proposed structure of a complete Fuzzy logic controller.

The proposed structure of a complete FLC is given in Fig. 6. The Fuzzy control rules are designed for a fuzzy set of the control input in each combination of fuzzy sets for V_{dc} and ΔV_{dc} through which a very small amount of real loss is required for voltage regulation taken as the output from the FLC. The input and output variables are converted into linguistic variables.

The direct axis load current is added to the fuzzy logic output and it is forwarded to reference feeder dqo current. The MSRF based currents are directly given to relay and it is senses a control signal to shunt VSC control circuit.

Instead of using conventional (PI) controller mentioned in references a FLC is being used for its transient response to make MC-UPQC very fast in reducing the total harmonic distortions on the source and load side voltages as well as currents on both the feeders. Here five labels of fuzzy subsets; negative large (NL), negative medium (NM), zero (ZR), positive medium (PM), positive large (PL). The control rule base table is shown in Tab. 2. In which the row and column represent the error and its changes respectively.

Tab. 2: Rule based for Voltage control.

		INPUT 2 [ΔV_{dc}]				
		NL	NM	ZR	PM	PL
INPUT 1 [V_{dc}]	NL	NL	NL	NM	NM	ZR
	NM	NL	NM	NM	ZR	PM
	ZR	NM	NM	ZR	PM	PM
	PM	NM	ZR	PM	PM	PL
	PL	ZR	PM	PM	PL	PL

5. Simulation Results and Discussion

In order to certify the control strategy as discussed above, the MATLAB simulation based proposed system as described in Fig. 2, is compared without and with MC-UPQC controller, with (PI) controller [2]. The proposed control system gives better simulation results as expected. The same model is re-defined with some modifications and also incorporated with FLC is realized using the MATLAB/Simulink software environment as shown in Fig. 7. The performance of the MC-UPQC along with FLC is connected with two feeder system to mitigate the voltage sag swell and interruptions. The load and source voltage unbalance are effectively reduced. Reactive power compensation also achieved with the help of series compensation and the THD of load current is reduced. The proposed control strategies for the series and shunt VSC's with FLC are discussed in below. The simulation circuit diagram for MSRF based shunt and series controller is shown in Fig. 8. Design of new source controller simulation circuit diagram is shown in Fig. 9. The controller parts are discussed in section 3.

5.1. Performance of the MC-UPQC when Connected to Feeder One

The performance of the MC-UPQC which is connected to feeder one with proposed control techniques simulation results is shown in Fig. 10, Fig. 11, Fig. 12 and Fig. 13.

The MC-UPQC is turned on at the time $t = 0.02$ s. The feeder one bus voltage has voltage sag between 0.1 to 0.2 sec and voltage swell between 0.2 to 0.3 sec. This voltage contains 24.9 % of sag and 119.9 % of swell. To compensate this voltage sag, swell using MSRF based series and shunt VSC controller with FLC is presented.

THD is reduced in bus 1 voltage from 22.01 % to 21.98 % and series compensation voltage from 65.91 % to 65.80 % simultaneously load 1 voltage from 38.90 % to 24.75 %. The simulation results and harmonic spectrum of bus 1 voltage, series compensation voltage and load 1 voltage in feeder one are shown in Fig. 10, Fig. 11 respectively. The feeder one current, shunt current, non-linear current, DC-capacitor voltages in feeder one simulation results are shown in Fig. 12. The THD is reduced in feeder one current from 2.73 % to 2.45 % and shunt filter current from 15.92 % to 13.96 % simultaneously non-linear current from 13.61 % to 12.07 %. The harmonic spectrum of feeder one, shunt filter, non-linear currents in feeder one is shown in Fig. 13. These currents are improving very well when compared to the

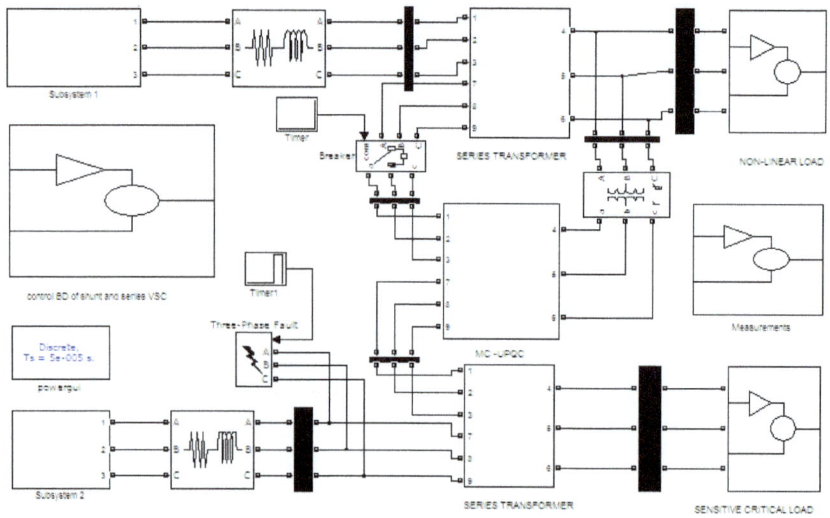

Fig. 7: Simulation circuit diagram of Multi Converter-UPQC is connected in a distribution system.

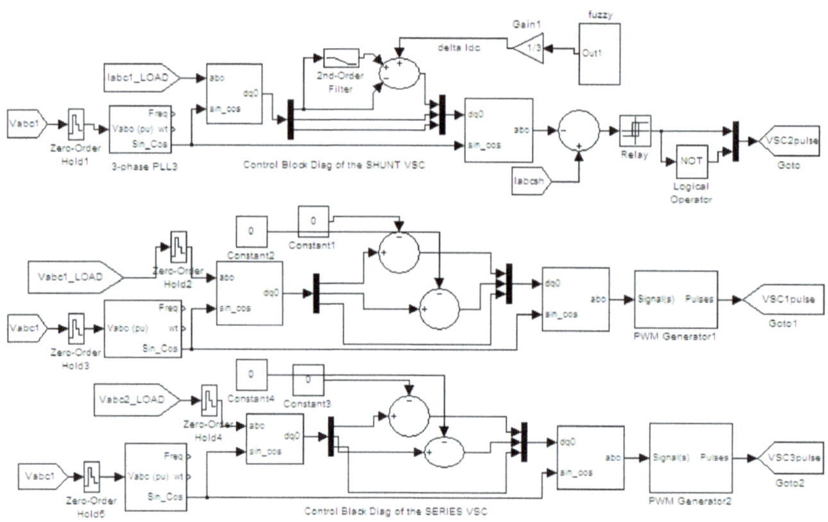

Fig. 8: Simulation control block diagram for MSRF based Shunt VSC and Series VSC controller.

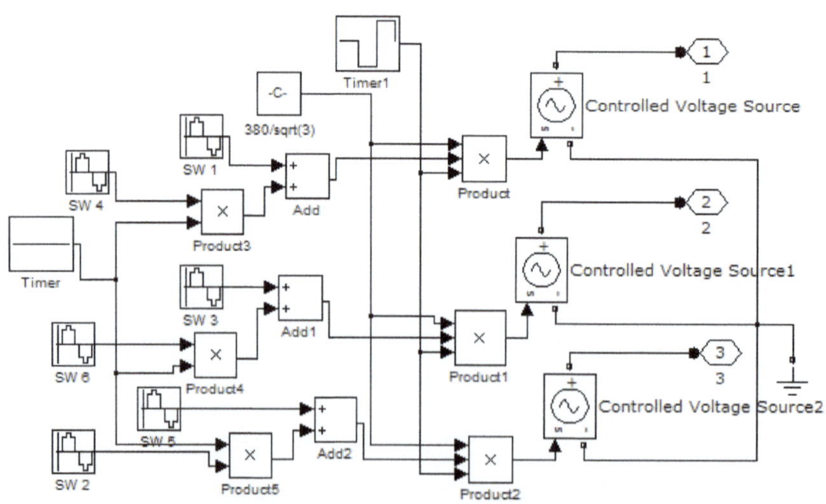

Fig. 9: Simulation circuit diagram of new source controller.

conventional system during the power quality problems in between 0.1 to 0.3 s.

5.2. Performance of the MC-UPQC when Connected to Feeder 2

The performance of the MC-UPQC connected to feeder two with proposed control techniques simulation results are shown in Fig. 14, Fig. 15, Fig. 16, Fig. 17 and Fig. 18. At the same time, the feeder two bus voltage having voltage sag between 0.15 to 0.25 s duration equal to 34.09 % is compensated using same proposed control system itself.

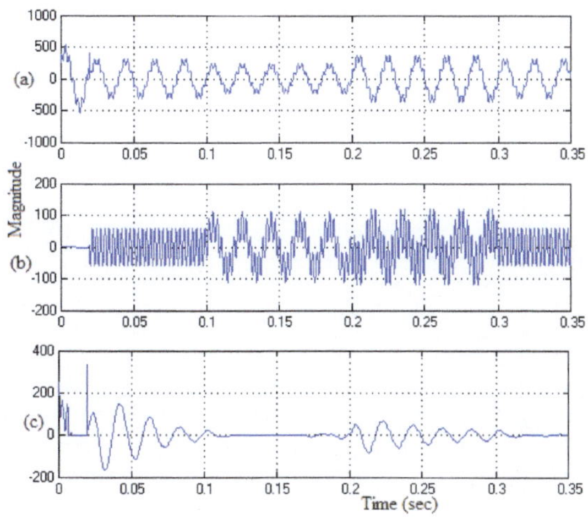

Fig. 10: Simulation results of (a) BUS 1, (b) series compensation, and (c) load 1 voltage in Feeder 1.

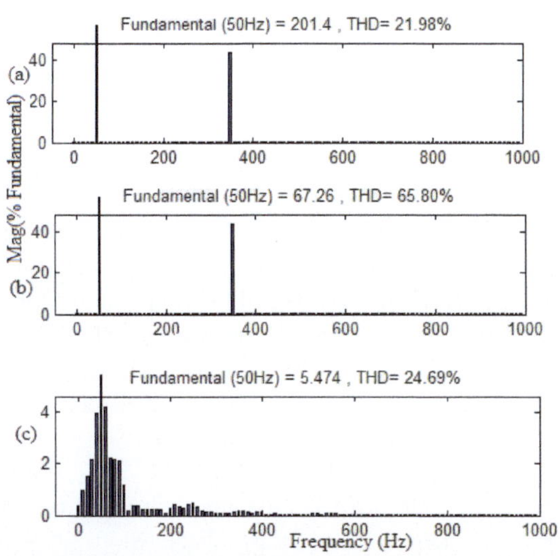

Fig. 11: Harmonic spectra of Harmonic spectrum of (a) Bus 1, (b) series compensation and (c) load 1 voltage in feeder 1.

Bus 2 having a voltage swell between 0.25 to 0.3 sec duration equal to 130 %, THD was reduced in bus 2

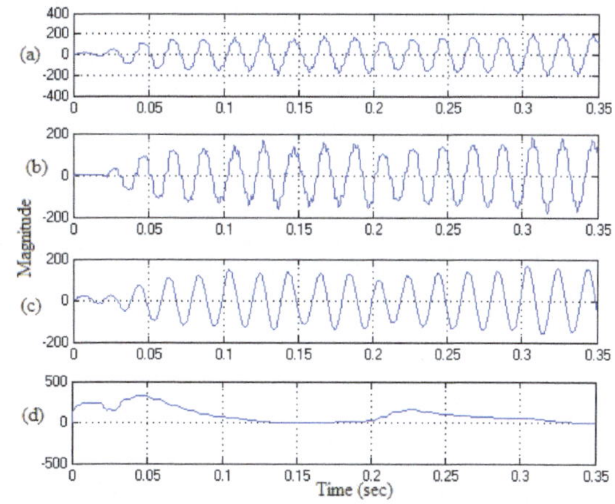

Fig. 12: Simulation results of (a) non-linear (b) shunt filter (c) feeder 1 currents and (d) DC capacitor voltage.

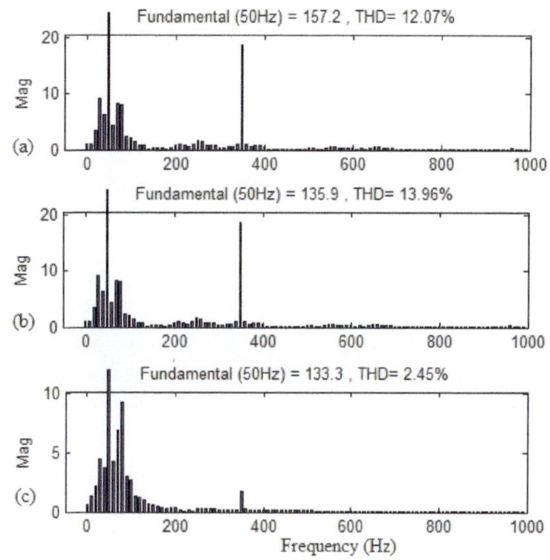

Fig. 13: Harmonic spectrum of (a) non-linear (b) shunt filter and (c) feeder 1 currents.

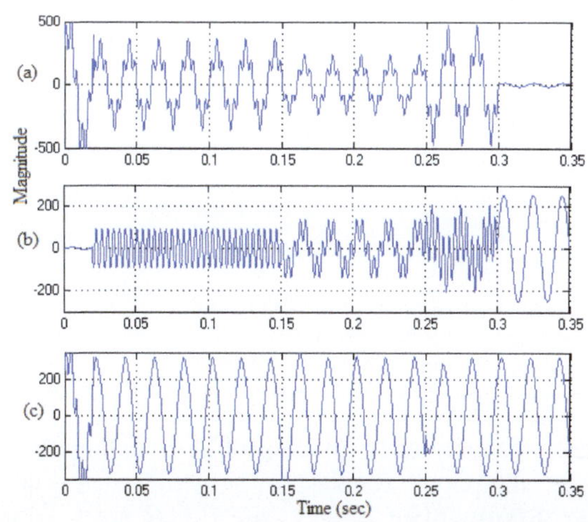

Fig. 14: Simulation results of bus 2, series compensation, load 2 voltages in feeder 2.

voltage from 34.97 % to 34.96 % and series compensa-

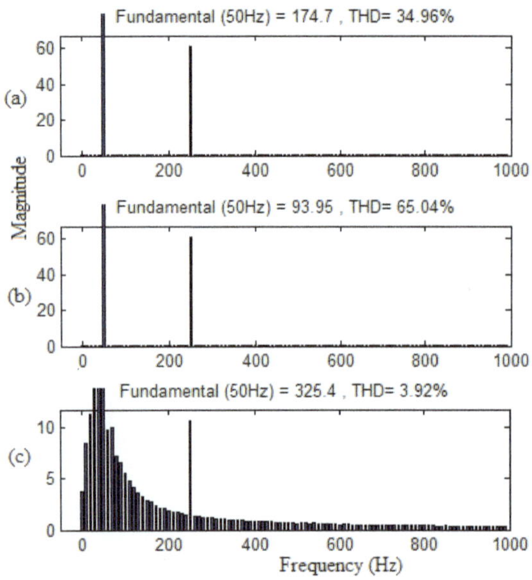

Fig. 15: : Harmonic spectra of (a) bus 2, (b) series compensation and (c) load 2 voltages in feeder 2.

tion voltage from 64.92 % to 65.04 % along with load 2 voltage from 4.30 % to 3.92 %.

The simulation results and harmonic spectrum of bus 2, series compensation and load voltages are shown in Fig. 14, Fig. 15 respectively. If any type of faults occurred in feeder two, the voltage across the linear load will be effected with sag, swell and interruptions.

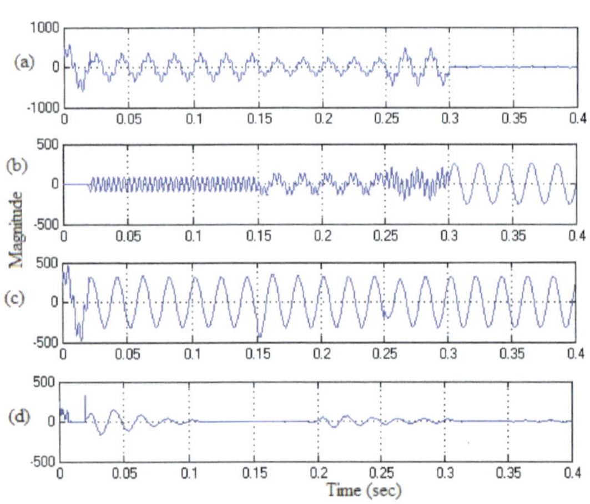

Fig. 16: Simulation results of Feeder 2 (a) bus 2, (b) series compensation, (c) load 1 and (d) load 2 voltages during upstream fault.

In order to compensate above problems a shunt VSC is used. In this paper, the system is tested with different types of faults like L-G, L-L and L-L-G faults. The L-L-L-G fault is applied to feeder two between the 0.3 to 0.4 s duration. It is observed that the bus 2 voltage is affected between 0.15 to 0.25 s duration with sag, 0.25 to 0.3 s duration with swell and 0.3 to 0.4 s duration momentary interruptions.

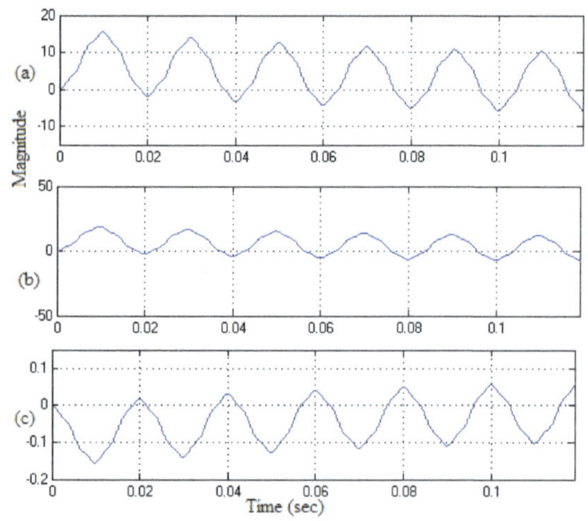

Fig. 17: Simulation results of (a) non-linear, (b) feeder 2 and (c) shunt filter current in feeder 2.

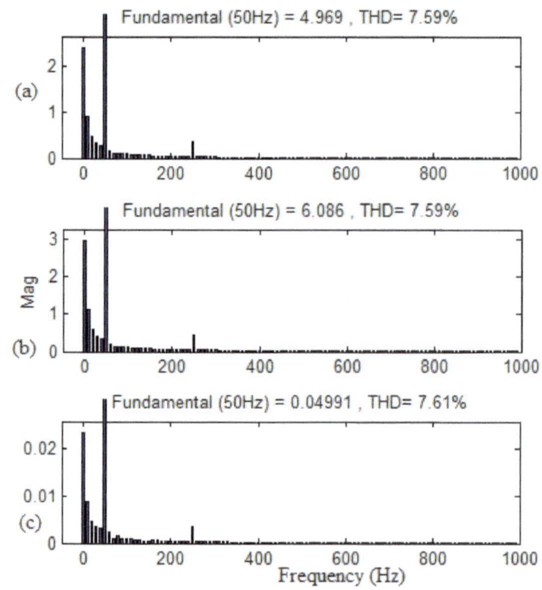

Fig. 18: Harmonic spectrum of (a) non-linear, (b) shunt filter, and (c) feeder 2 current in feeder 2.

The bus 2, series compensation, loads 1 and 2 voltages during the upstream fault conditions simulation results were shown in Fig. 16. The non-linear, feeder two and shunt currents in feeder two simulation results are also shown in Fig. 17. It is founded that THD are reduced in feeder two current from 8.92 % to 7.61 % and shunt filter current from 8.00 % to 7.59 % simultaneously non-linear currents from 8.00 % to 7.59 %. The harmonic spectrum of nonlinear, shunt filter and feeder two currents were shown in Fig. 18.

During the 0.3 to 0.4 sec, the bus 2 voltage has interruption up to 4.2 %. The above all results are compared with the conventional and proposed control system.

THD results were shown in Tab. 3, the voltage sags, swell, interruption is calculated according to IEEE

Tab. 3: % THD Results: Comparison between conventional [1] and proposed system.

System voltage/current	Without MC-UPQC	Conventional [1]	Proposed system
Bus 1 voltage(Ut1)	21.92	22.01	21.98
Compensation voltage	21.92	65.91	65.80
Load 1 voltage	21.95	38.90	24.75
Feeder 1 current	15.64	2.73	2.45
Shunt filter current	31.05	15.92	19.96
Non-linear current	15.64	13.61	12.07
Bus 2 voltage	35.08	34.97	34.96
Compensation voltage	14.11	64.92	65.04
Load 2 voltage	46.49	4.30	3.92
Feder 2 current	9.70	8.92	7.61
Shunt filter current	Undefined	8.00	7.59
Non-linear current	9.70	8.00	7.59

Tab. 4: % Results of Sag, Swell, Interruption According to IEEE 1159-1995 Std.

PQ issues [%]	Without MC-UPQC		Conventional [1]		Proposed system	
	Bus1	Bus2	Bus1	Bus2	Bus1	Bus2
Sag	25.01	35.01	25	35	24.9	34.09
Swell	119.96	130.5	120	130	119.9	130
Interruption		9.3		5		4.2

Tab. 5: % Results of Active (P), Reactive power (Q) on load and source side.

Active/reactive power	Without MC-UPQC		Conventional [1]		Proposed system	
	Load	Source	Load	Source	Load	Source
P [W]	13.82	1.745	466.6	45.23	1479	53.42
Q [var]	13.58	13.58	1170	-4.827	1792	130

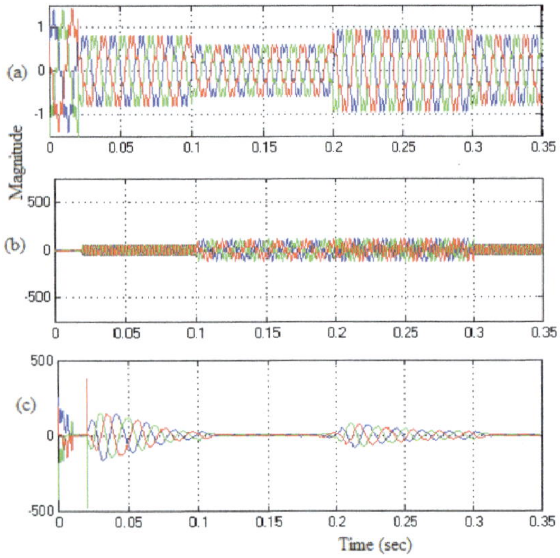

Fig. 19: Simulation results of the (a) bus1, (b) series compensation and (c) feeder 1 voltage in feeder 1 under unbalance source voltage.

1159-1995 standards as shown in Tab. 4. The active, reactive powers are calculated for both load and source side as given in Tab. 5. The simulation results for voltage at bus1, series compensated voltage and load voltage at feeder one are shown in Fig. 19. During the load changes, at the time $t = 0.5$ the load 1 voltage is doubled and it is reduced to half, reaming load is unchanged and DC bus voltage, non-linear current is compensated.

6. Conclusion

With the help of proposed MC-UPQC system the voltage /current harmonics, reactive power compensation was performed successfully. Additionally the voltage regulation was also done by protecting the loads against any distortions, sag, swell, and interruptions. By using FLC with MC-UPQC DC-link voltage controller it is observed that transient response is attained very fast. The proposed novel control strategy for a three phase three wire system utilizing with MC-UPQC is validated and compared with conventional controller.

References

[1] GHOSH, A. and G. LEDWICH. *Power Quality Enhancement Using Custom Power Devices.* Boston: Springer Verlag, 2012. ISBN 978-146-1354-185.

[2] MOHAMMADI, H. R., A. Y. VARJANI and H. MOKHTARI. Multiconverter Unified Power-Quality Conditioning System: MC-UPQC. *IEEE Transactions on Power Delivery.* 2009, vol. 24, iss. 3, pp. 1679–1686. ISSN 0885-8977. DOI: 10.1109/TPWRD.2009.2016822.

[3] NAIDU, P. V. and B. BASAVARAJA. Design of a SRF based MC UPQC used for load voltage control in Parallel distribution systems. In: *IEEE Fifth Power India Conference.* Murthal: IEEE, 2012, pp. 1–6. ISBN 978-1-4673-0763-5. DOI: 10.1109/PowerI.2012.6479578.

[4] JINDAL, A. K., A. GHOSH and A. JOSHI. Interline Unified Power Quality Conditioner. *IEEE Transactions on Power Delivery.* 2007, vol. 22, iss. 1, pp. 364–372. ISSN 0885-8977. DOI: 10.1109/TPWRD.2006.881581.

[5] BAHR ELDIN, S. M., K. S. RAMA RAO and N. PERUMAL. Generalized unified power quality conditioner for compensating current and voltage imperfections. In: *2011 IEEE Ninth International Conference on Power Electronics and Drive Systems.* Singapore: IEEE, 2011, pp. 83–88. ISBN 978-1-4577-0000-2. DOI: 10.1109/PEDS.2011.6147228.

[6] ZAVERI, T., B. BHAVESH and N. ZAVERI. Control techniques for power quality improvement in delta connected load using DSTATCOM. In: *2011 IEEE International Electric Machines & Drives Conference (IEMDC).* Niagara Falls: IEEE, 2011, pp. 1397–1402. ISBN 978-1-4577-0060-6. DOI: 10.1109/IEMDC.2011.5994811.

[7] ELDIN, S. M. B., K. S. R. RAO, R. IBRAHIM and N. PERUMAL. Convertible unified power quality conditioner to mitigate voltage and current imperfections. In: *2012 4th International Conference on Intelligent and Advanced Systems (ICIAS2012).* Kuala Lumpur: IEEE, 2012, pp. 473–478. ISBN 978-1-4577-1968-4. DOI: 10.1109/ICIAS.2012.6306240.

[8] FUJITA, H. and H. AKAGI. The unified power quality conditioner. *IEEE Transactions on Power Electronics.* 1998, vol. 13, iss. 2, pp. 315–322. ISSN 0885-8993. DOI: 10.1109/63.662847.

[9] BHARTI, A., R. VARSHNEY and S. K. SRINIVASTVA. A study of PI controller based unified power quality conditioner. *International journal of advanced research in computer science and software engineering.* 2012, vol. 2, iss. 9, pp. 85–88. ISSN 2277-128X.

[10] FAHMY, A., M. S. HAMAD, A. K. ABDELSALAM and A. LOTFY. Power quality improvement in three-phase four-wire system using a shunt APF with predictive current control. In: *IECON 2012 - 38th Annual Conference*

on IEEE Industrial Electronics Society. Montreal: IEEE, 2012, pp. 668–673. ISBN 978-1-4673-2420-5. DOI: 10.1109/IECON.2012.6388748.

[11] RASTOGI, M., R. NAIK and N. MOHAN. A comparative evaluation of harmonic reduction techniques in three-phase utility interface of power electronic loads. *IEEE Transactions on Industry Applications*. 1994, vol. 30, iss. 5, pp. 1149–1155. ISSN 1149-1155. DOI: 10.1109/28.315225.

[12] PENG, F. Z. Application issues of active power filters. *IEEE Industry Applications Magazine*. 1998, vol. 4, iss. 5, pp. 21–30. ISSN 1077-2618. DOI: 10.1109/2943.715502.

[13] AKAGI, H. New trends in active filters for power conditioning. *IEEE Transactions on Industry Applications*. 1996, vol. 32, iss. 6, pp. 1312–1322. ISSN 1312-1322. DOI: 10.1109/28.556633.

[14] GYUGYI, L., C. D. SCHAUDER, S. L. WILLIAMS, T. R. RIETMAN, D. R. TORGERSON and A. EDRIS. The unified power flow controller: a new approach to power transmission control. *IEEE Transactions on Power Delivery*. 1995, vol. 10, iss. 2. pp. 1085–1097. ISSN 0885-8977. DOI: 10.1109/61.400878.

[15] FUJITA, H. and H. AKAGI. The unified power quality conditioner: the integration of series and shunt-active filters. *IEEE Transactions on Power Electronics*. 1998, vol. 13, iss. 2, pp. 315–322. ISSN 0885-8993. DOI: 10.1109/63.662847.

[16] GHOSH, A. and G. LEDWICH. A unified power quality conditioner (UPQC) for simultaneous voltage and current compensation. *Electric Power Systems Research*. 2001, vol. 59, iss. 1, pp. 55–63. ISSN 0378-7796. DOI: 10.1016/S0378-7796(01)00141-9.

[17] AREDES, M., K. HEUMANN a E. H. WATANABE. An universal active power line conditioner. *IEEE Transactions on Power Delivery*. 2002, vol. 13, iss. 2, pp. 545–551. ISSN 0885-8977. DOI: 10.1109/61.660927.

[18] HU, M. and H. CHEN. Modeling and controlling of unified power quality compensator. In: *APSCOM 2000 - 5th International Conference on Advances in Power System Control, Operation and Management*. Hong Kong: IEEE, 2000, pp. 431–435. ISBN 0-85296-791-8. DOI: 10.1049/cp:20000437.

[19] BASU, M., S. DAS and G. DUBEY. Comparative evaluation of two models of UPQC for suitable interface to enhance power quality. *Electric Power Systems Research*. 2007, vol. 77, no. 7, pp. 821–830. ISSN 0378-7796.

[20] KOMATSU, Y. and T. KAWABATA. A control method of active power filter in unsymmetrical voltage system. In: *Proceedings of Second International Conference on Power Electronics and Drive Systems*. Singapore: IEEE, 1997, pp. 839–843. ISBN 0-7803-3773-5. DOI: 10.1109/PEDS.1997.627504.

About Authors

Chandra Babu PADUCHURI was born in 1987 in Thokalapalli, Porumamilla, and Andhra Pradesh, India. He received B. Tech from JNTU-K, AP and M. Tech from SRM UNIVERSITY Chennai in 2006 and 2012 respectively. He is currently pursuing his research at SRM University, Chennai, India, in the area power quality improvement using custom power devices. His area of Interests includes Power System operation, Control & Power Electronic Converters, Fuzzy, ANN controller, DSP and FACTS. He has published 8 international papers and 5 international journals.

Subhransu Sekha DASH was born in 1972. He received A.M.I.E graduation from I.E (India), Calcutta, India. The M.E Degree from U.C.E, Burla, Orissa, India, (with specialization in power systems) and the Ph.D. degree in Electrical Engineering from Anna University College of Engineering, Guindy, Chennai in 1994, 1996 and 2006 respectively. He has published more number of Papers in National and International reputed Journals. He is presently working as Professor and HOD (EEE) in SRM Engineering College, SRM University, Chennai, India and his areas of interests include FACTS, Power System Operation, Control & Stability, Power Electronics Drives and Intelligent Controlling Techniques.

Subramani CHINNAMUTHU was born in 1976. He is presently working as Associate professor in the Dept. of EEE at SRM University, Chennai, India. He received B.E graduation from Bharathiar University, Tamilnadu, India and M.E (Power System Engineering) from Anna University, Chennai in 2004 and 2006 respectively. He was completed his Ph.D. degree in Electrical and Electronics Engineering from SRM University in 2012. He has published 16 Papers in referred International Journals and more than 40 papers in International/National Conference. He having 15 years teaching and industrial experience. His areas of interest are power system stability, power system control and FACTS, electrical machine design.

21

Trust Based Algorithm for Candidate Node Selection in Hybrid MANET-DTN

Jan PAPAJ, Lubomir DOBOS

Department of Electronics and Multimedia Communications, Faculty of Electrical Engineering and Informatics, Technical University of Kosice, Letna 9, 042 00 Kosice, Slovak Republic

jan.papaj@tuke.sk, lubomir.dobos@tuke.sk

Abstract. *The hybrid MANET-DTN is a mobile network that enables transport of the data between groups of the disconnected mobile nodes. The network provides benefits of the Mobile Ad-Hoc Networks (MANET) and Delay Tolerant Network (DTN). The main problem of the MANET occurs if the communication path is broken or disconnected for some short time period. On the other side, DTN allows sending data in the disconnected environment with respect to higher tolerance to delay. Hybrid MANET-DTN provides optimal solution for emergency situation in order to transport information. Moreover, the security is the critical factor because the data are transported by mobile devices. In this paper, we investigate the issue of secure candidate node selection for transportation of the data in a disconnected environment for hybrid MANET-DTN. To achieve the secure selection of the reliable mobile nodes, the trust algorithm is introduced. The algorithm enables select reliable nodes based on collecting routing information. This algorithm is implemented to the simulator OPNET modeler.*

Keywords

Candidate node, delay tolerant network, mobile ad-hoc network, trust.

1. Introduction

Mobile ad-hoc network (MANET) is characterized by multi hop communication between mobile nodes by wireless links [1]. There are also no infrastructures and routing paths established by routing algorithms (Fig. 1). The traditional routing algorithms and protocols are based on routing schemes, which can find a path for a given node pair according to various metrics. Data packets are then transmitted from one intermediate relay node to the next specified relay based on physical condition of wireless channels [2], [4], [5]. The routing algorithm relies on the assumption that the network graph is fully connected and fails to route messages if there is not a complete route from source to destination at the time of sending [3].

The Delay/Disruption Tolerant Networks (DTN) or Opportunistic networks (OppNet) have been developed for intermittent communication between mobile terminals. The main feature of these networks is high propagation delays when transferring data between different terminals. DTN or OppNet can be deployed in environment, where high bit error rates and the long-term disconnections are experienced. These are proposed and designed to operate in hostile environments. The one is characterized by very long delay paths and frequent network partitions [6].

Security is an important issue in MANET, DTN and OppNet. In MANET, the security mechanisms are

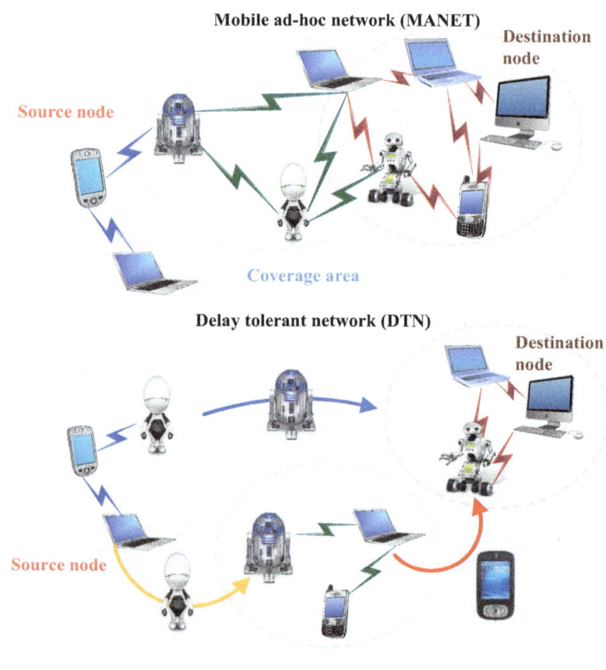

Fig. 1: Difference between MANET and DTN.

based on the assumption that there is/are a connection between source and target nodes (end-to-end connections) [2]. In OppNet, we require the security solutions, which provide security for all nodes, all services and application that participate in routing and transmitting process. There is a sporadic connectivity of nodes and we need to provide secure delivery of the messages from source node to the destination node.

In order to provide the effective communication between nodes, there is necessary to consider different aspects (disconnections, mobility, partitions, and norms instead of the exceptions). The mobility in opponent is used to provide efficient communication between unconnected groups of nodes. The mobile nodes forward data and also store data in the cache for a long time. This model is called store-carry-forward [2].

The selection of forwarding candidate node is the first key issue. There are a few works which deal with the selection of the candidate nodes. The firs access is based on the number of transitions (Expected Transmission Count (ETX), Expected Any-path Transmission (EAX)) which accounts for the specific characteristics of the opportunistic paradigm [6]. Other algorithms are based on link state algorithm for the unconstrained selection based on the Dijkstra algorithm [7]. Most methods are based on generalizing the Bellman-Ford algorithm and they prove its optimality [8], [9].

In this article the trust based candidate node selection algorithm is introduced. This mechanism enables the selection of the secure mobile node used for transportation of the data across the disconnected environment. Selected candidate nodes provide the secure mobile node selected to transport of the data. If the mobile node finds connection the DTN forwarding mechanism is activated.

1.1. Trust Mechanisms for Hybrid MANET-DTN

The term of trust is defined in different disciplines. Trust enables to combine many different aspects and then can be reflected by reliability, utilization, availability, reputation, risk, confidence, quality of services and other concepts. Trust enables to calculate a subjective assessment by a mobile node based on reliability and accuracy of information received from or traversing through that node in a given context.

In hybrid MANET-DTN, the key aspect is how these trust concepts can be applied in modeling trust [10], [11]. Trust calculation can be divided into the following groups (Fig. 2):

- Direct trust calculation - mobile nodes have implemented mechanisms for computing its own value of the trust on its neighbours.
- Indirect trust calculation - the trust is computed based on the recommendation of the neighbour mobile nodes.
- Hybrid trust calculation - combination of the direct and indirect method.

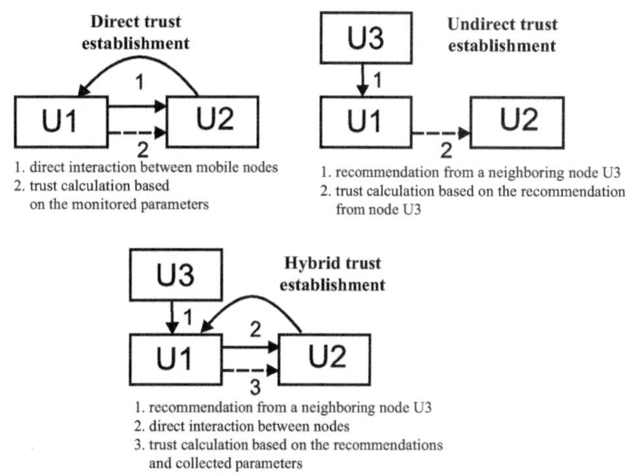

Fig. 2: Trust computation models for hybrid MANET-DTN.

2. Proposed Trust Algorithm for Hybrid MANET-DTN

The proposed algorithm is based on the direct model for the trust computation. This model is based on the assumption that each node in the network receives information about other nodes. Information is stored on each node and later used in the calculation of the resulting values of trust. The algorithm is designed on third layer so that all obtains from the node information throughout the entire operation of the network layer.

The data and routing streams are monitored. Figure 3 shows the flow diagram of the proposed algorithm for calculation of trust. The algorithm can be divided into two main phases:

- Phase of obtaining and storing information.
- Phase of trust computing.

These phases are carried out whenever the packet is received on the node. It is for this reason that the final value of trust is always up to date and it makes changes dynamically, as well as changing the network topology, which is related to the mobility of nodes. The final value of the trust reflects the current state of the parameters obtained from the network.

Fig. 3: Proposed algorithm for trust monitoring in hybrid MANET-DTN.

2.1. Phase of Obtaining and Storing Information

The algorithm for the trust calculation is triggered whenever a change occurs in the obtained parameters. The routing packet coming from the lower layer (physical and data link layer) is encapsulated and the network layer is an IP packet containing different information.

The one such important information is collected from the routing packets (DSR reactive routing protocol). This information tells what kind of packet goes. In the proposed algorithm are incoming packets used as parameters which enter into the final calculation of trust and the following parameters are used for calculation: total number of route request packets received in node, total number of route reply packets received in node, total number of route error packets received in node, total number of route acknowledgement packets received in node, total number of data packets received in node, number of route requests received from each node, number of route replies received from each node, number of route errors received from each node, number of route reply received from each node, number of data packets received from each node.

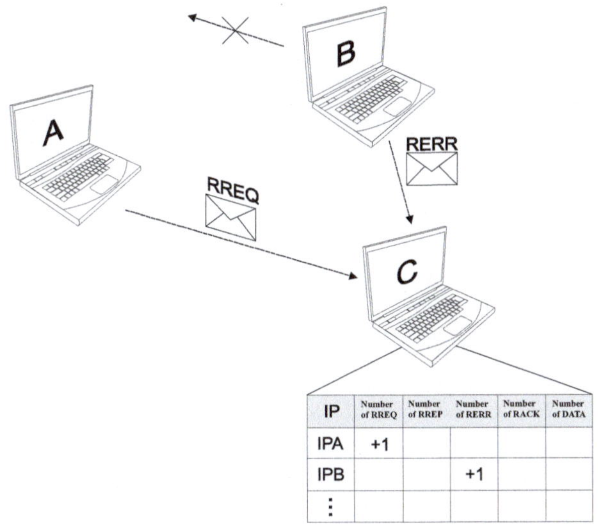

Fig. 4: Collecting of the routing data from hybrid MANET-DTN.

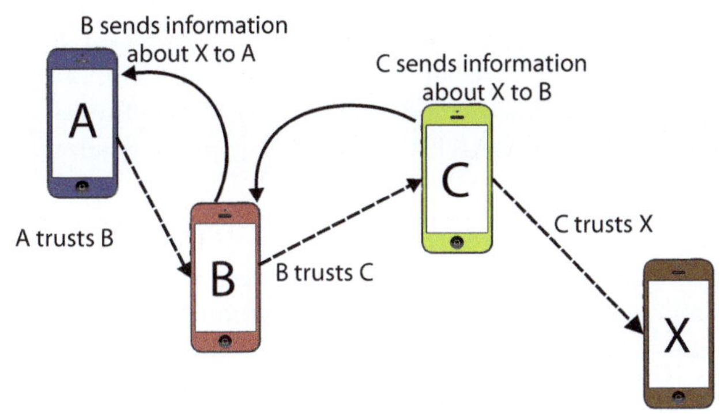

Fig. 5: Promotion calculated value confidentiality between nodes during computing of the trust in hybrid MANET-DTN.

This information from the phase of obtaining and saving of parameters are stored in the data structure in memory on each node separately. Each node contains information about nodes with whom he came in contact, and when they exchange routing information or data traffic between each other. Each node in the network is separate and calculates a resulting value of confidentiality to the other node. Information is stored in a format which has the form of a table. This structure is dynamic and its size can vary based on the number of nodes with which node communicates.

Figure 4 illustrates the phase of obtaining and saving parameters. In the picture we can see three communicating nodes A, B, C. In this scenario, node A sends a packet type Route Request to node C and node B sends a Route error packet to the node C (Fig. 4). From node C point of view, this node updates the parameter table and store new values into the structure. That structure is dynamic in size because it is not known in advance how many nodes in the network can be located and how many nodes can communicate with this node.

2.2. Phase of Computing Trust Parameter

After retrieving and updating the parameters in table of parameters for each node, it is necessary to calculate and update the value of trust. At this stage, the value of trust is calculated from the obtained parameters. Computing of trust is based on the direct model for the calculation of trust. This means that the resulting value was not sent further to other nodes as a recommendation. Serves locally on that node and helps him in deciding. In our case, the calculated value will serve by the decision on sending in a situation where there is a disconnection in the network and the node will not have a backup path to the destination node (Fig. 5). The actual calculation of confidentiality can be divided into two parts:

- Trust calculation from routing information.
- Trust calculation from data packets.

From these two sub-calculations we get the final value of trust, which is registered in the node data structure. Analyze individual fragment values are given in equations:

$$T_s = W_{rreq} \cdot \left(\frac{R_{req}}{R_{reqc}}\right) + W_{rrep} \cdot \left(\frac{R_{rep}}{R_{repc}}\right) + \\ + W_{rerr} \cdot \left(\frac{R_{rerr}}{R_{rerrc}}\right) + W_{rack} \cdot \left(\frac{R_{rack}}{R_{rackc}}\right), \quad (1)$$

$$T_d = W_d \cdot \left(\frac{D}{D_c}\right), \quad (2)$$

where T_s and T_d are mentioned partial value confidentiality. Furthermore W_{rreq}, W_{rrep}, W_{rerr}, W_{rack}, W_d are constants, which define a weight of trust value, and that resulting value will be in the selected range. Constants can be changed based on the types of attacks and on the basis of what we want to balance with the individual parameters entering into the calculation of trust. The values in the numerator of the equation represent the number of each type of packets from specific nodes in the network. Values R_{rreqc}, R_{rrepc}, R_{rerrc}, R_{rackc}, D_c represent the total number of packets received on the node from all nodes and R_{rreq}, R_{rrep}, R_{rerr}, R_{rack}, D are the values of packets received from each nodes. The value representing final trust is shown in Eq. (3):

$$T = T_s + T_d, \quad (3)$$

where T is the resulting value of confidentiality and T_s and T_d values is partial confidentiality mentioned above.

3. Simulations and Results

OPNET modeler simulator a professional tool was used for testing of the proposed mechanisms. OPNET provides a lot of useful tools for analyzing of the mobile networks and enables to simulate mobile ad-hoc networks with different routing protocol [12]. The Dynamic Source Routing protocol represents a reference value for analyzing of behaviour of the MANET [1]. In order to check the functionality of the DTN network, we have implemented a PRoPHET forwarding mechanism to the simulator OPNET modeler [12].

Tab. 1: Simulation setup for OPNET modeler.

Parameters	Values
Routing protocols	DSR, DSR_TRUST, PRoPHET
Routing parameters DSR	Default
Simulation area	2000 m×2000 m
Time of simulation	600 s
Transmission range	250 m
Number of nodes	20, 40, 60, 80, 100
Speed of nodes	$0 - 6$ m·s^{-1}
Model placement	Random
Mobility model	Default Random Waypoint with pause time 10 s
Transmitted power	1 mW
Values of trust	$W_{rreq} = 0.2$, $W_{rrep} = 0.5$, $W_{rerr} = 0.2$, $W_{rack} = 0.2$, $W_d = 0.3$
Number of values per simulation	1000

The following simulation scenarios were used to analyse the performance of the proposed mechanisms:

- MANET model - MANET network used standard DSR protocol without the possibility to find communication paths between mobile nodes.
- MANET model with TRUST - MANET network used the DSR protocol with the possibility to find candidate node based on trust.
- DTN model - network implements the PRoPHET protocol and mechanism enable to find a communication path if the links are disconnected during the transportation of the packets. The PRoPHET forwarding mechanism is implemented.

The simulation parameters for the OPNET modeler are summarized in the Tab. 1. During simulations, the average delay, average number of hops parameters, average routing traffic sent, average routing traffic received and an average number of salvaged packets, are used to check the functionalities of the proposed solution.

The average delay provides the average amount of the time that is necessary for useful transmission of the packet between source and destination nodes. The average numbers of routing traffic send and received inform how many data the routing protocol needs to send to the network until the communication paths will be found. The average numbers of the salvaged packets give us information how many packets were salvaged by routing protocol while routing or forwarding. Average number of hops represents number of transmissions necessary to find end-to-end connection between source and destination mobile node. During the analysis, the average values of the parameters were analyzed.

3.1. Comparison of the MANET and DTN

This experiment is focused on the situation when all communication paths are disconnected for a short time interval. The two routing protocols for MANET and DTN namely DSR and PRoPHET were analyzed. The disconnection of the communication paths was simulated with short transmission ranges and random mobility model. This mobility model enables to move mobile node across the simulation area randomly with randomly selected speed. Disconnections represent a short time interval when DSR routing protocol for MANET cannot find a communication path with other mobile nodes. Parameters average delay and average number of hops were analyzed. Collected results show why integration between MANET and DTN routing mechanisms are necessary.

Figure 6 shows the average delay for MANET and DTN with respect to the number of the mobile nodes with different speed. If the communication paths were disconnected, the values are higher for MANET than PRoPHET. Based on the idea of DSR, the routing mechanism sends the routing packets in order to find communication paths and on the other side PRoPHET protocol unicast only the packet for the nodes which are

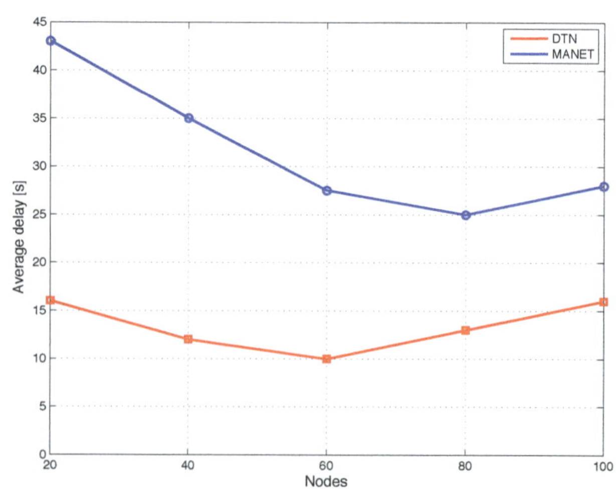

Fig. 6: The average delay for MANET and DTN.

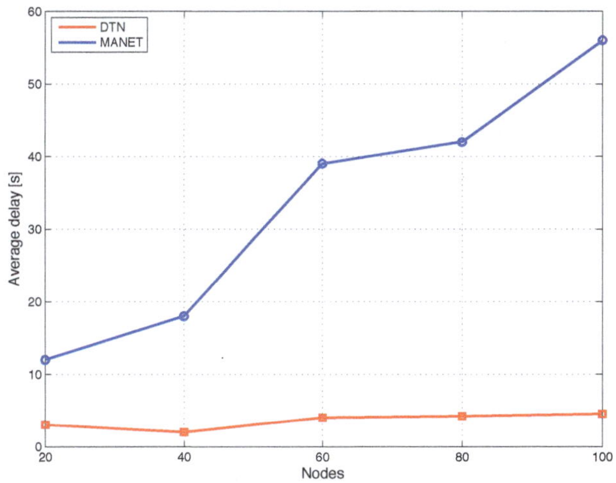

Fig. 7: The average number of the hops for MANET and DTN.

directly connected to the selected mobile node. The values of average delay for MANET are almost 2 times higher than values for DTN.

The second analyzed parameter is the average number of hops. The DSR protocol provides the routing mechanism if the communication paths are disconnected which is resulting in a significant number of the hops for networks with the highest number of the mobile nodes (Fig. 7). The PRoPHET protocol enables to find destination node with lower number of the hops based on the idea of the forwarding mechanism. If the communication paths are disconnected for a short time the PRoPHET shows the better performance than DSR routing protocol for MANET (Fig. 6, Fig. 7).

3.2. Analysis of the Routing Parameters in Disconnected Environment

This simulation is focused on the situation when the communication paths will be disconnected during transmission of the data. In this case the routing paths will be broken and trust algorithm will be activated in order to find the best candidate node with regards to trust.

During simulation, the average amount of routing traffic sent, average amount of routing traffic received, an average number of dropped packets are analyzed. Parameters give us information how the new algorithm will affect the routing of the data with regard to different speed of the mobile nodes.

Table 2 and Tab. 3 show impact of the speed of the mobile nodes on average routing data sent and received. In the case of disconnected communication paths, the results increase with a higher speed and number of the mobile nodes. DSR with trust has generally the highest average routing traffic sent and received as compared to DSR routing protocol without the trust mechanism. The reason for this behaviour is the existence of trust mechanism on the mobile nodes and for this reason the routing data are resend if the mobile node is untrusted.

Tab. 2: Average routing traffic sent vs number of mobile nodes [bit·s^{-1}].

Number of nodes	Model	Speed of mobile nodes		
		2 m·s^{-1}	4 m·s^{-1}	6 m·s^{-1}
20 nodes	DSR	708.15	752.55	650.50
	TRUST	700.70	826.34	947.68
40 nodes	DSR	1039.50	2003.12	1185.29
	TRUST	1989.60	2399.20	3204.34
60 nodes	DSR	2061.00	3298.55	2439.56
	TRUST	2924.66	2160.67	4957.57
80 nodes	DSR	4090.00	5931.42	3910.58
	TRUST	5982.50	6561.55	7625.53
100 nodes	DSR	8306.24	8784.08	10000.84
	TRUST	9122.49	11394.86	12694.57

Tab. 3: Average routing traffic received vs number of mobile nodes [bit·s^{-1}].

Number of nodes	Model	Speed of mobile nodes		
		2 m·s^{-1}	4 m·s^{-1}	6 m·s^{-1}
20 nodes	DSR	1097.7	1567.5	1663
	TRUST	1306	4135.8	1583.6
40 nodes	DSR	2067	3086	2185
	TRUST	6847	2851.2	9478
60 nodes	DSR	4999	7243	6594
	TRUST	12019	8843.5	9481.5
80 nodes	DSR	8316.5	10056	7897
	TRUST	14496.5	13421	12329
100 nodes	DSR	16014	16237	16104
	TRUST	28094	23258	32103

The average number of salvaged packets is another very important criterion. The one shows how many packets are salvaged during the sending of the routing packets in disconnected environments. These packets could be resend to the destination nodes. In Fig. 8, we compare the number of salvaged packets for MANET and MANET with implementing trust mechanisms versus speed of mobile nodes. Result shows, that trust algorithm to selection of candidate node enables salvage a lot of packets in comparison with DSR protocol.

4. Conclusion

In this article, we proposed the trust algorithm for selection of the best candidate node in the hybrid MANET-DTN. The trust algorithm allows select secure mobile node for transportation of the data in a disconnected environment. The one also works in a decentralized manner. Hybrid MANET-DTN networks provides the new way how the different application could be provided for end users. The main idea of hybrid MANET-DTN networks enables use the network not

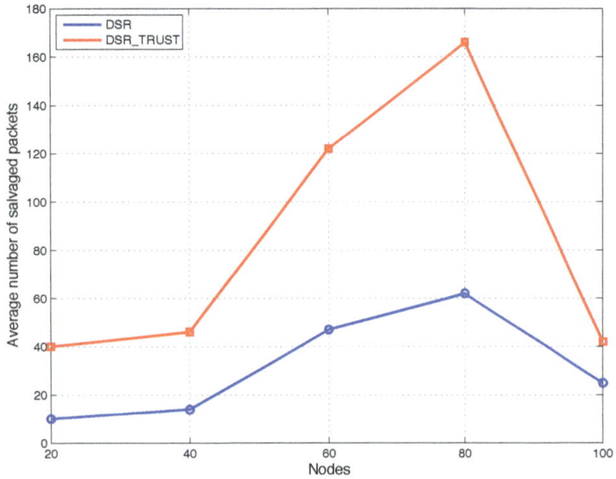

(a) 2 m·s^{-1}

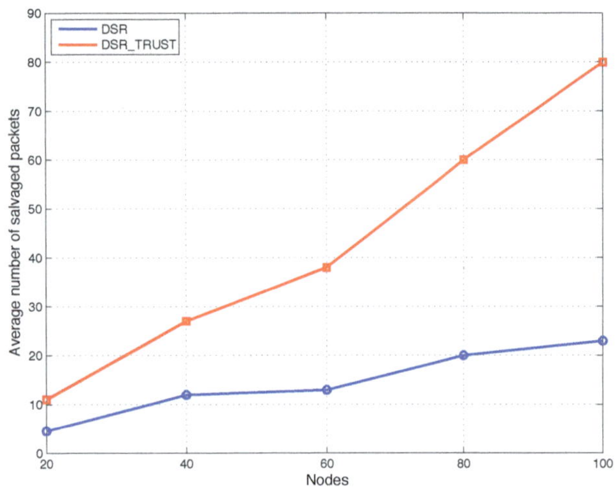

(b) 4 m·s^{-1}

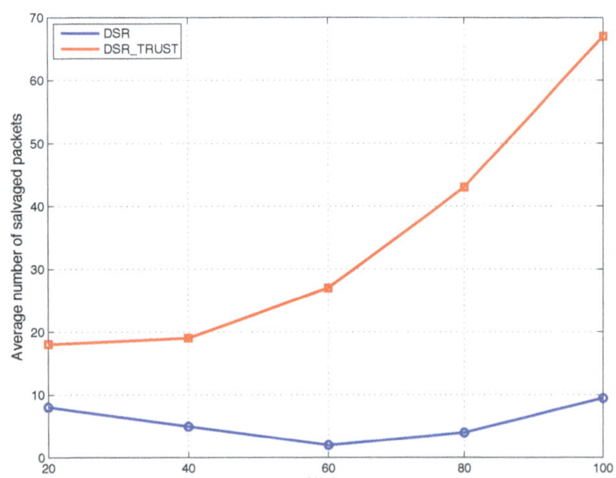

(c) 6 m·s^{-1}

Fig. 8: Average number of salvaged packets for different speed of mobile nodes.

only for personal usage but for emerging applications and services.

This article also gives us the basic performance analysis of the hybrid MANET-DTN networks. The models for MANET and DTN networks are implemented in OPNET modeler. Based on collecting results we can conclude trust algorithm for selection of the candidate node provides useful tool for selection of the optimal trusted path in the case of disconnected environment and also this algorithm allows enhance the performance of the hybrid MANET-DTN network from the routing point of view.

In the future we will focus on the design of the method of the candidate node selection based on game theory with regard to trust algorithm. The main idea of this algorithm is to combine routing properties of the MANET and DTN in order to increase the performance of the hybrid MANET-DTN network and also provides the secure transportation of the data across the disconnected environment.

Acknowledgment

This work has been performed partially in the framework of the Ministry of Education of Slovak Republic under research VEGA 1/0386/12 (20 %) and under research project ITMS-26220220155 supported by the Research & Development Operational Programme funded by the ERDF (80 %).

References

[1] CIZMAR, A., L. DOBOS and J. PAPAJ. Security and QoS Integration Model for MANETs. *Computing and Informatics*. 2012, vol. 31, no. 5, pp. 1025–1044. ISSN 1335-9150.

[2] PAPAJ, J., L. DOBOS and A. CIZMAR. Routing Strategies in Opportunistic Networks. *Journal of Electrical and Electronics Engineering*. 2012, vol. 5, no. 1, pp. 167–172. ISSN 1844-6035.

[3] CALEFFI, M. and L. PAURA. Opportunistic routing for disruption tolerant networks. In: *IEEE 23rd International Conference on Advanced Information Networking and Applications, AINA '09*. Bradford: IEEE, 2009, pp. 826–831. ISBN 978-1-4244-3999-7. DOI: 10.1109/WAINA.2009.201.

[4] BRIDA, P. and J. MACHAJ. A Novel Enhanced Positioning Trilateration Algorithm Implemented for Medical Implant In-Body Localization. *International Journal of Antennas and Propagation*. 2013, vol. 2013, no. 819695, pp. 1–10. ISSN 1687-5877. DOI: 10.1155/2013/819695.

[5] AMBROZIAK, S. and R. KATULSKI. An Empirical Propagation Model for Mobile Radio

Links in Container Terminal Environment. *IEEE Transactions On Vehicular Technology.* 2013, vol. 62, no. 9, pp. 4276–4287. ISSN 0018-9545. DOI: 10.1109/TVT.2013.2266618.

[6] LI, Y., W. CHEN and Z.-L. ZHANG. Optimal forwarder list selection in opportunistic routing. In: *IEEE 6th International Conference on Mobile Adhoc and Sensor Systems, 2009.* Macau: IEEE, 2009, pp. 670–675. ISBN 978-1-4244-5113-5. DOI: 10.1109/MOBHOC.2009.5336939.

[7] KURTH, J.-P., A. ZUBOW, and J.-P. REDLICH. Cooperative opportunistic routing using transmit diversity in wireless mesh networks. In: *The 27th Conference on Computer Communications.* Phoenix: IEEE, 2008, pp. 1310–1318. ISBN 978-1-4244-2025-4. DOI: 10.1109/INFOCOM.2008.188.

[8] LU, M., LI, F. and J. WU. Efficient opportunistic routing in utility-based ad hoc networks. *IEEE Transactions on Reliability.* 2009, vol. 58, iss. 3, pp. 485–495. ISSN 0018-9529. DOI: 10.1109/TR.2009.2020100.

[9] DUBOIS-FERRIERE, H., M. GROSSGLAUSER and M. VETTERLI. Least-cost opportunistic routing. In: *Proceedings of Allerton Conference on Communication, Control, and Computing, 2007.* Illinois: University of Illinois at Urbana-Champaign, 2007, pp. 1–8. ISBN 978-1605600864.

[10] GOVINDAN, K. and P. MOHAPATRA. Trust Computations and Trust Dynamics in Mobile Adhoc Networks: A Survey. *IEEE Communications Surveys & Tutorials.* 2012, vol. 14, iss. 2, pp. 279–298. ISSN 1553-877X. DOI: 10.1109/SURV.2011.042711.00083.

[11] TRIFUNOVIC, S., F. LEGENDRE and C. ANASTASIADES. Social trust in opportunistic networks. In: *INFOCOM IEEE Conference on Computer Communications Workshops.* San Diego: IEEE, 2010, pp. 1–6. ISBN 978-1-4244-6739-6. DOI: 10.1109/INFCOMW.2010.5466696.

[12] OPNET Modeler Simulation Software. *Riverbed* [online]. 2012. Available at: `http://www.opnet.com`.

About Authors

Jan PAPAJ was born in Liptovsky Mikulas, Slovak Republic in 1977. He is graduated at the Department of Computers and Informatics in 2001, Faculty of Electrical Engineering and Informatics at Technical University in Kosice. He obtained his Ph.D. degree in Telecommunications from the Faculty of Electrical Engineering, Technical University of Kosice in 2010. His research interests cover the fields of mobile ad-hoc networks, QoS and security, context and content routing protocols, opportunistic networks, delay tolerant networks, sensor networks.

Lubomir DOBOS was born in Vranov nad Toplou, Slovak Republic in 1956. He received the Ing. (M.Sc.) degree and Ph.D. degree in Radioelectronics from the Faculty of Electrical Engineering, Technical University of Kosice, in 1980 and 1989, respectively. He defended his habilitation work - Broadband Wireless Networks for Multimedia Services in 1999. His scientific research area is: adaptive noise cancellations in speech, broadband information and telecommunication technologies, multimedia systems, telecommunications networks and services, universal mobile communication systems (UMTS), 5G networks.

The Derivation of the Electrical Conductance/Temperature Dependency for Tin Dioxide Gas Sensor

Libor GAJDOSIK

Department of Telecommunications, Faculty of Electrical Engineering and Computer Science, VSB–Technical University of Ostrava, 17. listopadu 15/2172, 708 33 Ostrava-Poruba, Czech Republic

libor.gajdosik@vsb.cz

Abstract. *The dependency between the electrical conductivity and the temperature of the tin dioxide layer at the constant concentration of the detected gas is derived here. The derived equation is modified as a function of two variables describing the dependence of the electrical conductivity of the sensor on both the temperature and the concentration. The derived formulas were verified by approximation of the measured data. These formulas can be useful in practical applications.*

Keywords

Electrical conductance, gas sensor, modelling, tin dioxide.

1. Introduction

The tin dioxide sensors for the gas detection in air have been used for years. The detection process is based on the changes of the electrical conductivity of a tin dioxide layer heated to a sufficiently high temperature. Sensor response - it is a term used here to refer to the electrical conductance of the tin dioxide layer. The metal oxide detection layer is a polycrystalline structure. This is why the physical-chemical phenomena appearing on the surface of the layer are complex and it is very difficult to built the mathematical description of the behaviour of the sensor during the detection of the substance.

There are several methods for the description of the sensor behaviour. One group of the methods is based on the band theory. On its base the equations describing the electrical conductance of the sensor are derived in [1], [2], [3]. The advantage of the above-mentioned description is that it is general and universal. The disadvantage is its low objectivity and weak coherence with physical-chemical phenomena during the detection. The second possibility is using a pure mathematical approximation of the measured data. It is shown for example in [4], [5], [6]. The advantage of this approach is that it is not necessary to deal with the complex physical-chemical phenomena or the band theory. But the obtained equations are without any physical relation to the object being described. The third group deals with the physical-chemical phenomena. The examples of such an approach can be found in [7], [8], [9], [10], [11], [12], [13], [14], [15], [16], [17]. The advantages of this description are the objectivity and the relation to the physical-chemical phenomena. So the sensor behaviour may be better understood from the macroscopic point of view. On the other hand, the complexity of the obtained formulas and their accordance with the experimental values depends on the choice of the phenomena used for the behaviour of the sensor at the given conditions. This article describes the sensor response in terms of physical-chemical phenomena under simplified assumptions. The aim of this work is to conceive a simple model of the sensor behaviour, that fits in the sensor response and that can be useful for the sensor applications. This model extends the results described in [18].

2. Experiments

The commercial tin dioxide sensors of type TGS 813, TGS 822, SP 11 and MQR 1003 were chosen for the experiments. The sensors TGS 813, SP 11 and MQR 1003 are meant for detection of methane. TGS 822 is meant for the detection of ethanol, acetone and organic solvents. Different technologies were used for production of the sensors, so they have different levels of selectivity and sensitivity. The vapour of acetone, benzene, ethanol and hexane were used in the experiments, always only one compound in air. Each substance was tested by all sensors at the same time to ensure the equal concentration of the tested vapours and the equal value of the heating voltage for all sensors. The saturated vapour above the liquid was obtained at a constant ambient temperature of 25 °C. The desired concentration was prepared by injecting of a certain volume of saturated vapour into the tested chamber filled with the air of a known volume of 2700 ml according to the Antoine equation. The heating voltage was set to U=5 V and the sensors were inside the testing chamber in a flux of clean air before every measurement so as to remove the adsorbed undesired substances from the surface of SnO_2 [19], [20]. The used laboratory air was purged of dust and its relative humidity was decreased to RH=25 %. The recovery of the sensors was completed when the sensor response became practically independent of time. Related value of the sensor response never exceeded 10 µS. Afterwards, the heating voltage was decreased to $U = 2$ V by a step change to start sensing process, but to keep suitable temperature of the layer to avoid adsorption of water to the surface of SnO_2. According to the previous experience, the value of $U = 2$ V corresponds to the surface temperature of approximately 500 K. After that the tested vapour was injected into the chamber and the measurement was carried out. The heating voltage was increased step by step by 20 mV up to 5 V. The electrical conductance was sensed 10 s after the change of the heating voltage. The time of 10 s was established experimentally as the moment at which the value of the conductivity is practically independent of time. The electrical circuit for sensing the conductivity of the sensor was used according to the data sheet recommendations of the producers of the sensors. The resistor of $R = 2$ kΩ in series with the sensor and the supply voltage of $U = 5$ V were used. The temperature of the detection layer relevant to the range of the heating voltage of 2 V up to 5 V was estimated according to the temperature dependency of the alloy NiCr electrical resistance [21], which is widely used for the heating system of the given sensors. Furthermore, this estimated temperature was compared to the data from [22], where the temperature range for the detection of ethanol is described. Considering this, the temperature range related to the heating voltage $U = 2$ V up to 5 V corresponded to $T = 500$ K up to $T = 700$ K. The measurements were carried out at different values of concentration of the tested substances in air. The temperature characteristics of the sensor - it is a term used here to refer to the dependency of the electrical conductivity of the sensor on the temperature at the constant concentration.

3. Theory

Chemisorption of gases on the surface of a tin dioxide detection layer can be considered as monomolecular adsorption. So the process of chemisorption can be described by the Langmuir isotherm. On the basis of this, the electrical conductivity of the sensor Eq. (1) was derived in [18].

$$G(x) = G_0 + Z\frac{Kx}{1+Kx}, \quad \text{where } x = \frac{p_R}{p_T}10^6, \tag{1}$$

$$K = b\frac{p_T}{10^6}, \; Z = \frac{eN_Ak_1m_1a_{\max}}{U}, \tag{2}$$

where x is the relative concentration of the detected gas in ppm, G_0 is the basic electrical conductivity of the sensor in clean air, that is at $x = 0$, b is the adsorption coefficient in Pa^{-1}, p_T is the total pressure of the mixture in Pa, e is the electron electrical charge in C, N_A is the Avogadro constant in mol^{-1}, k_1 is the reaction rate constant in s^{-1}, m_1 is the mass of the detection layer SnO_2 in kg, $a_{\max}$ is the adsorption capacity of SnO_2 in $mol \cdot kg^{-1}$, U is the used measuring voltage in V and it is always kept constant during the experiments. It is assumed in Eq. (1) that the temperature of the detection layer is constant while the concentration x is a variable. In this article this equation is referred to as the transfer characteristics of the sensor. The examples of the reaction mechanisms of some substances on the surface during the detection are described in [23] ,[24], [25] ,[26] ,[27] ,[28]. Equation (1) is derived under the assumption that a system of subsequent reactions on the surface can be commonly expressed by one symbolic equation

$$\mathrm{R_{(g)} + O_{(ads)} \xrightarrow{k_1} D_{(g)} + }r\mathrm{V_O}, \tag{3}$$

where R represents the reducing gas being detected, O the adsorbed oxygen on the layer of SnO_2, D represents the final products formed during the detection, V_O are the oxygen vacancies (the released electrons changing the electrical conductivity of the sensor), r is a rational number.

The derivation described herein is based on the hypothesis, that the system of subsequent reactions during the detection can be represented by one symbolic reaction and the reactions during the detection in mentioned range of temperatures release free electron. This electron increases the electrical conductivity of the tin dioxide. Next hypothesis is, that the adsorption process of the detected substance and the desorption process of the reaction product are the features both for the transfer characteristic and for the temperature characteristic. All hypotheses and assumptions are supported with the experimental results presented herein.

Since there is a sufficient amount of the detected substance contained in the air surrounding of the sensor, the electrical conductivity of SnO_2 reaches the value that is relevant to the dynamic equilibrium at the given conditions. This value is practically independent of time.

It is necessary to modify Eq. (1) to be used for the derivation of temperature characteristics of the sensor. The concentration x will be constant and the temperature T of the detection layer of SnO_2 will be a variable. It means that coefficients G_0, Y and K become dependent on the temperature T. In general, the temperature dependency of the conductivity of a detection layer exists in clean air $x = 0$ as well [1]. The temperature dependencies of the conductivities of the tested sensors were found experimentally and they do not exceed the values of 10 µS at $x = 0$ in the range of $T = 500$ K up to 700 K. That value can be neglected compared to the usual values of the sensor response (100 µS up to 1000 µS) for used range of concentrations and range of temperatures. So the term G_0 in formula Eq. (1) can be neglected.

The temperature dependency of coefficients Z and K in Eq. (1) can be derived by using the kinetic theory of gases. Relative coverage of a surface is used for gas adsorption on a solid surface and it is defined by

$$\Theta = \frac{N}{N_t} = \frac{a}{a_{\max}}, \tag{4}$$

where N is the number of species adsorbed on the surface, N_t is the total number of the adsorption sites on the surface, a is the adsorption coefficient $\mathrm{mol \cdot kg^{-1}}$, $a_{\max}$ is the adsorption capacity of the surface $\mathrm{mol \cdot kg^{-1}}$. The adsorption process of the substance on the surface can be described by

$$\frac{\mathrm{d}\Theta}{\mathrm{d}t} = SF\sigma_0, \tag{5}$$

where S is the sticking coefficient, F is the number of the molecules striking the surface equal one per the time equal one, σ_0 is the area of one adsorption site. The Hertz Knudsen equation is valid for the ideal gas:

$$F = \frac{p}{\sqrt{2\pi m_1 kT}}, \tag{6}$$

where p is the gas pressure, m_1 is the mass of one gas molecule, k is the Boltzmann constant, T is the thermodynamic temperature. The sticking coefficient represents the probability that the struck molecule will remain adsorbed on the surface and will not get back to the gas phase. The coefficient is dependent on the temperature and on the coverage of the surface. In case that one molecule occupies only one adsorption site the formula is as follows:

$$S = S_0(1-\Theta)\mathrm{e}^{-\frac{E_A}{RT}}, \tag{7}$$

where S_0 is the probability of occupation of one site in case of a clean, still uncovered surface, that is $\Theta = 0$. S_0 usually equals 1. E_A is the activation energy of adsorption of gas in $\mathrm{J \cdot mol^{-1}}$, R is the universal gas constant, T is the thermodynamic temperature. For desorption of the substance out of the surface of coverage Θ, the Wigner Polanyi equation is used having the following form for the reaction of the first order:

$$\frac{\mathrm{d}\Theta}{\mathrm{d}t} = -k_2\Theta \quad \text{where} \quad k_2 = \nu_0 \mathrm{e}^{-\frac{E_d}{RT}}, \tag{8}$$

where E_d is the activation energy of desorption of the adsorbed substance in $\mathrm{J \cdot mol^{-1}}$, R is the universal gas constant, T is the thermodynamic temperature, k_2 is the rate constant of the reaction. ν_0 is the frequency factor in the Arrhenius equation. For the frequency factor the approximative formula according to the Frenkel equation [29] is valid:

$$\nu_0 \doteq \frac{kT}{h} \quad \frac{1}{\nu_0} = \tau_0, \tag{9}$$

where k is the Boltzmann constant, h is the Planck constant, T is the thermodynamic temperature, ν_0 represents the frequency of vibration of one molecule chemisorbed on the surface. If the energy delivered to amount-of-substance of 1 mol equals to or is greater than E_d, the bond between the surface and the molecule is broken and the molecule desorbs to the gas phase. Time τ_0 in [30] is referred to the time for which the molecule remains on the surface before leaving it. For the equilibrium state between adsorption and desorption according to Eq. (5) and Eq. (8) there is

$$SF\sigma_0 - k_2\Theta = 0. \tag{10}$$

By substitution from Eq. (6), Eq. (7) and Eq. (9) to Eq. (10) the following equation is obtained:

$$\frac{p(1-\Theta)\sigma_0}{\sqrt{2\pi k m_1 T}}\mathrm{e}^{-\frac{E_A}{RT}} = \frac{1}{\tau_0}\Theta \mathrm{e}^{-\frac{E_d}{RT}}. \tag{11}$$

The equation can be modified into the form of

$$K_A(1-\Theta)p = K_D\Theta \tag{12}$$

$$K_A = \frac{\sigma_0 \mathrm{e}^{-\frac{E_A}{RT}}}{\sqrt{2\pi mkT}} \qquad K_D = \frac{1}{\tau_0}\mathrm{e}^{-\frac{E_d}{RT}}. \tag{13}$$

From formula Eq. (12) the coverage of the surface Θ can be calculated:

$$\Theta = \frac{\frac{K_A}{K_D}p}{1+\frac{K_A}{K_D}p} = \frac{bp}{1+bp}. \tag{14}$$

Formula Eq. (14) is the Langmuir isotherm. Coefficient b can be calculated by substitution of K_A and K_B from Eq. (13) and we get

$$b = \frac{K_A}{K_D} = \frac{\tau_0\sigma_0}{\sqrt{2\pi m_1 kT}}\mathrm{e}^{\frac{E_d - E_A}{RT}} = \frac{\tau_0\sigma_0}{\sqrt{2\pi m_1 kT}}\mathrm{e}^{\frac{\Delta E}{RT}}, \tag{15}$$

by using the following formulas

$$k = \frac{R}{N_A} \qquad m_1 = \frac{M}{N_A}, \tag{16}$$

formula Eq. (15) can be modified to the form mentioned in [30], or in [31], that is to

$$b = \frac{\tau_0 \sigma_0 N_A}{\sqrt{2\pi MRT}} \mathrm{e}^{\frac{\Delta E}{RT}}, \tag{17}$$

where ΔE is the heat necessary for desorption of 1 mol of amount-of-mass of the substance (it practically equals the negative value of the adsorption heat), M is the molar mass of the adsorbed substance, N_A is the Avogadro constant, σ_0 is the area of one adsorption site, τ_0 is the time for the duration of which the adsorbed molecule stays on the surface, R is the universal gas constant, T is the thermodynamic temperature. At this moment the temperature dependency of coefficients Z and K in Eq. (2) can be derived. The Arrhenius equation is valid for the temperature dependency of the rate constant

$$k_1 = \nu_0 \mathrm{e}^{-\frac{E_1}{RT}} \quad \text{where} \quad \nu_0 = \frac{kT}{h}, \tag{18}$$

where E_1 in $\mathrm{J \cdot mol^{-1}}$ is the activation energy of the chemical reaction in Eq. (3). The following equation will be valid for Z after substitution of Eq. (18) into formula Eq. (2):

$$Z = \frac{eN_A kTm_1 a_{\max}}{hU} \mathrm{e}^{-\frac{E_1}{RT}}. \tag{19}$$

After substitution b from Eq. (17) and τ_0 from Eq. (9) and their inserting into Eq. (2) we obtain the following equation for K:

$$K = \frac{p_T h \sigma_0 N_A}{10^6 kT\sqrt{2\pi MRT}} \mathrm{e}^{-\frac{\Delta E}{RT}}. \tag{20}$$

And finally, after substitution Eq. (19) and Eq. (20) into Eq. (1), the equation for the temperature dependency of the detection layer will be in the following form:

$$G(T) \doteq Z\frac{Kx}{1+Kx} = C_0 T \mathrm{e}^{\frac{C_1}{T}} \frac{\frac{C_2}{T\sqrt{T}}\mathrm{e}^{\frac{C_3}{T}}}{1+\frac{C_2}{T\sqrt{T}}\mathrm{e}^{\frac{C_3}{T}}}, \tag{21}$$

where for coefficients C_0, C_1, C_2 and C_3 we get

$$C_0 = \frac{eN_A km_1 a_{\max}}{hU} \qquad C_1 = \frac{-E_1}{R}, \tag{22}$$

$$C_2 = \frac{\sigma_0 N_A h p_T x}{10^6 k\sqrt{2\pi MR}} \qquad C_3 = \frac{\Delta E}{R}. \tag{23}$$

Thus, from formulas Eq. (19) and Eq. (22) the following formulas emerge for coefficients Z and C_0:

$$\gamma_1 = \frac{Z}{C_0} = T\mathrm{e}^{\frac{-E_1}{RT}} = T\mathrm{e}^{\frac{C_1}{T}}, \tag{24}$$

and from formulas Eq. (20) and Eq. (23) the following formulas are valid for coefficients K and C_2:

$$\gamma_2 = \frac{K}{C_2} = \frac{1}{xT\sqrt{T}}\mathrm{e}^{\frac{\Delta E}{RT}} = \frac{1}{xT\sqrt{T}}\mathrm{e}^{\frac{C_3}{T}}. \tag{25}$$

4. Results

Equation (21) was used for approximation of the measured values of the electrical conductance of a tin dioxide layer at the concentration level $x = 600$ ppm at the temperature from 500 K up to 700 K. The computer programme XYMATH was used for the approximation by means of the least square method and the total error option was chosen as well. The correlation coefficient varied around $r = 0.98$ and the standard deviation was approximately 8 percent. That is acceptable. The example of approximation of the measured data for hexane and the used sensors are in Fig. 1, the example for different substances and sensor TGS 813 are in Fig. 2. The approximation of the measured data for different concentration values of acetone and sensor TGS 813 are in Fig. 3. These examples demonstrate good approximation ability of measured data by Eq. (21).

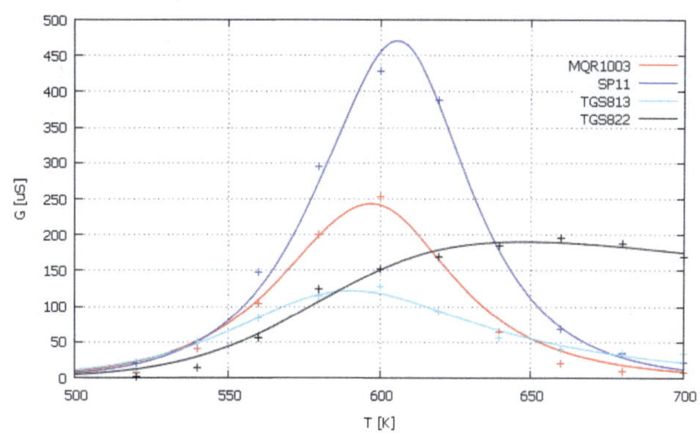

Fig. 1: The temperature dependencies of electrical conductivity G of the tested sensors for the concentration $x = 600$ ppm of hexane. The curves are approximations, the indicated points are the measured values.

The numerical values of the approximation coefficients of the temperature characteristics of the tested substances are in Tab. 1.

It is apparent from Tab. 1 that coefficients C_0 up to C_3 differ from each other. Coefficient C_0 is the highest and coefficient C_2 is the smallest one. The values of C_1 and C_3 are ordinarily comparable and the sign is opposite. C_1 is related to the activation energy of reaction Eq. (3) of the detected gas and oxygen. This value is always negative as, according to the Arrhenius equation, the temperature increases its value the reaction rate increases. The value of C_3 is related to the adsorption heat of the substance adsorbed on the surface of SnO_2. The value of the adsorption heat is negative because chemisorption is an exothermic process. It is

Tab. 1: The coefficients of approximations of the temperature dependency of the electrical conductance of the sensors for $x = 600$ ppm.

	C_0 (μS $\cdot$ K^{-1})	C_1 (K)	C_2 (K$^{3/2}$)	C_3 (K)
MQR 1003				
acetone	$3.07 \cdot 10^{10}$	$-1.36 \cdot 10^{4}$	$4.11 \cdot 10^{-18}$	$2.92 \cdot 10^{4}$
benzene	$1.05 \cdot 10^{13}$	$-1.78 \cdot 10^{4}$	$3.79 \cdot 10^{-25}$	$3.94 \cdot 10^{4}$
ethanol	$1.30 \cdot 10^{9}$	$-1.13 \cdot 10^{4}$	$1.43 \cdot 10^{-11}$	$1.92 \cdot 10^{4}$
hexane	$9.85 \cdot 10^{7}$	$-1.12 \cdot 10^{4}$	$1.02 \cdot 10^{-16}$	$2.79 \cdot 10^{4}$
SP 11				
acetone	$1.17 \cdot 10^{10}$	$-1.28 \cdot 10^{4}$	$4.06 \cdot 10^{-22}$	$3.52 \cdot 10^{4}$
benzene	$1.58 \cdot 10^{8}$	$-1.10 \cdot 10^{4}$	$2.47 \cdot 10^{-18}$	$3.00 \cdot 10^{4}$
ethanol	$1.29 \cdot 10^{6}$	$-7.61 \cdot 10^{3}$	$1.48 \cdot 10^{-14}$	$2.42 \cdot 10^{3}$
hexane	$3.22 \cdot 10^{9}$	$-1.31 \cdot 10^{4}$	$2.97 \cdot 10^{-20}$	$3.32 \cdot 10^{4}$
TGS 813				
acetone	$1.34 \cdot 10^{8}$	$-1.06 \cdot 10^{4}$	$3.35 \cdot 10^{-16}$	$2.70 \cdot 10^{4}$
benzene	$1.37 \cdot 10^{6}$	$-8.76 \cdot 10^{3}$	$4.64 \cdot 10^{-13}$	$2.30 \cdot 10^{4}$
ethanol	$5.03 \cdot 10^{4}$	$-5.94 \cdot 10^{3}$	$4.70 \cdot 10^{-12}$	$2.10 \cdot 10^{4}$
hexane	$1.09 \cdot 10^{7}$	$-1.00 \cdot 10^{4}$	$3.34 \cdot 10^{-10}$	$1.84 \cdot 10^{4}$
TGS 822				
acetone	$9.33 \cdot 10^{5}$	$-1.72 \cdot 10^{3}$	$1.88 \cdot 10^{-8}$	$1.70 \cdot 10^{4}$
benzene	$3.56 \cdot 10^{11}$	$-1.64 \cdot 10^{4}$	$1.92 \cdot 10^{-9}$	$1.78 \cdot 10^{4}$
ethanol	$1.71 \cdot 10^{9}$	$-1.14 \cdot 10^{4}$	$4.22 \cdot 10^{-9}$	$1.63 \cdot 10^{4}$
hexane	$8.41 \cdot 10^{8}$	$-1.26 \cdot 10^{4}$	$1.12 \cdot 10^{-6}$	$1.34 \cdot 10^{4}$

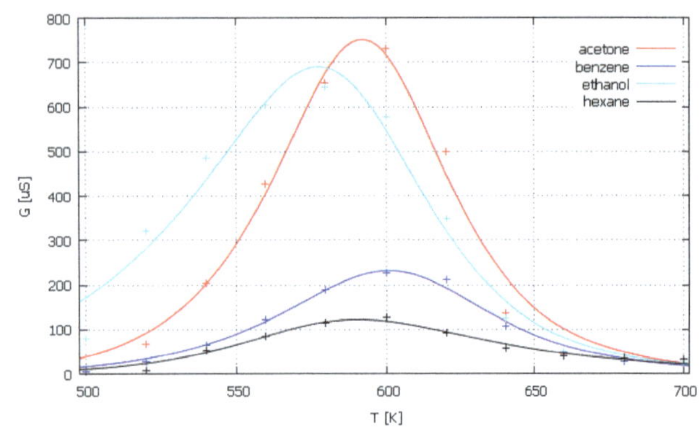

Fig. 2: The temperature dependencies of the electrical conductivity G of sensor TGS 813 for the concentration $x = 600$ ppm for different substances. The curves are approximations, the indicated points are the measured values.

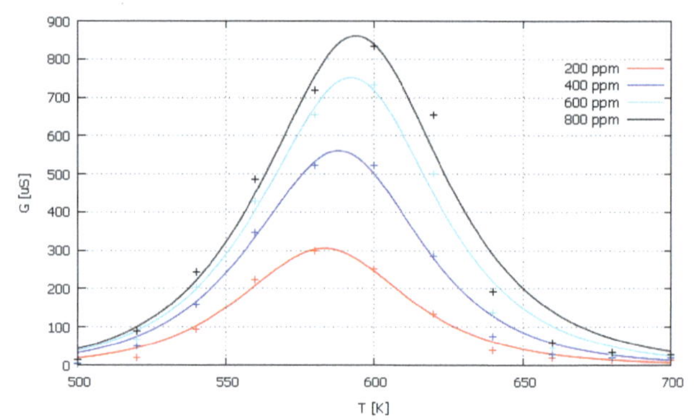

Fig. 3: The temperature dependencies of electrical conductivity G of sensor TGS 813 for different concentrations of acetone in air. The curves are approximations, the indicated points are the measured values.

necessary to supply this heat to the chemisorbed substance during the process of desorption to enable the substance to leave the surface. That is why coefficient C_3 is positive. Coefficients C_0 are ordinarily of high values and, on the other hand, coefficients C_2 are ordinarily of very low values. It is related to the values of the constants that occur in formulas Eq. (22) and Eq. (23). It emerges from Tab. 1 that the values of the coefficients depend both on the type of a sensor and on the type of a detected substance. The values of coefficients C_0 up to C_3 form a set of four values. This set is typical for the given substance, sensor and concentration. The set of four coefficients can help to discriminate the detected substance.

On the base of the derived formula Eq. (21) and its verification by the experimental results it is possible to suggest simplified behaviour of the sensor during the detection. If temperature of the layer increases its value the rate constant k_1 of reaction Eq. (3) increases. That is why the number of the released electrons increases. This leads to increasing the conductivity of the layer of SnO_2. When the maximum value of conductivity is reached, the sensor response starts to decrease together with the rise of the temperature. Behind the maximum value of the sensor response the desorption influence begins to prevail since the rate constant k_2 in Eq. (8) increases its value together with the rising temperature. The molecules of the adsorbed substance oscillate more intensively and it is more difficult for them

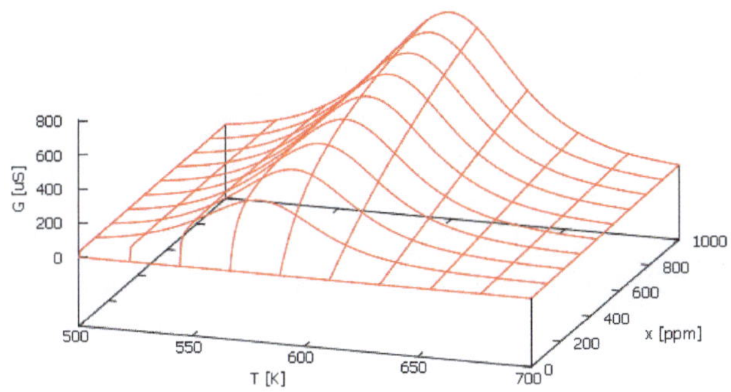

Fig. 4: The dependency of electrical conductance G of sensor TGS 813 on temperature T and gas concentration x for acetone.

to remain on the surface. Their number decreases with the rising temperature and due to this the amount-of-substance R in Eq. (3) decreases. That is why the number of the electrons released by chemical reaction Eq. (3) decreases as well and the conductivity of the layer decreases. These phenomena form the bell shape of the temperature characteristics, which is typical for most of the detected substances. Some substances have the increasing dependency only at the tested range of the temperature as apparent for hexane in Fig. 1. The hypothesis for explanation is, that the maximum of the detection curve of these substances is shifted towards the higher temperature that lies out of the tested temperature range. The maximum value of the response theoretically exists according the Eq. (21), but the observation of this maximum is limited by the maximum permissible value of the heating voltage so as not to damage the sensor.

Equation (21) manifests the typical features of the sensor. Then one equation of two variables can be formed from formulas Eq. (1) and Eq. (21):

$$G(T,x) = C_0 T e^{\frac{C_1}{T}} \frac{\frac{C_2^*}{T\sqrt{T}} x e^{\frac{C_3}{T}}}{1 + \frac{C_2^*}{T\sqrt{T}} x e^{\frac{C_3}{T}}}$$

$$\text{where} \quad C_2^* = \frac{C_2}{x_0}, \tag{26}$$

where x_0 is the reference value of the concentration. The dependency of the electrical conductivity of sensor TGS 813 for ethanol is in Fig. 4. The coefficients from Tab. 1 and the reference value $x_0 = 600$ ppm were used here.

5. Conclusion

The equation describing the temperature dependency of the electrical conductance of a tin dioxide layer for the detection of the reducing gas is derived here. The derivation is based on the physical-chemical phenomena. Next research can be focused to theoretical justification of the simplified assumptions used herein. The derived equation was successfully verified by comparing with the experimental data. Reached results can be useful in practical applications.

Acknowledgment

This article was enabled due to Project GAP 108/11/1057 entitled Synthesis, Structure and Properties of Nanocomposites Conducting Polymer/Phyllosilicate provided by GACR (The Czech Science Foundation). The research has been partially supported by the project No. CZ.1.07/2.3.00/20.0217 "The Development of Excellence of the Telecommunication Research Team in Relation to International Cooperation" within the frame of the operation programme "Education for competitiveness" financed by the Structural Funds and from the state budget of the Czech Republic.

References

[1] MORRISSON, S. R. Measurement of surface state energy levels of one-equivalent adsorbates on ZnO. *Surface science.* 1971, vol. 27, iss. 3, pp. 586–604. ISSN 0039-6028.

[2] MORRISSON, S. R. *The chemical physics of surfaces.* ed. 2. New York: Plenum Press, 1990. ISBN 03-064-3549-7.

[3] YAMAZOE, N. and K. SHIMANOE. Theory of power laws for semiconductor gas sensors. *Sensors and Actuators B: Chemical.* 2008, vol. 128, iss. 2, pp. 566–573. ISSN 0925-4005. DOI: 10.1016/j.snb.2007.07.036.

[4] HIROBAYASHI, S., M. A. KADIR, T. YOSHIZAWA and T. YAMABUCHI. Verification of a logarithmic model for estimation of gas concentrations in a mixture for a tin oxide gas sensor response. *Sensors and Actuators B: Chemical.* 2003, vol. 92, iss. 3, pp. 269–278. ISSN 0925-4005. DOI: 10.1016/S0925-4005(03)00311-3.

[5] WLODEK, S., K. COLBOW and F. CONSADORI. Signal-shape analysis of a thermally cycled tin-oxide gas sensor. *Sensors and Actuators B: Chemical.* 1991, vol. 1, iss. 1, pp. 63–68. ISSN 0925-4005.

[6] SEARS, W. M., K. COLBOW and F. CONSADORI. Algorithms to improve the selectivity

of thermally-cycled tin oxide gas sensors. *Sensors and Actuators.* 1989, vol. 19, iss. 4, pp. 333–349. ISSN 0925-4005.

[7] GARDNER, J. W. A non-linear diffusion reaction model of electrical conduction in semiconductor gas sensors. *Sensors and Actuators B: Chemical.* 1990, vol. 1, iss. 4, pp. 166–170. ISSN 0925-4005.

[8] WINDISCHMANN, H. P. and P. MARK. A model for the operation of a thin film SnO_2 conductance modulation carbon monoxide sensor. *Solid state science and technology.* 1979, vol. 126, iss. 4, pp. 627–633. ISSN 2162-8769. DOI: 10.1149/1.2129098.

[9] GULATI, S., N. MEHAN, D. P. GOYAL and A. MANSINGH. Electrical equivalent model for SnO_2 bulk sensors. *Sensors and Actuators B: Chemical.* 2002, vol. 87, iss. 2, pp. 309–320. ISSN 0925-4005. DOI: 10.1016/S0925-4005(02)00251-4

[10] HORN, M. A new theory of adsorption for the quantitative description of gas sensors. *Sensors and Actuators B: Chemical.* 1995, vol. 26, iss. 1–3, pp. 217–219. ISSN 0925-4005.

[11] SAKAI, G., N. MATSUNAGA, K. SHIMANOE and N. YAMAZOE. Theory of gas-diffusion controlled sensitivity for thin film semiconductor gas sensor. *Sensors and Actuators B: Chemical.* 2001, vol. 80, iss. 2, pp. 125–131. ISSN 0925-4005. DOI: 10.1016/S0925-4005(01)00890-5.

[12] YAMAZOE, N. and K. SHIMANOE. Theoretical approach to the gas response of oxide semiconductor film devices under control of gas diffusion and reaction effects. *Sensors and Actuators B: Chemical.* 2011, vol. 154, iss. 2, pp. 277–282. ISSN 0925-4005. DOI: 10.1016/j.snb.2010.01.018.

[13] NAKATA, S., K. TAKEMURA and K. NEYA. Non-linear dynamic responses of a semiconductor gas sensor: Evaluation of kinetic parameters and competition effect on the sensor response. *Sensors and Actuators B: Chemical.* 2001, vol. 76, iss. 1–3, pp. 436–441. ISSN 0925-4005. DOI: 10.1016/S0925-4005(01)00652-9.

[14] ZHANG, S., C. XIE, H. LI, Z. BAI, X. XIA and D. ZENG. A reaction model of metal oxide gas sensors and a recognition method by pattern matching. *Sensors and Actuators B: Chemical.* 2009, vol. 135, iss. 2, pp. 552–559. ISSN 0925-4005. DOI: 10.1016/j.snb.2008.10.021.

[15] AHLERS, S., G. MULLER and T. DOLL. A rate equation approach to the gas sensitivity of thin film metal oxide materials. *Sensors and Actuators B: Chemical.* 2005, vol. 107, iss. 2, pp. 587–599. ISSN 0925-4005. DOI: 10.1016/j.snb.2004.11.020.

[16] BECKER, T., S. AHLERS, C. BOSCH, V. BRAUNMUHL, G. MULLER and O. KIESEWETTER. Gas sensing properties of thin- and thick-film tin-oxide materials. *Sensors and Actuators B: Chemical.* 2001, vol. 77, iss. 1–2, pp. 55-61. ISSN 0925-4005. DOI: 10.1016/S0925-4005(01)00672-4.

[17] VILANOVA, X., E. LLOBET, R. ALCUBILLA, J. E. SUEIRAS and X. CORREIG. Analysis of the conductance transient in thick-film tin oxide gas sensors. *Sensors and Actuators B: Chemical.* 1996, vol. 31, iss. 3, pp. 175–180. ISSN 0925-4005. DOI: 10.1016/0925-4005(96)80063-3.

[18] GAJDOSIK, L. The derivation of the electrical conductance/concentration dependency for SnO_2 gas sensor for ethanol. *Sensors and Actuators B: Chemical.* 2002, vol. 81, iss. 2–3, pp. 347–350. ISSN 0925-4005. DOI: 10.1016/S0925-4005(01)00980-7.

[19] WATSON, J., K. IHOKURA and G. S. V. COLES. The tin dioxide gas sensor *Measurement Science and Technology.* 1993, vol. 4, iss. 7, pp. 711–719. ISSN 1361-6501. DOI: 10.1088/0957-0233/4/7/001

[20] CHAKRABORTY, S., I. MANDAL, I. RAY, S. MAJUMDAR, A. SEN and H. S. MAITI. Improvement of recovery time of nanostructured tin dioxide-based thick film gas sensors through surface modification. *Sensors and Actuators B: Chemical.* 2007, vol. 127, iss. 2, pp. 554–558. ISSN 0925-4005. DOI: 10.1016/j.snb.2007.05.005.

[21] *ASM speciality handbook: Nickel, cobalt, and their alloys.* Materials Park: ASM International, 2000. ISBN 978-0-87170-685-0.

[22] HAE-WON, C. and L. MAN-JONG. Sensing characteristics and surface reaction mechanism of alcohol sensors based on doped SnO_2. *Journal of Ceramic Processing Research.* 2006, vol. 7, iss. 3, pp. 183-191. ISSN 1229-9162.

[23] BERGER, F., J. B. SANCHEZ and O. HEINTZ. Detection of hydrogen fluoride using SnO_2-based gas sensors: Understanding of the reactional mechanism. *Sensors and Actuators B: Chemical.* 2009, vol. 143, iss. 1, pp. 152–157. ISSN 0925-4005. DOI: 10.1016/j.snb.2009.08.054.

[24] TOURNIER, G. and C. PIJOLAT. Influence of oxygen concentration in the carrier gas on the response of tin dioxide sensor under hydrogen and methane. *Sensors and Actuators B: Chemical.*

1999, vol. 61, iss. 1–3, pp. 43–50. ISSN 0925-4005. DOI: 10.1016/S0925-4005(99)00278-6.

[25] SBERVEGLIERI, G. *Gas sensors: principles, operation, and developments.* Boston: Kulwer Academic Publishers, 1992. ISBN 07-923-2004-2.

[26] KOHL, D., T. OCHS, W. GEYER, M. FLEISCHER and H. MEIXNER. Adsorption and decomposition of methane on gallium oxide films. *Sensors and Actuators B: Chemical.* 1999, vol. 59, iss. 2–3, pp. 140–145. ISSN 0925-4005. DOI: 10.1016/S0925-4005(99)00211-7.

[27] RUHLAND, B., T. BECKER. and G. MULLER. Gas-kinetic interactions of nitrous oxides with SnO_2 surfaces. *Sensors and Actuators B: Chemical.* 1998, vol. 50, iss. 1, pp. 85–94. ISSN 0925-4005.

[28] KOZIEJ, D., N. BARSAN, V. HOFFMANN, J. SZUBER and U. WEIMAR. Complementary phenomenological and spectroscopic studies of propane sensing with tin dioxide based sensors. *Sensors and Actuators B: Chemical.* 2005, vol. 108, iss. 1–3, pp. 75–83. ISSN 0925-4005. DOI: 10.1016/j.snb.2004.12.076.

[29] FRENKEL, J. Theorie der Adsorption und Verwandter Erscheinungen. *Zeitschrift fuer Physik.* 1924, vol. 26, iss. 1, pp. 117–138. ISSN 1434-6001. DOI: 10.1007/BF01327320.

[30] ADAMSON, A. W. *Physical chemistry of surfaces.* 5th ed. New York: Wiley, 1990. ISBN 0-471-61019-4.

[31] PROTODJAKONOV, I. O. and S. V. SIPAROV. *The mechanics of adsorption in the systems gas and solid phase (in russian).* 1st ed. Leningrad: Nauka, 1895.

About Authors

Libor GAJDOSIK graduated in telecommunication engineering from the Czech Technical University (Prague, Czech Republic) in 1983. He received his Ph.D. in 1998 from the Technical University of Ostrava (Ostrava, Czech Republic). His thesis was focused on the detection properties of tin dioxide gas sensors at the dynamic operating modes. He has been an assistant professor at the Technical University of Ostrava since 1989. His main interests are electronics, circuit theory and chemical sensors.

23

On Ladder Diagrams Compilation and Synthesis to FPGA Implemented Reconfigurable Logic Controller

Adam MILIK

Institute of Electronics, Silesian University of Technology, Akademicka 16, 44 100 Gliwice, Poland

adam.milik@polsl.pl

Abstract. *The paper presents synthesis process of a hardware implemented reconfigurable logic controller from a ladder diagram according to IEC61131-3 requirements. It is focused on the originally developed a high-performance LD processing method. It is able to process a set of diagrams restricted to logic operations in a single clock cycle independently from the number of processed rungs. The paper considers the compilation of the ladder diagram into an intermediate form suitable for logic synthesis process according to developed processing method. The enhanced data flow graph (EDFG) has been developed for the intermediate representation of an LD program. The original construction of the EDFG with attributed edges has been described. It allows for efficient representation and processing of logic and arithmetic formulas. The set of compilation algorithms that allow to preserve serial analysis order and to obtain massively parallel processing unit are presented. The overview of a hardware mapping concludes the presented considerations.*

Keywords

DFG, FPGA, high-level synthesis, IEC61131-3, LD, logic synthesis, PLC, reconfigurable hardware.

1. Introduction

The Programmable Logic Controllers (PLC) have been used since 1970s and first they were applied to relay control systems. Within years of fast development of electronic technology, the requirements given to PLC become higher all the time (operating speed, handling of analog objects, the increasing reliability, etc.). Today, the areas of PLC applications include small complexity processes as well as large manufacturing lines.

The general concept of a PLC is based on the microprogrammable circuits. It consists of two inseparable parts that are a hardware platform and software. The Hardware platform is able to execute given set of logic and arithmetic instructions. A control algorithm is created in the form of instructions sequence [1], [2], [8].

In contrast to software centric solutions, hardware offers intrinsic parallel execution of the tasks. It radically reduces the response time and offers better performance than software solutions. The implementation of the control algorithm with the use of reprogrammable and reconfigurable logic has been proposed by different research groups [3], [4], [5], [9], [11], [13], [15], [16], [19]. There have been proposed a custom FPGA architecture for direct mapping of the LD logic [17]. The significant limitation in wide use of reprogrammable digital circuits is a high design complexity of the implementation processes (in comparison to the instruction based standard approach).

A set of tools for creating a reconfigurable controller and its direct programming with well defined and commonly used ladder diagram has been developed. Presented work concentrates on transforming of control algorithm designed with the use of the ladder diagram into a form suitable for the entire process of hardware implementation incorporating: optimization, scheduling and hardware mapping. There has been considered details of the LD program execution according to IEC61131-3 requirements.

An intermediate form of the control program has been developed according to considered standard requirements. Algorithms, presented in this paper, are a part of the developed concept of reconfigurable logic controllers families and toolset for their programming according to the IEC61131-3 reference manual.

2. The LD Execution Model

The LD network is widely used method for describing control algorithms [1]. This method has been inherited from relay control systems. Contacts and coils represent logic dependencies between signals and function blocks. According to IEC61131-3 requirements a network is analyzed in a row based fashion that evaluates power flow through components rung by rung. In contrast to the electrical schematic diagram the power flow is unidirectional. The power is transmitted from the left rail to the right. There are implied limitations that prohibit of reverse power flow in ladder schematic [8]. Sequential analysis of schematic produces an ordered sequence of instructions for a PLC (Fig. 1). A switch is transformed into a logic AND operation. This operation is performed between current result coming from the predeceasing node and a signal that controls the switch. A junction merges power flow coming from several rungs. In some cases, there is a need for creating variables that enables storage of partial results for nested operations.

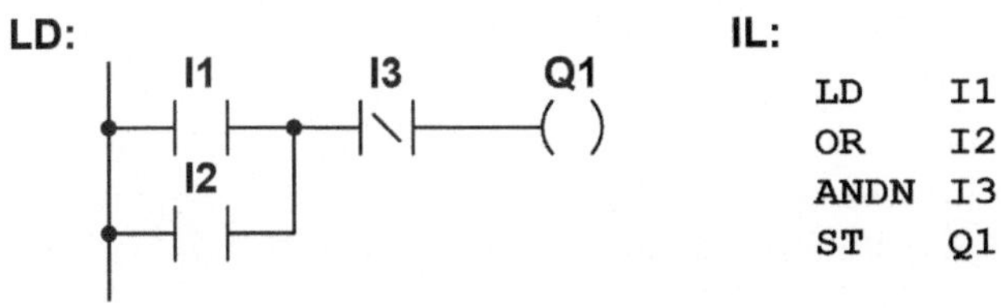

Fig. 1: The ladder diagram and its equivalent instruction sequence.

The LD network processing speed can be increased by parallel execution of the logic operation in programmable hardware. Transforming logic dependencies into combinatorial logic allows increasing performance several orders of magnitude. It is required to develop a universal method suitable for representing not only LD programs, but other programming methods specified by IEC61131. This method should be able to synthesize logic operation, but also other operations performed by PLCs (e.g. timers, counters, arithmetic operations).

2.1. Existing Synthesis Models for LD

An LD diagram is described by two sets of Boolean variables I and Q. The set I consists of variables associated with inputs while the set Q consists of variables associated with outputs and internal markers. The logic functions are defined by rungs and create an ordered sequence of Boolean expressions:

$$q_i = f_i(I, Q), i = 1...r, \tag{1}$$

where i is the rung index. Equation (1) defines the ordered sequence of processing according to the index i. This feature has been utilized in implementations proposed by [9], [10], [16].

An exemplary LD network and its implementation have been presented in the figure (Fig. 2). In this model each rung is processed in individual cycle. The controller response time is proportional to the number of rungs in a program. In comparison to the programmatic approach, this model reduces a computation time of logic functions. It can be noticed that calculations of some variables can be processed in parallel. Distributing calculation process for each rung (q variable) introduces redundant cycles. In considered diagram (Fig. 2) variables q_1 and q_3 do not depend on other q variables. The q_1 variable and the q_3 can be evaluated in the first cycle (t_1).

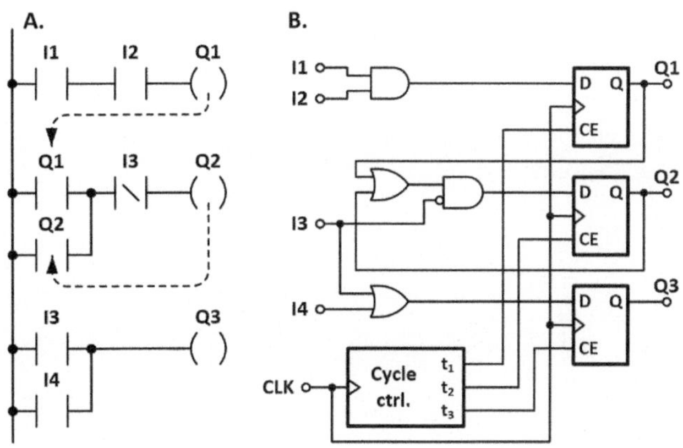

Fig. 2: The LD network (A) and its equivalent (B).

In order to reduce the number of calculation cycles dependencies between q_i variables have to be determined. In the paper [6] authors introduced an idea of using dependencies and simultaneities graphs for creating a sequential functional chart (SFC) from given LD. This idea has been employed in [3] for creating optimized hardware description. Similar idea has been employed in [15] for control algorithm partitioning. During the analysis of the LD, a dependencies graph is created. This is a directed graph that consists of nodes representing all q_i variables. The node v_i (representing variable q_i) is connected by directed edge with node v_j only if function f_i depends on variable q_j and $i > j$:

$$(v_j, v_i) \leftrightarrow f_i(q_i) \neq const. \tag{2}$$

The number of elementary cycles based on dependencies analysis is equal to:

$$T = p_{max} + 1, \tag{3}$$

where p_{max} is the longest path in the dependencies graph.

2.2. The LD High-Performance Synthesis Model

Presented dependency analysis of the ladder diagram is directly derived from sequential execution model. There are considered rungs as independent units delivering variables value. The early approaches consider each rung to be dependant of predeceasing rungs. The execution is performed in serial fashion a rung by a cycle. Applying rung dependencies analysis allows to determine the calculation dependencies in the form of the graph which is used for improved rungs scheduling.

The LD can be considered as a sequence of operations. Let assume that variables associated with inputs are updated before the start of the calculation process and remain constant during it. Let introduce the set of variables D that are assigned with a value of processed expressions. Value of the variable d_i is assigned to variable q_i at the end of calculation cycle ($q_i = d_i$). This approach allows to distinguish between two values that are calculated in the present cycle (d_i) and in the previous cycle (q_i). Equation (1) for m^{th} rung can be rewritten in following form:

$$d_m = f_m(I, d_0, ..., d_{m-1}, q_m, ..., q_n), \qquad (4)$$

$$q_m = d_m. \qquad (5)$$

Using proposed substitution of q variables allows to propagate calculation results through all functions bypassing registers (Fig. 3). The current value of control process is updated by single clock pulse after calculating all d_i values. In presented form the calculation process is fully parallel and completes in a single cycle that transfers values from d to respective q variables.

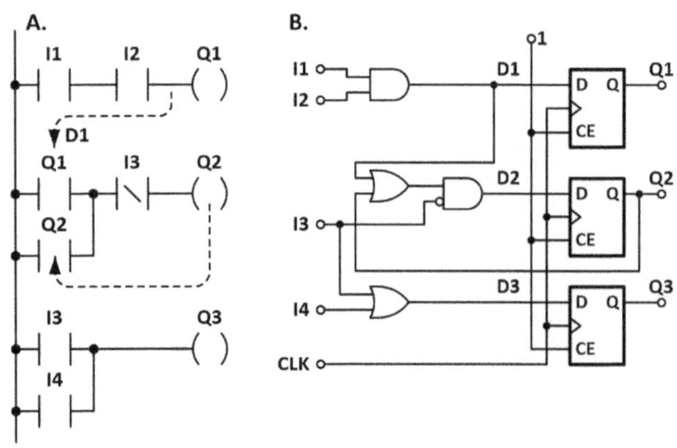

Fig. 3: The LD network (A) and its hardware equivalent obtained with proposed synthesis method (B).

3. Intermediate Representation with the Use of Data Flow Graphs

It is required to develop appropriate representation for an intermediate form of control algorithm that is suitable for high-level synthesis process. The intermediate form should be able to represent logic and arithmetic operations performed by PLCs maintaining operation sequence and reviling its dependencies. Commonly used form of intermediate representations for logic synthesis and compilers are data flow graphs [7]. A node of the graph represents elementary operation while directed edges indicate processing flow between nodes.

3.1. The EDFG Concept

For the purpose of recording PLC programs, the author has developed a form of enhanced data flow graph (EDFG). This has been inspired by concept of attributed edges used in BDD introduced by Minato [14]. In a similar way the EDFG handle unary operations like logic inversion and arithmetic complement. The other implemented extension is a multiple argument node for commutative operations. Presented modifications allow for efficient creating of data flow graph and reduces algorithmic complexity.

The Extended Data Flow Graph is given by $G = \langle V, E \rangle$ where V is a set of nodes representing elementary operations and E is a set of directed edges with attributes. The directed edge $e \in E$ is described by triple $e = \langle v_S, v_D, a \rangle$ where v_S is a predeceasing node and v_D is a successor node of the directed edge. The a is an attribute of the selected from the set A of allowed attributes.

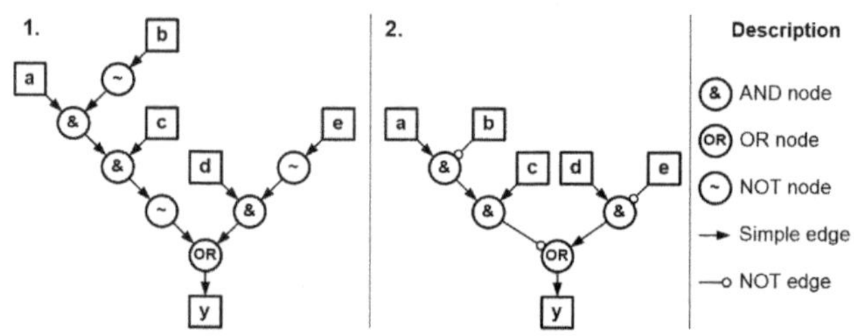

Fig. 4: Comparison of the general DFG (1) with EDFG (2).

An equivalent DFG to the Boolean formula $y = \overline{a \cdot \bar{b} \cdot c} + d \cdot \bar{e}$ is presented in the figure (Fig. 4). There have been considered two cases: a standard approach DFG (1) and with the use of the EDFG (2). The DFG (1) implements logic inversion by separate NOT nodes. Introducing attributed edges with logic inversion eliminates a NOT node EDFG (2). The attributed edge not only reduces the number of nodes in the diagram, but also allow to simplify logic operation handling.

Figure 5 shows the graph transformations for logic nodes. One of the common operation after compilation process is operation merge. Using EDFG simplifies the algorithms of node merging. Predeceasing node can be merged if it is connected with simple edge and both nodes implement the same logic operation. (shown in Fig. 5.1.). The reference node is marked with gray color. The operation is performed by modifying graph edges. There is exchanged v_D that equals v_1 to v_2. The merge rule can be extended with the use of de Morgan's laws. In the case (Fig. 5.2.) the predeceasing node is connected with inverted edge and implements opposite logic function (AND↔OR) to the reference node. The operation is performed by exchanging the v_D that equals v_2 to v_4 and the edge attribute is inverted. Finally, the v_2 node is removed The final EDFG is presented in Fig. 5.3.

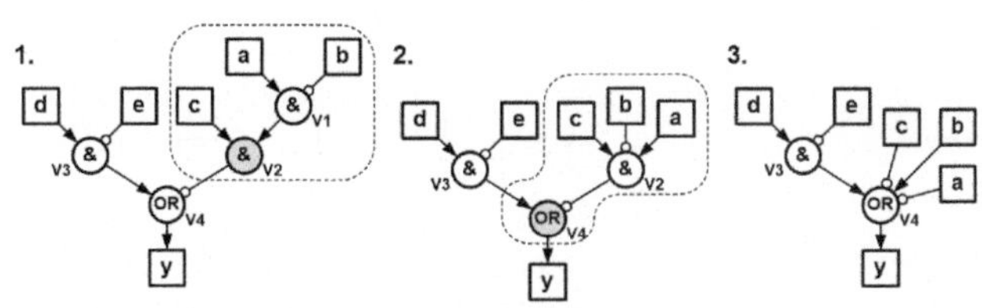

Fig. 5: Implementation of node merge (1) and de Morgan's laws (2) in DFG with attributed edges.

Thanks to attributed edges similar flexibility is achieved for arithmetic operations. In the domain of arithmetic operations, the subtraction node is replaced by an edge with complement value attribute. It reduces the set of arithmetic operations to: addition, multiplication and division. The figure (Fig. 6) shows the implementation of the expression: $y = a+b-c+d-e$ and compares the use of attributed edges for arithmetic operations. Using a standard approach with separate addition and subtraction nodes is shown in Fig. 6.A. Similar result is achieved using attributed edges (Fig. 6.B). The attributed edges simplify algorithms of operation merge and constant propagation as shown in Fig. 6.C.

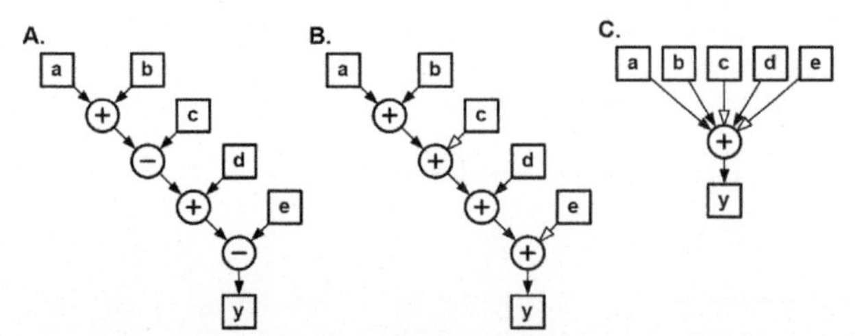

Fig. 6: Comparison of the general DFG (A) and the EDFG (B, C).

4. Converting LD to EDFG

The compilation process delivers basic items of a language [18]. Subsequent algorithms show systematic methods of translating those items into an EDFG suitable for hardware mapping.

4.1. Variables

The variables are declared at the beginning of a network according to IEC61131-3 requirements. For the purpose of the synthesis process the variables set is divided into three subsets. The variable classification is based on the signal association to input, output and internal marker areas. The variable reduction procedure takes into consideration variables membership. Variables associated with input signals are allowed to be read while value assignment is implied and made from input signals. Variables associated with outputs and markers are allowed for read and write access. The variables associated with markers can be eliminated when only write access is detected. There are two possible cases. The first one when a variable is used as a temporary storage for distributing the value and the other one when a variable is unused. The unused variable is distinguished as the only sink for the driving node. Variables associated with output signals and markers are not allowed to be read without value assignment. The moment of assignment is independent of the read access but must be completed at least once in entire calculation cycle.

The variable value access implementation assures sequential variable access according to LD description. In order to satisfy this requirement, a variable current value is accessed by following algorithm.

Algorithm 1: Let the x is a variable that the value is going to be read by node v, v_{xWR} is an assignment node to the x variable, v_{xDRV} is the node delivering value to the variable x. The variable x refers to EDFG nodes through the table pointing read and write nodes (Fig. 7). The v_{DRV} (if exists) is connected with a directed edge with v_{xWR}. The v_{xRD} is a value reading node of the x variable. Following two cases are possible depending on the value assignment sequence. If the variable x value is not assigned than the v_{xWR} node does not exist. The value of the x variable is read by creating the v_{xRD} node (this notifies that the value is coming from the previous cycle – e.g. forward coil reference). The v_{xRD} node becomes the argument of the v node (Fig. 7.1). If the variable x is assigned the v_{xWR} node exists. The directed edge connects it with driving node v_{DRV}. The v_{DRV} is used for driving the v node (Fig. 7.2). The attribute of the edge is inherited. It should be noticed that this mechanism follows recently assigned value to the x variable.

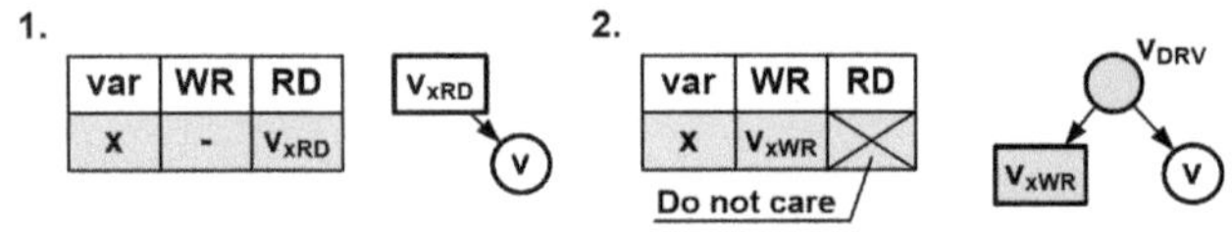

Fig. 7: The variable value access algorithm.

4.2. Nodes

The ladder diagram node merges power flow coming from multiple sources. It should deliver a logic sum of all connected signals. A general algorithm that accepts multiple drivers for a node is used. Process starts from the variable declaration. The variable name is inherited from network node name (automatically generated). It is included into a subset of internal signals.

Algorithm 2: Let the x is a variable associated with a schematic node (junction), v_{xWR} is the x variable value assignment node, v_C is the graph node that is going to drive the x variable, v_P is the node that currently drives the x variable. There are two possible cases. If the x variable is not assigned than the v_{xWR} node does not exist. The v_C node is connected with newly created v_{xWR} node (Fig. 8.1). If the x variable is already assigned then the v_{xWR} node exists. In this situation, the v_{OR} node is created. Nodes v_P and v_C are connected to v_{OR} node. The v_{OR} node becomes the only driver of v_{xWR} (Fig. 8.2). The described algorithm can be repeated iteratively for nodes with multiple driving sources.

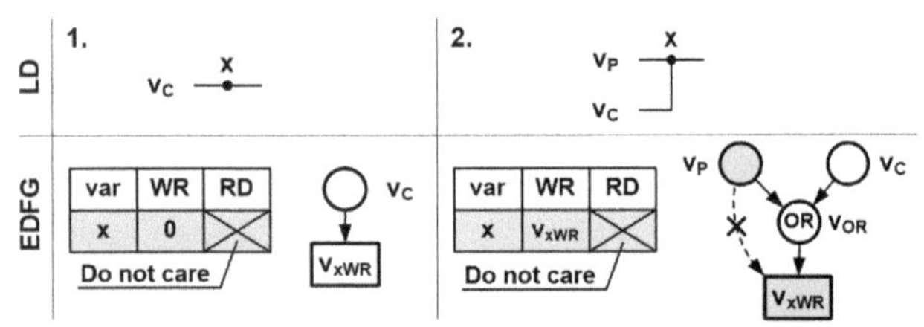

Fig. 8: The iterative conversion of the LD node into equivalent EDFG.

4.3. Switches

A switch is a basic component used for creating the logic AND operation between driving and input signals. The switch is converted into EDFG equivalent that has been shown in the Fig. 9. The EDFG procedure utilizes two previously described algorithms for variable access and node driving.

Algorithm 3: Let the x is a variable associated with an input node, a is a variable driving the switch and y is a variable associated with the output node. The v_{AND} node is created for considered switch. The value of the x and a variables are accessed with the use of the algorithm 1. The procedure returns respective driving nodes that are connected to the v_{AND} node. The attribute of the edge for variable a is set to a logic inversion for normally closed switch (Fig. 9.2). The v_{AND} node assigns value to the y variable. The assignment is performed according to the algorithm 2.

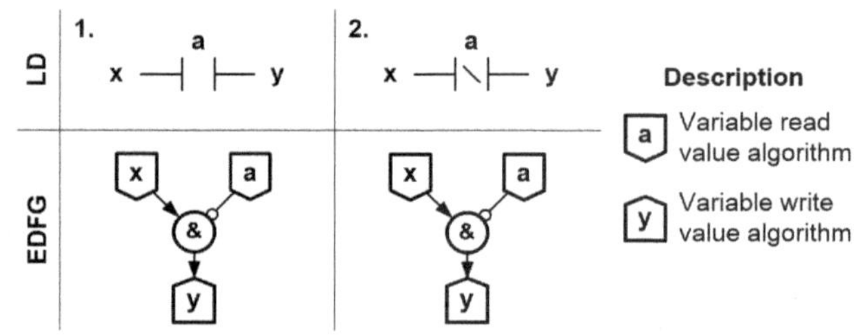

Fig. 9: The EDFG switch equivalent.

4.4. Coils

The coil assigns or reassigns value to a particular variable. Following algorithm is used for obtaining EDFG from a coil item. This algorithm is adopted to cooperate with remaining compilation algorithms, especially with variable value access.

Algorithm 4: Let the a is a driving signal, y is the variable associated with a signal driven by the coil. The variable value is read according to the algorithm 1 that returns driving node v_{aD}. Returned v_{aD} node is assigned to the variable y. If a value assignment node v_{yWR} does not exist it is created and linked with v_{aD} (Fig. 10.1). If the variable y is already assigned than the directed edge is reconnected to the recent driving node (Fig. 10.2). The edge attribute is set according to the coil type (e.g. inverted coil - Fig. 10.3).

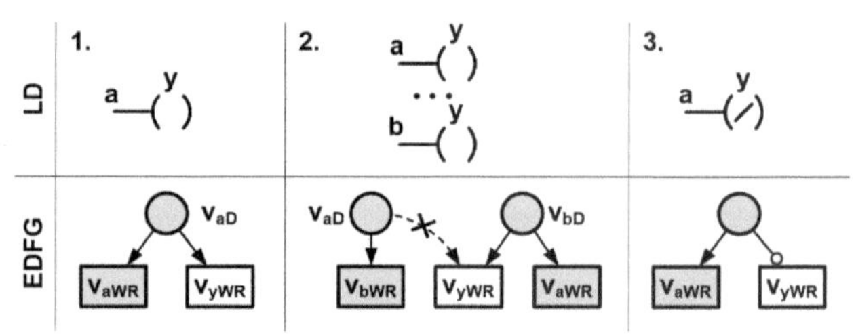

Fig. 10: The coil compilation scheme.

4.5. Functional Modules

The complex arithmetic or mixed arithmetic-logic functionality of the controller is implemented with the use of functional blocks. In the form of blocks are implemented timers, counters and arithmetic functions [1], [8]. Those blocks are controlled by logic and arithmetic signals. All logic signals are connected into ladder network while numeric variables are referenced by identifiers (names).

All these blocks perform conditional execution controlled by input logic signals. This implies the conditional execution of arithmetic operations. The EDFG requires introduction of the conditional selection node that implement selection and assures parallel operation execution). The conditional selection node shown in Fig. 11.1 reflects the high-level synthesis concept based on EDFG. It enables selection between arguments of same data type and can be used for flow control in logic and arithmetic paths. The optimization process can be separately applied to the data paths and control (selection input) path. The additional optimization rules are defined for selection nodes. This is important for optimization performed in arithmetic operations paths. When considered for logic paths the conditional node is transformed into multiplexer equation formed from logic nodes. This operation creates consistent logic EDFG that further can be optimized.

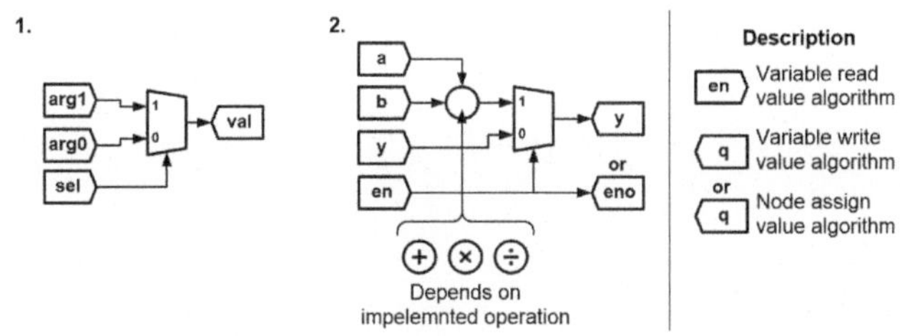

Fig. 11: The EDFG conditional selectin node (1) and general implementation EDFGimplementation of arithmetic blocks (2).

The exemplary general arithmetic module is shown in Fig. 11.2. This block performs calculations conditionally depending on the enable (*en*) signal. An arithmetic block is combined into EDFG between source and sink nodes. Conditional execution of the block forces an automatic variable implementation. It is achieved by executing algorithm 1 for the output variable before calling assignment algorithm 4 This variable is responsible for driving output when block is disabled ($en = 0$). The optimization procedures of selection nodes allow for eliminating covered by logic condition intermediate nodes. At the block level there are used two different value assignment procedures. For logic variables, the algorithm 2 is used while for numeric variables the algorithm 4 described for coils is used.

The figure (Fig. 12) shows an EDFG implementation of TON timer. This block is delivered in the form of sub EDFG that is incorporated into final EDFG during the compilation process. The timer is a specific implementation of a counter. It declares a hidden clocking signal t that is triggered with time base period. This signal enables counting of the timer unit. It is driven from the controller framework created during the implementation phase. Similarly to the arithmetic modules, there are declared internal variables responsible for storing elapsed time (et) and timer activity (q).

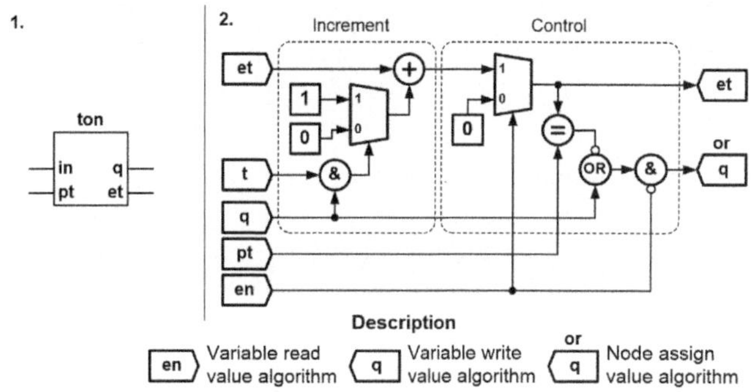

Fig. 12: The timer ton equivalent sub EDFG.

5. The EDFG Hardware Mapping

A synthesizable HDL model optimized for an FPGA target is obtained from described EDFG structure. The mapping procedure for an EDFG starts from the optimization process. The EDFG allows for limited optimization of the logic operations as presented in chapter 3. After initial operation merge and logic absorption, the Espresso minimization is used. This allows for further reduction of the logic operations and optimization of unused paths. In the domain of the arithmetic operation, a constant merge and common subexpression extraction are performed. Finally, multiple argument arithmetic nodes are expanded into a two argument nodes that can be directly mapped into arithmetic modules (adders and multipliers). The expansion process balances the propagation delay of operations in EDFG paths [12].

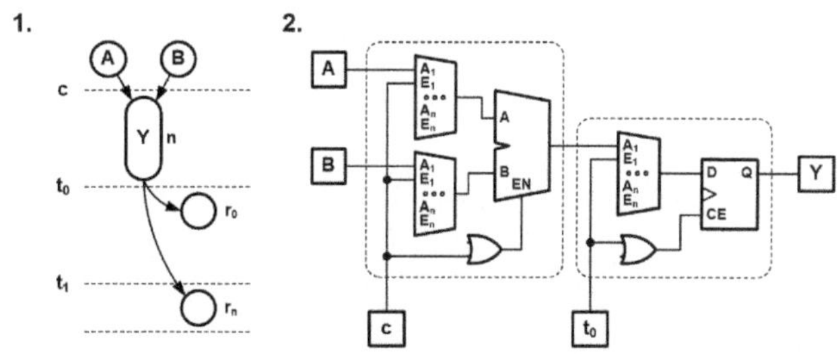

Fig. 13: The EDFG scheduling and mapping process.

The optimized EDFG is a subject of scheduling. It can be directly implemented with the use of greedy approaches with ALAP or ASAP scheduling methods [7]. In contrast to logic operations, arithmetic operations resource requirements are much higher. The greedy mapping approach will lead to quick resources run out. To overcome this limitation, a scheduling method based on list approach with original operation sorting is applied. The operation schedule takes into consideration the operation mobility and local dependencies. Scheduled nodes are mapped into a set of arithmetic resources. After operation schedule, the register allocation is made with the use of modified left edge algorithm. Schematically EDFG hardware mapping process is shown in (Fig. 13). There is shown a schedule result in the form of an EDFG (Fig. 13.1). On the graph are marked: operation start time (c) operation end time (t_0) and last access time to the result (t_1). Obtained mapping reflects the hardware structure shown in Fig. 13.2.

The mapping procedure optimizes the resource distribution by minimizing the cost of argument multiplexing. Structure of the controller and principles of its operation are shown schematically in Fig. 14. The specific EDFG depicts the allocated registers (small circles) and operations (large circles). The thick line connecting small circles denotes the variable lifetime. There are three computation stages. The calculation process starts from image registers update and internal variables exchange. After this operation, a calculation process takes place. The cycle is ended with a result write back.

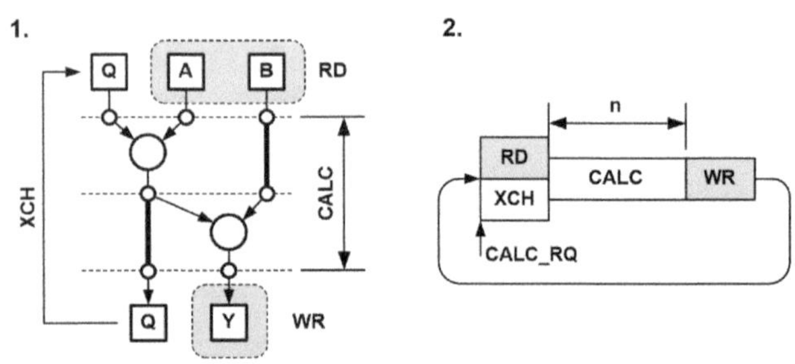

Fig. 14: The calculation cycle and its EDFG representation with marked variable lifetime.

The proposed method of implementation has been compared with direct EDFG mapping approach. The result are presented in Tab. 1. There has been selected 3 representative FPGA families that are Spartan II, Spartan 3 and Spartan 6. The Spartan II and the Spartan 3 are equipped with 4 input LUTs while the Spartan 3 is additionally equipped with combinatorial 18x18 multipliers. The Spartan 6 family is equipped with 6 input LUTs and DSP48A1 units. For illustrating implementation, two representative projects for automatic control have been chosen. The C2 project implements double PID controller with low pass filtering and hysteresis trigger. The T8 project implements the cascade of 8 timers controlling time dependant process. The optimization process can influence the controller response time by extending calculation time due to resource sharing. In the case of the C2 project there was used method of resource sharing that prohibits performance loss. This approach presents a non-redundant greedy assignment accommodated to FPGA architecture sharing. In general, the controller area has been reduced between 50 % - 60 % of the initial area. There can be observed a strong reduction of multipliers usage (50 %).

Tab. 1: The FPGA resource usage comparison.

FPGA	Proj.	Direct	Opt	Gain
		[LUT/MUL]		[%]
Spartan II (LUT4)	C2	1021/-	612/-	59.9 %
	T8	341/-	93/-	27.2 %
Spartan 3 (LUT4 + MUL)	C2	1252/4	717/2	57.2 %
	T8	341/-	93/-	27.2 %
Spartan 6 (LUT6 + DSP)	C2	1274/4	598/2	46.9 %
	T8	205/-	95/-	46.3 %

The T8 projects demonstrate the idea of resource sharing with an acceptable increase of response time. In the case of timers, the flat structure response time is 2 cycles. For this structure, the algorithm takes benefits form use of distributed RAMs that allows to reduce resource requirements about 3.67 times on expense of response time increase. This method is applicable when controller performs other calculation or performance reduction is acceptable.

6. Conclusion

The paper presents entire synthesis process of hardware implemented reconfigurable logic controller from a ladder diagram to hardware mapping. The paper is focused on the compilation of the ladder diagram into an intermediate form suitable for logic synthesis process. As it was presented, chosen intermediate form and methods of creating it has extremely high impact on the final result of the synthesis. The author has developed a method of intermediate representation based on the enhanced data flow graph that utilize attributed edges. It significantly simplifies the graph construction and processing. The intermediate form is created from the ladder diagram with the use of presented algorithms. Due to limited space only general overview of the LD compilation has been described. There are also developed method of representing arithmetic operations and complex functional blocks like timers and counters.

The graph representation is well suited for further processing oriented to FPGA implementation. Finally, a brief overview of mapping and implementation of synthesizable HDL description was given. Developed optimization methods allow to reduce controller size between 1.66 – 2.13 times without performance loss. Significant reduction of the controller size (about 3.6 times) is observed for implementation where specific features of FPGAs are used and little performance loss is acceptable.

Presented algorithms belong to originally developed a hardware PLC synthesis tool capable of synthesizing custom hardware implementation from LD, IL and SFC [11], [12]. The compilation and synthesis tool is subject of ongoing research and development. It is planned to extend arithmetic support to floating point numbers and improving scheduling and mapping processes.

References

[1] BOLTON, W. *Programmable Logic Controllers.* Burlington: Elsevier Newnes, 2009. ISBN 978-0-7506-8112-4.

[2] CHMIEL, M. and E. HRYNKIEWICZ. Concurrent operation of processors in the bit-byte CPU of a PLC. *Control and Cybernetics.* 2010, vol. 39, no. 2, pp. 559–579. ISSN 0324-8569.

[3] DAOSHAN, D., X. XIAODONG and K. YAMAZAKI. A study on the generation of silicon-based hardware PLC by means of the direct conversion of the ladder diagram to circuit design language. *International Journal of Advanced Manufacturing Technology.* 2010, vol. 49, no. 5, pp. 615–626. ISSN 1433-3015.

[4] ECONOMAKOS, C. and G. ECONOMAKOS. FPGA implementation of PLC programs using automated high-level synthesis tools. In: *IEEE International Symposium on Industrial Electronics.* Cambridge: IEEE, 2008, pp. 1908–1913. ISBN 978-1-4244-1665-3.

[5] ECONOMAKOS, C. and G. ECONOMAKOS. C-based PLC to FPGA translation and implementation: The effects of coding styles. In: *16th International Conference on System Theory, Control and Computing (ICSTCC).* Sinaia: IEEE, 2012, pp. 1–6. ISBN 978-1-4673-4534-7.

[6] FALCIONE, A. and B. H. KROGH. Design Recovery for Relay Ladder Logic. *IEEE Control Systems.* 1993, vol. 13, iss. 2, pp. 90–98. ISSN 1066-033X.

[7] GAJSKI, D., N. DUTT, A. WU and S. LIN. *High-Level Synthesis - Introduction to Chip and System Design.* New York: Kluwer Academic Publishers, 1992. ISBN 978-1-4613-6617-1.

[8] JOHN, K. H. and M. TIEGELKAMP. *IEC 61131-3: Programming Industrial Automation Systems: Concepts And Programming Languages, Requirements for Programming Systems, AIDS to Decision-making Tools.* Berlin: Springer Science & Business Media, 2001. ISBN 978-3540677529.

[9] ICHIKAWA, S., M. AKINAKA, R. KIEDA and H. YAMAMOTO. Converting PLC instruction sequence into logic circuit: A preliminary study. In: *IEEE International Symposium on Industrial Electronics.* Montreal: IEEE, 2006, pp. 2930–2935. ISBN 1-4244-0497-5. DOI: 10.1109/ISIE.2006.296082.

[10] LIU, Y., K. YAMAZAKI, M. FUJISIMA and M. MORI. Model-driven programmable logic controller design and FPGA-based hardware implementation. In: *ASME 2005 International Design Engineering Technical Conferences and Computers and Information in Engineering Conference.* Long Beach: ASME, 2005, pp. 81–88. ISBN 0-7918-3766-1. DOI: 10.1115/DETC2005-85119.

[11] MILIK, A. and E. HRYNKIEWICZ. Synthesis and implementation of reconfigurable PLC. *International Journal of Electronics and Telecommunications.* 2012, vol. 58, no. 1, pp. 85–94. ISSN 2300-1933.

[12] MILIK, A. On Mapping of DSP48 Units for Arithmetic Operation in Reconfigurable Logic Controllers. *Programmable Devices and Embedded Systems.* 2012, vol. 11, no. 1, pp. 249–254. ISSN 1474-6670. DOI: 10.3182/20120523-3-CZ-3015.00048.

[13] MILIK, A. On Mapping of DSP48 Units for Arithmetic Operation in Reconfigurable Logic Controllers. *Programmable Devices and Embedded Systems.* 2006, vol. 6, no. 1, pp. 14–16. ISSN 1474-6670.

[14] MINATO, S. *Binary Decision Diagrams and Applications for VLSI CAD.* Berlin: Springer, 1996. ISBN 978-1-4613-1303-8.

[15] MOCHA, J. and D. KANIA. Hardware Implementation of a control program in FPGA structures. *Electrical Review.* 2012, vol. 88, iss. 12, pp. 95–100. ISSN 0013-4384.

[16] WELCH, J. T. Translating Relay Ladder Logic for CCM Solving. *IEEE Transactions on Robotics and Automation.* 1997, vol. 13, iss. 1, pp. 148–153. ISSN 1042-296X. DOI: 10.1109/70.554356.

[17] WELCH, J. T. and J. CARLETTA. A direct mapping FPGA architecture for industrial process control applications. In: *International Conference on Computer Design.* Austin: IEEE, 2000, pp. 595–598. ISBN 0-7695-0801-4. DOI: 10.1109/ICCD.2000.878352.

[18] WIRTH, N. *Algorithms + Data Structures = Programs.* New Jersey: Prentice Hall, 1976. ISBN 978-0130224187.

[19] ZIEBINSKI, A., R. CUPEK and W. SROKA. Application in Java language realizing the function parser of pseudocode describing structure of a specialized coprocessor of PLC in VHDL. *Measurement Automation and Monitoring.* 2011, vol. 57, no. 8, pp. 148–153. ISSN 0032-4140.

About Authors

Adam MILIK received M.Sc. and Ph.D. degrees from Silesian University of Technology of Gliwice in 1997 and 2003 respectively. Since 2003 he is a professor assistant at Silesian University of Technology of Gliwice. His main interests and research areas are: high-level logic synthesis and implementation, algorithm implementation, technology mapping in FPGA devices, the hardware high-level modeling systems based on HDLs and its integration with other tools like MATLAB, Simulink or SystemVue.

24

An Experimental Study of Laboratory Hybrid Power System with the Hydrogen Technologies

Daniel MINARIK[1], *Bohumil HORAK*[2], *Petr MOLDRIK*[1], *Zdenek SLANINA*[2]

[1]Centre ENET—Research Centre of Energy Units for Utilization of Non Traditional Energy Sources, VSB–Technical University of Ostrava, 17. listopadu 15, 708 33 Ostrava, Czech Republic
[2]Department of Cybernetics and Biomedical Engineering, Faculty of Electrical Engineering and Computer Science, VSB–Technical University of Ostrava, 17. listopadu 15, 708 33 Ostrava, Czech Republic

daniel.minarik@vsb.cz, bohumil.horak@vsb.cz, petr.moldrik@vsb.cz, zdenek.slanina@vsb.cz

Abstract. *This paper presents very small laboratory hybrid photovoltaic-hydrogen power system. The system was primarily assembled to verify the operability of the control algorithms and practical deployment of available commercial hydrogen technologies that are directly usable for storage of electricity produced from renewable energy sources in a small island system. This energetic system was installed and tested in Laboratory of fuel cells that is located in the university campus of VSB–Technical University of Ostrava. The energetic system consists of several basic components: a photovoltaic field, accumulator's bank, water commercial electrolyzer and compact fuel cell system. The weather conditions recorded in two different weeks as model weather and solar conditions are used as case studies to test the energetic system and the results for two different cases are compared each other. The results show and illustrate selected behaviour curves of the power system and also average energy storage efficiency for accumulation subsystem based on hydrogen technologies or at the energetic system embedded components. On the basis of real measurement and its evaluation the ideal parameters of the photovoltaic field were calculated as well as the hydrogen technologies for supposed purpose and the power requirements.*

Keywords

Efficiency, electrolysis, energy conversion, energy storage, fuel cells, hybrid power system, hydrogen storage, photovoltaic, renewable energy sources.

1. Introduction

The planet Earth has no other resources besides those that evolved as a result of natural physical processes long before the arrival of human era. These materials include mainly fossil fuels, as the greatest contribution to the development of mankind, radioactive elements or hydrogen bonded in water. Since the deposits of the first two raw materials mentioned above are unfortunately limited and the whole world is still dependent on them, further technologies are appreciated for sustainable development and catering for increasing demand, while these enable utilization of other energy resources, including mainly energy from wind, water and the Sun. The problems associated with utilisation of the said energy resources address mainly their variability in time, substantial dependency on local conditions and low density of energetic flow from transformation equipment. Mankind has been using these energy resources for centuries, and the current state of the art equipment makes them a very prospective and interesting supply of energy. The technology of thermonuclear fusion, which is still in development, could solve most economic and environmental problems incidental to retrieval of energy. This technology would further enable a smooth transition to hydrogen technologies as it could be considered a hydrogen-based technology itself. Several optimistic prognoses say this technology should be launched in commercial practice within the "mere" period of 40–50 years. The transition period before full adoption of the thermonuclear fusion technology can be used for to proceed with retrieval of energy from the relatively familiar hydrogen technologies being developed and put into use. Just the combination of technologies for the utilization of energy from renewable sources and hydrogen technologies can help to remove mentioned deficiencies or at least improve the usability of renewable energies and generally expand the limits of acceptability of these sources into the power distribution system. Such hybrid power systems was also examined and tested by the authors in other studies [6], [7], [8], [9], [10], [11], [12], [13], [14], [15], [16], [17].

2. Description of Experimental Energy System

The basic idea of the whole project was to create a self-contained and self-sufficient source of electric energy to provide power supply into appliances within suitable premises with respect to the estimated output from the entire system. Premises suitable for this application would comprise a smaller leisure facility, remote buildings or facilities that are either impossible or very costly to connect into the grid. The system, designed here, could be also used as an alternate solution for power supply employing motor-driven generators. The system output should be ideally used to cover a certain permanent consumption of electric power for a defined period matching the duration of supply provided with an integrated hydrogen power supply unit as well.

The primary energy resource installed within the power system is, therefore, represented by an array of photovoltaic panels located on the laboratory roof. The photovoltaic system (further PVS only) is made of 12 pieces photovoltaic panels with the total installed power of 1.98 kWp.

The direct-current (DC) power is fed into the DC/DC converter (charger) and subsequently forwarded to the direct-current bus with nominal voltage of 48 V defined by the auxiliary accumulator storage bank linked with the bus. The nominal capacity of power storage bank that consist from four accumulators in series connection is 75 Ah, which defines its total power capacity of 3.6 kWh. The main function of the accumulator storage bank in the energetic system is accumulation of electrical energy produced by photovoltaic panels that does not directly cover consumption of a load (appliances connected by the user), consumption of the electrolyzer, transmission and also transformation losses of the system. Another benefit and unambiguously indispensable function of the accumulator power storage is the functionality as power "buffer" from which electrolyzer's consumption is covered under a defined temporal constraints of the photovoltaic system that are caused by immediate worsening of local meteorological situation.

The said direct-current bus is also provided with a stabilizing DC/DC converter for connection of one of the essential hydrogen technologies - a Polymer Electrolyte Membrane (PEM) type fuel cell. As the specific fuel cell was after several necessary adjustments its electronic systems used the Ballard's Nexa™ Power Module readily available on the market.

Production of hydrogen in the system is handled by an electrolyzer (ELCLZR in diagram), which operated during the period of power surplus from photovoltaic source to store the power in hydrogen. The electrolyzer

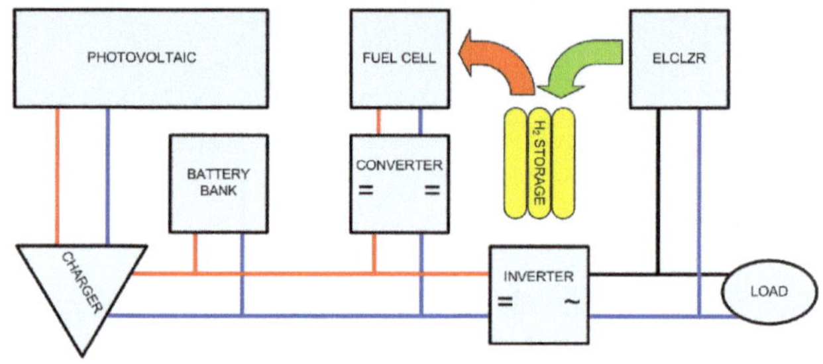

Fig. 1: Diagram of key components of tested energetic system [1].

available for production of hydrogen was the laboratory unit low temperature PEM type Hogen GC600 connected to the system via an alternating-current (AC) bus. The electrolyzer is able to supply approx. 0.6 Nl per minute of gaseous hydrogen at the pressure up to 13.8 bar.

The produced gaseous hydrogen was stored in three standard pressure vessels with water volume of 50 l used for capturing pure hydrogen gas (see Fig. 1). The total capacity of storage therefore equals to water volume of 150 l, which corresponds to approx. 2 Nm^3 (normal cubic meters; gas volume related to normal pressure 100 kPa and temperature 20 °C) of hydrogen gas in a bundle, i.e. which is approx. 650 Nl (normal liters) per one pressure vessel at the maximum pressure in electrolyzer. The system for storage and distribution of hydrogen is also equipped with a connection point for potential supply of hydrogen from an external supply of hydrogen gas from the laboratory hydrogen distribution. The fuel from external tanks is used in case the produced hydrogen stored in the primary and above mentioned tank for power system is depleted during the operation of the fuel cell and the whole power system faces the risk of complete shutdown.

The last essential element within the power system is a power inverter to ensure production of alternating voltage with standard parameters in the single-phase low voltage grid (230 V, 50 Hz) used to supply the alternating current bus. The alternating current bus was also employed to supply the load and electrolyzer as needed [1].

The power system also featured the control and independent measuring system. The measuring system was implemented to ensure measuring of non-electric and electric values of the whole power system, for digitalisation, processing, visualization and storage of data obtained by measurement and values monitored during operation of particular components of the power system. The whole power system is controlled and operated via a control system fitted with the main control unit in the form of a programmable logic controller (PLC) Siemens SIMATIC type S7-224XP with the relevant control and visualization peripheries. The user's interface and control algorithm were programmed to make the intended ideal operation of power system ensure mainly uninterrupted supply for the permanent design consumption of 200 W and the implemented hydrogen accumulation system with fuel cell was to balance overhang of the power produced by photovoltaic system or to supply electric power in case of production lack using renewable sources.

3. Analysis of Power System Operation Characteristics

Measurements further used as the basis for the relevant conclusions were run during first stage debugging of the power system during the autumn season. To serve the assessment purpose, model measuring was split into two periods differing mainly with respect to the nature of weather and solar radiation conditions. As far as sunshine power is concerned, the first period was below average, and the second period of measuring could be defined as very good [2].

The following figure (see Fig. 2) shows a fragment of data obtained by measurement over five days from 23.10. to 27.10.2010 representing behaviour of selected values within the power system. As the first behaviour curve clearly shows, depicting the power supplied by photovoltaic system (PPV value), this period was associated with solar power below average. The curve illustrates mainly two days with average solar power - 24. and 26.10 - plus one day with overcast condition (25.10.) and one model day (27.10.), when the electrolyzer was launched. Activation of the electrolyzer is evident from the second behaviour curve showing the course of power drain by inverter from the direct-current bus. As one can further see, besides deviations, the permanent load ranged around approx. 240 W corresponding with a permanently connected load of 200 W. The different in output values corresponds mainly with losses during transformation and transmissions losses in the distribution system.

To assess the operation of the power system, we will further deal with behaviour shown in Fig. 3 detailing the behaviour of six values in the power system. Values $I_{PV\ dc}$ and $U_{PV\ dc}$ relate to the photovoltaic system and define the current and voltage from the photovoltaic system. Measuring of values from the photovoltaic system was run on supply cables connected to input terminals of the charger. Further indicators of behaviour – $I_{ChAR\ dc}$ and $U_{BAT\ dc}$ - show the course of current at charger output and voltage from the power storage bank, which basically defines the voltage in direct-current bus. The values of $I_{ChAR\ dc}$ and $U_{BAT\ dc}$ are used to calculate the output power of the charger, which is further defined as the value $P_{ChAR\ dc}$. Values of $I_{ISL\ dc}$ and $U_{ISL\ dc}$ show the input current and voltage at inverter terminals. In case the system is connected to a single fuel cell only, values $I_{ISL\ dc}$ and $U_{ISL\ dc}$ show the current and voltage at the output from stabilizing converter fitted to the fuel cell.

Figure 3 shows ten time indicators, which highlight key moments in operation of the power system. In the beginning, the values shown to the left from indicator 1, the system was in operation, and the consumption of system and load was covered with the photovoltaic system. That was followed by gradual decrease of output from PVS up to the point, when the PVS was no longer able to cater for consumption. Coverage of power consumption was then launched with the assistance of the accumulator power storage bank. This is also reflected in the voltage behaviour captured in UBAT dc curve, which showing the clear drop in voltage from batteries. Once the insufficient capacity of batteries was assessed (indicator 2), the system was supported by the fuel cell to cover the system consumption in full. The fuel cell was in operation till dawn (indicator 3), when the supply from PVS was restored. However, the relatively low output of PVS during the particular day could not ensure sufficient re-charging of batteries resulting in inevitable re-connection to the fuel cell (indicator 4). The very next day was even less convenient in terms of solar power and indicators 5 and 6 show several attempts to shut the fuel cell down. The power system operation was similar on the subsequent day, with partial re-charging of the power storage bank. The fuel cell took over power production next night only. The last day in the description brought clear skies, which is also reflected by its operation data. As the behaviour indicators after assessment of sufficient power from the PVS (indicator 7) clearly show, the fuel cell has been disconnected. Batteries were re-charged to a sufficient level gradually to enable connection of the electrolyzer (indicator 8). It was in operation until cloudy conditions (indicator 9), when the control system interfered to disconnect it and regenerate the power storage bank, which is shown by increase of voltage within. Once the skies cleared up again, sufficient output from the photovoltaic system resumed and the production of hydrogen restarted. The production cycle terminated approx. 1 hour later and the power system began to utilize the power accumulated in batteries.

If we abstract from assessment of the very efficiency of photovoltaic panels, the first indicator in chain of power transformations may be represented by the efficiency of first converter (charger) that transforms power produced by PVS to parameters of the direct-current bus. This efficiency value has been calculated as a rate between the two values of energies i.e. energy measured on the charger output and energy value delivered on to charger input (9.756 kWh / 10.008 kWh)

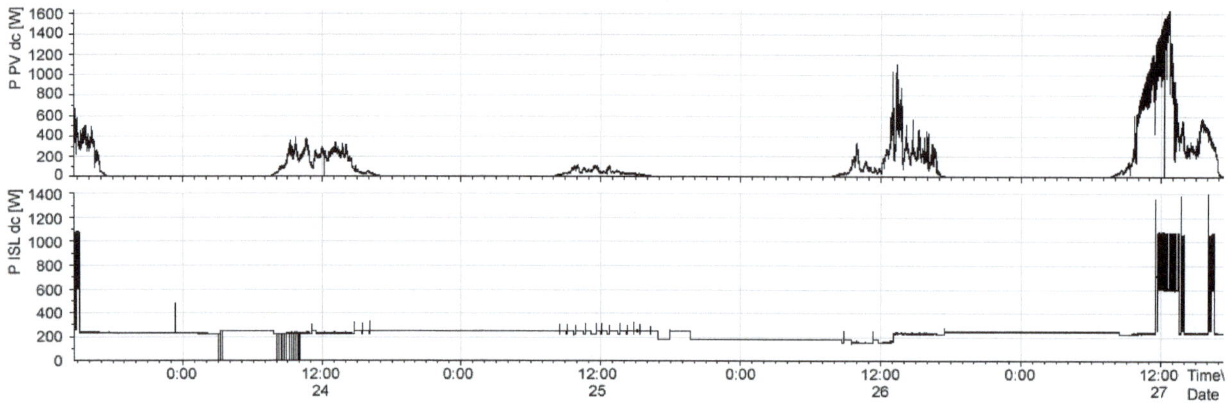

Fig. 2: Selected behaviour curves of the power system for a period of inconvenient solar and weather conditions. ($P_{PV\ dc}$ - power supplied by photovoltaic panels, $P_{ISL\ dc}$ - power drawn at inverter).

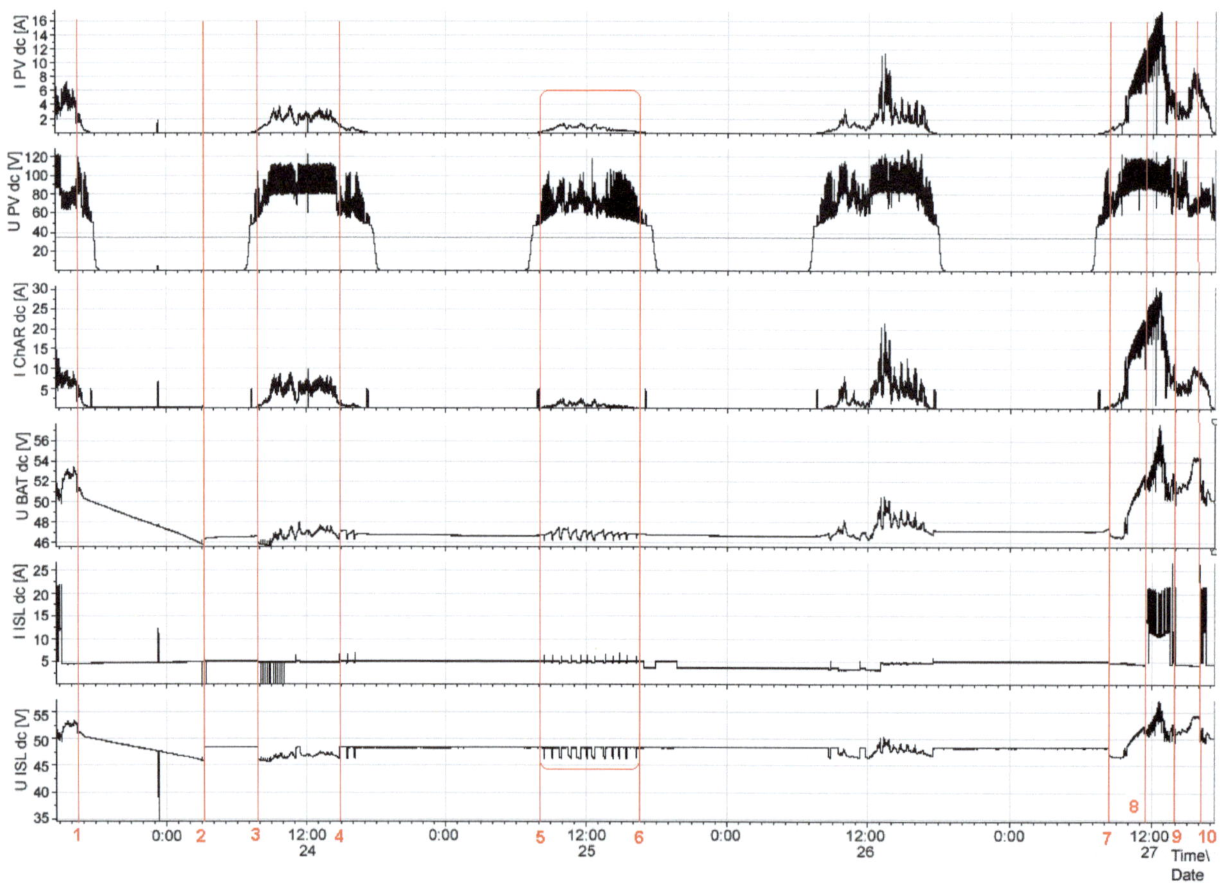

Fig. 3: Selected behaviour indicators of measured values in the power system for cloudy days ($I_{PV\ dc}$ - current in photovoltaic panels, $U_{PV\ dc}$ - voltage on photovoltaic panels, $I_{ChAR\ dc}$ - current at charger input, $U_{BAT\ dc}$ - voltage in power storage bank, $I_{ISL\ dc}$ - current at inverter input, $U_{ISL\ dc}$ - voltage at inverter input). All the power or efficiency indicators shown below have been calculated using the said behaviour indicators.

for overall operating time and it is equal to 97.48 %. Another important element in the system is the inverter whose overall average efficiency was calculated from the measured data as 81.18 %. Concretely, was for determine the efficiency of the inverter calculated the energies delivered to the inverter input with a value of 24.139 kWh and energy taken from the inverter output with value 19.596 kWh. This efficiency is relatively low and it can be justified by the fact that the inverter was not even near its operating scope of defined nominal output, which is 4.2 kW; and worked in the system under load from 0.24 kW to 1.1 kW only.

The most vital information about efficiency of the hydrogen power storage system is derived from data on consumption and production of hydrogen within the power system, resp. relevant amounts of electric power produced and consumed by hydrogen technologies. As far as these details are concerned, the total amount of hydrogen consumed by the fuel cell within the observed period was 12,631 Nl. Converted with respect to the maximum power utilization by PEM type fuel cell, this amount corresponded to 37.173 kWh of stored energy. The fuel cell supplied the system with 14.138 kWh, which corresponds with efficiency rating of transformation of power from hydrogen into electric power supplied to the direct-current bus equal to 38.03 %. This efficiency includes losses on the route from "stack" of fuel cell up to terminals on the inverter, i.e. losses in distribution lines, consumption within the fuel cell itself and losses during transformation of output using the stabilizing DC/DC converter, this is the net production efficiency of fuel cell in other words.

The reverse process - electrolysis - run in accordance with behaviour indicators described earlier, especially the data obtained on the last day (see Fig. 3 – between indicators 8 and 9, past 10 and partially even before indicator 1), consumed 1.575 kWh to supply the electrolyzer which produced 100.53 Nl of hydrogen. This amount of hydrogen gas corresponds to 0.296 kWh of energy usable by the fuel cell and the efficiency of electric power conversion into hydrogen equals to 18.8 %. Another fact worth mentioning is that the said efficiency value reflects rather the efficiency of subsystem for hydrogen production, not just the electrolyzer as a device, since this value is also affected by losses incurred during idle times, when no hydrogen was produced and the electrolyzer stayed in the operation initialization stage etc. The transformation efficiency during stable operation of the electrolyzer is equal to 24.8 %. In other words, the example of the electrolyzer operation shows a serious effect of settings made to the algorithm of the control system with impact to the overall efficiency of power system as a whole. The level of efficiency of transformation of the electric power is very low and that is mainly due to the design of implemented electrolyzer and definitely not due to the conversion efficiency of electrolysis as a electrochemical process. The most common value of the energy efficiency of electrolysis process which is achieved in practice on laboratory testing devices is in interval roughly between 60–80 % [5], [6], [8], depending on specific parameters of used equipment and its operational parameters.

Another measurement to be assessed has been taken with respect to the said period under the most convenient conditions possible. The first four days illustrated by the fragment of measured data shown in Fig. 4 brought absolutely clear skies, which is matched by achieved peaks of output from the photovoltaic system reaching values above 1.4 kW, i.e. 73 % (usage) of the installed capacity. As far as the operation characteristics of the power system are concerned, these four days are almost identical, and time indicators can be used to interpreter all the events observed, similarly to the previous case. The fifth day shown in behaviour indicators at the last one did not bring such convenient weather, compared to preceding days, and the course of output drawn by inverter shows the electrolyzer was not launched either. The power available from the photovoltaic system was used to re-charge the power storage bank only. Figure 5 shows almost ideal characteristics providing a clear demonstration of essential principles of the power system. The indicator 1 (as well as 8) shows the moment, when the capacity of batteries has been depleted, and the production of electric power in the fuel cell was initialized around one o'clock in the morning. The fuel cell was shut down, and the coverage of input for load was taken over by the photovoltaic system at the moment shown by indicator 2 (as well as 9). The time indicators 3 and 4 define a time interval (approx. 3 hrs.) of the electrolyzer operation. Termination of charging of the power storage bank and the beginning of its discharge is marked using the indicator 5 (as well as 7). The subsequent three days show repetition of the power system cycle at the same level of defined operation logic.

Power parameters and efficiency levels set for individual components of the power system determined using the data obtained by measurement during the said interval, i.e. from 27.10. to 1.11., have been summarized in Tab. 2.

Compared to the previous assessments, the converter efficiency has been determined at 89.68 % that is higher thanks to its improved utilization. The efficiency of the charger achieved a slightly lower level of 97.38 %, which is most likely due to measurement inaccuracy and subsequent calculation operations during processing of the measured data only. As far as the particular period is concerned, the amount of power produced by photovoltaic system reached 25.3 kWh and the charger supplied the system with 24.6 kWh. The amount of power supplied to the system by fuel cell was

Tab. 1: Summary of data on power system operation in the period of inconvenient solar and weather conditions.

Energy [kWh] Amount of H_2 [NL]	Supplied	Drawn	Efficiency of	%
Energy supplied from PVS [W_{PV}]	10.008	-		
Energy at charger output [W_{ChAR}]	9.756	-	Charger	97.48
Energy at inverter input [W_{ISL}]	24.139	-	Inverter	81.18
Energy drawn from the system [W_Z]	-	19.596		
Energy drawn by load [W_{LOAD}]	-	18.021		
Amount of hydrogen produced	-	100.53	Electrolyzer	18.78
Equivalent amount of energy in H_2	-	0.296		
Energy drawn by electrolyzer [W_{ELC}]	-	1.575		
Amount of hydrogen consumed	12 631.0	-	FC+Converter	38.03
Equivalent amount of energy in H_2	37.173	-		
Energy supplied by fuel cell [W_{FC}]	14.138	-		

Tab. 2: Summary of data on power system operation in the period of convenient solar and weather conditions.

Energy [kWh] Amount of H_2 [NL]	Supplied	Drawn	Efficiency of	%
Energy supplied from PVS [W_{PV}]	25.267	-		
Energy at charger output [W_{ChAR}]	24.605	-	Charger	97.38
Energy at inverter input [W_{ISL}]	34.081	-	Inverter	89.68
Energy drawn from the system [W_Z]	-	30.564		
Energy drawn by load [W_{LOAD}]	-	23.506		
Amount of hydrogen produced	-	465.12	Electrolyzer	19.39
Equivalent amount of energy in H_2	-	1.369		
Energy drawn by electrolyzer [W_{ELC}]	-	7.058		
Amount of hydrogen consumed	8 138.8	-	FC+Converter	37.20
Equivalent amount of energy in H_2	23.952	-		
Energy supplied by fuel cell [W_{FC}]	8.911	-		

8.9 kWh, which required more than 8 Nm^3 of hydrogen consumed. The generator block with fuel cell, i.e. the fuel cell and its stabilizing converter, worked with the total efficiency of fuel transformation equal to 37.2 %. The different between power supply efficiency of the fuel cell shown by the data obtained is not substantial and it is mainly caused by longer period, when the fuel cell operated outside the optimal operation conditions, compared to the total operation period. The previous assessment shows the fuel cell in operation basically throughout the entire period of measuring, which can be easily compared using behaviour indicators illustrated in Fig. 3 and Fig. 5.

Similar logic is implemented for reasoning of higher efficiency of hydrogen production reaching the level of 19.4 % in the second case of operation. This slight increase of operation efficiency of the electrolyzer is probably supported by the essentially stable operation of the electrolyzer without unnecessary warm-up stages compared to its total operation period. The amount of hydrogen produced throughout the entire period of measurement assessed was 465 Nl, which can be expressed by the power equivalent of 1.37 kWh.

4. Summary

The data and parameters mentioned above imply several important conclusions. It is absolutely evident that "the bottleneck" of the whole power system is represented by the specific electrolyzer employed. The very low efficiency (around 19 %) makes it totally unacceptable for similar installations. As already mentioned above, the minimum operation efficiency of regular PEM type electrolyzers available on the market ranges within the interval of 56–73 % [4], depending mainly on the volume of production capacity.

If the power system were to employ an electrolyzer technology with the same parameters (efficiency), the coverage of hydrogen consumption ensured by the fuel cell, properly quantified and determined for the first assessment period at approx. 12.6 Nm^3, would have to be catered by the total of 14 electrolyzers operated in parallel 5 hours a day for the total period of 5 days (all of the same type, i.e. Hogen GC 600). However, the power consumed by these devices would reach the level of 196 kWh, which in turn corresponds with the required absorbed power of approx. 7.8 kW. Considering the assessed efficiency of the charger, this would be matched by 60 % usage of photovoltaic producing 13 kWp, which requires an installation comprising 79 photovoltaic panels. Various variants of the abovementioned modification with said parameters are summarized in Tab. 3.

Similar ideas may be elaborated for the second case too as it is based on values analyzed under relatively convenient weather conditions. This case should be considered with values matching the data from Tab. 3, obviously with the different quantity of hydrogen required by fuel cell. As implied by values summarized

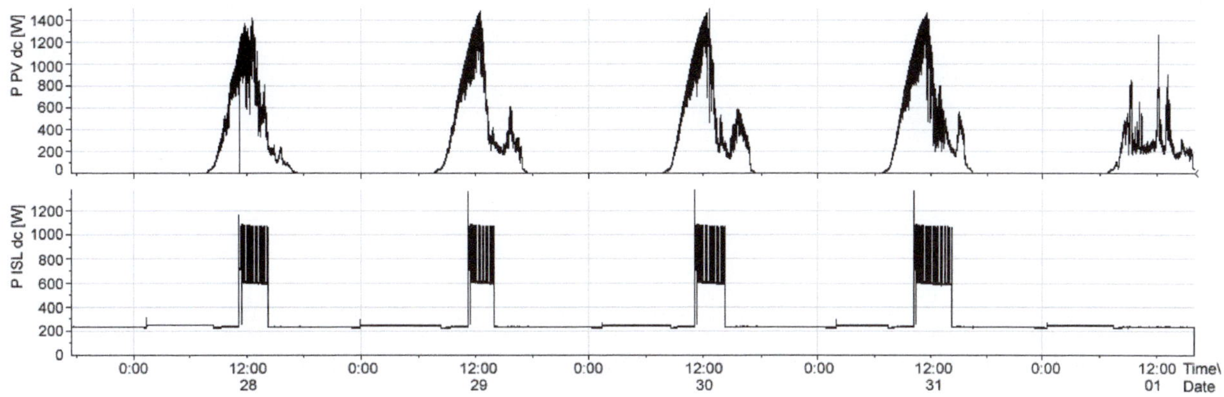

Fig. 4: Selected behaviour curves of the power system for a period of convenient solar and weather conditions. ($P_{PV\ dc}$ - power supplied by photovoltaic panels, $P_{ISL\ dc}$ - power drawn at inverter).

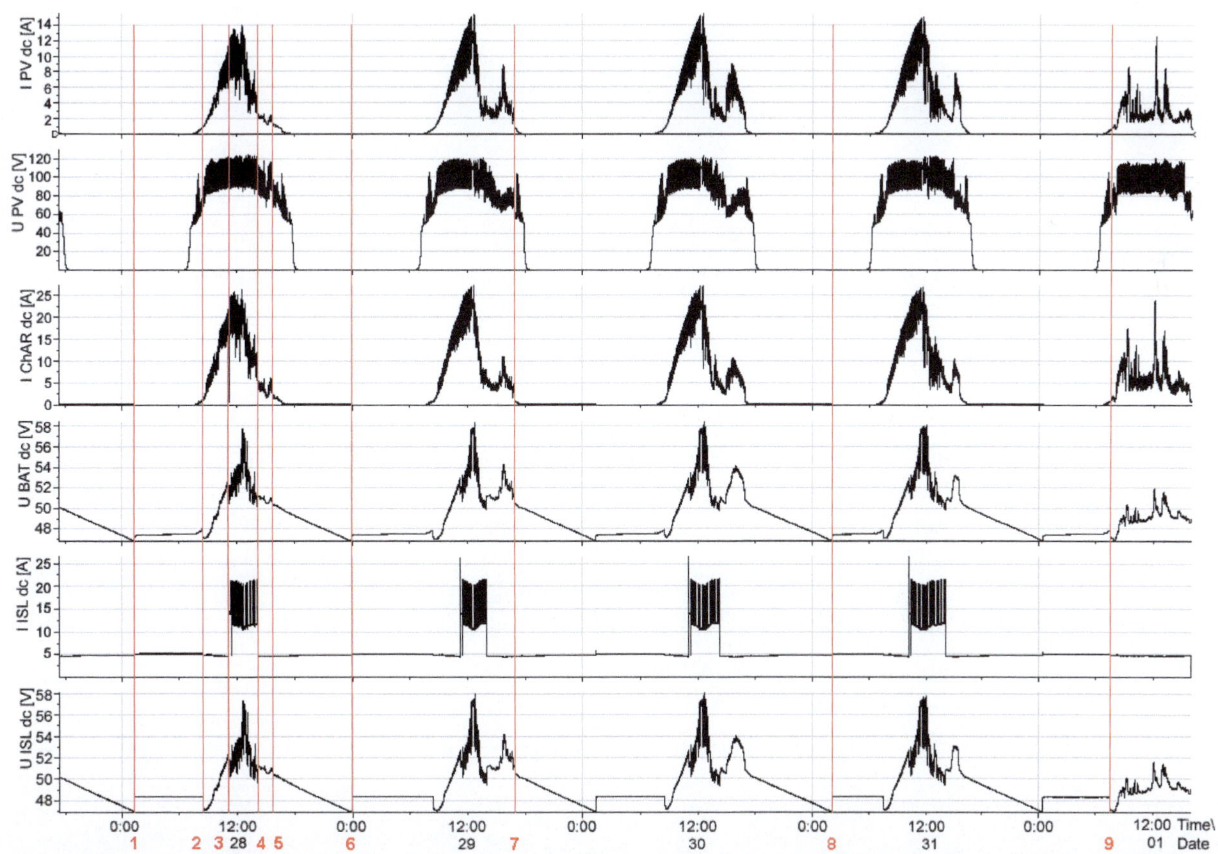

Fig. 5: Selected behaviour indicators of measured values in the power system for clear sky days. ($I_{PV\ dc}$ - current in photovoltaic panels, $U_{PV\ dc}$ - voltage on photovoltaic panels, $I_{ChAR\ dc}$ - current at charger input, $U_{BAT\ dc}$ - voltage in power storage bank, $I_{ISL\ dc}$ - current at inverter input, $U_{ISL\ dc}$ - voltage at inverter input .

Tab. 3: Summary of the intended substitution of equipment required to achieve the necessary hydrogen production levels in case of poor weather conditions.

Entered parameters		
Total amount of hydrogen required		12 631 Nl
Equivalent energy		37.173 kWh
Number of days of operation		5 days
Number of hours of operation per day		5 h
Efficiency of transformation		19.00 %
Charger efficiency		97.48 %
Electrolyzers production capacity		36 Nl/h
One PV panel output		165 Wp
Calculated parameters		
Number of electrolyzers		14 pcs
Energy consumed by electrolyzers		195.647 kWh
Electrolyzers power requirement		7.826 kW
Overall PVS system utilization	**Total installed capacity of PVS**	**Number of panels**
30%	26.086 kWp	158 pcs
40%	19.565 kWp	119 pcs
50%	15.652 kWp	95 pcs
60%	13.043 kWp	79 pcs
70%	11.180 kWp	68 pcs
80%	9.782 kWp	59 pcs
90%	8.695 kWp	53 pcs

Tab. 4: Summary of the intended substitution of equipment required to achieve the necessary hydrogen production levels in case of good weather conditions.

Entered parameters		
Total amount of hydrogen required		8 139 Nl
Equivalent energy		23.952 kWh
Number of days of operation		5 days
Number of hours of operation per day		5 h
Efficiency of transformation		19.00 %
Charger efficiency		97.48 %
Electrolyzers production capacity		36 Nl/h
One PV panel output		165 Wp
Calculated parameters		
Number of electrolyzers		9 pcs
Energy consumed by electrolyzers		126.063 kWh
Electrolyzers power requirement		5.043 kW
Overall PVS system utilization	**Total installed capacity of PVS**	**Number of panels**
30%	16.808 kWp	102 pcs
40%	12.606 kWp	76 pcs
50%	10.085 kWp	61 pcs
60%	8.404 kWp	51 pcs
70%	7.204 kWp	44 pcs
80%	6.303 kWp	38 pcs
90%	5.603 kWp	34 pcs

in the Tab. 4, the system would require 51 photovoltaic panels to produce approx. 8.1 Nm^3 of hydrogen with the power equivalent of 24 kWh, involving 9 electrolyzers used within the experimental power system.

Further components of the power system show very favourable efficiency parameters. The fuel cell worked as expected, with respect to the efficiency of stabilizing converter declared by its manufacturer at 90 %, with the average conversion efficiency of 41.8 %, which exactly corresponds with the value determined from the above mentioned independent measurements on the idle fuel cell within a power system.

The total efficiency of a completed hydrogen accumulation cycle within the experimental power system is determine when considering the average efficiency of key components. These partial efficiencies were determined for two observed periods with inconvenient and more convenient solar conditions and were calculated as its arithmetic averages, i.e. 37.62 % for generator production block with a fuel cell and 19.09 % for the electrolyzer. This total efficiency, therefore, reached the level of 7.18 %. Inclusion of the average charger efficiency of 97.43 % makes the overall efficiency of the hydrogen cycle equal to 7.0 %.

5. Conclusion

This paper deals with issues associated with integration of system for accumulation of electric power based on hydrogen technologies into power systems producing electric power from renewable energy sources.

The contents of this paper summarize results obtained through practical measurements on the built hybrid power system with accumulation of electric power by means of hydrogen technologies and especially the subsequent practical analysis of results produced by actual testing operation of this power system.

Acknowledgment

This work was mainly supported by the project „ENET - Energy Units for Utilization of non-Traditional Energy Sources“ (reg. No. CZ.1.05/2.1.00/03.0069, Operational Programme Research and Development for Innovations) and this paper was also supported by the project New creative teams in priorities of scientific research (reg. No. CZ.1.07/2.3.00/30.0055, Operational Programme Education for Competitiveness) cofinanced by the European Social Fund and the state budget of the Czech Republic.

References

[1] MINARIK, D. Experimental project of an energy system using accumulation of electric energy on the basis of hydrogen technologies. In: *Proceedings of 12th International Scientific Conference Electric Power Engineering 2011.* Ostrava: VSB–TUO, 2011, pp. 155–159. ISBN 978-80-248-2393-5.

[2] MINARIK, D. and P. MOLDRIK. Control and measurement systems for experimental standalone energetic system with accumulation of electrical energy based on hydrogen technologies. *IFAC postprint volumes IPPV / International Federation of Automatic Control.* 2013, vol. 12, iss. 1, pp. 371–377, ISSN 14746670. DOI: 10.3182/20130925-3-CZ-3023.00113.

[3] WANG, M., Z. WANG, X. GONG and Z. GUO. The intensification technologies to water electrolysis for hydrogen production – A review. *Renewable and Sustainable Energy Reviews.* 2014, vol. 29, iss. 1, pp. 573–588. ISSN 1364-0321. DOI: 10.1016/j.rser.2013.08.090.

[4] IVY, J. Summary of Electrolytic Hydrogen Production. In: *National Renewable Energy Laboratory* [online]. 2007. Available at: `http://www.nrel.gov/docs/fy04osti/36734.pdf`

[5] SRINIVASAN, S. *Fuel cells: from fundamentals to applications.* New York: Springer, 2006. ISBN 978-038-7251-165.

[6] URSUA, A., I. SAN MARTIN, E. L. BARRIOS and P. SANCHIS. Stand-alone operation of an alkaline water electrolyser fed by wind and photovoltaic systems. *International Journal of Hydrogen Energy.* 2013, vol. 38, iss. 35, pp. 14952–14967. ISSN 0360-3199. DOI: 10.1016/j.ijhydene.2013.09.085.

[7] AICHE-HAMANE, L., M. BELHAMEL, B. BENYOUCEF and M. HAMANE. Feasibility study of hydrogen production from wind power in the region of Ghardaia. *International Journal of Hydrogen Energy.* 2009, vol. 34, iss. 11, pp. 4947–4952. ISSN 0360-3199. DOI: 10.1016/j.ijhydene.2008.12.037.

[8] BARBIR, F. PEM electrolysis for production of hydrogen from renewable energy sources. *Solar Energy.* 2005, vol. 78, iss. 5, pp. 661–669. ISSN 0038-092X. DOI: 10.1016/j.solener.2004.09.003.

[9] GRIGORIEV, S., V. POREMBSKY and V. FATEEV. Pure hydrogen production by PEM electrolysis for hydrogen energy. *International Journal of Hydrogen Energy.* 2006,

vol. 31, iss. 2, pp. 171–175. ISSN 0360-3199. DOI: 10.1016/j.ijhydene.2005.04.038.

[10] SHAPIRO, D., J. DUFFY, M. KIMBLE and M. PIEN. Solar-powered regenerative PEM electrolyzer/fuel cell system. *Solar Energy.* 2005, vol. 79, iss. 5, pp. 544–550. ISSN 0038-092X. DOI: 10.1016/j.solener.2004.10.013.

[11] YILANCI, A., I. DINCER and H. K. OZTURK. A review on solar-hydrogen/fuel cell hybrid energy systems for stationary applications. *Progress in Energy and Combustion Science.* 2009, vol. 35, iss. 3, pp. 231–244. ISSN 0360-1285. DOI: 10.1016/j.pecs.2008.07.004.

[12] YILANCI, A., I. DINCER and H. OZTURK. Performance analysis of a PEM fuel cell unit in a solar–hydrogen system. *International Journal of Hydrogen Energy.* 2008, vol. 33, iss. 24, pp. 7538–7552. ISSN 0360-3199. DOI: 10.1016/j.ijhydene.2008.10.016.

[13] CETIN, E., A. YILANCI, Y. ONER, M. COLAK, I. KASIKCI and H. K. OZTURK. Electrical analysis of a hybrid photovoltaic-hydrogen/fuel cell energy system in Denizli, Turkey. *Energy and Buildings.* 2009, vol. 41, iss. 9, pp. 975–981. ISSN 0378-7788. DOI: 10.1016/j.enbuild.2009.04.004.

[14] ZINI, G. and P. TARTARINI. Hybrid systems for solar hydrogen: A selection of case-studies. *Applied Thermal Engineering.* 2009, vol. 29, iss. 13, pp. 2585–2595. ISSN 1359-4311. DOI: 10.1016/j.applthermaleng.2008.12.029.

[15] SOPIAN, K., M. Z. IBRAHIM, W. R. WAN DAUD, M. Y. OTHMAN, B. YATIM and N. AMIN. Performance of a PV–wind hybrid system for hydrogen production. *Renewable Energy.* 2009, vol. 34, iss. 8, pp. 1973–1978. ISSN 0960-1481. DOI: 10.1016/j.renene.2008.12.010.

[16] DA SILVA, E. P., A. J. MARIN NETO, P. F. P. FERREIRA, J. C. CAMARGO, F. R. APOLINARIO and C. S. PINTO. Analysis of hydrogen production from combined photovoltaics, wind energy and secondary hydroelectricity supply in Brazil. *Solar Energy.* 2005, vol. 78, iss. 5, pp. 670–677. ISSN 0038-092X. DOI: 10.1016/j.solener.2004.10.011.

[17] KAURANEN, P., P. LUND and J. VANHANEN. Development of a self-sufficient solar-hydrogen energy system. *International Journal of Hydrogen Energy.* 1994, vol. 19, iss. 1, pp. 99–106. ISSN 0360-3199. DOI: 10.1016/0360-3199(94)90181-3.

About Authors

Daniel MINARIK was born in 1982. He received his Ph.D. degree in the field of implementation of hydrogen technologies in stationary applications for utilization in combination with renewable sources at Department Electrical Power Engineering at VSB–Technical University of Ostrava on Faculty of Electrical Engineering and Computer Science in 2011. He is founder of new Laboratory of Hydrogen Technologies in research center ENET, which is focused to testing general hydrogen technologies for production of hydrogen or electricity in real operation, mainly with cooperation with renewable sources or/with other varied non-traditional energetic technologies for producing and accumulation of electricity. His research interests include practical aspects and impact of e-mobility or specific hydrogen mobility and also applied research of high temperature fuel cells, which will be pushed further possibility special using of mine gases or syn-gases rich in hydrogen for purposes of combined heat and power production.

Bohumil HORAK was born in Prerov. In 1990 - he graduated at the VSB Ostrava, branches – Mechanical technology and Robotics. He obtained Ph.D. degree at the VSB–Technical University of Ostrava at branch Electronics in the year 1998, he became an associate professor in 2008 at branch Technical cybernetics. Since 1991 works at the Department measurement and control (since 2010 Department of the Cybernetics and Biomedical Engineering) of the Faculty of Electrical Engineering and Computer Science. His research work is dedicated to signal analysis, sensors, measurement and control, robotics, measurement and control of the alternate and renewable energy sources, photovoltaic systems, hydrogen technologies, energy accumulation, smart technologies and electromobility.

Petr MOLDRIK was born in 1979, graduated from the faculty of electrical engineering and computer science, VSB–Technical university of Ostrava, Czech republic, from electrical power engineering branch, in 2003, and received the Ph.D. degree in electrotechnics, communication and computer engineering, in 2008. His research activities are mainly research of energy storage gained from renewable energy sources using hydrogen and other technologies, quality of electric power supplied by co-generation units with biogas and firedamp combustion, and application of parametric models of multi-criteria analysis in the field of electrical power engineering.

Zdenek SLANINA was born in 1977. He received his Ph.D. degree in the field of real-time operating systems applications remote monitoring at Department of measurement and control, Faculty of electrical engineering and computer science, VSB–Technical University of Ostrava in 2008. He is member of "Systems with alternative energy sources" group. His research interests including control systems generally, electromobility and alternative energy sources monitoring.

25

Comparative Method for Indirect Sensitivity Measurement of UHF RFID Reader with Respect to Interoperability and Conformance Requirements

Lukas KYPUS, Lukas VOJTECH

Department of Telecommunication Engineering, Faculty of Electrical Engineering,
Czech Technical University in Prague, Technicka 2, 166 27 Prague, Czech Republic

kypusluk@fel.cvut.cz, vojtecl@fel.cvut.cz

Abstract. *There is never-ending race for the competitive advantage that forces RFID technology service integrators to focus more on used technology qualitative aspects and theirs impacts inside RFID ecosystem. This paper contributes to UHF RFID reader qualitative parameters evaluation and assessment problematic. It presents and describes in details indirect method and procedure of sensitivity measurement created for UHF RFID readers. We applied this method on RFID readers within prepared test environment and confirmed long term intention and recognized trend. Due to regulations limitations, there is not possible to increase output power over defined limits, but there are possibilities to influence reader sensitivity. Our proposal is to use customized comparative measurement method with insertion loss compensation for return link. Beside the main goal achievement, results show as well the qualitative status of development snapshot of reader. Method and following experiment helped us to gain an external view, current values of important parameters and motivation we want to follow up on as well as compared developed reader with its commercial competitors.*

Keywords

Conformance, interoperability, middleware, quality of service, RFID reader sensitivity.

1. Introduction

Radio frequency identification technology as a part of Auto-ID carriers is available in different forms for industry usage for almost 30 years. By this time it has established a strong position mainly in logistic industry [1] for its great suitability in tracking of goods [2]. RFID utilizes four or five main blocks which create each general RFID system [3]: tag, reader and antenna(s), RF channel, and optionally RFID back-end called Middleware. Its main purpose is to collect, aggregate, filter, format, normalize and report captured and gathered data by read TAGs. RFID Middleware has indisputable role and impact on quality of final service. [4]. There are overall qualitative parameters with significant impact on service, which are combination of used technology capabilities and common readers' parameters and Middlewares [5].

RFID reader faces a unique set of requirements [6]. In this paper we focused on RFID reader as the main and bearing component of whole RFID ecosystem with its main parameters. The most important are transmission power and sensitivity. Sensitivity was chosen because it is possible to influence it within the defined range already in the process of design and development. Other typical UHF operating parameters [7] are:

- Reader transmission power P_r [W].
- System operating wavelength (m), $\lambda = 0.35$ m, for $f = 865.0$ MHz.
- Reader receiver sensitivity S_r [dBm].
- Reader antenna gain G_r [dBi].
- Tag power (sensitivity) requirement P_t [dBm].
- Tag antenna gain G_t [dBi].
- Tag backscatter efficiency $E_t = -20$ dB.

Operational parameters were reflected and are relevant to our method and experiment. The most important one, for the following experiment, was sensitivity.

There are other approaches beside the availability of Digital Signal Processing (DSP) methods suitable for use cases with very low Signal to Noise Ratio (SNR) [8]. The reader sensitivity is the subject worth studying and improving. According to the industry, the reader sensitivity has a strong impact on the performance, stability, security and quality of consequent functions and services within the whole RFID ecosystem.

The main challenge we identified is how to evaluate sensitivity of the developed reader (with not known parameters from vendor) with respect to existing best practices. Sensitivity together with the output power, the reading distance and read rate/speed, are considered as the operational parameters. Reader sensitivity is dependent on several design choices as well. These are HF components and circles sensitivity, antenna layout configuration, that become more important as tag power is forced to lower value [6]. And finally the power is limited by the regulatory body as in Eq. (1), Eq. (2). We still have a chance to influence and improve previously described and explained sensitivity of the reader with respect to certain limitations. Assumed that the local regulations [9] specify that we must not exceed a transmitting power level of:

$$P_{ERPmaxEU} = 2 \text{ W} = 33 \text{ dBm}, \quad (1)$$

$$P_{ERPmaxUS} = 4 \text{ W} = 36 \text{ dBm}, \quad (2)$$

for Europe and for US.

The approach and the structure of the experimental work where we tested the presented method and the procedure in main steps are:

1. Measurement devices calibration and initial setup.
2. RFID reader's configuration.
3. continuous reading tests for all three readers.
4. record and evaluation of captured data.

Our approach was based on comparison and math of known and measured powers. The motivation was to demonstrably compare important parameters of developed RFID reader with the values of same parameters of referred, commercially available ones. The two main outputs are measurement method and reader sensitivity comparison table. Values are results of applied method on readers in testbed.

In this paper there are demonstrated measurement method and cycle in the experimental RFID test bed in details in Section 3. You find here described steps in diagram, testbed schema and situation plan. Main section ends with results, where you find table with measured values, which are discussed. This section is followed by conclusions and proposal for future work in Section 4.

2. Related Works

There is not available straight forward way to assess and compare discussed performance parameters with competition but there are certified methodologies for static and dynamic tests of RFID TAGs' sensitivity measurement and testing available [10], [11]. We studied complementary problems in papers focused on sensitivity of transponders with TAG sensitivity measurement. Industry relies on EPCglobal approved TAG related testing methodologies as well, not only according to [12], [13]. From the RFID hardware certifications perspective it is important to mention two guides: Hardware Conformance Requirements [14] and Hardware Interoperability Test Requirements [15] valid for the Electronic Product Code (EPC) world.

Since there was no indirect method available for readers, we decided to build and test one.

3. Method and Experiment

We decided to confirm our hypothesis that we are capable to assess developed reader parameters such as the sensitivity by considering initial reader setup, with eliminated and compensated the inserted path loss combined with the knowledge of forward and return link budget within testing environment.

Assumptions were confronted with tests and comparison of results with two other readers by the continuous increasing insertion loss within each measurement cycle to find the right value of the sensitivity. The aim of the experiment was to find lowest power and highest level of attenuation by comparing parameters of known readers' parameters with developed reader. The baselines were set by competitive readers Metra and AMS where there are know the key values from product sheets.

The sensitivity of a receiver indicates how great the field strength or the induced voltage U must be at the receiver input for a signal to be received without errors [3]. Method relies on well-known Friis formula[6] customized to testbed needs and suitable to count unknowns in total link energy budget for our specific case. The backscatter communication radio link budget describes the amount of modulated power that is scattered from the RF tag to the reader [14].

Equation defines the initial link budget as:

$$P_{REC} = P_{Tag} + G_R + G_T - \sum Losses - FSL - E_T, \quad (3)$$

where P_{REC} is transmitting power, P_{TAG} is TAG power sensitivity, G_R is reader antenna gain and G_T is TAG antenna gain. FSL is Free Space Loss and E_T is backscatter efficiency.

3.1. Measurement Cycle Details

We prepared measurement cycle in order to test and verify our method as applicable procedure. Steps are depicted in the Fig. 1. It starts with the initial RFID reader Fig. 1(1) configuration and setup. We kept constantly set parameters for each reader Fig. 1(2): reader output transmission power, operational frequency, frequency hopping, swept frequency range and complex impedance match. Continuous reading started Fig. 1(3). And it is processed within testbed shown in the Fig. 3 established from blocks in Fig. 2.

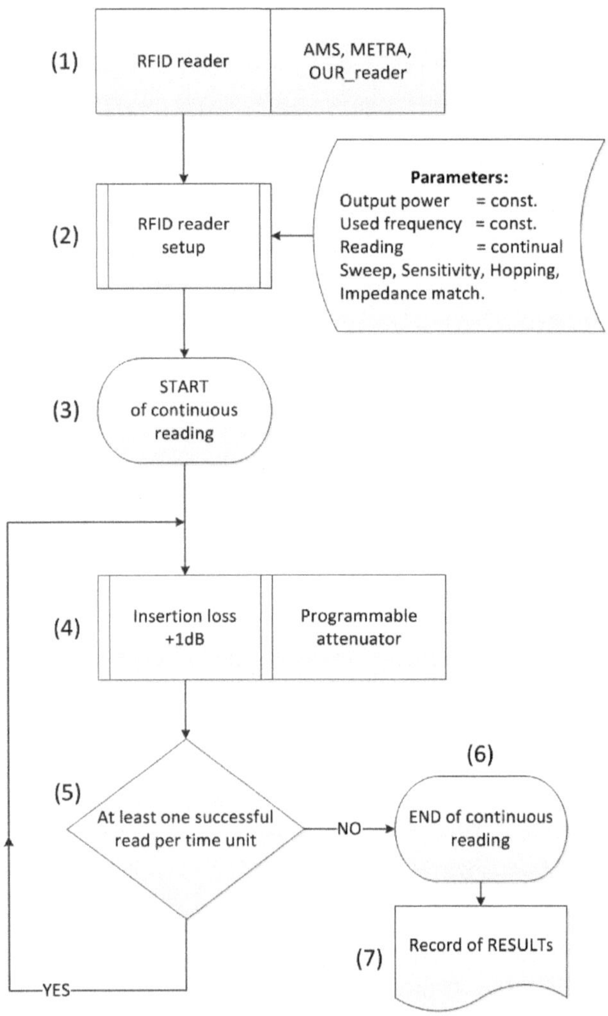

Fig. 1: Diagram of one measurement cycle.

We kept the power on the compensated level of 0 dBm, which allowed us to eliminate link path loss and then do the inter-reader sensitivity comparison, for each one of measurement cycles.

At each cycle we increased insertion loss for about 1 dB Fig. 1(4) and visually checked if the successful reading of TAG continued. We set fixed time period for about 10 s as a trigger Fig. 1(5). This is considered internally as an equivalent of BER Eq. (4), bit error rate test, which is one of the most qualitative parameter in area of digital data transmission:

$$\text{BER} = \frac{\text{number of errors}}{\text{number of transmitted bits}}. \quad (4)$$

We replaced bits with reads within defined time period. If the test failed the experiment relevant to one reader ends Fig. 1(6) and the valid value of inserted attenuation is the one from previous round. Cycle as described is the base stone of this indirect measurement method. After the failure we stopped continuous reading, recorded results Fig. 1(7), replaced reader and jumped back to the point Fig. 1(1).

3.2. RFID Testbed

Our RFID test bed environment in the Fig. 2 consists of PC Fig. 2(1) with installed reader relevant vendor applications for monitoring and management of each reader (METRA software and AMS software suitable as well for AMS reader and developed one). RFID reader itself Fig. 2(2). All readers are in impedance match with interconnected TEM-cell Fig. 2(7). Signal goes in forward and return link through two circulators Fig. 2(3) exactly according their link characteristics. Precise directional coupler Fig. 2(5) separates the power and differentiates forward and return signal direction with one gate connected as a power sensor. We had two options how to measure power of return link either on Power meter Fig. 2(6A) in [mW] or [dBm] or to use Spectrum analyzer Fig. 2(6B) as we can see in the Fig. 4. Chosen analyzer has the advantage of visualization of signal within certain frequency range and for its user friendliness as well.

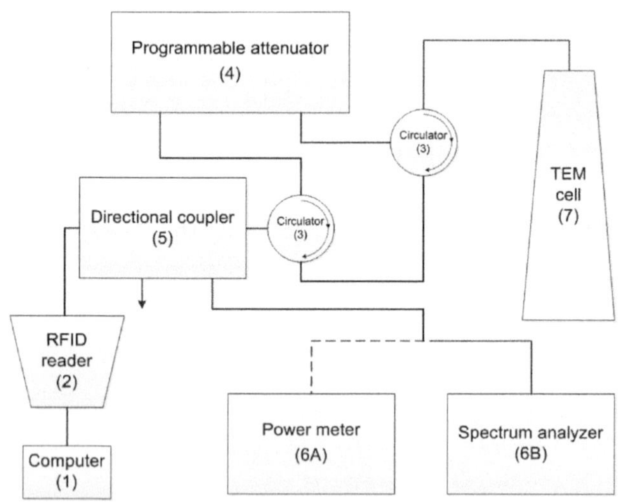

Fig. 2: Block schema of RFID testbed.

List of main components for the experiment:

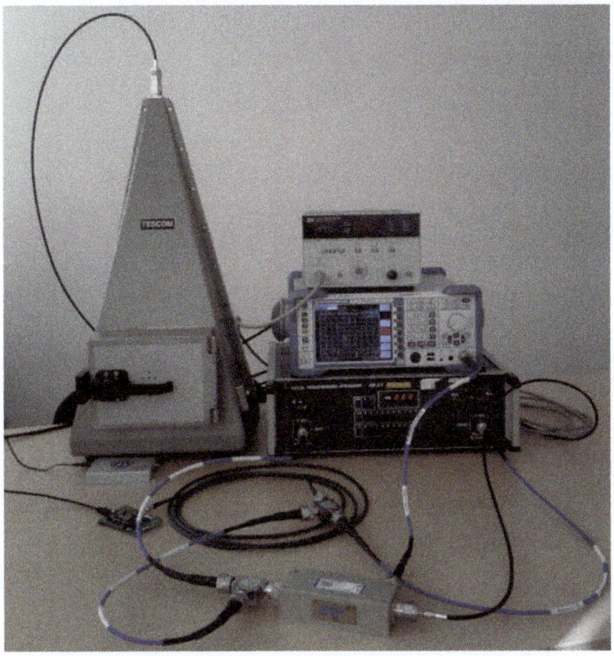

Fig. 3: Testbed situation plan.

- Certified TEM Cell Tescom TC-5060A/B UHF.
- Programmable attenuator TESLA BM577.
- Power meter Hewlet Packard 436A.
- Power sensor Hewlet Packard 8481A.
- Spectrum analyzer Rohde Schwarz 9 kHz – 6 GHz.
- Coaxial Circulator UIYCC 3232A800T900 NF.
- Ultra-broadband Directional coupler DC3010A.
- RFID UHF tags (ISO 18000-6C/EPC Class 1 Gen 2, ISO18000-6B anti-collision support).

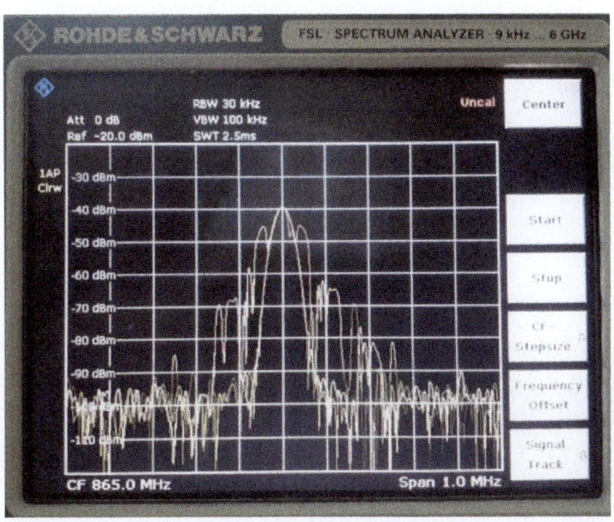

Fig. 4: Spectrum analyzer and return link power visualization.

3.3. Results

Table 1 shows parameters, relevant figures from data sheets, where applicable, and measured and counted results for each type of reader. Final values were achieved by repeated test and budget calculations for each reader during cycles with custom loss compensation and with respect to link power budget in return link direction.

Tab. 1: Comparison of data sheet values with values measured and counted results. DR=developed reader.

RFID Reader	Sensitivity datasheet [dBm]	TX power [dBm]	Sensitivity measurement [dBm]
AMS	−90	20	−90
METRA	−63	25	−63
DR	NA	20	$-63-3=-66$

Results of the practical experimental work confirmed the hypothesis from introduction as well as the real figures supported our indirect measurement method. During test with developed reader we were able to increase insertion loss for about −3 dBm more than at Metra reader. The appropriate link budget math with final sensitivity of −66 dBm is emphasized in Tab. 1.

4. Conclusion

The method and procedure together with experiment approach presented in this paper were developed and tested on three different UHF RFID readers. Results were intended to help to watch important parameters of RFID reader prototype during development and especially in level of achieved sensitivity as well as to give us direction how to compare developed reader with external commercial products with sensitivity focus. Interesting is the impact of fully managed sensitivity on overall quality of service to whole RFID ecosystem with Middlewares due to increasing demand for efficiency, speed and accuracy of reading. It is directly visible from the results that we are for about 24 dBm from the project goal with level of −90 dBm.

We experimented with placement of Directional coupler represented by block Fig. 2(6B) in Fig. 4 and at the end we decided to put it as close as possible to evaluated reader, to eliminate an impact of additional loss.

Requirements for future work were identified as need of improved triggers (from manual to automated) inside the "read error rate test" (RER). We want to be able to exactly recognize and record first fail period of time and incorporate statistical mechanisms.

The goal of each development team is to be able to influence positively the sensitivity which is supported by the visible trend at competitive readers. Achieved and measured sensitivity values at reader contributed to our motivation to continue in development.

Acknowledgment

This work was supported by grant in EUREKA Cluster programme with EUREKA7592 AutoEpcis project and by Student grant at Czech technical university in Prague SGS14/141/OHK3/2T/13.

References

[1] KEBO, V., P. STASA, F. BENES, J. SVUB. RFID Technology in Logistics Processes. In: *Proceedings of the 13th International Multidisciplinary Scientific GeoConference*. Albena: EBSCO Publishing, 2013, pp. 219–226. ISBN 978-954-91818-9-0.

[2] VACULIK, J., I. MICHALEK and P. KOLAROVSZKI. ADM - Principles of selection, implementation and utilization of RFID in supply chain management. *Promet - Traffic & Transportation scientific journal on traffic and transportation research*. 2009, vol. 21, no. 1, pp. 41–48. ISSN 0353-5320.

[3] FINKENZELLER, K. *RFID Handbook: Fundamentals and Applications in Contactless Smart Cards and Identification*. Third ed. Chichester: John Wiley & Sons, 2010. ISBN 978-0-470-69506-7.

[4] RIDA, A., L. YANG and M. TENTZERIS. *RFID-Enabled Sensor Design and Applications*. London: Artech House, 2010. ISBN 978-1-60783-981-1.

[5] KYPUS, L. *Research and evaluation of qualitative parameters of middleware for heterogeneous RFID networks: SGS14/141/OHK3/2T/13*. Prague, 2014. Available at: `http://www.fel.cvut.cz/en/en/13132/grants/2014.html`.

[6] DOBKIN, M. *The RF in RFID. Passive UHF RFID in Practice*. Oxford: Elsevier Inc., 2008. ISBN 978-0-7506-8209-1.

[7] LEHPAMER, H. *RFID Design Principles*. London: Artech House, 2008. ISBN 978-1-59693-194-7.

[8] VOJTECH, L. Feasibility of RFID Signal Denoising Using Neural Network. *Advances in Electrical and Electronic Engineering*. 2010, vol. 8, no. 5, pp. 111–113. ISSN 1804-3119. DOI: 10.15598/aeee.v11i5.889.

[9] PARET, D. *RFID at ultra and super high frequencies. Theory and application*. Chichester: John Wiley & Sons, 2009. ISBN 978-0-470-03414-9.

[10] ETSI EN 302 208-1. *Electromagnetic compatibility and Radio spectrum Matters (ERM); Radio Frequency Identification Equipment operating in the band 865 MHz to 868 MHz with power levels up to 2 W; Part 1: Technical requirements and methods of measurement*. Valbonne: ETSI, 2008.

[11] ISO/IEC 18000-6:2010. *Information technology – Radio frequency identification for item management: Part 6: Parameters for air interface communications at 860 MHz to 960 MHz*. Geneva: IEC, 2013.

[12] EPCglobal, Inc. Community. *Dynamic Test: Conveyor Portal Test Methodology For Applied Tag Performance Testing Rev 1.1.4 Test Methodology*. Brussels: EPCglobal, 2006.

[13] EPCglobal Technical Steering Committee and EPCglobal Business Steering Committee: *UHF Tag Performance Parameters and Test Methods Version 1.1.3*. Brussels: EPCglobal, 2008.

[14] EPCglobal: Powered by GS1. *EPC^{TM} Radio-Frequency Identity Protocols Class-1 Generation-2 UHF RFID Conformance Requirements V.1.0.6*. Brussels: EPCglobal, 2011.

[15] EPCglobal: Powered by GS1. *Interoperability Test System for EPC^{TM} Compliant Class-1 Generation-2 UHF RFID Devices, Interoperability test methodology, V.1.2.4*. Brussels: EPCglobal, 2006.

[16] TURCU, C. *Designing and deploying RFID applications*. Rijeka: InTech, 2011. ISBN 978-953-307-265-4.

About Authors

Lukas KYPUS is a Ph.D. student since 2013 in Center of Automatic Identification and RFID lab at the Czech Technical University in Prague. He received his M.Sc. degree on Faculty of Electrical Engineering and Communication at Brno University of Technology. He worked at different technical positions where he was responsible for design, implementation and operations of telecommunication protocols, devices and services. Currently his research interests are RFID/NFC technologies and information security. He works on thesis in area of low cost secure RFID identifiers.

Lukas VOJTECH received M.Sc. and Ph.D.

in Telecommunication Engineering at the Czech Technical University in Prague, Czech Republic in 2003 and 2010. He has been actively involved in several national and international projects founded by Ministry of Industry and Trade Czech Republic and European Commission such as EUREKA, RFID locator. He is a leader of RFID laboratory at the Czech Technical University in Prague since 2010. He is an author or co-author of more than 50 papers in international journals, conferences and results of applied research. His research interests are hardware prototyping and measurement especially in the field of RFID/NFC technology, textile antenna design and RFID localization.

Increasing Robustness of Multi-homed Systems in Heterogeneous Environment

Matej ROHLIK, Petr CHLUMSKY, Tomas VANEK

Department of Telecommunication Engineering, Faculty of Electrical Engineering, Czech Technical University in Prague, Technicka 2, 166 27 Praha 6, Czech Republic

matej.rohlik@fel.cvut.cz, petr.chlumsky@fel.cvut.cz, tomas.vanek@fel.cvut.cz

Abstract. *Nowadays, packet error rate in fixed networks can be considered as a negligible value. However, the increasing requirements for transmission speed of mobile devices, heterogeneous technology, and other high frequency sources cause interference growth within the electromagnetic spectrum. This affects the overall reliability and throughput of the network and may cause undesirable operation malfunction of application-level services. Higher speeds can be achieved by advanced modulation techniques, but at the price of lower resistance against the interference. On the other hand, error-correcting codes or higher-level protocols are utilized to correct the delivery failures. We introduce a novel method for increasing robustness of communication for multi-homed systems in heterogeneous environment. Furthermore, we propose a security measure to ensure confidentiality, integrity and availability of the transmitted data without influencing the transmission parameters. Finally, we show positive impact of the proposed method on transmission efficiency and effective throughput, especially in networks with high probability of error occurrence.*

Keywords

Heterogeneous packet network, multi-homed load balancing, network coding, security.

1. Introduction

Contemporary 4G mobile technology standards, such as the IEEE 802.16m-2011 (WirelessMAN-Advanced) or the LTE-Advanced (LTE-A) provide solid foundation for further development of the future 5G standard and related services. However, from its substance, wireless communication evinces variable parameters in the terms of signal to interference plus noise ratio (SINR), channel capacity, delay, jitter and throughput, which affect possible deployment of new potentially emerging services.

Clearly on a global scale, the volume of mobile data traffic continuously increases. To efficiently cover the demand of customers, one of the promising solutions seems to be to split the service requests among available communication ways, thus utilize a multiple path bonding of multi-homed devices and provide higher data rate to the end users.

The authors in [1] dealt with bandwidth aggregation techniques in heterogeneous multi-homed devices and surveyed the most prominent solutions that were designed to address effective bandwidth utilization of appliances connected to several computer networks, enhance overall communication reliability, or opportunistic load offloading. Because of several aspects, which needs to be taken into account, when designing multipath communication systems, the conclusion is ambiguous and does not provide one and only protocol that should be used. However, it outlines evolution trends challenges requiring further research and one of which we address in this paper.

Therefore, we present a general approach for heterogeneous multi-homed networks, which is based on inverse packet multiplexer (IPM) while it eliminates the original disadvantage of IPM, which represents an uncoordinated sorting of packets to multiple transmission paths, that can result into an uncontrolled growth of transmission delay [1], higher layer protocol instability and/or session breakdown.

The principle is achieved by application of a control polyline function that regulates the amount of packet data passing through the data channels. Moreover from end-to-end perspective, the system is transparent, independent on operating systems or network devices, increases stability of the connection by regulation of the delay value during the transmission, optimizes data coding to provide higher service reliability, and provides adequate security for the user data.

2. Towards Secure and Reliable Packet-Based Future Heterogeneous Systems

To incorporate techniques which substantially increase end-to-end throughput of user-data, stability of the connection and reliability of transferred data delivery into the future mobile networks, a transparent and architecture independent solution needs to be designed. Moreover, where efficient security is required [2], effective cryptographic and coding measures must be assured to provide adequate quality in the terms of end-to-end throughput and delay. Therefore, in this section, we briefly introduce current multi-homing architecture approaches in nowadays packet networks. Subsequently, we present enhancements towards secure, reliable and load-balanced packet delivery in future mobile systems.

2.1. Current IP-Related Multi-Homing Architectures

The network layer, as per the ISO/OSI reference model, or the so called the Internet layer, is widespread based on the Internet protocol (IPv4/IPv6) which relies on interconnection of independent network parts denoted as the autonomous systems (ASs). The natural dynamic behaviour is managed by border routers which utilize dynamic routing protocols, such as the border gateway protocol (BGP), to exchange the network topology modifications as the particular network availability changes. Even though the IPv4 and the IPv6 work in a different manner, they both rely on the AS principle and both are desired to provide redundancy and load-sharing [3]. However generally, the principles have not been completely resolved or standardized yet [4], and without proper end-to-end administrative control, the proportional load balancing (Fig. 1) is impossible to ensure for the returning path even while utilizing BGP [5].

In [1], the authors accomplished a complex survey of the bandwidth multiplexing methods in heterogeneous environment, such as the multiple radio access technology (Multiple-RAT), [6]. The contribution has been published lately thus we consider it comprehensive. Even though the analysed solutions do affect either the application layer, the transport layer, the network layer, or the link layer, none of them has been standardized, because none of them is universal enough to cover all heterogeneous networks (IP-, non-IP-based) and/or various physical connections under different application requirements while efficiently implementing load balancing with minimal header overhead, delay and providing adequate security. Therefore, we focus on a transparent approach that addresses them.

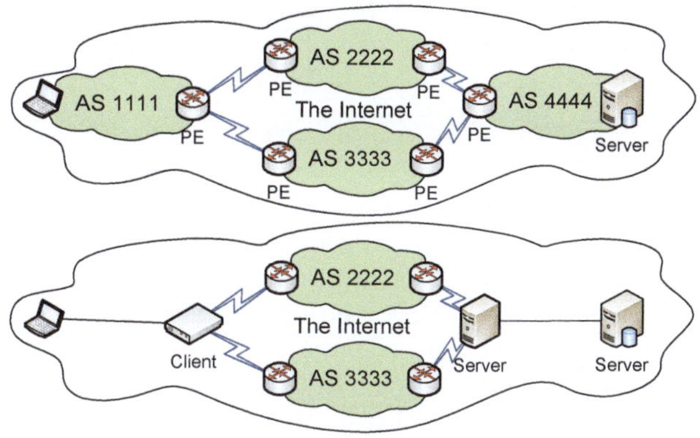

Fig. 1: BGP-based and non-BGP-based multi-homing example.

2.2. Transparent Transmission Environment Compliance

The elementary idea is to prevent congestion and deterioration of the transmission parameters due to classification of incoming packets according to current states of the individual outgoing interfaces the packet are about to be sent to, and to establish a stable data channel resistant to fluctuating transmission parameters, by which a wireless or a mobile environment is typically characterized. The solution adopts a packet (polyline) regulator principle that is based on inverse packet multiplexer [7].

However, the regulator principle does not address optimized reliability or security (Fig. 2). Therefore, within our approach, we focus on the network coding principles that reduce the overall processing delay and require a minimum buffers size while preserving adequate reliability in the terms of delivery of network packets.

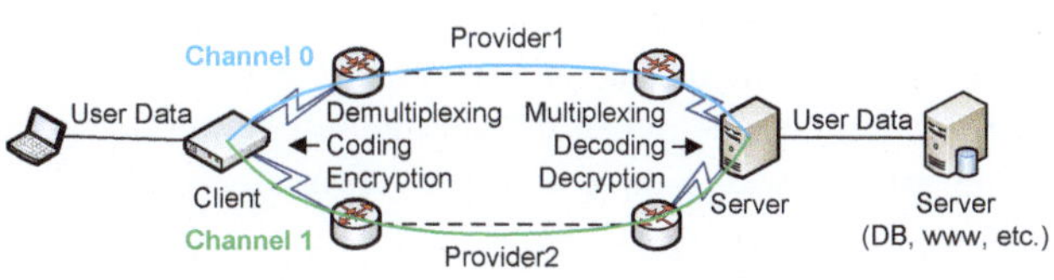

Fig. 2: Transparent transmission multi-homing principle.

Network coding is a novel technique introduced at the turn of the millennium with the intention to improve network throughput, robustness and wireless resources. In usual communication networks, every intermediate node works with an assumption that independent data streams may share resources, but the information itself is separate. Network coding breaks this assumption. Instead of simply forwarding data, nodes are enabled to recombine several messages from different input flows into one or several output packets [8]. The authors in [9] showed that we can allow intermediate nodes to process incoming streams, not just forward them. Benefits of network coding techniques in terms of throughput and robustness are described in e.g. [10], [11], [12].

Our system is based on the combination of network coding and path diversity principles. An appropriate linear combination of data, specifically with logical function, produces an additional redundancy for the transmission that creates an option of setting the ratio of resistance and transmission rate. Channel 1 is used for the combination of two consecutive messages. If a single message or more messages M_n get lost, is possible to reconstruct original data without any retransmissions in Channel 0 (Fig. 3(a)). Number of messages suitable for reconstruction is dependent on the size of buffer at the receiver side of transmission.

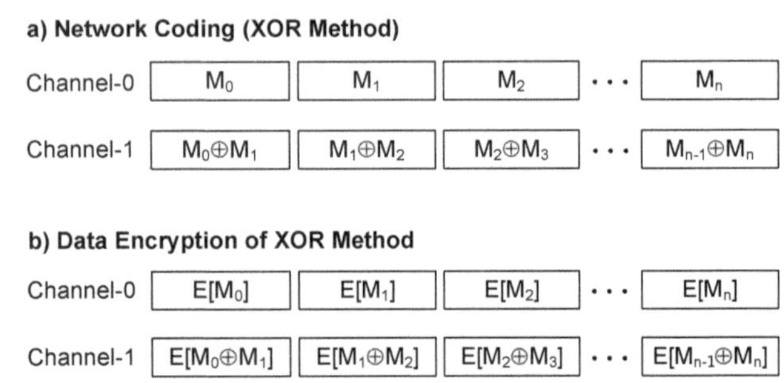

Fig. 3: Network encryption principle.

2.3. System Security Aspects

Undoubtedly, multiple transmission paths make a potential eavesdropping attack difficult. But on the other hand, considering that the attacker is able to intercept data on a single connection, most probably will be (technically) able to perform the same malicious activity on the rest of the links. Therefore, such measure cannot be considered as a data protection technique [13], hence we focus on a satisfactory security principle which does not affect the transmission parameters while preserves the system resources.

To be able to state that the system is secure, it is necessary to fulfil the three information security attributes: confidentiality, integrity and availability (CIA). The availability is assured by the network coding mechanism described in the previous section (referred to as reliability). To provide consistency of data, integrity mechanisms generally add redundant information to the original data. Therefore, considering that today's modern devices are endowed with HW support of block algorithm denoted as the Advanced Encryption Standard (AES), we utilized AES in Galois/counter mode of operation (AES-GCM), which is a confidentiality and authenticated encryption mechanism that includes integrity function while minimizing the extra information added to the user data thus minimizing latency and operation overhead [14]. Even though, stream ciphers are faster when implemented in HW, they are not considered in this case as they do not provide the integrity feature and vulnerable to the modification attacks. In [15] on the other hand, the authors outlined parameters (such as pipelined and parallelized implementations and minimal computational latency) which are requested that the mode must fulfil in order to be efficient for high-speed data rates. The counter mode has emerged as the best method.

Algorithm 1 Robustness increasing.

ASSUMPTION: a packet buffer B exists
CREATE an empty queue Q
GET packet P from B
ENQUEUE P onto Q
GET packet P from B
ENQUEUE P onto Q
while Q is NOT empty **do**
  $M_0 \leftarrow Q.$
  $M_1 \leftarrow Q.$
  E_0 $E\ M_0$
  E_1 $E\ M_0 \oplus M_1$
  SEND E_0 via Channel 0
  SEND E_1 via Channel 1
  GET packet P from B
  ENQUEUE P onto Q
end while

The principle of securing the multiple encoded data stream (Alg. 1) is based on encryption (M_n) of single messages (M_n) or encryption of coded pairs of consequent packets ($M_{n-1} \oplus M_n$) by application of AES (see Fig. 3b). The AES-GCM is an algorithm providing encryption and hashing mechanism while it minimizes the overall data overhead of the original message and maximizes throughput, especially, when efficiently combined with dedicated HW resources such as the field-programmable gate arrays (FPGAs),[16].

3. Performance Evaluation

The purpose of this paper is to improve the architecture of heterogeneous multi-homed devices or networks with high probability of packet loss, such as the wireless communication systems. While preserving adequate security for the user data, the proposed approach aims at increasing reliability and robustness of the connection by optimizing the data coding. We utilized the open-source OMNeT++ network simulation framework to demonstrate the proposed mechanism behaviour.

3.1. Transmission Efficiency

In the first simulation, we focused on the transmission efficiency, which was assumed as number of received to the number of retransmitted messages ratio. Compared to a load-balancing mechanism utilizing both data channels for data transmission; based on its nature, the proposed approach does not require as many messages to be resent as the original load-balancing mechanism. This is due to the feature, which enables the receiver to reconstruct specific amount of messages which were not delivered as a result of loss.

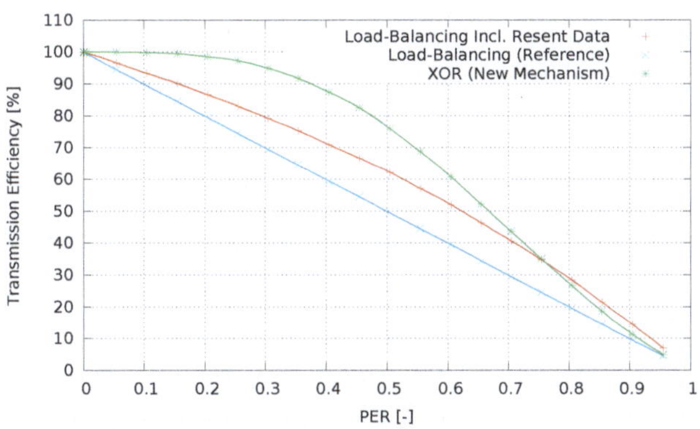

Fig. 4: Transmission efficiency.

The Fig. 4 demonstrates comparison of common load-balance method with and without retransmission with the proposed mechanism in terms of transmission efficiency. The load-balancing without resending lost data is the worst case in the entire monitored section because of the impact of packet error rate ( ) for both channels and the impossibility of any recovery. The mechanism is very efficient up to packet error rate (.), where the load-balancing becomes more efficient. This is caused by various options for data reconstruction of the mechanism. With the growing number of errors, the load-balance loses its natural advantage of two separate channels for the data transmission and the benefit of lost packet reconstruction in the new mechanism becomes more important. The efficiency turning point (.) describes situation where the new mechanism does not have enough data for the reconstruction because of extremely high error rate.

3.2. Effective Throughput

From the effective throughput perspective considering lossless environment, there is a clear benefit in utilizing both channels for data transmission in a load-balancing mode and not for redundancy as in the case of a new mechanism. Nevertheless with the increasing error rates (above .), it can be observed that the proposed mechanism becomes efficient.

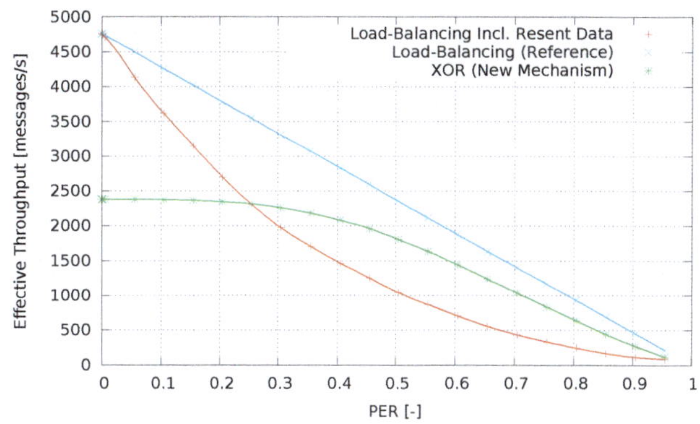

Fig. 5: Effective throughput.

Retransmissions of the data poses serious impact on effective throughput of the data; especially, shared wireless transmission technologies are sensitive to the amount of transmitted data and any data sent in addition has negative impact on the performance. In order to preserve generality of the result, we considered within the simulations that each lost message is resent only once and then assumed delivered. In other words, the probability a resent message will be lost again is put to zero. In a real world environment, such probability is not null, thus the effective throughput of the load-balancing mechanism with retransmissions would be far less effective (Fig. 5).

3.3. Practical Impact of Data Loss

By paying more attention to the fact, how does to lost data affects higher layer protocols, it can be seen that based on specific implementations of reliable protocols, such as the worldwide used transmission control protocol (TCP), can a single missing fragment of data cause retransmission of all consequent data or a more efficient method, such as the caching memory (buffer), is required to be implemented on both ends of the transmission path.

If the transmission systems operate in environment with higher value of error rate ( ), the size of the buffer is required to be increased with respect to the value. Otherwise with increasing value, the efficient throughput will further decrease as already received data will be required to be resent due to insufficient memory space in the buffer (Fig. 6).

The proposed method utilizes buffers of a constant size. Therefore, it is unaffected by the actual value. Moreover, size of these buffers is given by a required number of messages correction possibility and its size is negligible compared to the neces-

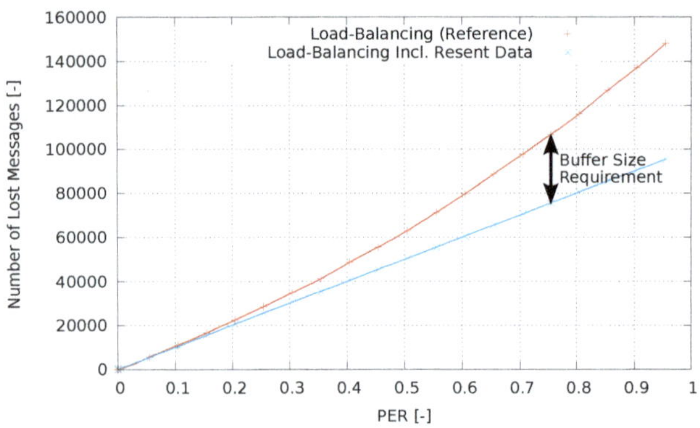

Fig. 6: Buffer size requirement.

sary buffer size in the case of the load-balance method including lost data retransmissions. Apparently, the extreme values of PER → were evaluated for simulation and theoretical research purposes. We assume that the kind reader considers the practical aspects of using networks within such ranges.

4. Conclusion

The paper is focused on a novel secure transparent multi-homing architecture. We have evaluated standardized and non-standardized approaches, which are available, and have concluded that a secure and transparent solution for higher-level protocols is currently unavailable. Since our proposal is not based on dynamic routing protocols, it can be utilized not only for network connectivity backup, but also as an efficient solution used by the end devices. Moreover, the approach we have designed is high-performance in environments evincing variable parameters due to the proposed coding scheme, as it significantly reduces number of resent messages if these get lost as a result of increased packet error rate.

Currently, the solution utilizes two proprietary linux-based devices (a client and a server appliance) which can be considered as a relative disadvantage. On the other hand, the client-side is planned to be programmed as a software application (e.g. for mobile devices). We believe that its redesign and utilization of more than two physical interfaces can prepare such solution for standardization, without which it can hardly consider wider application.

Acknowledgment

This work was supported by the Grant of the Ministry of the Interior of the Czech Republic, No. VG20122014095, "Development of adaptable and data processing systems for high-speed, secure and reliable communication in extreme conditions", and was researched in cooperation with CERTICON.

References

[1] HABAK, K., K. A. HARRAS and M. YOUSSEF. Bandwidth aggregation techniques in heterogeneous multi-homed devices: A survey. *Computer Science - Networking and Internet Architecture.* 2013, vol. abs/1309.0542. Available at: `http://arxiv.org/abs/1309.0542`.

[2] MACURA, L., J. ROZHON, F. REZAC, J. VYCHODIL and M. VOZNAK. Secured monitoring probe with highly redundant WAN connection. In: *Proceeding of the conference Knowledge in Telecommunication Technologies and Optics.* Ostrava: VSB–Technical University of Ostrava, 2010, pp. 118–121. ISBN 978-80-248-2330-0.

[3] ABLEY, J., B. BLACK and V. GILL. *Goals for IPv6 Site-Multihoming Architectures.* RFC 3582, 2003. Available at: `https://www.ietf.org/rfc/rfc3582.txt`.

[4] KUNTZ, R., J. MONTAVONT and T. NOEL. Multihoming in IPv6 mobile networks: progress, challenges, and solutions. *Communications Magazine.* 2013, vol. 51, iss. 1, pp. 128–135. ISSN 0163-6804.

[5] ABLEY, J., K. LINDQVIST, E. DAVIES, B. BLACK and V. GILL. *IPv4 Multihoming Practices and Limitations.* RFC 3582, 2005. Available at: `http://www.ietf.org/rfc/rfc4116.txt`.

[6] GHAMARI, S., M. SCHELLMANN, M. DILLINGER and E. SCHULZ. An approach for automated spectrum refarming for multiple radio access technologies. In: *Telecom World (ITU WT), 2011 Technical Symposium at ITU.* Geneva: IEEE, 2011, pp. 187–192. ISBN 978-1-4577-1148-0.

[7] KOCUR, Z., J. VODRAZKA, P. MACEJKO and V. MARIK. *Adaptive system for increasing the speed and reliability of data transmission in a packet network with optimized delay.* Utility Model no. 25772. August 2013.

[8] FRAGOULI, C., J. Y. LE BOUDEC and J. WIDMER. Network coding: An instant primer. *ACM SIGCOMM Computer Communication Review.* 2006, vol. 36, no. 1, pp. 63–68. ISSN 0146-4833. DOI: 10.1145/1111322.1111337.

[9] ALSHWEDE, R., N. CAI, S. R. LI and R. W. YEUNG. Network information flow. *IEEE Transactions on Information Theory.* 2000,

vol. 46, iss. 4, pp. 1204–1216. ISSN 0018-9448. DOI: 10.1109/18.850663.

[10] YEUNG, R., S. R. LI, N. CAI and Z. YHANG. Network coding theory. *Foundation and Trends in Communications and Information Theory.* 2005, vol. 12, iss. 4-5, pp. 241–381. ISSN 1567-2328.

[11] HO, T. and D. S. LUN. *Network Coding: An Introduction.* Cambridge: Cambridge University Press, 2008. ISBN 978-1-139-47018-6.

[12] KATTI, S., H. RAHUL, W. HU, D. KATABI, M. MEDARD and J. CROWCROFT. XORs in the Air: Practical Wireless Network Coding. *IEEE/ACM Transactions on Networking.* 2008, vol. 16, iss. 3, pp. 497–510. ISSN 1063-6692. DOI: 10.1109/TNET.2008.923722.

[13] LOU, W., W. LIU and I. FANG. Spread: improving network security by multipath routing. In: *Military Communications Conference.* Boston: IEEE, 2003, pp. 808–813. ISBN 0-7803-8140-8.

[14] DWORKIN, M. *Recommendation for block cipher modes of operation: Galois/Counter Mode (GCM) and GMAC.* U.S.: National Institute of Standards and Technology, 2006.

[15] MCGREW, D. A. and J. VIEGA. The Security and Performance of the Galois/Counter Mode (GCM) of Operation. In: *5th International Conference on Cryptology.* Berlin: Springer, 2005, pp. 343–355. ISBN 978-3-540-24130-0. DOI: 10.1007/978-3-540-30556-9_27.

[16] ABDELLATIF, K. M., R. CHOTIN-AVOT and H. MEHREZ. Efficient AES-GCM for VPNs using FPGAs. In: *56th International Midwest Symposium on Circuits and Systems.* Columbus: IEEE, 2013, pp. 1411–1414. ISBN 9781479900657. DOI: 10.1109/MWSCAS.2013.6674921.

About Authors

Matej ROHLIK received his M.Sc. and Ph.D. degree from the Czech Technical University in Prague, Faculty of Electrical Engineering, in 2008 and 2012 respectively. His research is focused on security of future mobile networks and broadcast security. He is actively involved in projects focused on security of the next generation mobile networks and high speed mobile transmission optimization.

Petr CHLUMSKY received his M.Sc. from the Czech Technical University in Prague in 2010. Since 2010 he has been studying Ph.D. degree. His research interests include wireless transmission, network coding and network simulation.

Tomas VANEK received his M.Sc. degree in telecommunication engineering from the Czech Technical University in Prague, Faculty of Electrical Engineering, in 2000. In 2008 he received the Ph.D. degree in applied cryptography from Czech Technical University in Prague, Faculty of Electrical Engineering. Currently, he works as an assistant professor at the Department of telecommunication engineering, CTU in Prague. His research interests include cryptography, advanced network protocols and VoIP security.

Extreme Production Conditions of Photovoltaic Power Plant Operated in Distribution Grid

Martin SMOCEK, Zdenek HRADILEK

Department of Electrical Power Engineering, Faculty of Electrical Engineering and Computer Science, VSB-Technical University of Ostrava, 17.listopadu 15/2172, 708 33 Ostrava-Poruba, Czech Republic

martin.smocek@vsb.cz, zdenek.hradilek@vsb.cz

Abstract. *Photovoltaic power plants are sources of electrical energy that are very dependent on weather conditions. This paper interests in description of stochastic production of power in photovoltaic power plants and subsequently its impact on the distribution power grid. Stochastic production of power from these sources is considerable and hardly predictable. The aim of this survey is to assess the influence of photovoltaic power plants operation on the daily load diagram with regard to output change difference of active power at the relevant substation. Evaluation is based on real-time synchronous data measuring both on-site photovoltaic power plant operation and on-site electric power line output from the substation. This measurement is used for suggestion of a general calculation methodology for the assessment of the difference of active power at a particular node of the grid under various power changes from the photovoltaic power plant. Statistical methods have been employed to process a methodology in order establish extreme conditions of production power for photovoltaic power plant.*

Keywords

Calculation methodology, photovoltaic power plant, real measurement, statistical methods, stochastic effect.

1. Introduction

Dependence operation of both photovoltaic power and wind power plant is considerable on meteorological conditions [3], [4]. This paper contains a description of the assessment focused on the issues of calculation methodology. Most studies deal with the prediction of power as [6], but the purpose of this methodology is to determine differences in active power induced by operation of photovoltaic power plants (PVP). The survey is based on actual long-term monitoring of a specific power plant. The data available has been obtained by measurement at the PVP operation site as well as the point of electric power output from substation plant, where the PVP is connected. Both measurements were conducted on synchronised basis.

The evaluation of PVP production is conducted by means of several individual tasks that can be split into two stages. The first stage determines methodology for set-up of extreme thresholds for production generated by the PVP and the second stage concerns analysis of the impact of these extreme threshold values on the magnitude of active power flows at the place of measurement within substation plant. The reached results from the analysis of PVP are use to establish a methodical calculation model to consider the impact of stochastic production of active power in the PVP on the distribution grid at various nominal values of peak power within various time intervals throughout the year.

The operators of such grids consider the stochastic nature of the power supplied undesirable. Such conditions result in higher demand for reserve capacity in terms of Ancillary Services to ensure the required balance between immediate production and consumption of electric power. Potential changes to power flow represent just another negative consequence. That affects the proper functioning of means for grid protection. There are also problems with keeping the required stable operating grid parameters. This calculation model is established with respect to the potential impact on preserving the magnitude of capacity Ancillary Services [5].

2. Area of a Measured Grid

The data was obtained by measurement both at PVP operation and on the relevant power line output of the substation plant, where this PVP is connected [1].

2.1. Database of Measured Data

1) Photovoltaic Power Plant

The peak output of this PVP is 1.1 MWp. The measurement was conducted on the low voltage side on a continuous basis for one year. Accurately that was from 30.6.2010 till 29.6.2011. The data was recorded with time step set at one-minute increments.

2) Output Substation Plant

The measurement at the power line output was conducted on the high voltage end and ran within various time periods from 26.7.2010 till 23.6.2011. Same as for PVP, the measurement was performed at one-minute increments. The measurement process produced many operating values and parameters. This survey deals with the analysis of the active power P over time path t. Percentage evaluation and mutual comparison of data measured on PVP and at the substation point have been illustrated in Fig. 1. Columns in charts represent individual months. The last column shows an average of all months monitored. In total, there are 98 % of the data from PVE and 64 % of the data from the substation, where:

- p - percent availabilily,
- x-axis - individual months,
- su - average sum of individual months.

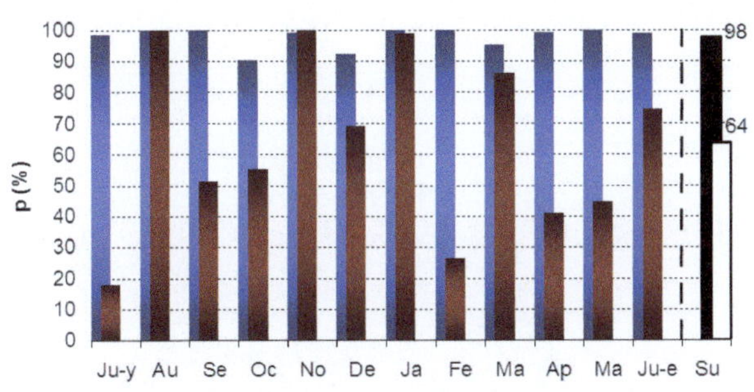

Fig. 1: Percentage comparison of data measured [1].

2.2. Network Topology

The network topology has a loop system, yet it is operated in the radial system. When this measurement was conducted, the PVP was the only electric power source with significant output on a given power line, where:

- PVP - photovoltaic power plant,
- SS - section switch (disconnected),
- 110/22 kV - substation plant 110/22 kV.

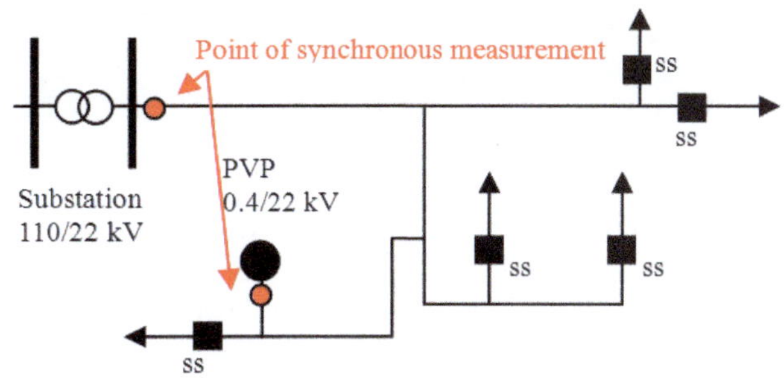

Fig. 2: Measure grid topology [1].

The first measurement was performed at the outgoing feeder from the substation 110/22 kV using the side of lower voltage level of 22 kV. The second measurement was done at the connection point of the photovoltaic plant to the grid at the voltage level of 0.4 kV. Both measurements were conducted in a synchronous manner. Points of measurement are flagged red in Fig. 2.

3. Ancillary Services

The Transmission Grid Code (GC), part II, as amended by revision 14 duly approved by the Energy Regulatory Office on 1.1.2014 in Czech Republic, defines the matters relevant to the utilisation of GC with respect to Ancillary Services. The document contains methodology for determination of summary control backups of SS for the power grid in Czech Republic. One of the prerequisites for determination of summary control backups of SS is the so-called deviation $OD_{OZE(t)}$. This is a parameter to establish the increment of balancing deviation caused by production from newly installed renewable resources of electric power. As stated here (1), this increment is defined by the sum of the increase generated by operation of wind $OD_{VtE(t)}$ and photovoltaic $OD_{FVE(t)}$ power plants. The deviation $OD_{OZE(t)}$ is then used to determine volumes of SS required for the subsequent period, based on existing information from preparations for the operation, e.g.:

$$OD_{OZE(t)} = OD_{VtE(t)} + OD_{FVE(t)}. \qquad (1)$$

This statement, described equation (1), implies that the effect of installed capacity of photovoltaic power plants has a certain impact on the calculation of SS. Aim of this survey is to assess if and how much the PVP production has influence to total calculation of SS with respect to individual months. The similar approach of evaluation is also described in [7].

4. Extreme Conditions

The consistency and comprehensibility of this article are supported by the results from own survey to define a methodology for determination of extreme production conditions of photovoltaic power plant. These matters were presented in detail at the conference [2]. Essential results will be described.

4.1. Photovoltaic Power Plant

Figure 3 shows three curves as a sample of analyzed data in this paper. Individual curves correspond as follows:

- producend power from PVP (green),
- power flowing through the substation (blue),
- loads (black).

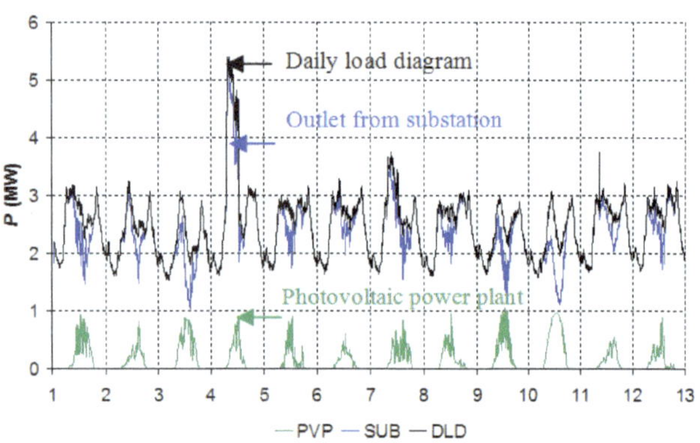

Fig. 3: Sample of examined data.

The power difference of PVP has a negative impact on immediate power balance between production and consumption within the power grid. For this reason, the analysis of extreme thresholds is necessary. Based on factors described in [2], one-minute output series was converted into five-minute increments. Deformation of the data information with respect to further processing and evaluation target is negligible. As an example, further steps described in this article analyse data in April 2011.

Figure 4 shows extreme areas of PVP production (the green curve from Fig. 3. The thresholds of these extreme conditions are defined by two regression polynomial Eq. (2) and Eq. (3), which form the top and bottom envelope curves of stochastic changes in the produced power. The equations are as follows:

$$P_{FVEmax} = -5700,51 + 63525,7 \cdot Time \\ -246642 \cdot Time^2 + 400745 \cdot Time^3 \\ -279206 \cdot Time^4 + 66861,7 \cdot Time^5, \quad (2)$$

$$P_{FVEmin} = 4554,05 - 68932,4 \cdot Time \\ +414328 \cdot Time^2 - 1,25591 \cdot 10^6 \cdot Time^3 \\ +2,0145 \cdot 10^6 \cdot Time^4 - 1,62904 \cdot 10^6 \cdot Time^5 \\ +523013 \cdot Time^6. \quad (3)$$

Determination of these extreme conditions was conducted based on statistical methods that utilize the relevant frequency categories using the so-called Sturges rule. This is a rule for optimal determination of frequency categories. Next step is to evaluate daily load

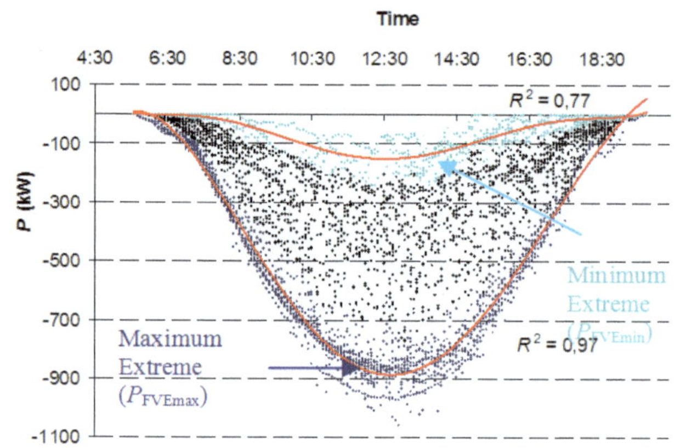

Fig. 4: Extreme conditions of PVP.

diagram. Correlation fields of daily load diagram (the black curve from Fig. 3) can be subject to regression analysis applied to enable determination of differential power thresholds of loads. It is described in Fig. 5. Extreme threshold conditions, Eq. (4) and Eq. (5), have been determined using 95 % prediction levels. For more information on this issue refer to [2]. The equations are as follows:

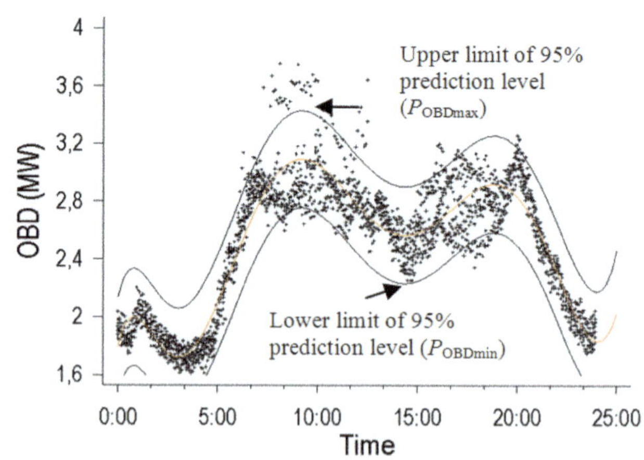

Fig. 5: Extreme conditions of daily diagram.

Previous analyses have produced the extreme difference of output generated by PVP and determination of thresholds for daily load diagram. Extreme conditions that might occur at the substation plant outlet correspond with the sum of extreme conditions both PVP

$$P_{OBDmax}(MW) = 2,13445 + 0,000174187 \cdot Time - 4,6908 \cdot 10^{-8} \cdot Time^2 + 4,32657 \cdot 10^{-12} \cdot Time^3 \\ -1,78877 \cdot 10^{-16} \cdot Time^4 + 3,85326 \cdot 10^{-21} \cdot Time^5 - 4,49778 \cdot 10^{-26} \cdot Time^6 \\ +2,69318 \cdot 10^{-31} \cdot Time^7 - 6,46948 \cdot 10^{-37} \cdot Time^8, \quad (4)$$

$$P_{OBDmin}(MW) = 1,45615 + 0.000176551 \cdot Time - 4,72509 \cdot 10^{-8} \cdot Time^2 + 4,35116 \cdot 10^{-12} \cdot Time^3 \\ -1,79852 \cdot 10^{-16} \cdot Time^4 + 3,87564 \cdot 10^{-21} \cdot Time^5 - 4,52732 \cdot 10^{-26} \cdot Time^6 \\ +2,71398 \cdot 10^{-31} \cdot Time^7 - 6,52996 \cdot 10^{-37} \cdot Time^8. \quad (5)$$

and the daily load diagram. They have been defined as follows:

1) Maximum Extreme Conditions:

matching the maximum extreme condition that may occur. It is equal to the difference between the maximum extreme of DLD (upper prediction level 95 %) and the minimum extreme of PVP:

$$P_{ROZ(max)} = P_{OBD(max)} - P_{FVE(min)} \; (MW). \quad (6)$$

2) Minimum Extreme Conditions:

matching the minimum extreme conditions that may occur. It is equal to the difference between the minimum extreme of DLD (lower prediction level 95 %) and the maximum extreme of PVP:

$$P_{ROZ(min)} = P_{OBD(min)} - P_{FVE(max)} \; (MW). \quad (7)$$

These curves Eq. (6) and Eq. (7) are portrayed in Fig. 6. They demonstrate the maximum extreme conditions that may occur for a relevant outlet of the grid at the outgoing feeder from the substation. In order to control of developed methodology, the measured data of flowing power at the point of the substation (the blue curve from Fig. 3) are plotted in calculated competent thresholds of curves describing the minimum and maximum extreme power (described in Fig. 5) which could occur at the outlet from the substation plant. This situation shows Fig. 7. All the data and evaluation methodology were considered with the reliability level of 95 %. This particular case concerns 95.54 % of values situated within the defined output extreme conditions. This fact justifies the correctness of the process.

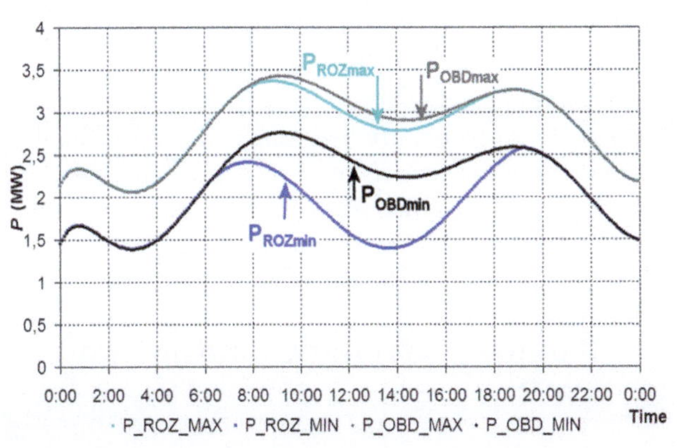

Fig. 6: Substation outlet extreme conditions.

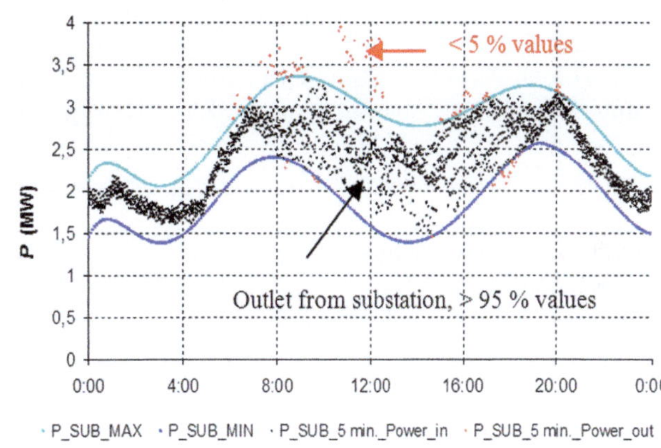

Fig. 7: Control of adjustment of power output difference tresholds.

5. Suggested Methodical Models

5.1. Assessment Methodology

The process of assessment methodology is described by the block diagram in Fig. 8. The diagram is divided into three vertical and three horizontal sections. Vertical columns are separated by a dashed vector, and horizontal lines are highlighted by blue background.

1) Vertical Sections

The first column represents measured data fundamental for the establishment of this methodology.

The second column provides a description of suggested calculation methodology. The first step required processing of data, or the so-called Post-Processing (block P.-Process.). The next step was to enumerate extreme conditions for power output of photovoltaic power plants (block Extr. PVP), extreme conditions of the daily load diagram (Extr. DLD) and extreme conditions at the outgoing feeder from the substation (Extr. SUB). Determination of Extr. SUB firstly required obtaining the data without impact of production from the PVP. This data represents so-called the row data (only loads) for daily load diagram (block DLD). The row data was determined by subtraction of the immediate output data during operation of PVP from the immediate output data measured at the substation. This subtraction is subject to accurate synchronising (block Synch.) of both sets of data measured. Subsequently, extreme conditions of DLD were evaluated. Suitable sum of Extr. DLD and Extr. PVP can help with assessment of extreme conditions occurring at the substation (block SUM Extr.).

The last column contains differences of active power of the PVP (block P diff. PvP) and the data measured at outgoing feeder from the substation (block P diff. SUB and P diff. DLD).

2) Horizontal Sections

As previously mentioned, the diagram is also divided into three horizontal lines.

The first line comprises blocks linked by means of red arrows. These describe the steps for assessment of power differences on the PVP itself only. This indicator does not include the impact of daily load diagram. This assessment is then used to determine the model situations for individual months with further potential for flexible alterations according to the panel type already known and its nominal peak power. This step is the most important and difficult of all evaluation.

The last horizontal line contains blocks linked with green arrows. The data measured at substation includes the impact of PVP operation too. Determination of the difference of active power in this manner is correct, although dedicated only. It cannot be applied to another model situation in a broader sense, being limited to these specific data measured at this particular substation and the relevant PVP.

These issues do not also deal with a case, when the difference of active power would change with the construction of a new PVP within the particular location.

The second line proves itself as an optimal solution, and it is linked to blocks using blue arrows. We need to realize that extreme conditions of power at the substation occur due to extreme conditions on both the PVP as well as a daily load diagram. As mentioned previously, the data measured at outgoing feeder from the substation must be reduced by subtracting the data pertaining to power produced by the PVP. This stage already enables to determine the extreme conditions of daily load diagram as well as PVP for any types of panels with varying peak power figures.

Model situations of extreme conditions of PVP may also be useful for various model or simulation processes, where the change to load diagram is already anticipated.

5.2. Calculation Model

The suggested assessment methodology can be used to set up a calculation model. The calculation model contains blocks linked with red arrows.

The main advantage of the very assessment of extreme states to PVP and DLD has been described above. This idea has been also transferred into the calculation model. Another irrefutable advantage brought by analysis of such separated data is a more accurate mathematical description. It has to be realized that the DLD and PVP production curves differ greatly in nature that is also one of the reasons, why examining either extreme separately is more convenient before calculation of model situation of extreme conditions of PVP (block PVP) requires knowledge of the panel type and their nominal peak power figures. Further requirements to be met include the precondition assuming static position of the panel in order to maximize profit (inclination at approx. 38° with respect to the horizontal plane and southwards orientation). The red arrow in Fig. 8 represents the integration of model extreme conditions of PVP into the calculation process and the point of assessment. The point for assessment (block 110/22), which is to be used for assessment of the power difference, must be identified with the nature of power load - the daily load diagram. These load curves are usually monitored by distributors at certain vital points. Data originating from various model or simulation processes can be also used.

5.3. Annual Extreme Conditions of PVP

Maximum extreme conditions of production are described in Fig. 9 and Fig. 10. Figure 9 covers months within the period from December till June, i.e. the periods of winter and summer solstice. One can no-

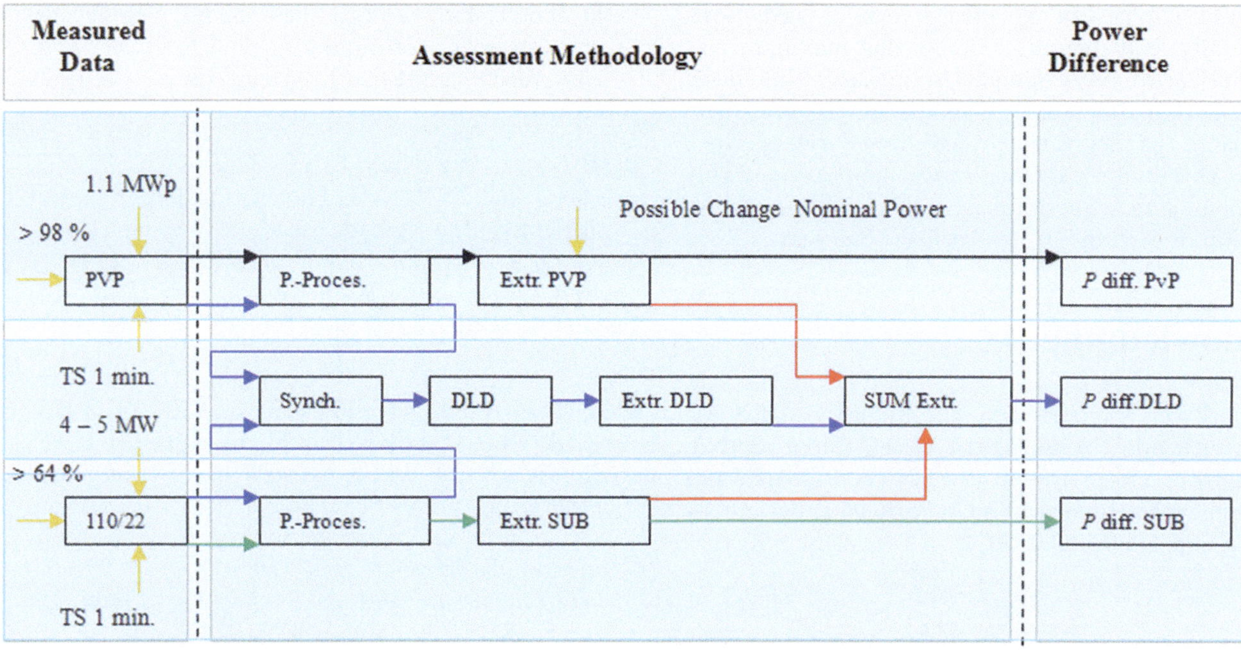

PVP – photovoltaic power plant,
110/22 – substation 110/22 kV,
P. - Process. – post-processing of data file,
Synch. – synchronising of measured data,
DLD – data of the daily load diagram,
Extr. PVP – extreme conditions of PVP,
Extr. DLD – extreme conditions of the daily load diag.,
Extr. SUB – extreme conditions at the substation 110/22 kV,
SUM Extr. – resultant extreme conditions,
P diff. PVP – active power difference of Extr. PVP,
P diff. DLD – active power difference of Extr. SUB, with inclusion of model Extr. PVP,
P diff. SUB – active power difference of Extr. SUB, without inclusion of model Extr. PV.

Fig. 8: Block diagram of assessment methodology.

tice that the dispersion of production time is extended proportionally to the increasing duration of sunshine within particular months, and the amplitude is rising with increasing intensity of solar radiation. The maximum occurs in May rather that during the summer solstice, as might be expected. This case therefore leads to the establishment of an optimal ratio between the intensity of solar radiation and average temperature during the particular month (previous months indicate lack of solar radiation and the subsequent months bring higher ambient temperature with a negative impact on panel efficiency). Figure 10 shows maximum curves

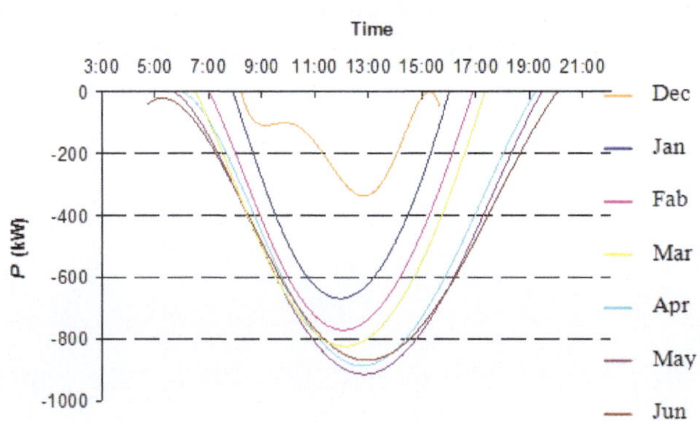

Fig. 9: Extreme conditions of PVP, December to June.

from July till December, i.e. between the summer and winter solstice.

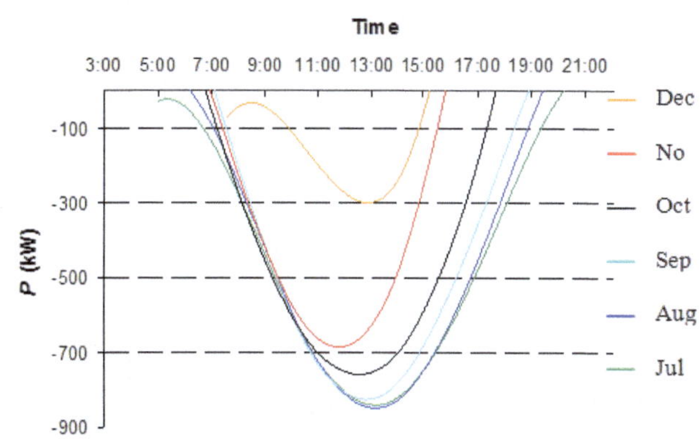

Fig. 10: Extreme conditions of PVP, July to December.

6. Conclusion

Briefly speaking, this measurement is used for suggestion of a general calculation methodology for assessment of difference o active power at a particular node of the grid under various fictional changes to power from the photovoltaic power plant or the whole solar farm, or even in case of expected changes or predictions in the daily load diagram. Comparing maximum extremes of each month indicates significant differences. It is clear that different periods during the year have influence to the assessment of Ancillary Services. Next Steps of this study will focus on the application of this methodology on a full database and evaluation the relationship of each month to Ancillary Services.

Acknowledgment

This work was supported by the Ministry of Education, Youth and Sports of the Czech Republic (No. SP2014/187) and by the project ENET (Research and Development for Innovations Operational Programmes No. CZ.1.05/2.1.00/03.0069).

References

[1] VSB-TU Ostrava, Department of Electrical Power Engineering. "Data source". VSB-TU Ostrava 2013: unpublished.

[2] SMOCEK, M. and Z. HRADILEK. Methodology for Evaluation Extreme Power Conditions of Photovoltaic Power Plants. In: *14th International Scientific Conference on Electric Power Engineering, EPE 2013.* Kouty nad Desnou: VSB-Technical University Ostrava, 2013, pp. 20–23. ISBN 978-80-248-2988-3.

[3] HRADILEK, Z. and T. SUMBERA. Simulator of Power Forecasting Gained from Wind Power Plants. *Przeglad Elektrotechniczny.* 2010, vol. 86, no. 8, pp. 196–199. ISSN 0033-2097.

[4] HRADILEK, Z. and T. SUMBERA. Reliability and Predictions of Power Supplied by Wind Power Plants. In: *International Conference on Renewable Energies and Power Quality.* Las Palmas de Gran Canaria: University of La Coruna, 2011, pp. 254–259. ISBN 978-84-614-7527-8.

[5] PIOTROWSKI, P. Analiza statystyczna danych do prognozowania ultrakrotkoterminowego produkcji energii elektrycznej w systemach fotowoltaicznych. *Przeglad Elektrotechniczny.* 2014, vol. 90, no. 4, pp. 1–4. ISSN 0033-2097.

[6] JIE, S., L. WEI-JEN, L. YONGQIAN, Y. YONGPING and P. WANG. Forecasting power output of photovoltaic system based on weather classification and support vector machine. In: *Industry Applications Society Annual Meeting (IAS).* Orlando: IEEE, 2011, pp. 1–6. ISBN 978-1-4244-9498-9. DOI: 10.1109/IAS.2011.6074294.

[7] YONA, A., T. SENJYU, A. SABER, T. FUNABASHI, H. SEKINE and K. CHUL-HWAN. Application of neural network to 24-hour ahead generating power forecasting for PV system. In: *Power and Energy Society General Meeting - Conversion and Delivery of Electrical Energy in the 21st Century.* Pitsburgh: IEEE 2008, pp. 1–6. ISBN 978-1-4244-1905-0. DOI: 10.1109/PES.2008.4596295.

About Authors

Martin SMOCEK was born in Hranice. In 2011 he graduated VSB-Technical University of Ostrava, Faculty of Electrical Engineering and Computer Science, electrical power engineering. Today he is Ph.D. student in the Department of Electrical Power Engineering, VSB-Technical University of Ostrava and he applies himself to the issue of photovoltaic power plant.

Zdenek HRADILEK was born in Brno. After graduation of college education at Faculty of Electrical Engineering and Computer Science Brno University of Technology in 1962 he worked as a technician in company Southern Moravian power plants in Brno, than he worked as a major power-supply director in Heat-supply Ostrava and from 1966 until now he is at the VSB-Technical University Ostrava. His scientific preparation graduated by his candidate dissertation defending at the Brno University of Technology in 1972. He defended his doctoral thesis at the Czech Technical University in Prague in 1988 and was appointed as professor.

28

A Model-Based Fuzzy Control of an Induction Motor

Daniela PEDRUKOVA , Pavol FEDOR

Department of Electrical Engineering and Mechatronics, Faculty of Electrical Engineering and Informatics, Technical University of Kosice, Letna 9, 042 00 Kosice, Slovak Republic

daniela.perdukova@tuke.sk, pavol.fedor@tuke.sk

Abstract. *The paper presents a method of obtaining a simplified fuzzy model of an induction motor from measured data, without the necessity of preliminary knowledge of its internal structure and parameters. With the aim of avoiding a heuristic search for linguistic control rules, the paper presents one of the possibilities of the application of this method for an inverse fuzzy model based control. The proposed simplified fuzzy model of an induction motor was applied in the control of the desired torque of the drive with induction motor. Obtained results were first verified by simulation in programme Matlab and finally experimentally validated by measurements on an IC inverter–induction motor system. Simulation results and experimental measurements confirmed the correctness of the proposed fuzzy modelling and control method and its applicability also to other nonlinear dynamic systems.*

Keywords

Fuzzy logic control, fuzzy modelling, induction motor, inverse fuzzy model.

1. Introduction

The fuzzy approach in the description and control of dynamic systems has recently undergone considerable development. One of the most successful fuzzy system applications is fuzzy logic control (FLC), which has become an alternative to the use of conventional control techniques in control systems that are often impossible to describe analytically. A very widely used FLC type is control based on fuzzy modelling methods that offer an alternative approach to describing complex nonlinear systems [1], [2], [3], [4] and also reduce the number of rules in modelling higher order nonlinear systems [5]. Fuzzy dynamic models [6], [7], also provide a basis for thr development of systematic approaches to fuzzy controller design in view of powerful conventional control theory and techniques [8]. Fuzzy model based control methods have many modifications that depend on the particular application [9], [10], [11], [12], [13] and [14], while the quality of the fuzzy model of the controlled system is also of significant importance. This may often be a non-linear system without availability of prior knowledge of its structure and parameters. The problem in the development of fuzzy models of these systems lies in obtaining their qualitative properties on the basis of measured experimental data, which, having no prior knowledge on the parameters and structure of these systems, often results in an inconsistent database, in problems with covering the entire possible space of the fuzzy system inputs, etc. The fuzzy model obtained in this way then often shows unusable in practice, although, when a different method of data collection is used, or a suitable selection of qualitative data from the database is made, it is possible to construct a corresponding fuzzy model of the unknown non-linear dynamic system.

The philosophy of systems description on basis of their qualitative parameters enables the finding of relatively simple and practically applicable models even in cases of complex non-linear systems, of course, at the cost of a certain degree of inaccuracy in their description [15]. A typical representative of such a strongly non-linear system in the area of electric drives is the three-phase induction motor, whose analytical description, subject to certain simplifying presumptions, is represented by a set of five non-linear differential equations with three inputs. Modern artificial intelligence methods (fuzzy approach, neural networks) enable the finding of its substantially more simple and practically applicable models based on the description of its significant qualitative properties [16], [17], [18], [19] and [20].

The paper deals with the method of designing a simplified fuzzy model of an induction motor only on basis of external information (i.e. measured relations between inputs and outputs). Obtained results are ap-

$$\frac{\mathrm{d}}{\mathrm{d}t}\begin{bmatrix} i_{1x} \\ i_{1y} \\ \psi_{2x} \\ \psi_{2y} \end{bmatrix} = \begin{bmatrix} -\omega_0 & \omega_1 & -\mathrm{K}_{12}\cdot\omega_g & -\mathrm{K}_{12}\cdot n_p\cdot\omega_m \\ -\omega_1 & -\omega_0 & \mathrm{K}_{12}\cdot n_p\cdot\omega_m & -\mathrm{K}_{12}\cdot\omega_g \\ M_n\cdot\omega_g & 0 & -\omega_g & \omega_2 \\ 0 & M_n\cdot\omega_g & -\omega_2 & -\omega_g \end{bmatrix}\cdot\begin{bmatrix} i_{1x} \\ i_{1y} \\ \psi_{2x} \\ \psi_{2y} \end{bmatrix} + \begin{bmatrix} \mathrm{K}_{11} & 0 \\ 0 & \mathrm{K}_{11} \\ 0 & 0 \\ 0 & 0 \end{bmatrix}\cdot\begin{bmatrix} u_{1x} \\ u_{1y} \\ 0 \\ 0 \end{bmatrix}. \tag{1}$$

$$n_p = \frac{M_n}{L_2}\left(\psi_{2x}\cdot i_{1y} - \psi_{2y}\cdot i_{1x}\right) - M_z = J\frac{\mathrm{d}\,\omega_m}{\mathrm{d}t}. \tag{2}$$

plied in the torque control of the drive with induction motor and experimentally validated by measurements on an IM inverter–induction motor system. In this case, the amount of a model based fuzzy control methods is extended.

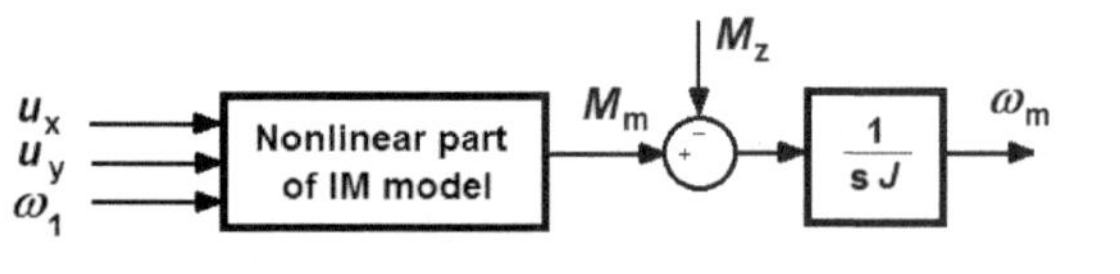

Fig. 1: Structure of the system with induction motor.

2. Method of Determination of an Induction Motor Simplified Fuzzy Model

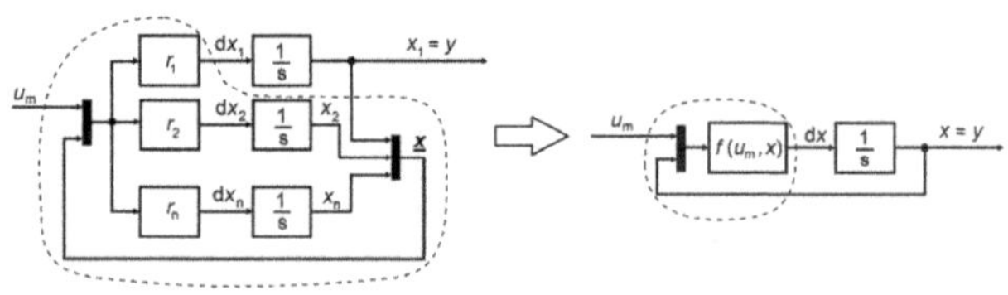

Fig. 2: Procedure of substituting a nonlinear higher order dynamic system by a first order dynamic system.

The induction motor (IM) represents a fifth-order nonlinear system described by state equations in the form Eq. 1, Eq. 2,[21], [22].

It is a current-flux induction motor model with structure as shown in Fig. 1, where u_x, u_y are components of stator voltage vector U_1 in a rotating coordinate system x, y and $\omega = 2\cdot\pi\cdot f_1$ is its angular frequency. The parameters of the model correspond to the parameters of the real motor used in experimental measurements and are specified in the Appendix.

The nonlinear part of the IM represents a fourth-order system for which we need to design a fuzzy model applying the method based on fuzzy linearization of non-linear dynamic system.

In general we want to substitute a nonlinear higher order dynamic system:

$$\mathrm{d}\vec{x} = \mathbf{R}\left(u_m, x\right), \tag{3}$$

$$y = x_1, \tag{4}$$

$$\mathbf{R} = \left[r_1, r_2, \ldots, r_n\right]^{\mathrm{T}}, \tag{5}$$

where function $\mathbf{R} = \left[r_1, r_2, \ldots, r_n\right]^{\mathrm{T}}$ is an unknown nonlinear continuous function, $\vec{x}$ is a state vector and n is the system order (the measured output variable of the system is the first state variable, i.e. $y = x_1$) by a simplified dynamic first-order system in the form:

$$\mathrm{d}x = f\left(u_m, x\right), \tag{6}$$

$$y = x, \tag{7}$$

in which $f\left(u_m, x\right)$ is an unknown nonlinear function obtained from suitable measurements in the system Eq. 3. The best possible similarity to responses of systems Eq. 3 and Eq. 6 for identical input variables um is the criterion for substitution. The procedure is illustrated in Fig. 2.

The basic premise for the identification of system Eq. 3 is that it is stable and at constant input u_m its output settles, after fading of the transient performances, at a constant value y. This means that the transition of some measured system Eq. 3 from one steady state A to another steady state B at constant input u_m determines one relation among the relevant three values $[u_m, x, \mathrm{d}x]$, i.e. one point of function f where $y = x$.

The course of the IM torque for the desired speed step from 0 to 100 % is shown in Fig. 3. When substituting the fourth-order system by a first-order system, the undergoing fast electromagnetic processes in the motor are ignored, which means that the IM torque course is simplified and is presented as a curve representing its medium value as is shown in Fig. 3.

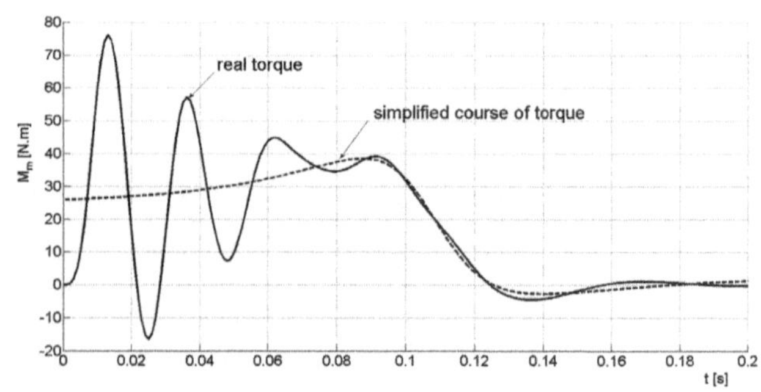

Fig. 3: IM real torque course and its simplified course.

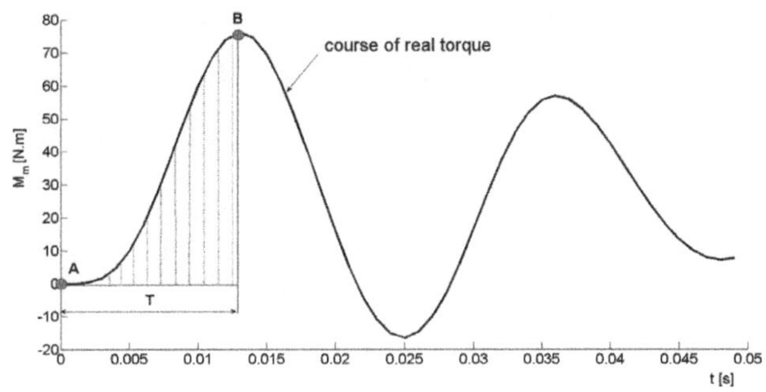

Fig. 4: Determination of starting point.

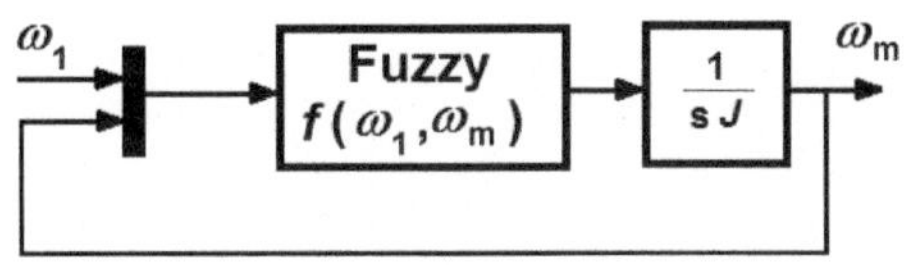

Fig. 5: Fuzzy model structure of the drive with induction motor.

For the purpose of identification of function f in Eq. 6 we need to determine the starting value dx (starting point) for each pair of inputs $[u_m, x]$, which is the initial value of the simplified course of its torque. It can be calculated e.g. as the medium value of the real torque course in the range from zero to the first maximum, as illustrated in Fig. 4.

The mathematical representation is:

$$M_{\text{start}} = \frac{1}{T}\int_A^B M_m \mathrm{d}t. \tag{8}$$

The database of measured data for identification of function f should cover the whole work space of pairs $[u_m, x]$, i.e. it is necessary to measure the desired transitions between the steady states in this space. If we split the input space um and thus also the relevant space of steady outputs Eq. 3 into n parts, then the number of desired measured responses is equal to the number of variations of n elements in groups of 2, i.e.:

$$V_2(n) = \frac{n!}{(n-2)!}. \tag{9}$$

For creating the complete fuzzy model of IM, that structure is shown in Fig. 5, we divided the work space into 6 regular levels. In this case in accordance with relation Eq. 9 it is necessary to measure 30 transitions between steady states and from these to define the identification points for function f. The result of the transition processes between each of the levels is one starting point identified according to the relation Eq. 9. The complete database of starting points for IM fuzzy model generation is shown in Tab. 1. Points on the diagonal represent complemented values of the system steady states.

Tab. 1: Identified points of function f.

$\frac{\omega_1}{\omega_m}$	0 %	20 %	40 %	60 %	80 %	100 %
0 %	0	−18	−34	−44	−50	−54
20 %	7.4	0	−20	−37	−48	−53
40 %	15	25	0	−24	−46	−55
60 %	27	27	28	0	−26	−50
80 %	27	29	34	32	0	−27
100 %	25	23	29	35	39	0

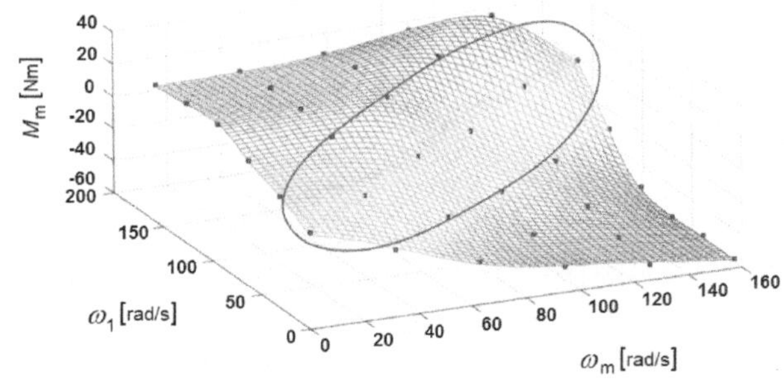

Fig. 6: Characteristic surface of induction motor fuzzy substitute.

This table is the basis for constructing a fuzzy description (model) of function f obtained by Anfisedit tool of the MATLAB programme package. The method used for the fuzzy system generation was the subclustering method, having the following parameters:

- *Range of influence* = 0.5,
- *Squash factor* = 1.25,
- *Accept ratio* = 0.5,
- *Reject ratio* = 0.15.

Function f is represented by Sugeno type fuzzy system with 7 rules, the characteristic surface of which is shown in Fig. 6.

It has to be pointed out that this is a simplified fuzzy model of an induction motor where we substituted a fifth order system by a first order system.

3. Inverse Fuzzy Model Based Torque Control of an Induction Motor

The described method of obtaining a simplified fuzzy model of an induction motor based in general on fuzzy linearization of a nonlinear dynamic system can be used for the control of such systems using a control structure based on an inverse fuzzy model, the basic idea of which is demonstrated in Fig. 7. Nonlinearity of the fuzzy

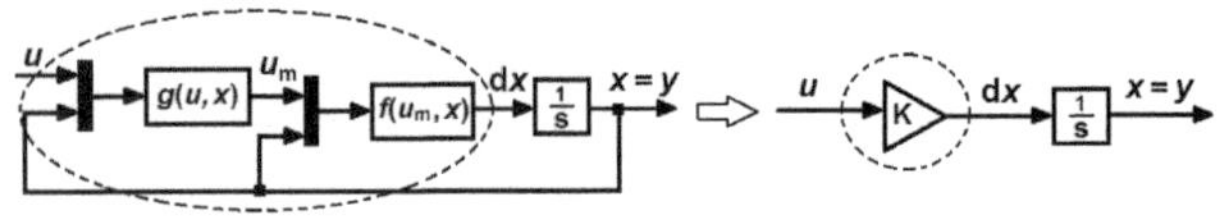

Fig. 7: A control structure based on an inverse fuzzy model.

system $f(u_m, x)$ is compensated by an inverse function $g(u, x)$ that satisfies the following condition:

$$\mathrm{d}y = f(g(u,x)\,x) = \mathrm{K} \cdot u, \tag{10}$$

where K is a constant representing the resulting linearized system (Fig. 7).

3.1. Example

Consider a simple first-order nonlinear dynamic system in the form:

$$\dot{x} = u_m - sin(x). \tag{11}$$

In this case it is obvious that $r_1 = f = (u_m - sin(x))$. When choosing function g in the form $g = \mathrm{K} \cdot u + sin(x)$, it applies that:

$$\dot{x} = g(u,x) - sin(x) = \\ \mathrm{K} \cdot u + sin(x) - sin(x) = \mathrm{K} \cdot u, \tag{12}$$

which obviously accurately satisfies condition Eq. 10. In this simple example we are dealing with linearization of a first-order system, the problem can be solved analytically and linearization is complete.

The role of inverse function g is to adapt input um to the nonlinear system in such a way that the first derivative of the output value corresponds, as much as possible, to the control input u. In the preceding chapter we illustrated the procedure of obtaining a fuzzy description of the function $\mathrm{d}x = f(u_m, x)$ on the basis of experimental measurements. From this relation it is then possible to determine for each particular desired value $\mathrm{d}x$ and for given system output $y = x$ the corresponding value of u_m. For linearization of system Eq. 3 the desired $\mathrm{d}x$ values are determined by equation Eq. 10, in which the amplification K is usually chosen such that it regulates the range of real $\mathrm{d}x$ values to the range of the control signal u.

Described idea was used for the control of the desired torque of an induction motor (M_{mref}) in the control structure based on an inverse fuzzy model, as shown in Fig. 8.

Function g is the function inverse to function f. When creating an inverse function, it is necessary to define correctly the work space for which all the points of the inverse function g are uniquely determined [23], [24] and [25].

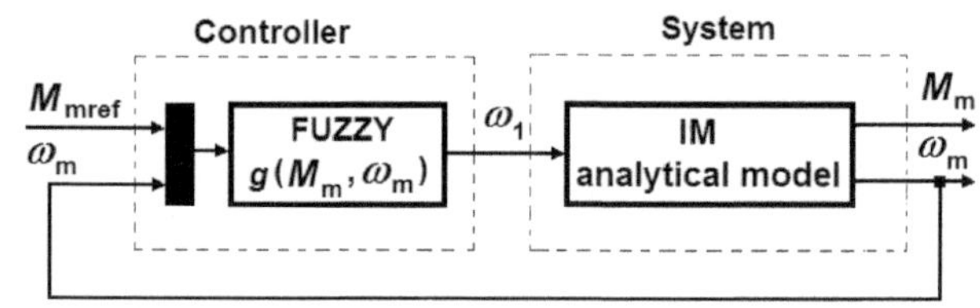

Fig. 8: Control structure based on inverse fuzzy model.

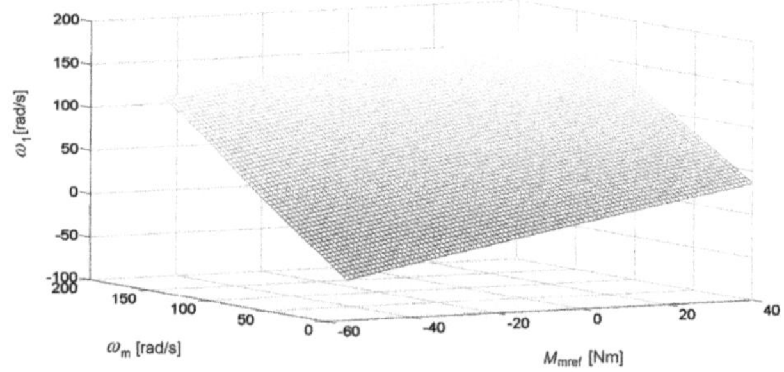

Fig. 9: Characteristic surface of inverse function g used in the control structure.

It is known from the torque characteristic that the IM behaves in operational space as a quasi linear system. In Fig. 6 the black line defines the operational space in the vicinity of zero torque, which represents the transition of the IM from motor to generator mode. From this it follows that if we want to control the IM by means of an inverse model, this model has to be generated only from the database of points that correspond to this operational space. The remaining spaces do not need to be considered, as the controller will not be entering these spaces. The range of M_m will then be $\langle -60, 40 \rangle$ and the range of ω_m can be assumed as being $\langle 0, 200 \rangle$. The matrix of corresponding points for the delimited operational space will be used for generation of the fuzzy description of function g by means of the Anfisedit tool and the subclustering method with standardly pre-set parameters. Function g is represented by a Sugeno type fuzzy system with 8 rules. Its characteristic surface is shown in Fig. 9.

The properties of the control structure based on the inverse fuzzy model (Fig. 8) were validated by simulation for various steps of the desired torque Mmref (Fig. 10). Because of the unexcited motor, it is not possible to achieve the desired torque value at the start of the simulation. With rotations of approximately 30 to 40 rad/s the motor is already excited and the motor torque reaches the desired value. At the time of approx. 2, 3 seconds we can see that the desired torque Mmref is at the value of 20 Nm, but the real torque has already fallen to zero. This is because mechanical angular velocity has reached the nominal value.

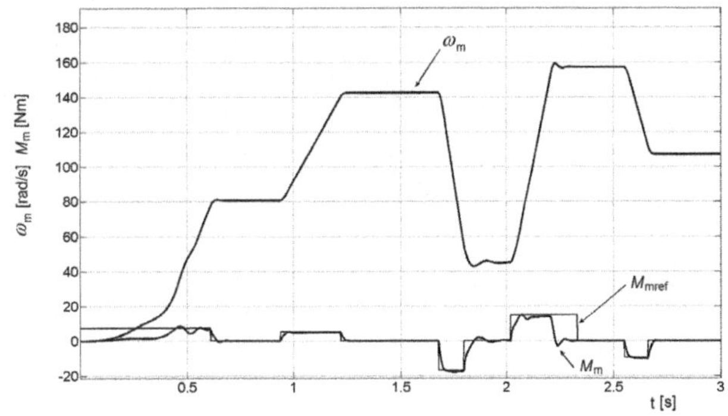

Fig. 10: Properties of control structure based on inverse fuzzy model for various steps of desired torque.

4. Experimental Results

Control based on an inverse fuzzy model was applied to a system consisting of a scalar control AC inverter without revolutions control and a three-phase asynchronous motor with parameters as specified in the Appendix.

The procedure of designing the fuzzy model of the controlled system (function f) was identical to the one described in Chapter 2. Identification measurements for the generation of the database of points (see Tab. 1) for the purpose of generating the controlled systems fuzzy model were carried out by means of the *Drive Monitor* software tool. The aim of the identification was to obtain the motors torque course and to subsequently use this for determining starting point values. We used the Drive Monitor tool to record transition actions of mechanical angular velocity. Motor torque was determined indirectly from the equation of motion:

$$M_m - M_z = J\frac{\mathrm{d}\,\omega}{\mathrm{d}\,t} \qquad (13)$$

ignoring load torque M_z which is generated by the resistance of the motor bearings and fan and is negligible compared to the motor torque.

The resulting fuzzy systems describing functions f and g were identical with the results obtained by simulation in Chapter 3.

The inverse model based control was implemented by means of the Real Time system OPAL RT – LAB. This real time system makes it possible to convert a model created in Simulink into a simulation running in real time. The layout of the implementation workplace is shown in Fig. 11.

The typical measured control responses of angular speed and torque are shown in Fig. 12. The smaller dynamics of the real torque course to the first step in its reference value is caused by the non-excited status of the motor. The resulting quality of control strongly depends upon the quality of the fuzzy model of the system, where, due to the application of the described procedure, the fifth-order system becomes a first-order fuzzy system, which in IM causes neglect of fast electromagnetic activities. The most important issue with respect to the design of the fuzzy model of the system is the correct identification of the smoothed course of its torque, as illustrated in Fig. 3 and Fig. 4.

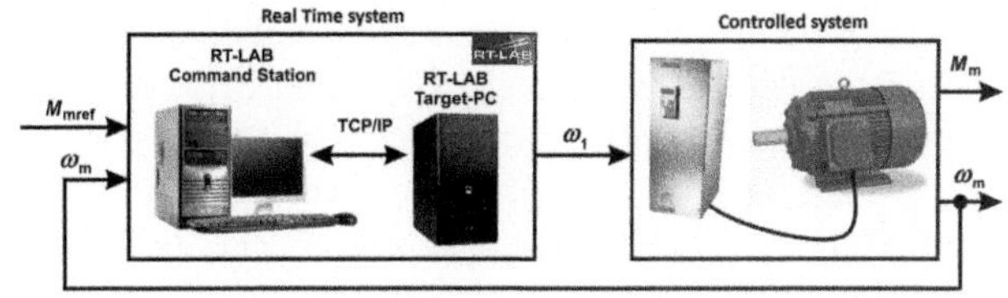

Fig. 11: Layout of the implementation workplace.

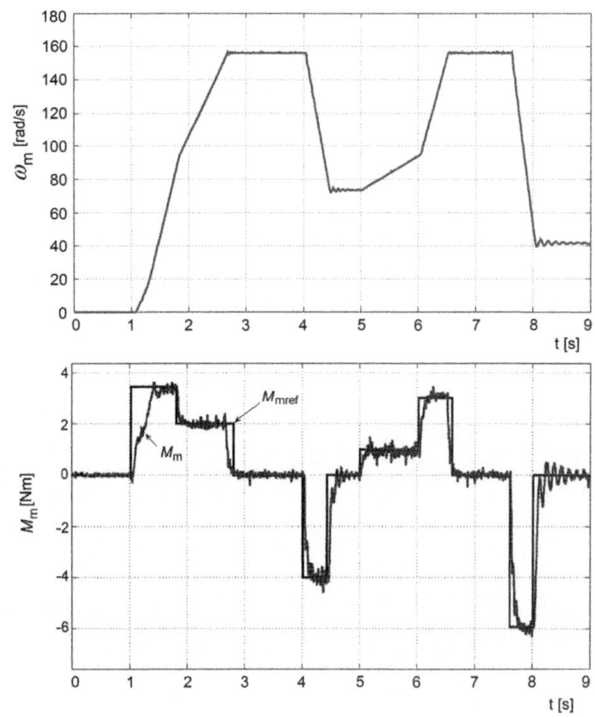

Fig. 12: Measured control responses of angular speed and torque of IM.

5. Conclusion

The paper deals with the setup of a simplified fuzzy model of an induction motor using method based on fuzzy linearization of non-linear dynamic system. For the setup of this model it is necessary to carry out several standard measurements of step responses on the motor, and no prior knowledge of the motor parameters is required.

Simulation results and experimental measurements have confirmed that the fuzzy model obtained by the described method can be exploited also for relatively good control of complex nonlinear systems, such as e.g. that of a drive with induction motor with as little previous knowledge as possible and with no the heuristic search for a rule-based controller, while the quality of the control structure based on an inverse fuzzy model is strongly dependent on good-quality identification of induction motor properties necessary for the construction of its fuzzy model and the correct definition of the operational space.

Comparing fuzzy to conventional modelling and control it must be stressed that in some cases the achieved results are better and in other cases they are not. It depends on the system or a problem to be solved which methods fits best. In this spirit the proposed method of obtaining a simple fuzzy model of a nonlinear dynamic system for which only external information is available (i.e. the measured dependencies between the inputs and outputs) and its exploitation for an inverse fuzzy model based control has to be understood as an enhancement of the wide range of fuzzy modelling and control methods. Because the article deals with fuzzy IM torque control based on the simplified fuzzy model, the resulting dynamics of the controlled system cannot be compared with the much more precise and complex vector control.

In the future the method described could also be applied to other types of nonlinear systems with similar properties that render their identification and control using conventional methods difficult, if not even impossible.

Appendix A AC Drive Parameters

- $P_N = 3$ kW
- $U_{1N} = 220$ V
- $I_{1N} = 6.4$ A
- $n_N = 1420$ rev/min
- $M_N = 20$ Nm
- $J = 0.022$ kg·m^2
- Stator phase resistance: $r_1 = 1.83\ \Omega$
- Rotor phase resistance: $r_2 = 2.26\ \Omega$
- Main inductance: $L_h = 0.216$ H
- Mutual inductance: $M = (3/2)L_h = 0.324$ H
- Leakage inductance $L_{S1} = L_{S2} = 12.036$ mH
- $R_1 = (2/3)r_1 = 1.22\ \Omega$
- $R_2 = (2/3)r_2 = 1.506\ \Omega$
- $L_2 = 2/3(L_{S2} + L_h) = 152.31$ mH
- $K_{11} = 3/2(L_{S1} + (L_{S2}L_h/(L_h + L_{S2}))^{-1} = 63.999\ H^{-1}$
- $K_{12} = -3/2(L_{S1} + L_{S1}L_{S2}^2/L_h)^{-1} = \check{}60.627\ H^{-1}$
- Slip angular speed $\omega_2 = \omega_1 - n_p\omega_m$
- $\omega_0 = K_{11}(R_1 + (M_2/L_2)\omega_g) = 164.723\ s^{-1}$
- $\omega_g = R_2/L_2 = 9.902\ s^{-1}$
- Mechanical angular speed of the motor: ω_m
- Angular frequency of the stator voltage: $\omega_1 = 2 \cdot \pi \cdot f_1 = 314\ s^{-1}$
- Number of pole pairs: $n_p = 2$

Acknowledgment

This work was supported by project KEGA 011TUKE–4/2013.

References

[1] CAO, S. G., N. W. REES and G. FENG. Analysis and design for a class of complex control systems Part I: Fuzzy modelling and identification. *Automatica*. 1997, vol. 33, iss. 6, pp. 1017–1028. ISSN 0005-1098. DOI: 10.1016/S0005-1098(97)00010-1.

[2] JOHANSEN, T. A., R. SHORTEN and R. MURRAY-SMITH. On the interpretation and identification of dynamic Takagi-Sugeno fuzzy models. *IEEE Transactions on Fuzzy Systems*. 2000, vol. 8, iss. 3, pp. 297–313. ISSN 1063-6706. DOI: 10.1109/91.855918.

[3] TANAKA, K. and H. O. WANG. *Fuzzy control systems design and analysis: a linear matrix inequality approach*. New York: Wiley, 2001. ISBN 04-713-2324-1.

[4] ZHANG, J. Modeling and optimal control of batch processes using recurrent neuro-fuzzy networks. *IEEE Transactions on Fuzzy Systems*. 2005, vol. 13, iss. 4, pp. 417–427. ISSN 1063-6706. DOI: 10.1109/TFUZZ.2004.841737.

[5] SUGENO, M. On stability of fuzzy systems expressed by fuzzy rules with singleton consequents. *IEEE Transactions on Fuzzy Systems*. 1999, vol. 7, iss. 2, pp. 201–224. ISSN 1063-6706. DOI: 10.1109/91.755401.

[6] CAO, S. G. and N. W. REES. Analysis and design of fuzzy control systems. In: *IEEE International Conference on Fuzzy Systems.* Yokohama: IEEE, 1995, pp. 317-324. ISBN 0-7803-2461-7. DOI: 10.1109/FUZZY.1995.409698.

[7] CAO, S. G., N. W. REES and G. FENG. Analysis and design for a class of complex control systems part II: Fuzzy controller design. *Automatica.* 1997, vol. 33, iss. 6, pp. 1029–1039. ISSN 0005-1098. DOI: 10.1016/S0005-1098(97)00011-3.

[8] FENG, G. A Survey on Analysis and Design of Model-Based Fuzzy Control Systems. *IEEE Transactions on Fuzzy Systems.* 2006, vol. 14, iss. 5, pp. 676–697. ISSN 1063-6706. DOI: 10.1109/TFUZZ.2006.883415.

[9] XU, W.-L., N.-Y. ZHANG and K. ZENG. A comparative study on sufficient conditions for Takagi-Sugeno fuzzy systems as universal approximators. *IEEE Transactions on Fuzzy Systems.* 2000, vol. 8, iss. 6, pp. 773–780. ISSN 1063-6706. DOI: 10.1109/91.890337.

[10] TANG, K. S., K. F. MAN, G. CHEN and S. KWONG. An optimal fuzzy PID controller. *IEEE Transactions on Industrial Electronics.* 2001, vol. 48, iss. 4, pp. 757–765. ISSN 0278-0046. DOI: 10.1109/41.937407.

[11] TAO, C. W. and J. S. TAUR. Robust fuzzy control for a plant with fuzzy linear model. *IEEE Transactions on Fuzzy Systems.* 2005, vol. 13, iss. 1, pp. 30–41. ISSN 1063-6706. DOI: 10.1109/TFUZZ.2004.839653.

[12] BATURONE, I., F. J. MORENO-VELO, S. SANCHEZ-SOLANO and A. OLLERO. Automatic Design of Fuzzy Controllers for Car-Like Autonomous Robots. *IEEE Transactions on Fuzzy Systems.* 2004, vol. 12, iss. 4, pp. 447–465. ISSN 1063-6706. DOI: 10.1109/TFUZZ.2004.832532.

[13] CHANG, Y.-CH. and CHEN B.-S. Intelligent robust tracking controls for holonomic and nonholonomic mechanical systems using only position measurements. *IEEE Transactions on Fuzzy Systems.* 2005, vol. 13, iss. 4, pp. 491–507. ISSN 1063-6706. DOI: 10.1109/TFUZZ.2004.840125.

[14] SHAFEI, S. E. and S. SEPASI. Incorporating Sliding Mode and Fuzzy Controller with Bounded Torques for Set-Point Tracking of Robot Manipulators. *Elektronika ir elektrotechnika.* 2010, no. 8, iss. 104, pp. 3-8. ISSN 2029-5731.

[15] BABUSKA, R. *Fuzzy Modeling for Control.* Dordrecht: Springer, 1998. ISBN 978-940-1148-689.

[16] FENG, G. A Survey on Analysis and Design of Model-Based Fuzzy Control Systems. *IEEE Transactions on Fuzzy Systems.* 2006, vol. 14, iss. 5, pp. 676–697. ISSN 1063-6706. DOI: 10.1109/TFUZZ.2006.883415.

[17] JOHANSEN, T. A. Fuzzy model based control: stability, robustness, and performance issues. *IEEE Transactions on Fuzzy Systems.* 1994, vol. 2, iss. 3, pp. 221–234. ISSN 1063-6706. DOI: 10.1109/91.298450.

[18] KIRIAKIDIS, K. Fuzzy model-based control of complex plants. *IEEE Transactions on Fuzzy Systems.* 1998, vol. 6, iss. 4, pp. 517–529. ISSN 1063-6706. DOI: 10.1109/91.728444.

[19] PEDRYCZ, W. *Fuzzy control and fuzzy systems.* New York: Wiley, 1989. ISBN 04-719-2311-7.

[20] PARK, Y., M.-J. TAHK and H. BANG. Design and Analysis of Optimal Controller for Fuzzy Systems With Input Constraint. *IEEE Transactions on Fuzzy Systems.* 2004, vol. 12, iss. 6, pp. 766–779. ISSN 1063-6706. DOI: 10.1109/TFUZZ.2004.836089.

[21] VITTEK, J,, P. BRIS, P. MAKYS and M. STULRAJTER. Forced dynamics control of PMSM drives with torsion oscillations. *COMPEL - The international journal for computation and mathematics in electrical and electronic engineering.* 2010, vol. 29, iss. 1, pp. 187–204. ISSN 0332-1649. DOI: 10.1108/03321641011008046.

[22] BRANDSTETTER, P. Sensorless Control of Induction Motor Using Modified MRAS. *International Review Of Electrical Engineering.* 2012, no. 7, iss. 3, pp. 4404–4411. ISSN 1827-6660.

[23] ABONYI, J., H. ANDERSEN, L. NAGY and F. SZEIFERT. Inverse fuzzy-process-model based direct adaptive control. *Mathematics and Computers in Simulation.* 1999, vol. 51, iss. 1-2, pp. 119–132. ISSN 0378-4754. DOI: 10.1016/S0378-4754(99)00142-1.

[24] ABONYI, J., L. NAGY and F. SZEIFERT. Indirect Model Based Control using Fuzzy Model Inversion. In: *International Conference on Intelligent Systems.* Vienna, Austria: IEEE, 1998. pp. 951–956. ISBN 0-7803-4465-0.

[25] ABONYI, J. *Fuzzy Model Identification for Control.* Boston, MA: Birkhäuser Boston, 2003. ISBN 978-146-1200-277.

About Authors

Daniela PERDUKOVA works as a Professor at the Department of Electrical Engineering and Mechatronics at the Faculty of Electrical engineering and Informatics, Technical University of Kosice. Her research and educational activities are focused mostly on AI techniques and their applications in the field of non-linear systems; a special attention is also paid to modelling of technological processes, their monitoring and technological process visualization.

Pavol FEDOR works as a Professor at the Department of Electrical Engineering and Mechatronics at the Faculty of Electrical Engineering and Informatics, Technical University of Kosice. He has extensive experience in the installation of control systems in industry. In the area of theory, his current research activities are focused on the development of control structures by means of Lyapunovs second method for electrical drives and multi-input multi-output systems.

Impact of H.264/AVC and H.265/HEVC Compression Standards on the Video Quality for 4K Resolution

Miroslav UHRINA[1], *Jaroslav FRNDA*[2], *Lukas SEVCIK*[2], *Martin VACULIK*[1]

[1] Department of Telecommunications and Multimedia, Faculty of Electrical Engineering, University of Zilina, Univerzitna 8215/1, 010 07 Zilina, Slovakia
[2] Department of Telecommunications, Faculty of Electrical Engineering and Computer Science, VSB–Technical University of Ostrava, 17. listopadu 15/2172, 708 33 Ostrava-Poruba, Czech Republic

miroslav.uhrina@fel.uniza.sk, jaroslav.frnda@vsb.cz, lukas.sevcik@vsb.cz, martin.vaculik@fel.uniza.sk

Abstract. *This article deals with the impact of H.264/AVC and H.265/HEVC compression standards on the video quality for 4K resolution. In the first part a short characteristic of both compression standards is written. The second part focuses on the well-known objective metrics which were used for evaluating the video quality. In the third part the measurements and the experimental results are described.*

Keywords

H.264/AVC, H.265/HEVC, PSNR, SSIM, VQM.

1. Introduction

In the last years the demand of the multimedia services that means the broadcasting, transmission and receiving the video, audio and other data in one stream, the multimedia stream has increased. Because of this progress, the video quality measuring as one part of the multimedia technology has become an important role. The video quality is affected by:

- the resolution of the scanning part of the camera,
- the processing of the television signal in the studio,
- the compression technology,
- the transmission link imperfection.

The compression technology can be considered as one of the main factors that influence the video quality. Nowadays many new compression standards are being developed, e.g. H.265/HEVC or VP9. Because of that video quality assessment plays still a big role in research area. Today some institutes and research teams deal with video quality evaluation but most of them focuse on the FullHD resolutions [20], [21] and [22]. While in this paper the impact of H.264/AVC and H.265/HEVC compression standards on the video quality for 4K resolution is measured and evaluated.

2. MPEG Compression Standards

MPEG, which stands for Moving Picture Experts Group, is the name of a family of standards used for coding audio-visual information (e.g. movies, video, music) in a digital compressed format [1].

2.1. H.264/AVC

H.264/AVC is currently the most used compression standard. It has been designed for a wide range of applications, ranging from video for mobile phones through web applications to TV broadcasting (HDTV). Some of the feature enhancements in H.264/AVC standard over the earlier codecs are:

- DCT algorithm works at 4×4 pixels instead of 8×8, but also supports 8×8,
- DCT is layered using Hadamard transforms,
- colour sampling is supported at 4:2:2 and 4:4:4,
- up to 12 bits per pixel are possible,
- motion compensation blocks are variable sizes,
- arithmetic variable-length coding,
- built-in de-blocking filter and hinting mechanism,

- rate-distortion optimizer,
- weighted bi-directional prediction,
- redundant pictures,
- flexible macroblock ordering,
- direct mode for B-frames,
- multiple reference frames,
- sub-pixel motion compensation.

H.264/AVC also defines profiles and levels. There are only three profiles currently defined: Baseline, Main and Extended [2], [3], [4], [5] and [6].

2.2. H.265/HEVC

The High Efficiency Video Coding (HEVC) standard is the most recent joint video project of the ITU-T video Coding Experts Group (VCEG) and the ISO/IEC Moving Picture Experts Group (MPEG) standardization organizations, working together in a partnership known as the Joint CollaborativeTeam on Video Coding (JCT-VC).

H.265/HEVC has the same basic structure as previous standards such as MPEG-2 Video and H.264/AVC. However, H.265/HEVC contains many incremental improvements such as:

- More flexible partitioning, from large to small partition sizes.
- Greater flexibility in prediction modes and transform block sizes.
- More sophisticated interpolation and deblocking filters.
- More sophisticated prediction and signalling of modes and motion vectors.
- Features to support efficient parallel processing.

The result is a video coding standard that can enable better compression, at the cost of potentially increased processing power. With H.265/HEVC, it should be possible to store or transmit video more efficiently than with earlier technologies such as H.264/AVC. This means:

- At the same picture size and quality, an HEVC video sequence should occupy less storage or transmission capacity than the equivalent H.264 video sequence.
- At the same storage or transmission bandwidth, the quality and/or resolution of an HEVC video sequence should be higher than the corresponding H.264 video sequence.

The H.265/HEVC standard is designed to achieve multiple goals, including coding efficiency, ease of transport system integration and data loss resilience, as well as implementability using parallel processing architectures. The video coding layer of H.265/HEVC employs the same hybrid approach (inter/intrapicture prediction and 2-D transform coding) used in all video compression standards since H.261. In the following, the various features involved in hybrid video coding using H.265/HEVC are highlighted as follows:

- coding tree units and coding tree block (CTB) structure,
- coding units (CUs) and coding blocks (CBs),
- prediction units and prediction blocks (PBs),
- TUs and transform blocks,
- motion vector signaling,
- motion compensation,
- intrapicture prediction,
- quantization control,
- entropy coding,
- in-loop deblocking filtering,
- sample adaptive offset (SAO) [7], [8] and [9].

3. Objective Video Quality Assessment Methods

The video quality evaluation can be differentiated into objective and subjective assessment. The subjective assessment consists of the use of human observers (people) who score the video quality. It is the most reliable way how to determine the video quality. The disadvantage of these methods is that they are time consuming and human resources are needed. Because of this fact, the objective methods are mostly used. They consist of the use of computational methods called "metrics" that produce values that score the video quality. They measure the physical characteristics of a video signal such as the signal amplitude, timing, signal-to-noise ratio. The big advantage of them is their repeatability. The well-known and mostly used objective metrics are peak signal-to-noise ratio (PSNR), video quality metric (VQM) and structural similarity index (SSIM).

3.1. PSNR (Peak Signal-to-Noise Ratio)

The PSNR in decibels is defined as:

$$\mathrm{PSNR} = 10log\frac{m^2}{\mathrm{MSE}}\,[\mathrm{dB}]\,, \qquad (1)$$

where m is the maximum value that pixel can take (e.g. 255 for 8-bit image) and MSE (Mean Squared Error) is the mean of the squared differences between the gray-level values of pixels in two pictures or sequences I and $\tilde{I}$:

$$\mathrm{MSE} = \frac{1}{TXY}\sum_t\sum_x\sum_y\left[I(t,x,y) - \tilde{I}(t,x,y)\right]^2 \qquad (2)$$

for pictures of size $X \times Y$ and T frames.

Technically, MSE measures image difference, whereas PSNR measures image fidelity. The biggest advantage of the PSNR metric is that it can be computed easily and fast [2].

3.2. SSIM (Structural Similarity Index)

The SSIM metric measures three components, the luminance similarity, the contrast similarity and the structural similarity and combines them into one final value, which determines the quality of the test sequence (Fig. 1). This method differs from the methods described before, from which all are error based, using the structural distortion measurement instead of the error one. It is due to the human vision system that is highly specialized in extracting structural information from the viewing field and it is not specialized in extracting the errors. Owing to this factor, SSIM metric achieves good correlation with subjective impression [10].

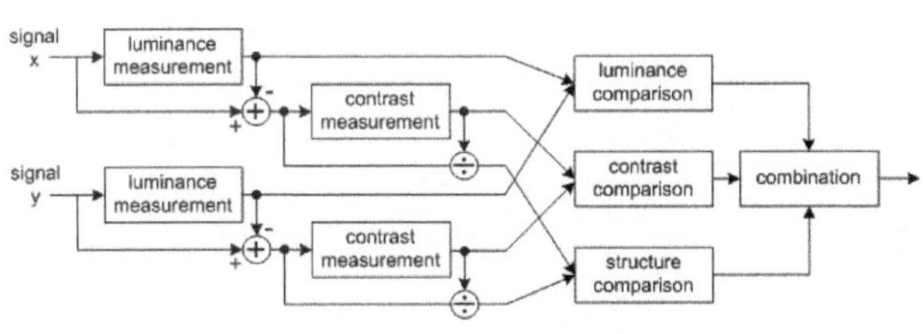

Fig. 1: The block diagram of SSIM metric.

The results are in interval [0,1], where 0 is for the worst and 1 for the best quality.

3.3. VQM (Video Quality Metric)

The VQM metric computes the visibility of artefacts expressed in the DCT domain. Figure 2 shows the

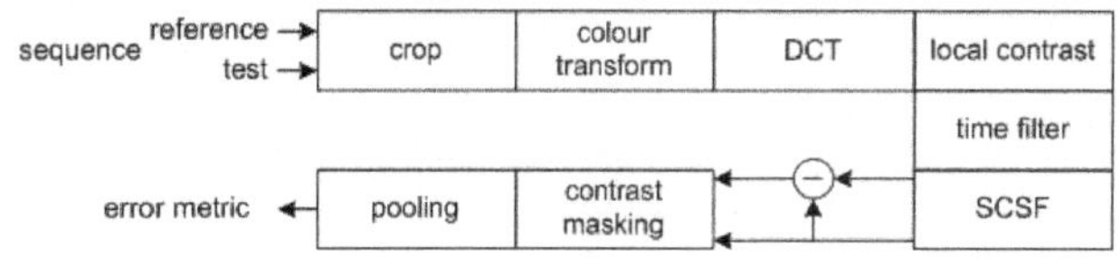

Fig. 2: The block diagram of VQM metric.

block diagram of this metric, which can be divided into 9 steps.

The input of the metric is a pair of colour image sequences, the reference one and the test one. Both sequences are cropped, then converted from the input colour space to the YOZ colour space, then transformed to blocked DCT and afterwards converted to units of local contrast. In the next step the input sequences are subjected to temporal filtering, which implements the temporal part of the contrast sensitivity function. The DCT coefficients, expressed in a local contrast form, are then converted to just-noticeable-differences (jnds) by dividing their respective spatial thresholds. This implements the spatial part of the contrast sensitivity function. In the next step, after the conversion to jnds, the two sequences are subtracted to produce a difference sequence. In the following step the contrast masking operation to the difference sequence is performed. Finally the masked differences are weighted and pooled over all dimensions to yield summary measures of visual error [11]. The output value of the VQM metric indicates the amount of distortion of the sequence, for no impairment the value is equal to zero and for rising level of impairment the output value rises.

4. Measurements

Nowadays some free databases containing video sequences used for video quality evaluation are available. Many of them include video sequences in FullHD or 4K resolution in uncompressed form (yuv) or compressed by H.264/AVC or H.265/HEVC standards [12], [13], [14], [15] and [16]. In our experiments four types of test sequences were used:

- "Beauty": Closeup on female face, hair waving around. Black background (Fig. 4).
- "Bosphorus": Zoomed in luxury yacht, huge bridge on background. Panning right (Fig. 5).
- "Jockey": Horse racing with camera panning to the left to follow (Fig. 6).
- "ReadySteadyGo": Horse racing track, riders getting ready for launch. The gates open and horses are running to the left (Fig. 7).

All sequences were in the 4K resolution (3840×2160 px) and 16:9 aspect ratio with 30 fps (frames per second). The length of these sequences was 600 frames, i.e. 20 seconds. The measurement procedure consists of four steps:

- First, both sequences were downloaded from [12] in the uncompressed format (*.yuv) and used as the reference sequences.
- Afterwards, they were encoded to both MPEG compression standards, H.264/AVC and H.265/HEVC, using the x264 [17], x265 tool respectively [18]. The target bitrates were in range from 2 Mbps to 30 Mbps, changed in 2 Mbps steps. The parameters of the encoded sequences were set to High Profile, Level 5.2 for H.264/AVC compression standard and Main Profile, Level 5.2 for H.265/HEVC compression standard. The GOP parameter was set to $N = 12$ and $M = 3$ which means that GOP length was 12 and two B frames between two successive P frames were stored. The command line settings of both x264 and x265 tools are shown in Tab. 1.
- Then, the sequences using the same tools (x264, x265) were decoded back to the format *.yuv.
- Finally, the quality between these sequences and the reference (uncompressed) sequence was compared and evaluated. This was done using the MSU Measuring Tool Pro version 3.0 [19]. PSNR, SSIM and VQM objective metrics for the measurements were used.

Tab. 1: Command line settings of x264 and x265 tools.

Command line options	x265, x264, command line settings
Input options	– input Beauty_3840×2160 _30fps_420_8bit.yuv – input-res 3840×2160 – input-csp i420 – input-depth 8 – fps 30
Slice decision	– no-open-gop
Options	– keyint 12 – min-keyint 12 – no-scenecut – bframes 2 – b-adapt 0
Quality, rate control and rate distortion options	– bitrate 4000 – vbv-maxrate 4000 – vbv-bufsize 4000
Debugging options	– recon Beauty_3840×2160 _30fps_420_8bit_GOP12-BF2_4M_x256.yuv Beauty_3840×2160_30fps _420_8bit_GOP12-BF2 _2M_x256.mp4

The whole process of measuring of both sequences is shown in Fig. 3. Figure 8, Fig. 9, Fig. 10, Fig. 11, Fig. 12, Fig. 13, Fig. 14, Fig. 15, Fig. 16, Fig. 17, Fig. 18 and Fig. 19 show the measurements results of the H.264/AVC and H.265/HEVC compression standards impact on the video quality for various test sequences.

According to the graphs the H.265/HEVC compression standard yields better compression efficiency than H.264/AVC compression standard, at the same resolution and bitrate the compression quality of H.265/HEVC standard is better than H.264/AVC. Bigger difference in quality between these two compression standards is in lower bitrates, with increasing bitrate the quality of H.264/AVC standard approach

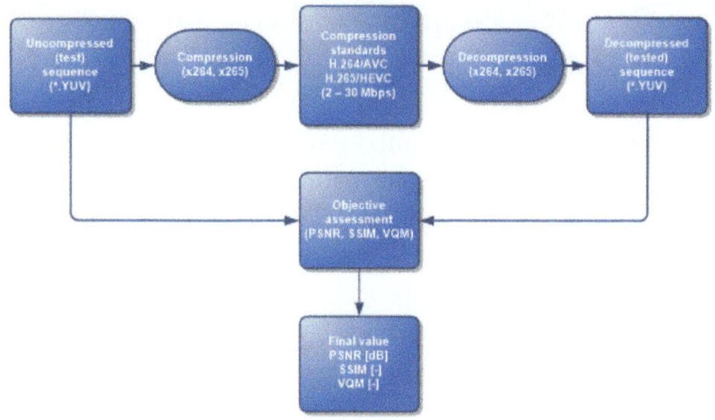

Fig. 3: The process of measuring the impact of H.264/AVC and H.265/HEVC compression standards on the video quality.

Fig. 4: The “Beauty” test sequence.

Fig. 5: The “Bosphorus” test sequence.

Fig. 6: The "Jockey" test sequence.

Fig. 7: The "ReadySetGo" test sequence.

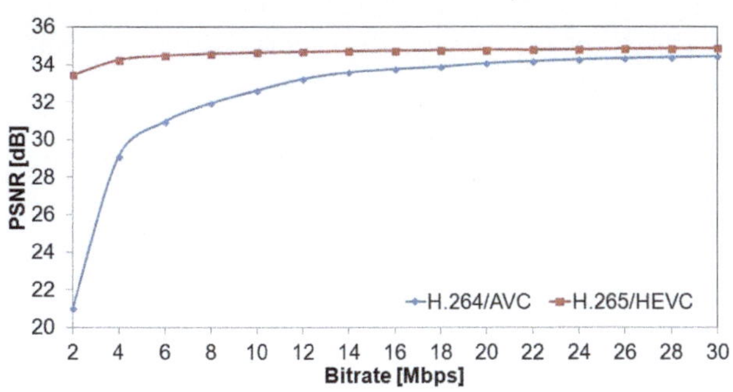

Fig. 8: The relationship between video quality (PSNR) and bitrate of both compression standards for "Beauty" test sequence.

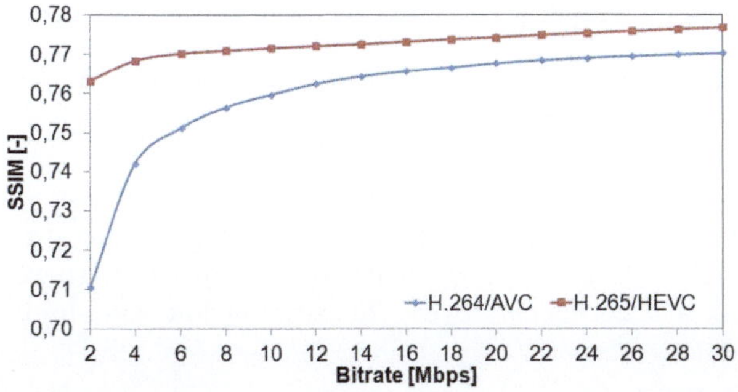

Fig. 9: The relationship between video quality (SSIM) and bitrate of both compression standards for "Beauty" test sequence.

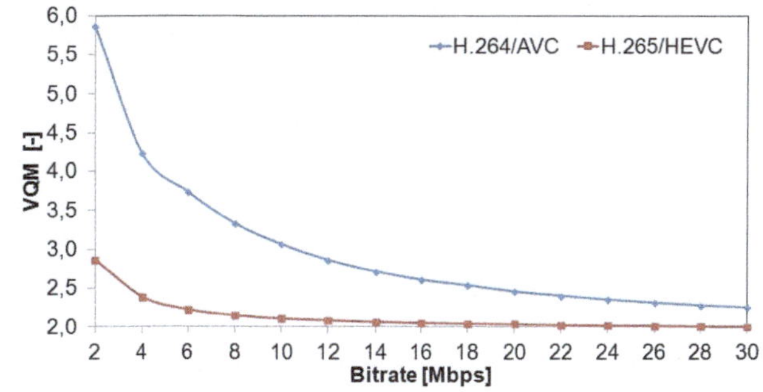

Fig. 10: The relationship between video quality (VQM) and bitrate of both compression standards for "Beauty" test sequence.

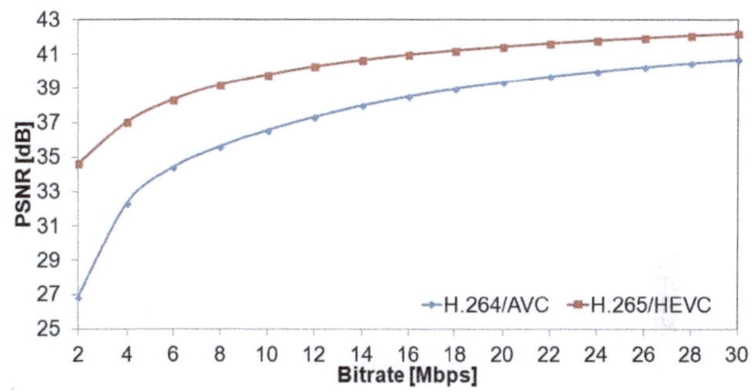

Fig. 11: The relationship between video quality (PSNR) and bitrate of both compression standards for "Bosphorus" test sequence.

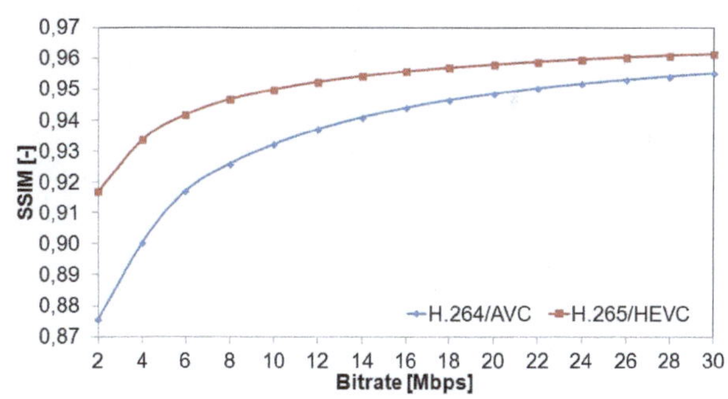

Fig. 12: The relationship between video quality (SSIM) and bitrate of both compression standards for "Bosphorus" test sequence.

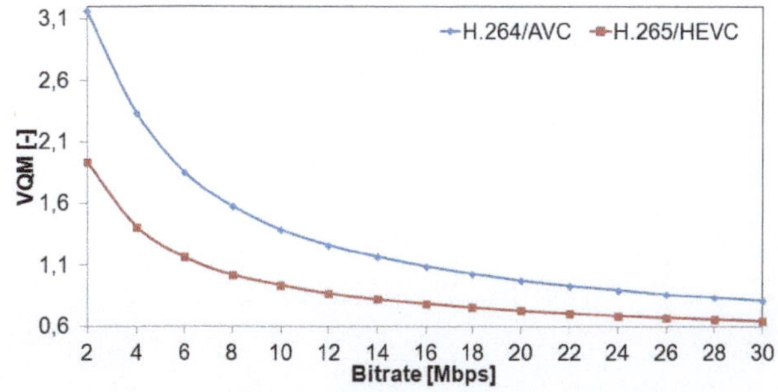

Fig. 13: The relationship between video quality (VQM) and bitrate of both compression standards for "Bosphorus" test sequence.

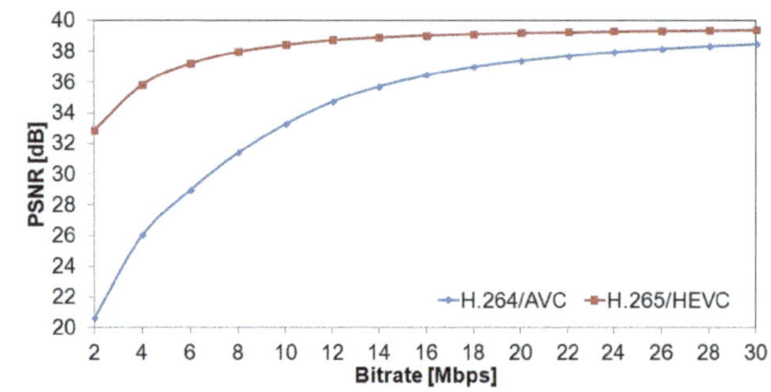

Fig. 14: The relationship between video quality (PSNR) and bitrate of both compression standards for "Jockey" test sequence.

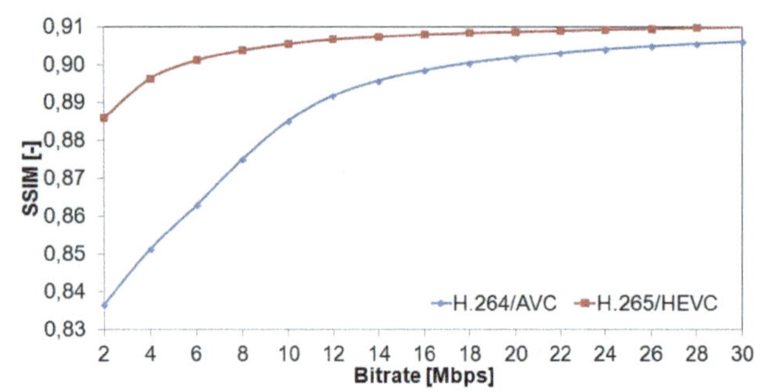

Fig. 15: The relationship between video quality (SSIM) and bitrate of both compression standards for "Jockey" test sequence.

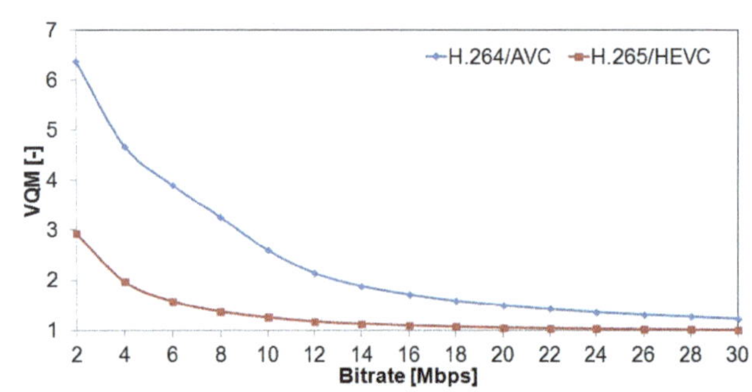

Fig. 16: The relationship between video quality (VQM) and bitrate of both compression standards for "Jockey" test sequence.

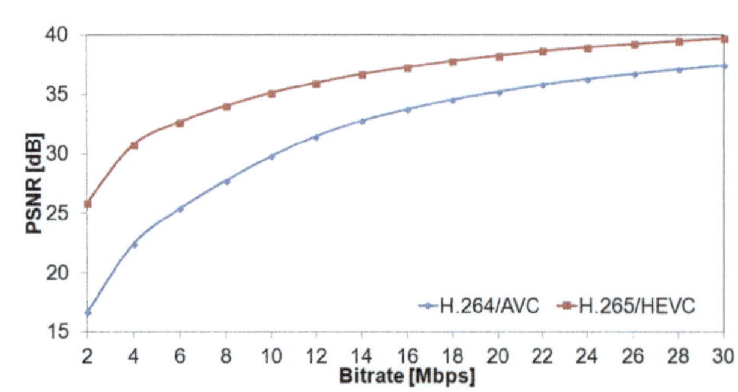

Fig. 17: The relationship between video quality (PSNR) and bitrate of both compression standards for "ReadySteadyGo" test sequence.

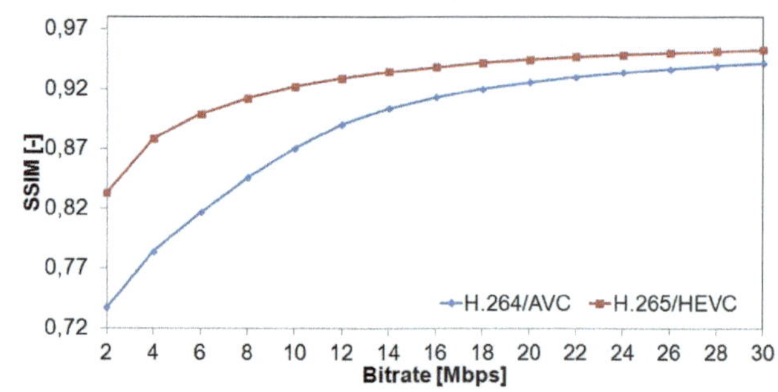

Fig. 18: The relationship between video quality (SSIM) and bitrate of both compression standards for "ReadySteadyGo" test sequence

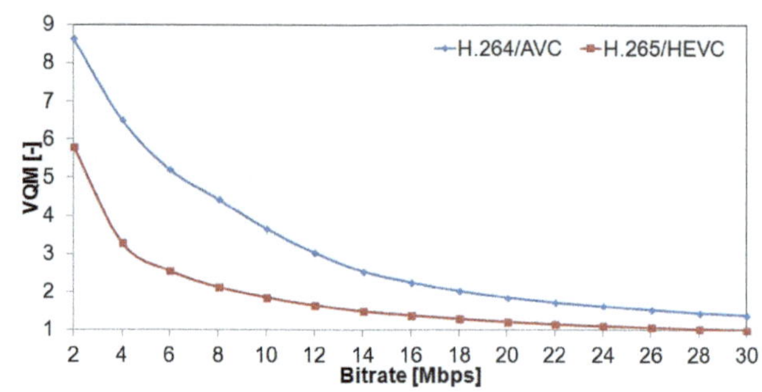

Fig. 19: The relationship between video quality (VQM) and bitrate of both compression standards for "ReadySteadyGo" test sequence

the H.265/HEVC standard. The effectiveness of compression depends on the type of sequence. Figure 20, Fig. 21, Fig. 22 and Fig. 23 show compression efficiency of both compression standards for various test sequences.

As it can be seen from the graphs, the compression efficiency depends on the types of test sequences. In consideration of measurements results we can say that:

- in sequences where is a slightly movement of one object on a static background as by "Bosphorus" sequence (Zoomed in luxury yacht, huge bridge on background. Panning right), not only best quality is achieved but also settled, there is no big difference between quality by low and high bitrates,
- in sequences where is a quick movement of more objects on a static background as by "ReadySteadyGo" sequence (Horse racing track, riders getting ready for launch. The gates open and horses are running to the left), only in higher bitrates very good quality is reached (not settled), there is a difference between quality by low and high bitrates,
- in sequences where is a quick movement of one object on a static background as by "Jockey" sequence (Horse racing with camera panning to the left to follow), very good quality is achieved and

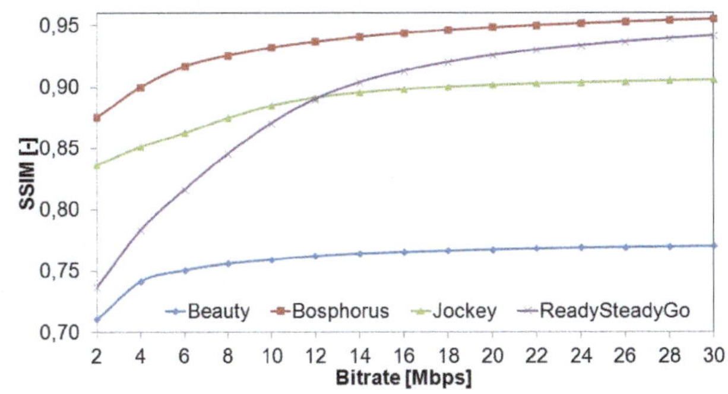

Fig. 20: The relationship between video quality (SSIM) and bitrate of all test sequences for H.264/HEVC compression standard.

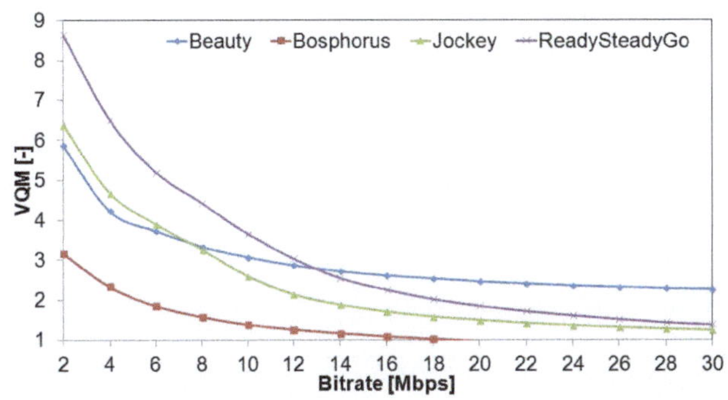

Fig. 21: The relationship between video quality (VQM) and bitrate of all test sequences for H.264/HEVC compression standard.

also settled, there is no big difference between quality by low and high bitrates,

- in sequences which are almost static as by "Beauty" sequence, where are only a few movements of female face (Closeup on female face, hair waving around. Black background), not so good quality is reached but settled, there is no big difference between quality by low and high bitrates.

5. Conclusion

In this article the impact of H.264/AVC and H.265/HEVC compression standards on the video quality for 4K resolution was tested. First a short characteristic of both compression standards was written, then the well-known objective metrics for evaluation of video quality were described and finally the measurements with experimental results were done and evaluated. Two types of experiments were done, the compression efficiency of both standards and the impact of the type of sequence on the video quality. According to the results, the H.265/HEVC compression standard yield better compression efficiency than H.264/AVC compression standard. The bigger difference in quality between these two compression standards is in lower bitrates with increasing bitrate the quality of H.264/AVC standard approaches the H.265/HEVC standard. It can be also stated that the effectiveness of compression depends on the type of sequence.

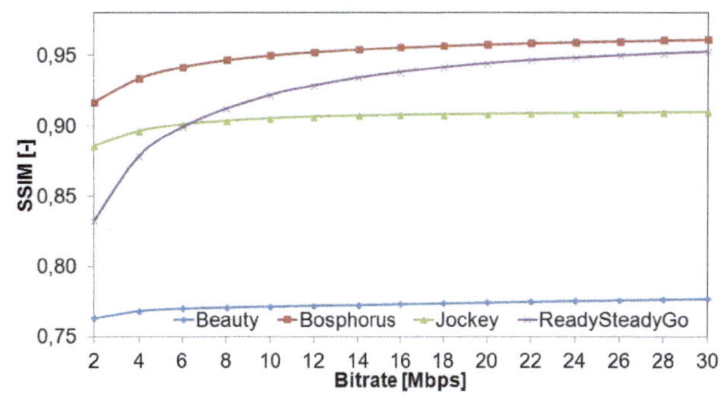

Fig. 22: The relationship between video quality (SSIM) and bitrate of all test sequences for H.265/HEVC compression standard.

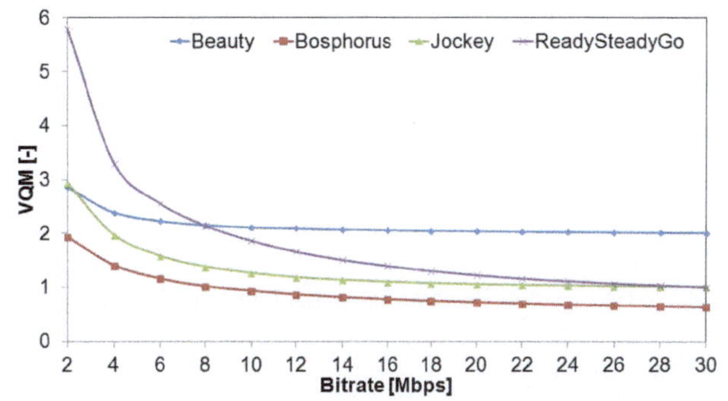

Fig. 23: The relationship between video quality (VQM) and bitrate of all test sequences for H.265/HEVC compression standard.

Acknowledgment

This contribution/publication is the result of the project implementation: Centre of excellence for systems and services of intelligent transport II., ITMS 26220120050 supported by the Research & Development Operational Programme funded by the ERDF.

References

[1] MPEG. Available at: https://www.mpeg.org.

[2] WINKLER, S. *Digital video quality: vision models and metrics.* Chichester: John Wiley, 2005. ISBN 04-700-2404-6.

[3] WOOTTON, C. *A practical guide to video and audio compression: from sprockets and rasters*

to macroblocks. Amsterdam: Focal Press, 2005. ISBN 02-408-0630-1.

[4] RICHARDSON, I. E. *H.264 and MPEG-4 video compression: video coding for next-generation multimedia*. Chichester: Wiley, 2003. ISBN 04-708-4837-5.

[5] RICHARDSON, I. E. *The H.264 advanced video compression standards*. Chichester: John Wiley, 2003. ISBN 978-0-470-51692-8.

[6] SCIENTIFIC ATLANTA. *MPEG-4 Part 10 AVC (H.264) Video Encoding*. Atlanta, 2005. Available at: http://pdf.aminer.org/000/322/892/reference_picture_selection_in_an_already_mpeg_encoded_bitstream.pdf.

[7] SULLIVAN, G. J., J.-R. OHM, W.-J. HAN and T. WIEGAND. Overview of the High Efficiency Video Coding (HEVC) Standard. *IEEE Transactions on Circuits and Systems for Video Technology*. 2012, vol. 22, iss. 12, pp. 1649–1668. ISSN 1051-8215. DOI: 10.1109/TCSVT.2012.2221191.

[8] WINKEN, M., P. HELLE, D. MARPE, H. SCHWARZ and T. WIEGAND. Transform codinginthe HEVC Test Model. In: *18th IEEE International Conference on Image Processing*. Brussel: IEEE, 2011, pp. 3693–3696. ISBN 978-1-4577-1303-3. DOI: 10.1109/ICIP.2011.6116521.

[9] RICHARDSON, I. VCODEX. *HEVC: An Introduction to High Efficiency Video Coding*. 2013. Available at: http://www.vcodex.com/images/uploaded/342512928230717.pdf.

[10] WU, H and K. RAO. *Digital video image quality and perceptual coding*. Boca Raton: Taylor, 2006. ISBN 08-247-2777-0.

[11] LOKE, M., E. ONG, W. LIN, Z. LU and S. YAO. Comparison of Video Quality Metrics on Multimedia Videos. In: *International Conference on Image Processing*. Atlanta: IEEE, 2006, pp. 457–460. ISBN 1-4244-0480-0. DOI: 10.1109/ICIP.2006.312492.

[12] Test sequences. ULTRA VIDEO GROUP. Available at: http://ultravideo.cs.tut.fi/#testsequences.

[13] Test media. XIPH. Available at: http://media.xiph.org.

[14] 4K Test Sequences. EMENTAL. Available at: http://www.elementaltechnologies.com/resources/4k-test-sequences..

[15] SJTU 4K Video Sequences. SJTU MEDIA LAB. Available at: http://medialab.sjtu.edu.cn/web4k/index.html.

[16] Free 4K Demo Footage. HARMONIC. Available at: http://www.harmonicinc.com/resources/videos/4k-video-clip-center.

[17] X264: free software library and application. VIDEO LAN ORGANIZATION. Available at: http://generator.citace.com/dok/yRN7548jgt2cGPOJ.

[18] x265: free software library and application. VIDEO LAN ORGANIZATION. Available at: http://www.videolan.org/developers/x265.html.

[19] MSU Quality Measurement Tool: PRO Version. EVERYTHING ABOUT THE DATA COMPRESSION. Available at: http://compression.ru/video/quality_measure/vqmt_pro_en.html#start.

[20] GROIS, D., D. MARPE, A. MULAYOFF, B. ITZHAKY and O. HADAR. Performance comparison of H.265/MPEG-HEVC, VP9, and H.264/MPEG-AVC encoders. In: *Picture Coding Symposium (PCS)*. San Jose: IEEE, 2013, pp. 394–397. ISBN 978-1-4799-0292-7. DOI: 10.1109/PCS.2013.6737766.

[21] KIM, I.-K., S. LEE, Y. PIAO and J. CHEN. Coding efficiency comparison of new video coding standards: HEVC vs VP9 vs AVS2 video. In: International Conference on Multimedia and Expo Workshops (ICMEW). Chengdu: IEEE, 2014, pp. 1–6. ISSN 1945-7871. DOI: 10.1109/ICMEW.2014.6890700.

[22] NGUYEN, T. and D. MARPE. Objective Performance Evaluation of the HEVC Main Still Picture Profile. *IEEE Transactions on Circuits and Systems for Video Technology*. 2013, vol. pp, iss. 99, pp. 1–8. ISSN 1051-8215. DOI: 10.1109/TCSVT.2014.2358000.

About Authors

Miroslav UHRINA was born in 1984 in Zilina, Slovakia. He received his M.Sc. and PhD. degrees in Telecommunications at the Department of Telecommunications and Multimedia, Faculty of Electrical Engineering, at the University of Zilina in 2008 and 2012, respectively. Nowadays he is an assistant professor at the Department of Telecommunications and Multimedia at the University of Zilina His research interests include audio and video compression, TV broadcasting (IPTV, DVB-T, DVB-H) and IP networks.

Jaroslav FRNDA was born in 1989 in Martin, Slovakia. He received his M.Sc. at the Department of Telecommunications, Faculty of Electrical Engineering and Computer Science, at the VSB–Technical University of Ostrava in 2013. Nowadays he is a Ph.D. student at the Department of Telecommunications at the VSB–Technical University of Ostrava. His research interests include quality of Triple play services.

Lukas SEVCIK was born in 1989. He received his M.Sc. at the Department of Informatics, Faculty of Management Science and Informatics at the University of Zilina in 2013. Nowadays he is a Ph.D. student at the Department of Telecommunications at the VSB–Technical University of Ostrava. His research interests include Quality of Triple play services.

Martin VACULIK was born in 1951. He received his M.Sc. and Ph.D. in Telecommunications at the University of Zilina , Slovakia in 1976 and 1987 respectively. In 2001 he was habilitated as associate professor of the Faculty of Electrical Engineering at the University of Zilina in the field of Telecommunications. Currently he works as a head of Department of Telecommunications and Multimedia University of Zilina . His interests cover switching and access networks, communication network architecture, audio and video applications.

30

Speech Quality Measurement of GSM Infrastructure Built on USRP N210 and OpenBTS Project

Marcel FAJKUS[1], *Martin MIKULEC*[1], *Miroslav VOZNAK*[1], *Martin TOMIS*[1], *Peppino FAZIO*[2]

[1]Department of Telecommunications, Faculty of Electrical Engineering and Computer Science, VSB–Technical Universit of Ostrava, 17. listopadu 2172/15, 708 33 Ostrava-Poruba, Czech Republic
[2]Department of Computer Science, Modeling, Electronics and Systems Engineering, University of Calabria, Edificio 42C - piano 5, 870 36 Arcavacata di Rende (CS), Italy

marcel.fajkus@vsb.cz, martin.mikulec@vsb.cz, miroslav.voznak@vsb.cz, martin.tomis@vsb.cz, pfazio@dimes.unical.it

Abstract. *The paper deals with the methodology for speech quality measuring in GSM networks using Perceptual Evaluation of Speech Quality (PESQ). The paper brings results of practical measurement of own GSM network build on the Universal Software Radio Peripheral (USRP N210) hardware and OpenBTS software. This OpenBTS station was installed in open terrain, and the speech quality was measured from different distances from the transmitter. The limit parameters of OpenBTS station with USRP N210 were obtained.*

Keywords

Asterisk, GSM, OpenBTS, PESQ, USRP.

1. Introduction

GSM and UMTS technologies have become a common part of almost every aspect of a human activity during its expansion of the last decade. Nowadays people consider using mobile phones for calling, texting and browsing the Internet as natural service like electricity or drinkable water at home. All these services require reliable infra-structure and control mechanism to ensure quality of the provided service. This paper deals with one aspect of control mechanism, measuring and controlling the quality of speech. In GSM/UMTS environment we can take advantage of algorithms that have been invented for IP-based networks. There are two basic evaluations of speech quality, objective and subjective. We have aimed at the objective method of measuring because of the zero influence of human factor and because of the possibility to repeat the same methodology for measuring speech quality in every GSM or UMTS infrastructure. The mobile terminals were connected to GSM infrastructure provided by testing OpenBTS station; build on Universal Software Radio Peripheral. We have aimed at measurement of speech quality in open terrain according to the distance of the mobile terminal from the base transceiver station and according to the number of simultaneous calls.

2. State of the Art

The speech quality evaluation is measured in Mean Opinion Score (MOS), which is five-degree scale developed by ITU-T. The objective methods are trying to be as precise as possible to gain adequate MOS value as it would be obtained by subjective methods with a sufficient number of participants for adequate statistical analysis. The objective methods are divided to intrusive and non-intrusive methods. The intrusive methods use the original sound sample as it has entered the communication channel and compare it with the degraded signal in the output. The most known intrusive methods are:

- PSQM (Perceptual Speech Quality Measurement),
- PAMS (Perceptual Analysis Measurement System),
- PESQ (Perceptual Evaluation of Speech Quality),
- P.OLQA (Perceptual Objective Listening Quality Assessment).

According to the fact, that PESQ algorithm is currently widely used, we have decided to use this algorithm as the keystone of our measurement. The intrusive method is used to determine the quality of the speech, when the test system compares the original signal $x(t)$ with the degraded signal $y(t)$ taken at the other end of the transmission chain. Subsequently, reached values are evaluated in MOS-PESQ scale and then transferred using complementary recommendation P.862.1 on the values of the MOS-LQO scale, as shown in Fig. 1, [1], [2].

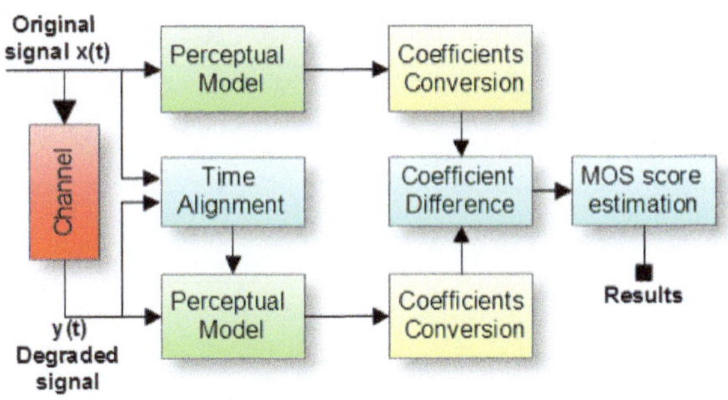

Fig. 1: Basic diagram of PESQ algorithm.

The PESQ method first computes several series of delays between the original and degraded signal, where each of these series corresponds to one interval of signal. The delay for each interval is different due to ensure proper functioning of the PESQ algorithm. For each interval, it is also determined start point and end point of time. Based on these series of delays, PESQ algorithm compares the original and degraded signal using the perceptual model.

The resulting PESQ-MOS score is expressed as a range of values from −0.5 to 4.5. This score has to be converted to more accurate scope, more accurate for human subjective evaluation. Therefore, it is necessary to use complementary ITU-T P.862.1, which will provide scale transfer from MOS-PESQ to MOS-LQO. Scale MOS-LQO provides a range of values from 1 to 5. Conversion from PESQ MOS to MOS LQO is defined by Eq. (1), [3], [4]:

$$y = 0.999 + \frac{4.999 - 0.999}{1 + e^{-1,4945x + 4.6607}}, \qquad (1)$$

where the variable x represents the value of the MOS-PESQ scale, and the y represent MOS-LQO score. Inverse score (MOS-PESQ) from LQO is shown in Eq. (2):

$$x = \frac{4.6607 - \ln\frac{4.999 - y}{y - 4.999}}{1.4945}. \qquad (2)$$

3. Testing Platforms

We have designed the measuring platform, which is able to generate GSM calls automatically in regular intervals and analyze the voice sample according to the P.862 specification. The results were logged for future purposes.

As the measured platform, we have used own GSM infrastructure provided by testing OpenBTS station build on Universal Software Radio Peripheral used in our laboratory.

3.1. Measuring Platform Details

The measuring platform was implemented on the low-energy consumption embedded device. According to this fact, we were able to use power supply from car and get high mobility of end mobile devices. The low-cost Huawei K3765 modems were used as end mobile devices for originating and receiving testing calls. These modems were connected directly with embedded device by USB ports and controlled by Asterisk PBX system installed inside the embedded device. As an operating system and core, we have decided to use a special Linux based system developed in our laboratory called BESIP [5].

The calls with original speech sample were generated automatically by Asterisk over SIP (Session Initiation Protocol) and RTP (Real-time Transport Protocol), as shown in Fig. 2. The modems were providing SIP/GSM translation of the outgoing and incoming calls. The Asterisk PBX was recording the incoming voice data to a separate WAV file with adequate timestamp of the measurement. After the end of the call, the PESQ algorithm was applied to the original and recorded degraded signal. The result, speech quality in MOS, was stored in the database for future evaluation.

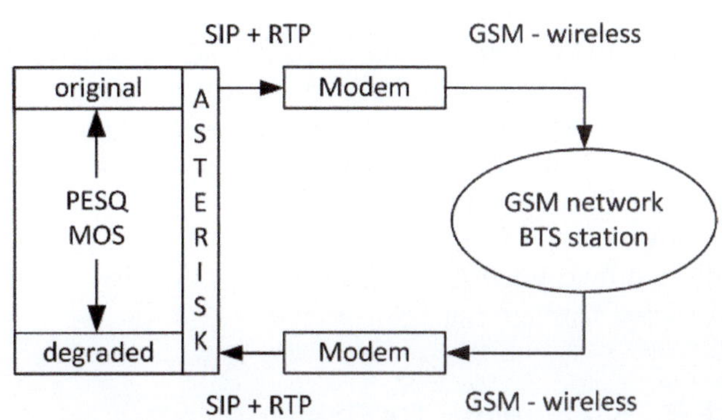

Fig. 2: BESIP measuring tool.

3.2. Measured Platform Details

The tested GSM infrastructure was built on Universal Software Radio Peripheral (USRP) hardware in connection with OpenBTS (Open Base Transceiver Station) software, which is open-source Linux-based application, that provides management of USRP to create wireless GSM interface, as shown in Fig. 3. We have used USRP N210 from Ettus Re-searchTM in combination with Daughterboard WBX [6].

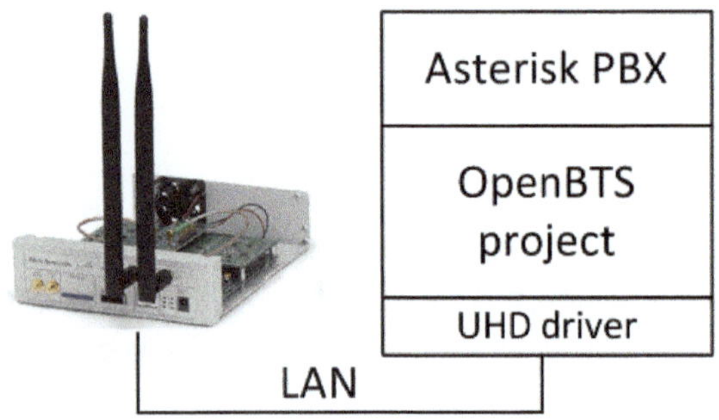

Fig. 3: USRP N210 connected with host OpenBTS server.

The USRP N210 architecture includes Xilinx Spartan 3A-DSP 3400 FPGA kit. In combination with WBX 50–2200 MHz Rx/Tx Daughterboard provides adequate architecture for creating cell phone base station or other applications.

There is UHD (USRP Hardware Driver), which provides API and drivers to the host computer for communication with USRP device. OpenBTS software is written in the C++ programming language and ensures the following functions of Um radio interface, which provides the radio interface for GSM standard [7], [8]:

- GMSK modulation with 13/48 MHz modulation rate and 200 kHz distance – supports GSM850, PGSM900, DCS1800 and PCS1900.
- Multiplexing and coding.
- Management of network resources.

Asterisk PBX is implemented inside the OpenBTS project and provides mobility management, authentication of the user and routing of the calls between registered users. Registered modems are identified by IMSI number of SIM card presented in the modems. The main advantage of this solution is that the calls can't be affected by commercial mobile operator; whole traffic is routed through our low-cost infrastructure.

3.3. Measuring Realization

We have installed USRP N210 with OpenBTS server statically outside the laboratory in an open terrain to ensure minimal influence of surrounding condition on a radio signal spreading.

Figure 4 depicts the measuring platform, represented by embedded BESIP with modems, was installed in the car and placed nearby our OpenBTS station.

Fig. 4: Scheme of measuring – BESIP + USRP N210 connected with host OpenBTS server.

When the modems were registered to our OpenBTS station, there were 3 calls originated on that place. The reason of the three calls was to get statistically reasonable results of speech quality and strength of the incoming signal to the modems. Then the measurement was repeated with increasing distance from the OpenBTS station.

We were interested in a few measuring parameters:

- The maximum number of simultaneous call, that is our OpenBTS able to provide.
- Dependence of receive signal strength on the distance from our OpenBTS.
- Dependence of a speech quality on the distance from our OpenBTS.
- Dependence of a speech quality on the number of simultaneous calls.
- Time period from originating to answering the call and its dependence on distance from OpenBTS and number of simultaneous calls.

4. Results of Measurement

After many repeated measurements in increasing distance, we have reached a sufficient amount of data to statistically evaluate limit parameters of our OpenBTS system. The results were following.

4.1. Maximum Number of Simultaneous Calls

OpenBTS in a basic non-commercial version uses one logical channel. GSM technologies use time multiplex-

Tab. 1: Success rate of established calls.

Number of simultaneous calls	Number of originated calls	Number of successfully answered calls	Probability of successful answered calls [%]
1	35	31	88.6
2	34	23	67.6
3	31	4	12.9

ing, one channel can be separated to 8 single time channels. We have found out that in our configuration the maximum number of simultaneous calls was 3 simultaneous calls. The probability of successful call establishing was only 12 % (4 successful calls from 31 tries).

4.2. Dependence of Receive Signal Strength on the Distance

There was a received signal strength (RSSI) measured during every active call. Measured and average values are depicted in Fig. 5. As the picture shows, the RSSI decreases −1.95 dB to 1 meter in average.

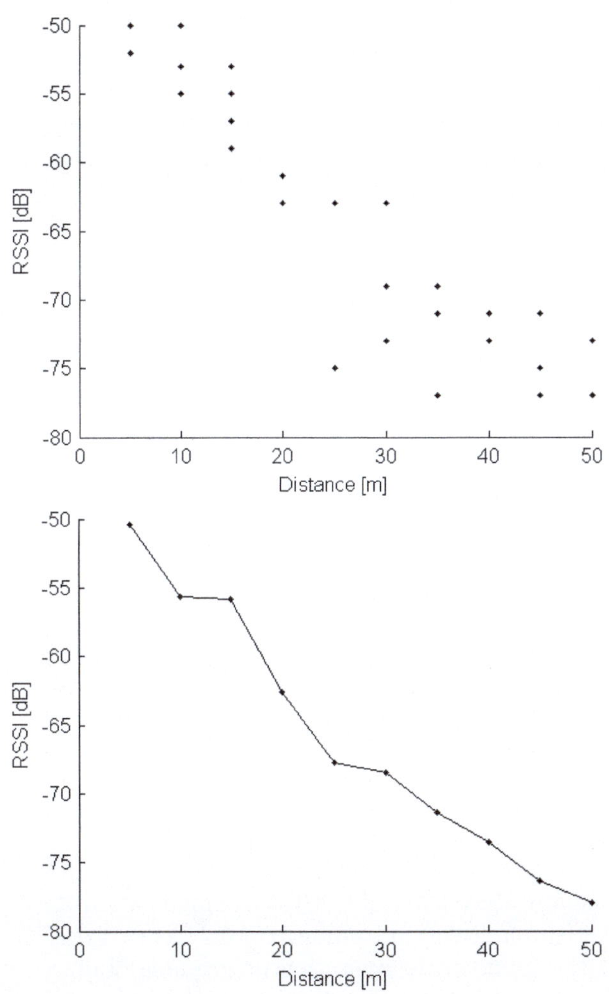

Fig. 5: RSSI of GSM signal in different distance from OpenBTS.

4.3. Dependence of a Speech Quality on the Distance from OpenBTS

The measured MOS values of speech quality are depicted in Fig. 6. We have removed values smaller than 2.2 MOS, because of the poor quality according to the ITU-T P.800 recommendation.

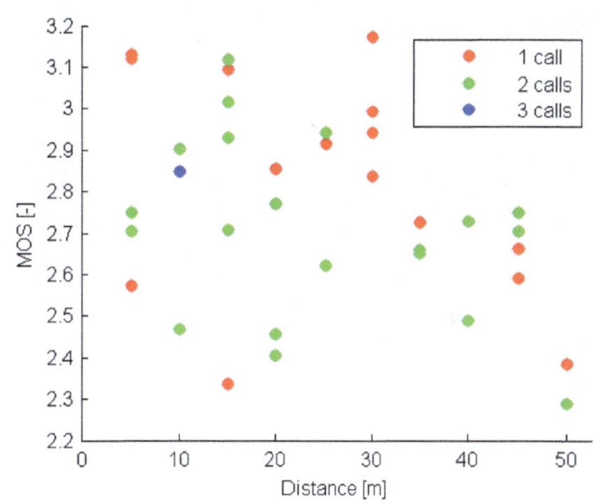

Fig. 6: Speech qualitz in MOS in different distance from OpenBTS.

We have made regression analysis from the measured data. The results are depicted in Fig. 7, Fig. 8, Fig. 9. According to the low count of the successful calls for 3 simultaneous calls, we have added the MOS values < 2.2 to show regression progress for 3 simultaneous calls in Fig. 9.

4.4. Time Period from Originating to Answering the Call

The results in Fig. 10 showed that time period of establishing call do not depend on the distance from the OpenBTS transmitter, but does depend on the count of active simultaneous calls.

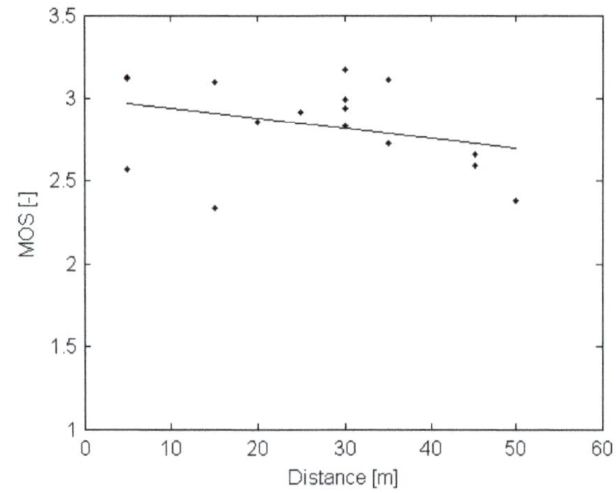

Fig. 7: Linear regression MOS value – 1 call.

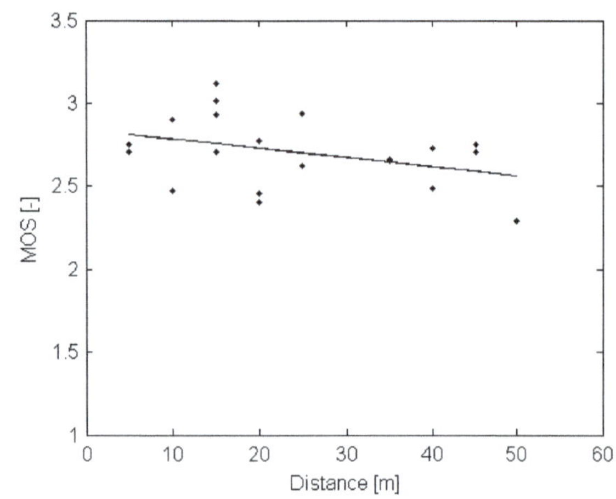

Fig. 8: Linear regression MOS value – 2 call.

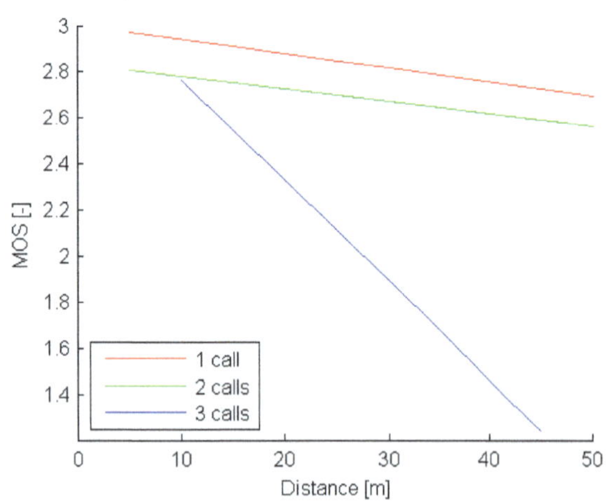

Fig. 9: Linear regression MOS value for all calls.

5. Conclusion

The results of the measurement showed that our OpenBTS station built on the USRP N210 is not suitable for practical application, only for laboratory purposes. The maximum of simultaneous calls were 3 calls, optimal load of the OpenBTS station is 1–2 simultaneous calls from 5 to 50 meters from the transmitter. The poor speech quality and connection failures are detected out of the measured limits. The measurement approved theoretical prerequisite that the speech quality decreases with increasing distance from the OpenBTS station and number of simultaneous calls.

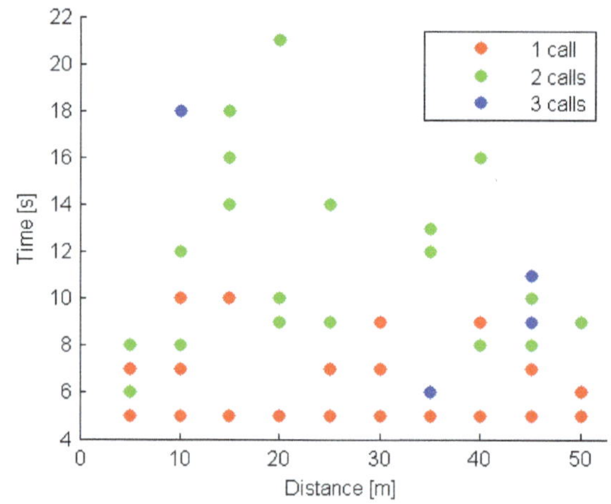

Fig. 10: Time period of establishing calls.

The main benefit of the paper is a methodology for automatic speech quality measurement of a general GSM infrastructure. The methodology uses low-cost embedded hardware, modems and open-source software tools.

Our next step will be to install new OpenBTS station built on more powerful USRP hardware from National Instruments, repeat the same measurement on it and compare the results.

Acknowledgment

This paper has been elaborated in the framework of the project "Support research and development in the Moravian-Silesian Region 2013 DT 2, reg. no. 02540/2013/RRC, financed from the budget of the Moravian-Silesian Region. This work has been partially supported by Grant of SGS No. SP2014/72 and by the project No. CZ.1.07/2.3.00/20.0217 "The Development of Excellence of the Telecommunication Research Team in Relation to International Cooperation" within the frame of the operation programme "Education for competitiveness" financed by the Structural Funds and from the state budget of the Czech Republic.

References

[1] ROZHON, J. and M. VOZNAK. Development of a speech quality monitoring tool based on ITU-T P.862. In: *34th International Conference on Telecommunications and Signal Processing (TSP).* Budapest: IEEE, 2011, pp. 62–66. ISBN 978-145771411-5. DOI: 10.1109/TSP.2011.6043771.

[2] ITU-T Recommendation G.107. *The E-model: A computation model for use in transmission planning.* Geneva: ITU-T, 2012.

[3] ITU-T Rec. P.862. *Perceptual Evaluation of Speech Quality (PESQ): An Objective Method for End-to-End Speech Quality Assessment of Narrow-Band Telephone Networks and Speech Codecs.* Geneva: ITU-T, 2001.

[4] ZHAO, J.-Y., F. CAO, M.-Y. LIANG and F. LIU YA. Very Low Bit Rate Speech Coding Algorithm and Its Implementation Based on DSP. *Computer Engineering.* 2011, vol. 37, iss. 21, pp. 261–263. ISSN 1000-3428. DOI: 10.3969/j.issn.1000-3428.2011.21.089.

[5] TOMALA, K., L. MACURA, M. VOZNAK and J. VYCHODIL. Monitoring the quality of speech in the communication system BESIP. In: *35th International Conference on Telecommunications and Signal Processing (TSP).* Prague: IEEE, 2012, pp. 255–258. ISBN 978-146731118-2. DOI: 10.1109/TSP.2012.6256293.

[6] ETTUS RESEARCH. *USRP N200/N210* Avalaible at: `http://www.ettus.com/content/files/07495_\Ettus_N200-210_DS_Flyer_HR_1.pdf`

[7] APVRILLE, A. *OpenBTS for dummies.* 2013. Available at: `http://gnuradio.org/redmine/attachments/\420/fordummies.pdf`

[8] NATALIZIO, E., V. LOSCRI, G. ALOI, N. PAOLI and N. BARBARO. The practical experience of implementing a GSM BTS through open software/hardware. In: *3rd International Symposium on Applied Sciences in Biomedical and Communication Technologies (ISABEL 2010).* Rome: IEEE, 2010, pp. 1–5. ISBN 978-1-4244-8131-6. DOI: 10.1109/ISABEL.2010.5702870.

About Authors

Marcel FAJKUS was born in 1987 in Ostrava. In 2009 received Bachelor's degree on VSB–Technical University of Ostrava, Faculty of Electrical Engineering and Computer Science, Department of Telecommunications. Two years later he received on the same workplace his Master's degree in the field of Telecommunications. He is currently Ph.D. student, and he works in the field of optical communications.

Martin MIKULEC received the M.Sc. degree from VSB–Technical University of Ostrava in 2011. Currently, he is working toward the Ph.D. degree at the Deptartment of Telecommunications, VSB–Technical University of Ostrava. His research is focused on advanced services in IP telephony and VoIP.

Miroslav VOZNAK is an Associate Professor with Department of Telecommunications, VSB–Technical University of Ostrava, Czech Republic. Topics of his research interests are Next Generation Networks, IP telephony, speech quality and network security.

Martin TOMIS was born in 1987. He received his master degree from Department of Telecommunications, Faculty of Electrical Engineering and Computer Science, VSB–Technical University Ostrava in 2011. His research include radio-communication technologies.

Peppino FAZIO was born in 1977. He received the degree in computer science engineering in May 2004. Since November 2004 he has been a Ph.D. student in Electronics and Communications Engineering at the University of Calabria and he has got the Ph.D. in January 2008; at the moment he is a research fellow at Department of Electronics, Computer Science and Systems of University of Calabria, after many collaborations with Department of "Universidad Politecnica de Valencia". His research interests include mobile communication networks, QoS architectures and interworking wireless and wired networks, mobility modeling for WLAN environments and mobility analysis for prediction purposes.

Permissions

All chapters in this book were first published in AEEE, by Technical University of Ostrava; hereby published with permission under the Creative Commons Attribution License or equivalent. Every chapter published in this book has been scrutinized by our experts. Their significance has been extensively debated. The topics covered herein carry significant findings which will fuel the growth of the discipline. They may even be implemented as practical applications or may be referred to as a beginning point for another development.

The contributors of this book come from diverse backgrounds, making this book a truly international effort. This book will bring forth new frontiers with its revolutionizing research information and detailed analysis of the nascent developments around the world.

We would like to thank all the contributing authors for lending their expertise to make the book truly unique. They have played a crucial role in the development of this book. Without their invaluable contributions this book wouldn't have been possible. They have made vital efforts to compile up to date information on the varied aspects of this subject to make this book a valuable addition to the collection of many professionals and students.

This book was conceptualized with the vision of imparting up-to-date information and advanced data in this field. To ensure the same, a matchless editorial board was set up. Every individual on the board went through rigorous rounds of assessment to prove their worth. After which they invested a large part of their time researching and compiling the most relevant data for our readers.

The editorial board has been involved in producing this book since its inception. They have spent rigorous hours researching and exploring the diverse topics which have resulted in the successful publishing of this book. They have passed on their knowledge of decades through this book. To expedite this challenging task, the publisher supported the team at every step. A small team of assistant editors was also appointed to further simplify the editing procedure and attain best results for the readers.

Apart from the editorial board, the designing team has also invested a significant amount of their time in understanding the subject and creating the most relevant covers. They scrutinized every image to scout for the most suitable representation of the subject and create an appropriate cover for the book.

The publishing team has been an ardent support to the editorial, designing and production team. Their endless efforts to recruit the best for this project, has resulted in the accomplishment of this book. They are a veteran in the field of academics and their pool of knowledge is as vast as their experience in printing. Their expertise and guidance has proved useful at every step. Their uncompromising quality standards have made this book an exceptional effort. Their encouragement from time to time has been an inspiration for everyone.

The publisher and the editorial board hope that this book will prove to be a valuable piece of knowledge for researchers, students, practitioners and scholars across the globe.

List of Contributors

Stefan BORIK
Department of Electromagnetic and Biomedical Engineering, Faculty of Electrical Engineering, University of Zilina, Univerzitna 1, 010 01 Zilina, Slovak Republic

Ivo CAP
Department of Electromagnetic and Biomedical Engineering, Faculty of Electrical Engineering, University of Zilina, Univerzitna 1, 010 01 Zilina, Slovak Republic

Suwat MANEEWAN
Department of Technology Electronics, Faculty of Industrial Technology, Buriram Rajabhat University, Buriram, 310 00, Thailand

Nawee UDORN
Department of Electronic Technology, Faculty of Industrial Technology, Nakhon Phanom University, Nakhon Phanom, 480 00, Thailand

Danupat DUANGMALAI
Department of Electronic Technology, Faculty of Industrial Technology, Nakhon Phanom University, Nakhon Phanom, 480 00, Thailand

Phamorn SILAPAN
Department of Electrical Computer and Industrial, Faculty of Industrial Technology, Uttaradit Rajabhat University, Uttaradit, 530 00, Thailand

Winai JAIKLA
Department of Engineering Education, Faculty of Industrial Education, King Mongkut's Institute of Technology Ladkrabang, Bangkok, 105 20, Thailand

Misel BATMEND
Department of Electrical Engineering and Mechatronics, Faculty of Electrical Engineering and Informatics, Technical University of Kosice, Letna 9, 042 00 Kosice, Slovak Republic

Daniela PERDUKOVA
Department of Electrical Engineering and Mechatronics, Faculty of Electrical Engineering and Informatics, Technical University of Kosice, Letna 9, 042 00 Kosice, Slovak Republic

Biplab BHATTACHARYYA
Department of Electrical Engineering, Indian School of Mines, Dhanbad, 826004 Jharkhand, India

Vikash Kumar GUPTA
Department of Electrical Engineering, Indian School of Mines, Dhanbad, 826004 Jharkhand, India

Sanjay KUMAR
Department of Electrical Engineering, Indian School of Mines, Dhanbad, 826004 Jharkhand, India

Fang LUO
Department of Electronic and Computer Engineering, College of Electrical and Computer Engineering, National Taiwan University of Science and Technology, No. 43, Sec. 4, Keelung Road, Taipei 106, Taiwan

Yu-Feng CHOU
Department of Electronic and Computer Engineering, College of Electrical and Computer Engineering, National Taiwan University of Science and Technology, No. 43, Sec. 4, Keelung Road, Taipei 106, Taiwan

Ray-Guang CHENG
Department of Electronic and Computer Engineering, College of Electrical and Computer Engineering, National Taiwan University of Science and Technology, No. 43, Sec. 4, Keelung Road, Taipei 106, Taiwan

Marek KALIKA
Department of Applied Informatics in Transport, Faculty of Transportation Sciences, Czech Technical University in Prague, Konviktska 20, 110 00 Prague, Czech Republic

Radek HOLY
Department of Applied Informatics in Transport, Faculty of Transportation Sciences, Czech Technical University in Prague, Konviktska 20, 110 00 Prague, Czech Republic

Jana KALIKOVA
Department of Applied Informatics in Transport, Faculty of Transportation Sciences, Czech Technical University in Prague, Konviktska 20, 110 00 Prague, Czech Republic

Lukas VOJTECH
Department of Telecommunication Engineering, Faculty of Electrical Engineering,
Czech Technical University in Prague, Technicka 2, 166 27 Prague, Czech Republic

Radoslav BORTEL
Department of Circuit Theory, Faculty of Electrical Engineering, Czech Technical University in Prague, Technicka 2, 166 27 Prague, Czech Republic

Marek NERUDA
Department of Telecommunication Engineering, Faculty of Electrical Engineering, Czech Technical University in Prague, Technicka 2, 166 27 Prague, Czech Republic

Milos KOZAK
Department of Telecommunication Engineering, Faculty of Electrical Engineering, Czech Technical University in Prague, Technicka 2, 166 27 Prague, Czech Republic

Stanislav KLUCIK
Institute of Telecommunications, Faculty of Electrical Engineering and Information Technology, Slovak University of Technology in Bratislava, Ilkovicova 3, 812 19 Bratislava, Slovakia

Martin LACKOVIC
Institute of Telecommunications, Faculty of Electrical Engineering and Information Technology, Slovak University of Technology in Bratislava, Ilkovicova 3, 812 19 Bratislava, Slovakia

Leos BOHAC
Department of Telecommunications Engineering, Faculty of Electrical Engineering, Czech Technical University in Prague, Technicka 2, 160 00 Prague, Czech Republic

Jiri VODRAZKA
Department of Telecommunications Engineering, Faculty of Electrical Engineering, Czech Technical University in Prague, Technicka 2, 160 00 Prague, Czech Republic

Hassan BARATI
Department of Electrical Engineering, Dezful Branch, Islamic Azad University, Dezful, Iran

Abdol Rasool FATTHI
Department of Electrical Engineering, Dezful Branch, Islamic Azad University, Dezful, Iran

Daniel HLADEK
Department of Electronics and Multimedia Communications, Faculty of Electrical Engineering and Informatics, Technical University of Kosice, Park Komenskeho 13, 042 00 Kosice, Slovakia

Jan STAS
Department of Electronics and Multimedia Communications, Faculty of Electrical Engineering and Informatics, Technical University of Kosice, Park Komenskeho 13, 042 00 Kosice, Slovakia

Jozef JUHAR
Department of Electronics and Multimedia Communications, Faculty of Electrical Engineering and Informatics, Technical University of Kosice, Park Komenskeho 13, 042 00 Kosice, Slovakia

Witthaya MEKHUM
Department of Industrial Management, Faculty of Industrial Technology, Suan Sunandha Rajabhat University, Ratchasima 295, 103 00 Bangkok, Thailand

Winai JAIKLA
Department of Engineering Education, Faculty of Industrial Education, King Mongkut's Institute of Technology Ladkrabang, Chalongkrung Road, 105 20 Bangkok, Thailand

Stanislav MISAK
Department of Electrical Power Engineering, Faculty of Electrical Engineering and Computer Science, VSB–Technical university of Ostrava, 17. listopadu 15, 708 33 Ostrava-Poruba, Czech Republic

Jindrich STUCHLY
Department of Electrical Power Engineering, Faculty of Electrical Engineering and Computer Science, VSB–Technical university of Ostrava, 17. listopadu 15, 708 33 Ostrava-Poruba, Czech Republic

Jakub VRAMBA
Department of Electrical Power Engineering, Faculty of Electrical Engineering and Computer Science, VSB–Technical university of Ostrava, 17. listopadu 15, 708 33 Ostrava-Poruba, Czech Republic

Lukas PROKOP
Department of Electrical Power Engineering, Faculty of Electrical Engineering and Computer Science, VSB–Technical university of Ostrava, 17. listopadu 15, 708 33 Ostrava-Poruba, Czech Republic

Marian UHER
Department of Electrical Power Engineering, Faculty of Electrical Engineering and Computer Science, VSB–Technical university of Ostrava, 17. listopadu 15, 708 33 Ostrava-Poruba, Czech Republic

Petr CHLEBIS
Department of Electronics, Faculty of Electrical Engineering and Computer Science, VSB–Technical University of Ostrava, 17. listopadu 15, 708 33 Ostrava-Poruba, Czech Republic

Martin TVRDON
Department of Electronics, Faculty of Electrical Engineering and Computer Science, VSB–Technical University of Ostrava, 17. listopadu 15, 708 33 Ostrava-Poruba, Czech Republic

Ales HAVEL
Department of Electronics, Faculty of Electrical Engineering and Computer Science, VSB–Technical University of Ostrava, 17. listopadu 15, 708 33 Ostrava-Poruba, Czech Republic

Katerina BARESOVA
Department of Electronics, Faculty of Electrical Engineering and Computer Science, VSB–Technical University of Ostrava, 17. listopadu 15, 708 33 Ostrava-Poruba, Czech Republic

Zdenek KINCL
Department of Radio Electronics, Faculty of Electrical Engineering and Communication, Brno University of Technology, Technicka 12, 616 00 Brno, Czech Republic

Zdenek KOLKA
Department of Radio Electronics, Faculty of Electrical Engineering and Communication, Brno University of Technology, Technicka 12, 616 00 Brno, Czech Republic

Chandra Babu PADUCHURI
Department of Electrical and Electronics Engineering, Faculty of Engineering and Technology, SRM University, SRM Nagar, Kattankulathur, 603 203 Chennai, India

Subhransu Sekhar DASH
Department of Electrical and Electronics Engineering, Faculty of Engineering and Technology, SRM University, SRM Nagar, Kattankulathur, 603 203 Chennai, India

Subramani CHINNAMUTHU
Department of Electrical and Electronics Engineering, Faculty of Engineering and Technology, SRM University, SRM Nagar, Kattankulathur, 603 203 Chennai, India

Ahmed SALHI
Laboratory of Electrical Engineering (LGEB), Department of Electrical Engineering, University Mohamed Khider of Biskra, BP 145, 070 00 Biskra, Algeria

Djemai NAIMI
Laboratory of Electrical Engineering (LGEB), Department of Electrical Engineering, University Mohamed Khider of Biskra, BP 145, 070 00 Biskra, Algeria

Tarek BOUKTIR
Department of Electrical Engineering, Faculty of Technology, University Ferhat Abbas of Setif, Cite Mabouda, 190 00 Setif, Algeria

Saeed MASOUMI KAZRAJI
Electrical Power Department, Faculty of Electrical and Computer Engineering, University of Tabriz, 29 Bahman Blvd, 5166616471 Tabriz, Iran

Mohammad Bagher BANNAE SHARIFIAN
Electrical Power Department, Faculty of Electrical and Computer Engineering, University of Tabriz, 29 Bahman Blvd, 5166616471 Tabriz, Iran

Hector Hugo SILVA LOPEZ
Laboratorio de Detectores, Instituto de Ciencias Nucleares, Universidad Nacional Autonoma de Mexico, Circuito exterior s/n Ciudad Universitaria AP, 70-543 Coyoacan CP. 04510, Distrito Federal, Mexico

Gustavo MEDINA-TANCO
Laboratorio de Detectores, Instituto de Ciencias Nucleares, Universidad Nacional Autonoma de Mexico, Circuito exterior s/n Ciudad Universitaria AP, 70-543 Coyoacan CP. 04510, Distrito Federal, Mexico

Lauro SANTIAGO CRUZ
Instrumentacion, Instituto de Ingenieria, Universidad Nacional Autonoma de Mexico, Circuito exterior s/n Ciudad Universitaria AP, 70-543 Coyoacan CP. 04510, Distrito Federal, Mexico

Mudathir Funsho AKOREDE
Department of Electrical & Electronics Engineering, Faculty of Engineering and Technology, University of Ilorin, P.M.B. 1515 Ilorin, Nigeria

Edris POURESMAEIL
Centre for Smart Energy Solutions, University of Southern Denmark, Campusvej 55, DK-5230 Odense, Denmark

Hashim HIZAM
Electrical & Electronic Engineering Department, Faculty of Engineering, Universiti Putra Malaysia, 43400 UPM Serdang, Selangor Darul Ehsan, Malaysia

Ishak ARIS
Electrical & Electronic Engineering Department, Faculty of Engineering, Universiti Putra Malaysia, 43400 UPM Serdang, Selangor Darul Ehsan, Malaysia

Mohd Zainal AB-KADIR
Electrical & Electronic Engineering Department, Faculty of Engineering, Universiti Putra Malaysia, 43400 UPM Serdang, Selangor Darul Ehsan, Malaysia

Chandra Babu PADUCHURI
Department of Electrical and Electronics Engineering, Faculty of Engineering and Technology, SRM University, SRM Nagar, Kattankulathur, 603 203 Chennai, India

Subhransu Sekhar DASH
Department of Electrical and Electronics Engineering, Faculty of Engineering and Technology, SRM University, SRM Nagar, Kattankulathur, 603 203 Chennai, India

Subramani CHINNAMUTHU
Department of Electrical and Electronics Engineering, Faculty of Engineering and Technology, SRM University, SRM Nagar, Kattankulathur, 603 203 Chennai, India

Jan PAPAJ
Department of Electronics and Multimedia Communications, Faculty of Electrical Engineering and Informatics, Technical University of Kosice, Letna 9, 042 00 Kosice, Slovak Republic

Lubomir DOBOS
Department of Electronics and Multimedia Communications, Faculty of Electrical Engineering and Informatics, Technical University of Kosice, Letna 9, 042 00 Kosice, Slovak Republic

Libor GAJDOSIK
Department of Telecommunications, Faculty of Electrical Engineering and Computer Science, VSB–Technical University of Ostrava, 17. listopadu 15/2172, 708 33 Ostrava Poruba, Czech Republic

Adam MILIK
Institute of Electronics, Silesian University of Technology, Akademicka 16, 44 100 Gliwice, Poland

Daniel MINARIK
Centre ENET—Research Centre of Energy Units for Utilization of Non Traditional Energy Sources, VSB–Technical University of Ostrava, 17. listopadu 15, 708 33 Ostrava, Czech Republic

Bohumil HORAK
Department of Cybernetics and Biomedical Engineering, Faculty of Electrical Engineering and Computer Science, VSB–Technical University of Ostrava, 17. listopadu 15, 708 33 Ostrava, Czech Republic

Petr MOLDRIK
Centre ENET—Research Centre of Energy Units for Utilization of Non Traditional Energy Sources, VSB–Technical University of Ostrava, 17. listopadu 15, 708 33 Ostrava, Czech Republic

Zdenek SLANINA
Department of Cybernetics and Biomedical Engineering, Faculty of Electrical Engineering and Computer Science, VSB–Technical University of Ostrava, 17. listopadu 15, 708 33 Ostrava, Czech Republic

Lukas KYPUS
Department of Telecommunication Engineering, Faculty of Electrical Engineering, Czech Technical University in Prague, Technicka 2, 166 27 Prague, Czech Republic

Lukas VOJTECH
Department of Telecommunication Engineering, Faculty of Electrical Engineering, Czech Technical University in Prague, Technicka 2, 166 27 Prague, Czech Republic

Matej ROHLIK
Department of Telecommunication Engineering, Faculty of Electrical Engineering, Czech Technical University in Prague, Technicka 2, 166 27 Praha 6, Czech Republic

Petr CHLUMSKY
Department of Telecommunication Engineering, Faculty of Electrical Engineering, Czech Technical University in Prague, Technicka 2, 166 27 Praha 6, Czech Republic

Tomas VANEK
Department of Telecommunication Engineering, Faculty of Electrical Engineering, Czech Technical University in Prague, Technicka 2, 166 27 Praha 6, Czech Republic

Martin SMOCEK
Department of Electrical Power Engineering, Faculty of Electrical Engineering and Computer Science, VSB-Technical University of Ostrava, 17.listopadu 15/2172, 708 33 Ostrava Poruba, Czech Republic

Zdenek HRADILEK
Department of Electrical Power Engineering, Faculty of Electrical Engineering and Computer Science, VSB-Technical University of Ostrava, 17.listopadu 15/2172, 708 33 Ostrava Poruba, Czech Republic

Daniela PEDRUKOVA
Department of Electrical Engineering and Mechatronics, Faculty of Electrical Engineering and Informatics, Technical University of Kosice, Letna 9, 042 00 Kosice, Slovak Republic

Pavol FEDOR
Department of Electrical Engineering and Mechatronics, Faculty of Electrical Engineering and Informatics, Technical University of Kosice, Letna 9, 042 00 Kosice, Slovak Republic

Miroslav UHRINA
Department of Telecommunications and Multimedia, Faculty of Electrical Engineering, University of Zilina, Univerzitna 8215/1, 010 07 Zilina, Slovakia

Jaroslav FRNDA
Department of Telecommunications, Faculty of Electrical Engineering and Computer Science, VSB–Technical University of Ostrava, 17. listopadu 15/2172, 708 33 Ostrava-Poruba, Czech Republic

Lukas SEVCIK
Department of Telecommunications, Faculty of Electrical Engineering and Computer Science, VSB–Technical University of Ostrava, 17. listopadu 15/2172, 708 33 Ostrava-Poruba, Czech Republic

Martin VACULIK
Department of Telecommunications and Multimedia, Faculty of Electrical Engineering, University of Zilina, Univerzitna 8215/1, 010 07 Zilina, Slovakia

Marcel FAJKUS
Department of Telecommunications, Faculty of Electrical Engineering and Computer Science, VSB–Technical Universit of Ostrava, 17. listopadu 2172/15, 708 33 Ostrava-Poruba, Czech Republic

Martin MIKULEC
Department of Telecommunications, Faculty of Electrical Engineering and Computer Science, VSB–Technical Universit of Ostrava, 17. listopadu 2172/15, 708 33 Ostrava-Poruba, Czech Republic

Miroslav VOZNAK
Department of Telecommunications, Faculty of Electrical Engineering and Computer Science, VSB–Technical Universit of Ostrava, 17. listopadu 2172/15, 708 33 Ostrava-Poruba, Czech Republic

Martin TOMIS
Department of Telecommunications, Faculty of Electrical Engineering and Computer Science, VSB–Technical Universit of Ostrava, 17. listopadu 2172/15, 708 33 Ostrava-Poruba, Czech Republic

Peppino FAZIO
Department of Computer Science, Modeling, Electronics and Systems Engineering, University of Calabria, Edificio 42C - piano 5, 870 36 Arcavacata di Rende (CS), Italy